ISBN 978-0-243-17139-2
PIBN 10605058

English
Français
Deutsche
Italiano
Español
Português

www.forgottenbooks.com

Mythology Photography **Fiction**
Fishing Christianity **Art** Cooking
Essays Buddhism Freemasonry
Medicine **Biology** Music **Ancient
Egypt** Evolution Carpentry Physics
Dance Geology **Mathematics** Fitness
Shakespeare **Folklore** Yoga Marketing
Confidence Immortality Biographies
Poetry **Psychology** Witchcraft
Electronics Chemistry History **Law**
Accounting **Philosophy** Anthropology
Alchemy Drama Quantum Mechanics
Atheism Sexual Health **Ancient History**
Entrepreneurship Languages Sport
Paleontology Needlework Islam
Metaphysics Investment Archaeology
Parenting Statistics Criminology
Motivational

COMPLETE

BUSINESS ARITHMETIC

BY

GEORGE H. VAN TUYL

TEACHER OF BUSINESS ARITHMETIC, HIGH SCHOOL OF COMMERCE
NEW YORK
FORMERLY TEACHER OF ARITHMETIC IN THE ALBANY BUSINESS COLLEGE
ALBANY, N. Y., AND IN THE PACKARD COMMERCIAL SCHOOL
NEW YORK

NEW YORK ∴ CINCINNATI ∴ CHICAGO
AMERICAN BOOK COMPANY

PREFACE

In the preparation of this book the following specific objects have been kept in view:

A training that leads to *facility* and *accuracy* in handling the fundamental operations.

The placing of emphasis on the *fundamental principles* of arithmetic rather than upon set rules for the solution of problems.

Clearness and fullness of *explanation*.

The providing of problems that have an *informational* value.

Attention is directed to the following features:

The chapter on aliquot parts, as applied to billing, trade discount, and simple interest, is placed early in the text.

Common and decimal fractions are treated together, as is the case in business.

A great many problems are provided for mental work.

Many of the problems are taken from the business affairs of corporations, cities, states, and nations of the world.

Many calculation tables are illustrated and applied to the solution of problems.

There are six sets of examinations (each set consisting of a speed test and a written test) to determine the student's mastery of the subject at various stages of the work.

The author is indebted to Head Master James E. Downey, of the High School of Commerce, Boston, Mass., and to Mr. G. J. Raynor, of the Business Department of the Commercial High School, Brooklyn, N.Y., for reading the manuscript and making helpful suggestions; to Mr. F. B. Hess, Head of the Business Department

3

of the Heffley Business Institute, Brooklyn, N.Y., for reading and criticizing the proof; and to the many business men in various lines of industry in the chief cities of the United States, who have so courteously and freely given invaluable advice and information concerning the latest and most approved methods of handling business calculations.

CONTENTS

CONTENTS

BUSINESS ARITHMETIC

——∘∘°⊙°∘∘——

READING AND WRITING NUMBERS

1. Numbers are expressed in three ways: (1) by written words, (2) by letters (Roman system), and (3) by figures (Arabic system).

ROMAN NOTATION

2. In the **Roman system** of writing numbers seven letters are used. They are:

I	V	X	L	C	D	M
1	5	10	50	100	500	1000

3. Writing numbers by the Roman system is based on the following principles.

PRINCIPLES: **1** *Repeating a letter repeats its value.*

2. *The value of a letter written after a letter of greater value is to be added to the larger value.*

3. *The value of a letter written before a letter of greater value is to be subtracted from the greater value.*

4. *A horizontal bar over a letter increases its value one thousand times.*

NOTES. (1) Only whole numbers can be written in the Roman system of notation.

(2) Roman notation is of little practical value, being used only to record dates and chapters in books, dates on buildings, numbers on a watch face, etc.

4.   ROMAN NOTATION ILLUSTRATED

II	= 2	XIV	= 14	LIX	= 59	CCCLXVI	= 366
III	= 3	XVI	= 16	LXXIV	= 74	DCXCVI	= 696
IV or IIII*	= 4	XIX	= 19	LXXXIX	= 89	MCXIX	= 1119
VI	= 6	XXIII	= 23	XC	= 90	MCM	= 1900
IX	= 9	XXXIV	= 34	CX	= 110	$\overline{\text{MCMX}}$	= 1,910,000

5. 1. Write in Roman notation:

12	85	336	598	10,000
18	94	472	1077	15,000
21	111	505	1384	100,000
47	128	666	1527	150,000
63	247	872	1898	1,000,000

2. Read the following:

XVIII	LXIX	CDLIX	$\overline{\text{MC}}$
XXIX	XCIX	$\overline{\text{D}}$CD	MDCCCXCIX
XXXI	CIX	DCCCLXXXIX	MDCCXXVIII
XLIV	CLXXII	CM	$\overline{\text{V}}$CCLXXX
XLIX	CCCLX	MCMIX	$\overline{\text{L}}$

ARABIC NOTATION

6. The Arabic system of notation makes use of ten characters or figures, and a period or decimal point. They are 0, 1, 2, 3, 4, 5, 6, 7, 8, 9, and .

7. With these figures and the decimal point any required number can be written.

8. For convenience in reading numbers, the figures are sometimes divided into groups of three, in this way:

2,564,879,598.

The reason for this grouping is best understood by observing the following table:

* IIII is used on timepieces.

TABLE SHOWING HOW TO READ NUMBERS

ten-trillions	trillions	hundred-billions	ten-billions	billions	hundred-millions	ten-millions	millions	hundred-thousands	ten-thousands	thousands	hundreds	tens	units
8	9,	7	2	4,	5	6	3,	9	0	0,	1	0	6.

The number is read, "eighty-nine trillion, seven hundred twenty-four billion, five hundred sixty-three million, nine hundred thousand, one hundred six."

9. In like manner read the following numbers:

1. 21,460		**7.** 962,024		**13.** 27,250,820	
2. 34,829		**8.** 872,463		**14** 20,300,600	
3. 127,724		**9.** 900,047		**15.** 65,897,249	
4. 306,200		**10.** 1,456,789		**16.** 185,496,728	
5 425,002		**11.** 3,248,764		**17.** 593,972,847	
6. 800,005		**12.** 5,002,005		**18.** 15,647,989,721	

10. Write the following numbers, using the Arabic system:

1. Seven hundred forty-three.

2. Nine thousand fourteen.

3. Eleven thousand one hundred four.

4. Fifteen thousand twenty.

5 Seventy-six thousand ninety-nine.

6. Fourteen hundred eighty-eight.

7 One hundred sixty thousand four hundred.

8. Eight hundred thousand nine hundred ninety.

9. Two million sixteen thousand seventeen.

10. Six hundred eleven million fourteen thousand eighty-six.

UNITED STATES MONEY

11. Any number of cents less than one hundred is read as cents : thus, $ .45 or 45 ¢ is read " forty-five cents," but any value equal to or greater than one hundred cents is read as dollars, or as dollars and cents: thus, $ 8.75 or 875 ¢ would be read " eight dollars, seventy-five cents." A period, or point, separates the number of dollars from the number of cents.

12. Read the following:

$ 3.17	$ 12.40	$ 826.52	$ 1476.83
$ 4.28	$ 16.50	$ 1624.43*	$ 1982.76
$ 9.15	$ 85.90	$ 1837.16	$ 14,728.41
$ 10.20	$ 104.26	$ 2487.49	$ 146,789.09

13. Write the following in figures:

1. One hundred sixteen dollars, fifty-nine cents.

2. Three hundred nine dollars, seventy cents.

3. Twelve hundred twenty-one dollars, eighty-eight cents.

4. Twenty-four hundred forty-three dollars, nine cents.

5. Fifty-seven hundred six dollars, thirty-three cents.

14. Money is the generally accepted measure of value.

15. Each nation has its own system of money or coinage.

The principal nations of the world have adopted gold as the standard of monetary value.

16. United States money consists of the coins, notes, and certificates authorized by Congress to be used as money. It is a decimal system.

TABLE

10 mills = 1 cent, ¢
10 cents = 1 dime, d.
10 dimes = 1 dollar, $
10 dollars = 1 eagle
(The mill is not a coin.)

17. There are ten different kinds of money in circulation in the United States: gold coins, silver dollars, subsidiary silver, nickels,

---

* In reading $ 1624.43 say " sixteen hundred twenty-four dollars, forty-three cents," not " one thousand six hundred twenty-four dollars, forty-three cents." One reason for saying it and writing it as suggested is that it takes less time and space.

pennies, gold certificates, silver certificates, Treasury notes, United States notes (greenbacks), and national bank notes.

18. The gold dollar, containing 25.8 grains of standard gold .9 fine, is the unit of value.

19. The silver dollar weighs 15.988 times as much as the gold dollar. That is, the ratio of gold to silver is 15.988 to 1 (commonly called 16 to 1).

20. DENOMINATIONS, WEIGHT, AND FINENESS OF THE COINS OF THE UNITED STATES

DENOMINATION	FINE GOLD CONTAINED (GRAINS)	ALLOY CONTAINED (GRAINS)	WEIGHT (GRAINS)
GOLD			
One dollar (not coined since 1890)	23.22	2.58	25.80
Quarter eagle ($ 2.50)	58.05	6.45	64.5
Half eagle ($ 5.00)	116.10	12.90	129.
Eagle ($ 10.00)	232.20	25.80	258.
Double eagle ($ 20.00)	464.40	51.60	516.
SILVER			
	PURE SILVER CONTAINED		
One dollar	371.25	41.25	412.5
Half dollar	173.61	19.29	192.9
Quarter dollar	86.805	9.645	96.45
Dime	34.722	3.858	38.58
MINOR			
	PURE COPPER CONTAINED		
Five cents.* (nickel)	57.87	19.29	77.16
One cent † (copper)	45.60	2.40	48.

NOTE. The alloy in gold and silver coins neither adds to nor detracts from the value of the coins. Its purpose is to harden the coin, thus reducing the loss by abrasion.

* 75 % copper and 25 % nickel.   † 95 % copper and 5 % tin and zinc.

ALIQUOT PARTS

21. An aliquot part of a number is a number that is contained in it an integral number of times.

Thus, 2, 4, 5, 10, 20, 25, and 50 are aliquot parts of 100; that is, $2 = \frac{1}{50}$ of 100, $25 = \frac{1}{4}$ of 100, etc.

NOTE. Other fractional parts of 100 are called aliquant parts of 100. For convenience, all fractional parts of 100 will be treated as aliquot parts.

HALVES AND QUARTERS

22. 1. What part of 100 cents, or $1.00, is 25¢? 50¢? 75¢?

2. At 25¢ a yard, how many yards can be bought for $1.00?

3. If 4 yd. cost $1.00, how much will 8 yd. cost? 12 yd.? 16 yd.?

4. How much will 20 yd. cost at 25¢ a yard? at 50¢ a yard? at 75¢ a yard?

5. Compare 25¢ with 50¢; with 75¢. 50¢ with 75¢; with 25¢. 75¢ with 25¢; with 50¢.

6. Compare the number of yards that can be bought for 50¢ and for 25¢; for 25¢ and for 75¢; for 75¢ and for 50¢.

7. State a rule for finding the cost of any number of articles at 25¢ each; at 50¢ each; at 75¢ each.

23. Make all extensions mentally, and find the total value of each of the following:

1.	2.	3.
16 yd. @ 25¢	32 yd. @ 50¢.	120 yd. @ 25¢.
24 yd. @ 25¢.	48 yd. @ 25¢.	140 yd. @ 50¢.
30 yd. @ 50¢.	54 yd. @ 50¢.	150 yd. @ 25¢.
16 yd. @ 75¢	60 yd. @ 75¢.	160 yd. @ 75¢.
40 yd. @ 50¢.	80 yd. @ 75¢.	126 yd. @ 25¢.

4.	5.	6.
13 yd. @ 25 ¢.	3486 lb. @ 25 ¢.	128 lb. @ 75¢.
21 yd. @ 50 ¢.	1984 lb. @ 50 ¢.	324 lb. @ 50 ¢.
39 yd. @ 25 ¢	4821 lb. @ 25 ¢.	233 lb. @ 25 ¢.
31 yd. @ 75 ¢.	8448 lb. @ 75 ¢.	199 lb. @ 75 ¢.
48 yd. @ 75 ¢	1331 lb. @ 50 ¢.	247 lb. @ 25 ¢.

Make up and find the total value of each of two or more groups of quantities like the above as directed by the teacher.

EIGHTHS

24. 1. $12\frac{1}{2}$¢ is what part of 25 ¢ ?   **2.** $\frac{1}{2}$ of $\frac{1}{4}$ is how much ?

3. $12\frac{1}{2}$¢ is what part of $1 ?

4. 25¢ is how many eighths of a dollar ?

5. $25¢ + 12\frac{1}{2}¢ =$ how many cents ?

6. $37\frac{1}{2}$¢ is what part of a dollar ?

7. What is the sum of $\frac{1}{2}$ and $\frac{1}{8}$? $50¢ + 12\frac{1}{2}¢ = ?$

8. $62\frac{1}{2}$¢ is what part of a dollar ?   **9.** $75¢ + 12\frac{1}{2}¢ = ?$

10. $\$\frac{7}{8} =$ how many cents ?

11. What part of a dollar is $12\frac{1}{2}$¢? $37\frac{1}{2}$¢? $62\frac{1}{2}$¢? $87\frac{1}{2}$¢?

12. What is the cost of 16 lb. of raisins at $12\frac{1}{2}$¢ a pound ? of 16 baskets of potatoes at $37\frac{1}{2}$¢ a basket ? of 16 yd. of cloth at $62\frac{1}{2}$¢ per yard ? of 16 bu. of apples at $87\frac{1}{2}$¢ a bushel ?

25. Using the aliquot-part method, find the total value of each of the following:

1.	2.	3.
24 lb. @ $37\frac{1}{2}$¢.	48 lb. @ $12\frac{1}{2}$¢.	120 yd. @ $12\frac{1}{2}$¢.
32 lb. @ $12\frac{1}{2}$¢.	64 lb. @ 25¢.	164 yd. @ $12\frac{1}{2}$¢.
40 lb. @ $62\frac{1}{2}$¢.	72 lb. @ $37\frac{1}{2}$¢.	200 yd. @ $37\frac{1}{2}$¢.
56 lb. @ $37\frac{1}{2}$¢.	88 lb. @ $87\frac{1}{2}$¢.	240 yd. @ $12\frac{1}{2}$¢

Make two or more groups like the above and find the total value of each.

Sixteenths

26. 1 $6\frac{1}{4}$¢ is what part of $12\frac{1}{2}$¢? of 25¢? of 50¢? of \$1?

2. 3 times $6\frac{1}{4}$¢ is how many cents?

3. $18\frac{3}{4}$¢ is what part of a dollar?

4. 25¢ or $\$\frac{1}{4}$ is how many sixteenths of a dollar?

5. 25¢ $+6\frac{1}{4}$¢ are how many cents?

6. $31\frac{1}{4}$¢ is what part of a dollar?

7. $\frac{1}{2}-\frac{1}{16}=$? $43\frac{3}{4}$¢ is what part of a dollar?

8. $\frac{1}{2}+\frac{1}{16}=$? $56\frac{1}{4}$¢ $=$ what part of a dollar?

9. In like manner find how many cents there are in $\$\frac{11}{16}$, $\$\frac{13}{16}$, and $\$\frac{15}{16}$ by comparing $6\frac{1}{4}$¢ ($\$\frac{1}{16}$) with 75¢ and \$1.

27. Complete the following table of sixteenths, eighths, quarters, and halves.

Compare one part with another

	Eighths	Quarters	Halves	Cents		Eighths	Quarters	Halves	Cents
$\$\frac{1}{16}$				$6\frac{1}{4}$	$\$\frac{9}{16}$				
$\$\frac{2}{16}$	$\$\frac{1}{8}$			$12\frac{1}{2}$	$\$\frac{10}{16}$				
$\$\frac{3}{16}$				$18\frac{3}{4}$	$\$\frac{11}{16}$				
$\$\frac{4}{16}$	$\$\frac{2}{8}$	$\$\frac{1}{4}$		25	$\$\frac{12}{16}$				
$\$\frac{5}{16}$					$\$\frac{13}{16}$				
$\$\frac{6}{16}$					$\$\frac{14}{16}$				
$\$\frac{7}{16}$					$\$\frac{15}{16}$				
$\$\frac{8}{16}$					$\$\frac{16}{16}$				

28. Make the extensions mentally, and find the total value of each of. the following:

1.	**2.**	**3.**
32 yd. @ $6\frac{1}{4}$¢.	80 yd. @ $31\frac{1}{4}$¢.	47 yd. @ 25¢.
48 yd. @ $12\frac{1}{2}$¢.	128 yd. @ $37\frac{1}{2}$¢.	53 yd. @ $12\frac{1}{2}$¢.
64 yd. @ $18\frac{3}{4}$¢.	144 yd. @ $56\frac{1}{4}$¢	58 yd. @ $6\frac{1}{4}$¢.
96 yd. @ $6\frac{1}{4}$¢.	96 yd. @ $93\frac{3}{4}$¢.	63 yd. @ 50¢.
160 yd. @ $43\frac{3}{4}$¢.	72 yd. @ $37\frac{1}{2}$¢.	75 yd. @ $37\frac{1}{2}$¢.

Note. In the second item, in No. 3, 53 yd. @ $12\frac{1}{2}$ ¢ = $6\frac{5}{8}$ = $6.62\frac{1}{2}$; write the extension as $6.63.

Write two or more groups like the above and find the total value of each.

Thirds and Sixths

29. 1. What part of a dollar is $33\frac{1}{3}$ ¢? $66\frac{2}{3}$ ¢?

2. What is $\frac{1}{2}$ of $\frac{1}{3}$? $\frac{1}{2}$ of $33\frac{1}{3}$ ¢ is how many cents?

3. $16\frac{2}{3}$ ¢ = what part of a dollar?

. 3 times $16\frac{2}{3}$ ¢ = how many cents? what part of a dollar?

. 4 times $16\frac{2}{3}$ ¢ = how many cents? what part of a dollar?

6. $1 − $16\frac{2}{3}$ ¢ = how many cents? what part of a dollar?

30. Complete the following table. State the relation of each part to the parts that follow and precede it.

	Thirds	Halves	Cents
$\frac{1}{6}			$16\frac{2}{3}$
$\frac{2}{6}	$\frac{1}{3}		$33\frac{1}{3}$
$\frac{3}{6}		$\frac{1}{2}	50
$\frac{4}{6}			
$\frac{5}{6}			
$\frac{6}{6}			

31. Find total value of each group:

1. 48 lb. @ $16\frac{2}{3}$ ¢
54 lb @ $66\frac{2}{3}$ ¢.
30 lb. @ $83\frac{1}{3}$ ¢.

2. 66 yd. @ $33\frac{1}{3}$ ¢.
76 yd. @ 25 ¢.
76 yd. @ $12\frac{1}{2}$ ¢.

3.
64 lb. @ $6\frac{1}{4}$ ¢.
96 lb. @ $33\frac{1}{3}$ ¢.
88 lb. @ $37\frac{1}{2}$ ¢.
91 lb. @ $66\frac{2}{3}$ ¢.

4.
48 qt. @ $33\frac{1}{3}$ ¢.
72 qt. @ $16\frac{2}{3}$ ¢.
96 qt. @ $12\frac{1}{2}$ ¢.
60 qt. @ $6\frac{1}{4}$ ¢.

5.
120 yd. @ $37\frac{1}{2}$ ¢.
126 yd. @ $16\frac{2}{3}$ ¢.
132 yd. @ $83\frac{1}{3}$ ¢.
144 yd. @ $33\frac{1}{3}$ ¢.

Write and find the total value of two or more groups like the above

Twelfths

32. 1. $\frac{1}{3}$ of 25 ¢ = how many cents?

2. $\frac{1}{3}$ of $\frac{1}{4}$ = ? $8\frac{1}{3}$ ¢ = what part of a dollar?

3. 50 ¢ − $8\frac{1}{3}$ ¢ = how many cents?

4. $\frac{1}{2} − \frac{1}{12}$ = ? $41\frac{2}{3}$ ¢ = what part of a dollar?

5. 50 ¢ + $8\frac{1}{3}$ ¢ = how many cents? what part of a dollar?

6. $1 − $8\frac{1}{3}$ ¢ = how many cents? what part of a dollar?

33. Complete the following table. Compare each part with the parts that follow and precede it.

34. Find total value of each of the following groups, making all extensions mentally:

	Sixths	Thirds	Quarters	Cents
$\frac{1}{12}$				$8\frac{1}{3}$
$\frac{2}{12}$	$\frac{1}{6}$			$16\frac{2}{3}$
$\frac{3}{12}$			$\frac{1}{4}$	25
$\frac{4}{12}$		$\frac{1}{3}$		$33\frac{1}{3}$
$\frac{5}{12}$				
$\frac{6}{12}$				
$\frac{7}{12}$				
$\frac{8}{12}$				
$\frac{9}{12}$				
$\frac{10}{12}$				
$\frac{11}{12}$				
$\frac{12}{12}$				

1.

36 yd. @ $8\frac{1}{3}$¢.

48 yd. @ $16\frac{2}{3}$¢.

24 yd. @ $41\frac{2}{3}$¢.

60 yd. @ $58\frac{1}{3}$¢.

72 yd @ $83\frac{1}{3}$¢.

2.

66 yd. @ $33\frac{1}{3}$¢.

84 yd. @ $41\frac{2}{3}$¢.

120 yd. @ $91\frac{2}{3}$¢.

42 yd @ $58\frac{1}{3}$¢.

54 yd. @ $16\frac{2}{3}$¢.

NOTE. Many articles are bought by the dozen and sold by the piece. A knowledge of the "twelfths" is essential in determining the price of one.

3.

75 lb. @ $8\frac{1}{3}$¢.

88 lb. @ $16\frac{2}{3}$¢.

90 lb @ $41\frac{2}{3}$¢.

96 lb. @ $58\frac{1}{3}$¢

84 lb. @ $66\frac{2}{3}$¢.

4.

24 yd. @ $58\frac{1}{3}$¢.

36 yd. @ $41\frac{2}{3}$¢

60 yd. @ $91\frac{2}{3}$¢.

48 yd. @ $16\frac{2}{3}$¢

72 yd. @ $33\frac{1}{3}$¢.

5.

140 lb. @ $33\frac{1}{3}$¢.

160 lb. @ $91\frac{2}{3}$¢.

240 lb. @ $58\frac{1}{3}$¢.

360 lb. @ $41\frac{2}{3}$¢.

118 lb. @ $8\frac{1}{3}$¢.

Write two or more groups like the above and find the total value of each. Endeavor to use the shortest method of solution in each exercise. Omit actual multiplication wherever possible.

35. OTHER SUGGESTIONS FOR SHORTENING THE WORK

1. Find the cost of 45 lb. at 18 ¢.

$18 ¢ = \$\frac{1}{5} - (\frac{1}{10} \text{ of } \$\frac{1}{5})$.

$\frac{1}{5}$ of $45 = 9$

$\frac{1}{10}$ of $9 = .90$

8.10; Cost = **$8.10**.

2. Find the cost of 72 yd. at $17\frac{1}{2}$ ¢.

$17\frac{1}{2} ¢ = \$\frac{1}{10} + (\frac{1}{2} \text{ of } \$\frac{1}{10}) + \frac{1}{2}(\frac{1}{2} \text{ of } \$\frac{1}{10})$ or, $10 ¢ + 5 ¢ + 2\frac{1}{2} ¢$. Hence 72 yd. @ $17\frac{1}{2}$ ¢ cost $7.20 + $3.60 + $1.80, or $12.60.

3. Formulate a short method for finding the cost of articles @ 24 ¢ ;
@ 27½ ¢ ; @ 45 ¢.

$$24 ¢ = \$ \tfrac{1}{4} - 1 ¢.$$
$$27\tfrac{1}{2} ¢ = \$ \tfrac{1}{4} + (\tfrac{1}{10} \text{ of } \$ \tfrac{1}{4}).$$
$$45 ¢ = \$ \tfrac{1}{2} - (\tfrac{1}{10} \text{ of } \$ \tfrac{1}{2}).$$

The student is urged to devise short methods for himself, and to use them on all possible occasions.

36. Find total value of each group:

1.	2.	3.
48 yd. @ 7½¢.	90 yd. @ 18¢.	7½ yd. @ 36¢.
56 yd. @ 9¢.	96 yd. @ 22¢.	11½ yd. @ 48¢.
64 yd. @ 13½¢.	110 yd. @ 22½¢.	16 yd. @ 45¢.
80 yd. @ 11½¢.	116 yd. @ 27½¢.	17½ yd. @ 62¢.
88 yd. @ 17½¢.	124 yd. @ 35¢.	18 yd. @ 65¢.

4.	5.	6.
55 yd. @ 44¢.	132 yd. @ 22¢.	219½ yd. @ 33¢.
55 yd. @ 36¢.	141 yd. @ 22½¢.	313¼ yd. @ 35¢.
65 yd. @ 33¢.	123 yd. @ 17½¢.	317½ yd. @ 44¢.
51 yd. @ 27 ¢.	163 yd. @ 27½¢.	273½ yd. @ 45¢.
39 yd. @ 55¢.	172 yd. @ 27 ¢.	124¾ yd. @ 55 ¢.

How did you calculate by 13½¢ in No. 1? by 22½ ¢ in No. 2? by 48¢ in No. 3? by 36¢ in No. 4? etc.

REVIEW

37. Using the shortest solution, find the total cost of:

1.	2.	3.
54 yd. @ 16⅔¢.	120 yd. @ 62½¢.	33 yd. @ 12½¢.
40 yd. @ 25¢.	240 yd. @ 87½¢.	41 yd. @ 8⅓¢.
36 yd. @ 33⅓¢.	160 yd. @ 66⅔¢.	52 yd. @ 6¼¢.
84 yd. @ 50¢.	360 yd. @ 91⅔¢.	63 yd. @ 12½¢.
75 yd. @ 66⅔¢.	164 yd. @ 37½¢.	89 yd. @ 33⅓¢.
96 yd. @ 12½¢.	66 yd. @ 41⅔¢.	113 yd. @ 75¢.
48 yd. @ 8⅓¢.	68 yd. @ 43¾¢.	97 yd. @ 62½¢.
36 yd. @ 16⅔¢.	75 yd. @ 58⅓¢.	116 yd. @ 87½¢.
42 yd. @ 25¢.	111 yd. @ 33⅓¢.	121 yd. @ 56¼¢.

4.	5.	6.
76 yd. @ 37½¢.	78 lb. @ 18¾¢.	87 lb. @ 6¼¢.
92 yd. @ 31¼¢	93 lb. @ 25¢.	94 lb. @ 8⅓¢.
83 yd. @ 43¾¢.	82 lb. @ 33⅓¢.	104 lb. @ 12½¢.
78 yd. @ 41⅔¢.	13 lb. @ 31¼¢.	231 lb. @ 16⅔¢.
107 yd. @ 6¼¢.	9 lb. @ 37½¢.	321 lb. @ 18¾¢.
113 yd. @ 62½¢.	27 lb. @ 41⅔¢.	88 lb. @ 93¾¢.
114 yd. @ 83⅓¢.	32 lb. @ 43¾¢.	136 lb. @ 91⅔¢.
131 yd. @ 16⅔¢.	84 lb. @ 50¢.	316 lb. @ 83⅓¢.
125 yd. @ 25¢.	79 lb. @ 56¼¢.	28 lb. @ 81¼¢.
143 yd. @ 6¼¢	97 lb. @ 58⅓¢	7 lb. @ 68¾¢

7.	8.	9.
117 lb. @ 93¾¢.	13 lb. @ 56¼¢.	67 lb. @ 6¼¢.
104 lb. @ 91⅔¢.	27 lb. @ 50¢.	74 lb. @ 8⅓¢.
116 lb. @ 87½¢.	4 lb. @ 43¾¢.	83 lb. @ 12½¢.
203 lb. @ 83⅓¢.	9 lb. @ 41⅔¢.	99 lb. @ 16⅔¢.
308 lb. @ 81¼¢.	17 lb. @ 37½¢.	199 lb. @ 18¾¢.
27 lb. @ 75¢.	37 lb. @ 33⅓¢.	127 lb. @ 25¢.
96 lb. @ 68¾¢.	45 lb. @ 31¼¢.	217 lb. @ 31¼¢.
84 lb. @ 66⅔¢.	83 lb. @ 25¢.	134 lb. @ 33⅓¢.
119 lb. @ 62½¢.	107 lb. @ 18¾¢.	229 lb. @ 37½¢.
17 lb. @ 58⅓¢.	119 lb. @ 16⅔¢.	168 lb. @ 41⅔¢.
45 lb. @ 12½¢.	128 lb. @ 18¾¢.	39 lb. @ 31¼¢.

SEVENTHS, NINTHS, AND FIFTEENTHS

38. In addition to the aliquot parts of a dollar already found, there are a few others of which use is occasionally made: viz. sevenths, ninths, and fifteenths.

1. What is $\frac{1}{7}$ of a dollar? $\frac{2}{7}$? $\frac{3}{7}$?

2. What is $\frac{1}{9}$ of a dollar? $\frac{2}{9}$? $\frac{4}{9}$?

3. What is $\frac{1}{15}$ of a dollar? $\frac{2}{15}$?

NOTE. These lists will not be made complete, as there is seldom any use for others than those given above.

39. Make the extensions mentally, and find total value of each group:

1.	2.	3.
21 lb. @ $14\frac{2}{7}$ ¢.	42 lb. @ $42\frac{6}{7}$ ¢.	33 lb. @ $14\frac{2}{7}$ ¢.
28 lb. @ $28\frac{4}{7}$ ¢.	54 lb. @ $44\frac{4}{9}$ ¢.	45 lb. @ $28\frac{4}{7}$ ¢.
36 lb. @ $11\frac{1}{9}$ ¢.	60 lb. @ $13\frac{1}{3}$ ¢.	83 lb. @ $6\frac{2}{3}$ ¢.
45 lb. @ $6\frac{2}{3}$ ¢.	84 lb. @ $14\frac{2}{7}$ ¢.	96 lb. @ $11\frac{1}{9}$ ¢.
48 lb. @ $6\frac{1}{4}$ ¢.	91 lb. @ $28\frac{4}{7}$ ¢.	87 lb. @ $12\frac{1}{2}$ ¢.

Write two or more groups and find total value of each.

40. In the preceding exercises the price has been an aliquot part of $1. The same principle applies when the quantity is an aliquot part of 100 units of the article involved.

The cost of 48 yd. @ $12\frac{1}{2}$ ¢ = the cost of $12\frac{1}{2}$ yd. @ 48 ¢.

(a) 48 yd. @ $12\frac{1}{2}$ ¢ = $\frac{1}{8}$ of 48 = 6, hence $6, cost.

(b) 100 yd. @ 48 ¢ = $48.
$12\frac{1}{2}$ yd. = $\frac{1}{8}$ of 100 yd.
Hence, $12\frac{1}{2}$ yd. @ 48 ¢ cost $\frac{1}{8}$ of $48 = $6, cost.

41. Find total value of each of the following groups:

1.	2.	3.
25 yd. @ 32 ¢.	$16\frac{2}{3}$ yd. @ 42 ¢.	$6\frac{1}{4}$ lb. @ 16 ¢.
$33\frac{1}{3}$ yd. @ 45 ¢.	$18\frac{3}{4}$ yd. @ 56 ¢.	$8\frac{1}{3}$ lb. @ 24 ¢.
50 yd. @ 78 ¢.	$31\frac{1}{4}$ yd. @ 80 ¢.	$12\frac{1}{2}$ lb. @ 32 ¢.
$37\frac{1}{2}$ yd. @ 48 ¢.	$62\frac{1}{2}$ yd. @ 72 ¢.	$16\frac{2}{3}$ lb. @ 42 ¢.
75 yd. @ 80 ¢.	$93\frac{3}{4}$ yd. @ 96 ¢.	$18\frac{3}{4}$ lb. @ 48 ¢.

4.	5.	6.
25 lb. @ 43 ¢.	78 yd. @ $33\frac{1}{3}$ ¢.	43 yd. @ $66\frac{2}{3}$ ¢.
$31\frac{1}{4}$ lb. @ 56 ¢	$16\frac{2}{3}$ yd. @ 33 ¢.	$68\frac{3}{4}$ yd. @ 32 ¢.
$33\frac{1}{3}$ lb. @ 58 ¢.	$56\frac{1}{4}$ yd. @ 44 ¢.	$81\frac{1}{4}$ yd. @ 16 ¢.
$37\frac{1}{2}$ lb. @ 68 ¢.	$58\frac{1}{3}$ yd. @ 66 ¢.	$83\frac{1}{3}$ yd. @ 72 ¢.
$43\frac{3}{4}$ lb. @ 96 ¢.	38 yd. @ $62\frac{1}{2}$ ¢.	80 yd. @ $93\frac{3}{4}$ ¢.

7.	8.	9.
89 lb. @ $6\frac{2}{3}$ ¢.	160 yd. @ 75 ¢.	1148 yd. @ $33\frac{1}{3}$ ¢.
93 lb. @ $6\frac{1}{4}$ ¢.	200 yd. @ $87\frac{1}{2}$ ¢.	2682 yd. @ $37\frac{1}{2}$ ¢.
87 lb. @ $12\frac{1}{2}$ ¢.	240 yd. @ $93\frac{3}{4}$ ¢	$81\frac{1}{4}$ yd. @ 80 ¢.
76 lb. @ $16\frac{2}{3}$ ¢.	25 yd. @ $1.20.	$68\frac{3}{4}$ yd. @ 96 ¢.
116 lb. @ 25 ¢.	50 yd. @ $1.60.	75 yd. @ 72 ¢.

APPLICATION OF THE PRINCIPLES OF ALIQUOT PARTS

42. 1. Multiply 48 by 10; by 100; by 1000.

2. What did you do to multiply by 10? by 100? by 1000?

3. $2\frac{1}{2}$ is what part of 10? 25 is what part of 100? 250 is what part of 1000?

4. Multiply 48 by $2\frac{1}{2}$; by 25; by 250.

First multiply 48 by 10: that is, annex one cipher, thus, 480. Since the given multiplier is $\frac{1}{4}$ of 10, the required product is $\frac{1}{4}$ of 480, or 120.

In like manner to multiply by 25, first annex two ciphers to 48, thus, 4800, and divide by 4, because 25 is $\frac{1}{4}$ of 100, etc.

5. $3\frac{1}{3}$ is what part of 10?

$33\frac{1}{3}$ is what part of 100?

$333\frac{1}{3}$ is what part of 1000?

6. Multiply 144 by $3\frac{1}{3}$.

Since $3\frac{1}{3}$ is $\frac{1}{3}$ of 10, first annex one cipher (multiply by 10), thus, 1440, and divide by 3. The required product is 480.

7.	**8.**
$1\frac{1}{4}$ is what part of 10?	$1\frac{2}{3}$ is what part of 10?
$12\frac{1}{2}$ is what part of 100?	$16\frac{2}{3}$ is what part of 100?
125 is what part of 1000?	$166\frac{2}{3}$ is what part of 1000?

43. Copy and complete the following table:

	1	10	100	1000
$\frac{1}{8}$'s	$.12\frac{1}{2}$	$1\frac{1}{4}$	$12\frac{1}{2}$	125
$\frac{1}{4}$'s				
$\frac{1}{6}$'s				
$\frac{1}{3}$'s				

44. Make the extensions mentally and find the total value of the items in each group:

1. 24 A. @ $25.	**2.** 25 A. @ $96.
18 cows @ $25.	$33\frac{1}{3}$ bu. @ $1.08.
72 sheep @ $2\frac{1}{2}$.	$166\frac{2}{3}$ bu. @ $.84.
16 plows @ $12\frac{1}{2}$.	1250 bu. @ $.64.
32 horses @ $125.	2500 bu. @ $.78.
160 bu. @ $1\frac{1}{4}$.	1600 bu. @ $1.25.
330 bu. @ $1.66\frac{2}{3}$.	$16\frac{2}{3}$ bu. @ $1.44.
48 wagons @ $125.	3800 lb. @ $.12\frac{1}{2}$.

ALIQUOT PARTS APPLIED TO DIVISION

45. **1.** At 25¢ a yard, how many yards of cloth can be bought for $1? for $2? for $10? for $25?

2. At $37\frac{1}{2}$¢ a yard, how many yards can be bought for $1? for $2? for $5? for $15?

$37\frac{1}{2}$¢ $= \$\frac{3}{8}$. $\$1 \div \$\frac{3}{8} = \frac{8}{3}$; $\frac{8}{3}$ yd., or $2\frac{2}{3}$ yd., can be bought for $1.

Since $\frac{8}{3}$ yd., or $2\frac{2}{3}$ yd., can be bought for $1, 2 times $2\frac{2}{3}$ yd., or $5\frac{1}{3}$ yd., can be bought for $2, and 5 times $2\frac{2}{3}$ yd., or $13\frac{1}{3}$ yd., can be bought for $5, etc.

NOTE. $1 \div \frac{3}{8} = \frac{8}{3}$. $\frac{8}{3}$ is called the reciprocal of $\frac{3}{8}$. The reciprocal of any fraction is the fraction inverted.

Hence, to divide by an aliquot part, multiply by the reciprocal of the fractional equivalent.

46. Calculate the number of yards that can be bought in each case:

	Total Cost	Cost of One Yard		Total Cost	Cost of One Yard
1.	$15.00	$$.33\frac{1}{3}$	**16.**	$91.00	$$.58\frac{1}{3}$
2.	36.00	.75	**17.**	100.00	$.62\frac{1}{2}$
3.	75.00	$.62\frac{1}{2}$	**18.**	110.00	$.66\frac{2}{3}$
4.	25.00	$.31\frac{1}{4}$	**19.**	121.00	$.68\frac{3}{4}$
5.	13.00	$.14\frac{2}{7}$	**20.**	129.00	.75
6.	16.00	$.08\frac{1}{3}$	**21.**	130.00	$.81\frac{1}{4}$
7.	27.00	$.18\frac{3}{4}$	**22.**	135.00	$.83\frac{1}{3}$
8.	29.00	.25	**23.**	59.50	$.87\frac{1}{2}$
9.	35.00	$.31\frac{1}{4}$	**24.**	66.00	$.91\frac{2}{3}$
10.	41.00	$.33\frac{1}{3}$	**25.**	75.00	$.93\frac{3}{4}$
11.	42.00	$.37\frac{1}{2}$	**26.**	45.00	1.25*
12.	47.50	$.41\frac{2}{3}$	**27.**	93.00	1.50
13.	49.00	$.43\frac{3}{4}$	**28.**	90.00	$1.87\frac{1}{2}$
14.	57.00	.50	**29.**	105.00	2.50
15.	81.00	$.56\frac{1}{4}$	**30.**	440.00	$3.33\frac{1}{3}$

TRADE DISCOUNT

47. All the information about aliquot parts in the preceding pages is directly applicable to Trade Discount.

48. The expression "per cent" (symbol %) means hundredths. That is, $25\% = \frac{25}{100} = \frac{1}{4}$. It is also written .25, in its decimal form, the figures being the same as are used to write 25¢ decimally. It follows, that for every price quoted in the problems in Aliquot Parts,

* $1.25 = \frac{5}{4}$, the reciprocal of which is $\frac{4}{5}$. Hence the number of yards $= \frac{4}{5}$ of 45.

the fractional part would be the same if the sign % were substituted for " cents."

Therefore, $12\frac{1}{2}\%$, $33\frac{1}{3}\%$, $56\frac{1}{4}\%$, etc. represent the same common fractions, or aliquot parts, as do $12\frac{1}{2}¢$, $33\frac{1}{3}¢$, $56\frac{1}{4}¢$, etc.

49. Find the net cost of the following, allowing a discount of $37\frac{1}{2}\%$:

$$66 \text{ lb. @ } 66\frac{2}{3}¢ = \$44$$
$$80 \text{ lb. @ } 40¢ = \$32$$
$$120 \text{ lb. @ } 75¢ = \$90$$
$$240 \text{ lb. @ } 37\frac{1}{2}¢ = \$90 \qquad \$256$$

Less $37\frac{1}{2}\%$ $(\frac{1}{4} + \frac{1}{2}$ of $\frac{1}{4})$ 96 $160

The items are extended mentally by taking $\frac{2}{3}$ of 66; $\frac{2}{3}$ of 80; $\frac{3}{4}$ of 120; and $\frac{3}{8}$ of 240. By adding, the gross cost is found to be $256. $37\frac{1}{2}\%$ discount = a discount of $\frac{3}{8}(\frac{1}{4} + \frac{1}{8})$. $\frac{3}{8}$ of $256 = $96, which, deducted, leaves $160, the net cost.

50. Find the net cost of each of the following:

1.	2.	3.
72 lb. @ $16\frac{2}{3}¢$	72 lb. @ $25¢$.	36 doz. @ $75¢$.
48 lb. @ $6\frac{1}{4}¢$	96 lb. @ $33\frac{1}{3}¢$.	48 doz. @ $12\frac{1}{2}¢$.
32 lb. @ $12\frac{1}{2}¢$	90 lb. @ $44\frac{4}{9}¢$.	60 doz. @ $25¢$.
56 lb. @ $37\frac{1}{2}¢$.	80 lb. @ $87\frac{1}{2}¢$.	72 doz. @ $37\frac{1}{2}¢$.
Less $12\frac{1}{2}\%$.	Less $18\frac{3}{4}\%$.	Less $16\frac{2}{3}\%$.

4.	5.	6.
48 doz. @ $6\frac{1}{4}¢$	64 lb. @ $12\frac{1}{2}¢$.	125 lb. @ $48¢$.
64 doz. @ $1.25.	240 lb. @ $33\frac{1}{3}¢$.	375 lb. @ $72¢$.
24 doz. @ $84¢$.	480 lb. @ $56\frac{1}{4}¢$.	320 lb. @ $62\frac{1}{2}¢$.
$37\frac{1}{2}$ doz. @ $2.40.	320 lb. @ $6\frac{1}{4}¢$.	250 lb. @ $36¢$.
Less $33\frac{1}{3}\%$.	Less 10 %.	Less $12\frac{1}{2}\%$.

7.	8.	9.
196 doz. @ $1.25.	326 doz. @ $62\frac{1}{2}¢$.	841 lb. @ $12\frac{1}{2}¢$.
26 doz. @ $75¢$.	112 doz. @ $75¢$.	963 lb. @ $16\frac{2}{3}¢$.
86 doz. @ $12\frac{1}{2}¢$.	84 doz. @ $62\frac{1}{2}¢$.	1047 lb. @ $25¢$.
50 doz. @ $74¢$.	2500 lb. @ $6\frac{1}{4}¢$.	141 lb. @ $33\frac{1}{3}¢$.
Less 25%.	Less $37\frac{1}{2}\%$.	Less 20 %.

FURTHER SUGGESTIONS FOR SHORTENING MULTIPLICATION

51. In addition to the so-called Aliquot Parts of 100, the use and application of which have been explained, there are many other numbers that are a fractional part more or less than the aliquot parts. By making use of this fact the work of multiplying by numbers that are not aliquot parts of 100 may be shortened. For example, $13\frac{1}{3}$ ¢ is not an aliquot part of a dollar, yet to multiply by $.13\frac{1}{3}$ requires no actual multiplication.

1. Multiply 120 by $.13\frac{1}{3}$.

$$.13\frac{1}{3} = .10 + (\tfrac{1}{3} \text{ of } .10).$$

Therefore, to multiply 120 by $.13\frac{1}{3}$, point off one place in 120, thus, 12.0, and add $\frac{1}{3}$ of 12, which gives a result of 16.

2. Multiply 36 by $.23\frac{1}{3}$.

$$.23\tfrac{1}{3} = .20 + (\tfrac{1}{6} \text{ of } .20) = \tfrac{2}{10} + (\tfrac{1}{6} \text{ of } \tfrac{2}{10}).$$
$$3.6 = \tfrac{1}{10} \text{ of } 36.$$
$$7.2 = \tfrac{2}{10} \text{ of } 36.$$
$$1.2 = \tfrac{1}{6} \text{ of } \tfrac{2}{10} \text{ of } 36.$$
$$8.4 = .23\tfrac{1}{3} \text{ of } 36.$$

Moving the point one place to the left in 36 gives $\frac{1}{10}$ of 36; multiplying by 2 gives $\frac{2}{10}$ of 36. To find the remaining part take $\frac{1}{6}$ of 7.2, which is 1.2. Add this to 7.2, giving 8.4, the desired result.

3. To multiply by $.02\frac{1}{2}$, point off 1 place and divide by 4.

4. To multiply by $.03\frac{1}{3}$, point off 1 place and divide by 3.

5. To multiply by $.05$, point off 1 place and divide by 2.

6. To multiply by $.07\frac{1}{2}$, point off 1 place and deduct $\frac{1}{4}$ of the result.

7. To multiply by $.11\frac{1}{4}$, point off 1 place and add $\frac{1}{8}$ of the result.

8. To multiply by $.13\frac{3}{4}$, point off 1 place and add $\frac{1}{4}$ of the result, plus $\frac{1}{2}$ of the $\frac{1}{4}$. Etc.

EXERCISE

52. Devise a short method of multiplying by each of the following:

1.	.09	**7.**	$.17\frac{1}{2}$	**13.**	$.23\frac{3}{4}$	**19.**	$.28\frac{3}{4}$	**25.**	$.37\frac{1}{2}$
2.	.10	**8.**	.18	**14.**	.24	**20.**	$.31\frac{1}{4}$	**26.**	.44
3.	.11	**9.**	.20	**15.**	.26	**21.**	.33	**27.**	.45
4.	$.11\frac{1}{4}$	**10.**	.22	**16.**	$.26\frac{2}{3}$	**22.**	$.33\frac{3}{4}$	**28.**	$112\frac{1}{2}$
5.	$.13\frac{1}{3}$	**11.**	$.22\frac{1}{2}$	**17.**	.27	**23.**	.35	**29.**	425
6.	.15	**12.**	$.23\frac{1}{3}$	**18.**	$.27\frac{1}{2}$	**24.**	.36	**30.**	375

53. Write by inspection the product of each of the following, and find the total of all the products in each group:

1.	2.	3.
24 lb. @ 2½¢.	123 lb. @ 10¢.	84 lb. @ 15¢.
24 lb. @ 3⅓¢.	113 lb. @ 11¢.	92 lb. @ 17½¢.
48 lb. @ 5¢.	120 lb. @ 11¼¢.	64 lb. @ 22 ¢.
44 lb. @ 7½¢.	140 lb. @ 12¢.	72 lb. @ 23¾¢.
56 lb. @ 9¢.	150 lb. @ 13⅓¢.	44 lb. @ 24¢.

4.	5.	6.
54 lb. @ 26⅔¢.	37 by 11.	140 by 31¼.
88 lb. @ 27½¢.	28 by 13¾.	150 by 37½.
144 lb. @ 31¼¢.	44 by 18¾.	160 by 43¾.
160 lb. @ 33¾¢.	48 by 22.	180 by 45.
176 lb. @ 35¢.	56 by 22½.	190 by 47½.

54. To find the cost of goods sold by the 100, 1000, or ton.

1. Find the cost of 168 lb. of feed @ $1.10 per cwt.

$1.68 = cost at $1   per cwt.   First find the number of hundred-
 .17 = cost at   .10 per cwt.   weight by pointing off 2 places in the
$1.85 = cost at 1.10 per cwt.   number of pounds, thus, 168 lb. = 1.68

cwt. At $1 per hundredweight, the cost would be $1.68. At $.10 per hundredweight, the cost would be $.168, or $.17. Hence, at $1.10 per hundredweight, the cost would be the sum of $1.68 and $.17, or $1.85.

2. At $1.87 per C, find the cost of 634 bolts.

634 = 6.34 C.   634 bolts = 6.34 hundreds. Since 1 hundred
6.34 × $1.87 = $11.86   costs $1.87, 6.34 hundreds cost 6.34 times $1.87, or $11.86.

3. At $15 per thousand, find the cost of 648 ft. of lumber.

$6.48 = cost at 1¢ per foot. $15 per M = 1½¢ per foot. At 1¢ a
 3.24 = cost at ½¢ per foot.   foot, 648 ft. cost $6.48, and at ½¢ a foot,
$9.72 = cost at $15 per M.   the cost is ½ of $6.48, or $3.24. Hence,

at 1½¢ a foot or $15 per M, the cost is $6.48 + $3.24, or $9.72.

4. At $16.40 per M, find the cost of 13,400 bricks.

13400 = 13.4 thousands.   The number of M is found by pointing off
13.4 × $16.40 = $219.76   3 places in the number 13,400, giving 13.4 M.

At $16.40 per M, 13.4 M cost 13.4 × $16.40, or $219.76.

5. At $6.50 per ton, how much will 13,480 lb. of coal cost?

$6.50 per ton = $3.25 per 1000 lb.

13.48 × $3.25 = $43.81

Since there are 2000 lb. in a ton, the price of 1000 lb. is ½ of the price of a ton, or $3.25.

13,480 lb. equal 13.48 thousand pounds. Since 1 thousand pounds cost $3.25, 13.48 thousands cost 13.48 × $3.25, or $43.81.

6. What is the cost of 24,680 lb. of hay at $18 per ton?

$18 per ton = $9 per 1000 lb.

24680 = 24.68 thousands.

$246.80 = cost at $10 per 1000 lb.

  24.68 = cost at   1 per 1000 lb.

$222.12 = cost at   9 per 1000 lb.

At $18 per ton, the cost of ½ of a ton, or 1000 lb., is $9. The number of thousands is found by pointing off 3 places in the number of pounds, thus, 24,680 lb. = 24.680 thousand pounds. If the price were $10 per 1000 lb., the cost would be $246.80 (point moved 1 place to the right). Deducting the cost at $1 per 1000 lb., or $24.68, leaves $222.12, the required result.

Note. The student should study each problem to see if there is not some short, easy solution. Make solutions mentally as far as possible, using pen or pencil only to write the results. Many of the everyday problems of business admit of short solutions which can be performed mentally, and with great rapidity and accuracy. In the preceding pages many suggestions have been made. The learner must not be content with using these alone. He should seek to make new short methods for himself, for only by so doing will he become efficient in the handling of numbers.

55. Solve mentally, if possible, the following:

Find the cost of:

1.	220 lb. bran	@ $ .90	per cwt. ;	@ $ 1.10	per cwt.
2.	350 lb. feed	@ 1.25	per cwt. ;	@ 1.15	per cwt.
3.	675 lb. nails	@ 3.12½	per cwt. ;	@ 3.37½	per cwt.
4.	875 lb. salt	@ .80	per cwt. ;	@ .88	per cwt.
5.	950 posts	@ 6.25	per C ;	@ 6.75	per C.
6.	1100 posts	@ 7.25	per C ;	@ 7.70	per C.
7.	1250 posts	@ 8.00	per C ;	@ 7.20	per C.
8.	1550 lb. lead	@ 2.50	per cwt. ;	@ 3.37½	per cwt.
9.	38,440 lb. pig iron	@ 1.50	per cwt. ;	@ 1.62½	per cwt.
10.	24,680 lb. sugar	@ 5.50	per cwt. ;	@ 5.75	per cwt.
11.	1875 ft. lumber	@ 16.00	per M ;	@ 18.00	per M.

12.	1450 ft. lumber	@	$ 25.00	per M ;	@	$ 22.00 per M.
13.	23,360 brick	@	8.75	per M ;	@	9.90 per M.
14.	14,500 shingles	@	5.62½	per M ;	@	5.25 per M.
15.	346 ft. lumber	@	33.00	per M ;	@	31.25 per M.
16.	23,465 ft. lumber	@	37.50	per M ;	@	33.75 per M.
17.	47,860 brick	@	11.25	per M ;	@	12.50 per M.
18.	12,375 brick	@	13.75	per M ;	@	15.00 per M.
19.	7850 shingles	@	6.75	per M ;	@	7.50 per M.
20.	319 ft. lumber	@	45.00	per M ;	@	55.00 per M.
21.	2660 lb. hay	@	18.00	per T. ;	@	22.50 per T.
22.	3474 lb. hay	@	26.00	per T. ;	@	24.00 per T.
23.	3096 lb. straw	@	14.50	per T. ;	@	13.75 per T.
24.	7276 lb. straw	@	13.75	per T. ;	@	12.50 per T.
25.	1680 lb. feed	@	24.60	per T. ;	@	27.90 per T.
26.	4288 lb. feed	@	26.00	per T. ;	@	27.00 per T.
27.	13,890 lb. coal	@	6.25	per T. ;	@	5.50 per T.
28.	47,976 lb. coal	@	6.75	per T. ;	@	6.60 per T.
29.	1890 lb. coal	@	4.50	per T. ;	@	4.40 per T.
30.	560 lb. coal	@	6.40	per T. ;	@	6.25 per T.

ALIQUOT PARTS APPLIED TO SIMPLE INTEREST

56. Simple interest * is treated (briefly) in this part of the text for two reasons: first, because it is required early in the course by all students who are at the same time pursuing the study of bookkeeping; and second, to correlate it with the work in "Aliquot Parts."

57. Interest is the price paid for the use of money.

58. The principal is the money for the use of which interest is paid.

59. For ordinary interest calculations, a year is considered as having 12 mo. of 30 da. each, or 360 da.

At 6 % for one year the interest on $1 is $.06.
2 mo., or 60 da., is ⅙ of a year.
Hence, at 6 % for 2 mo. or 60 da., the interest

on $1 is $.01
on $2 is $.02
on $15 is $.15
on $125 is $1.25
on $750.25 is $7.5025 = $7.50.

* For further work in interest see pages 270–291.

60. Find the interest on $450.50 for 60 da. at 6 %.

$$\$ 4|50.50 = \$ 4.51, \text{ int. at } 6 \% \text{ for } 60 \text{ da.}$$

Since at 6% for 60 da. the interest on $1 is $.01, the interest on $450.50 is $.01 for each dollar, or $ 4*5050, which equals $4.51. (Answers in interest problems are always given in the nearest whole cent.)

A vertical line is used in place of a decimal point in solutions of this kind.

61. Find the interest at 6 % on each of the following for 60 da.:

1. - $630	**5.** $846	**9.** $2465	**13.** $824.75
2. $750	**6.** $932	**10.** $3675	**14.** $968.66
3. $365	**7.** $1036	**11.** $4270	**15.** $732.45
4. $975	**8.** $1145	**12.** $362.50	**16.** $1248.72

62. In the preceding pages on aliquot parts 100 was the basis of comparison. In interest solutions 60 is the basis. Memorize the following aliquot parts of 60. (Compare with the parts of an hour.)

$30 = \frac{1}{2}$ of 60	$12 = \frac{1}{5}$ of 60	$5 = \frac{1}{12}$ of 60
$20 = \frac{1}{3}$ of 60	$10 = \frac{1}{6}$ of 60	$4 = \frac{1}{15}$ of 60
$15 = \frac{1}{4}$ of 60	$6 = \frac{1}{10}$ of 60	$3 = \frac{1}{20}$ of 60

63. What combinations of 60 da. and aliquot parts of 60 da. will best give (a) 45 da.? (b) 27 da.? (c) 85 da.?

(a) 60 da.	(b) 60 da.	(c) 60 da.
15 da. $= \frac{1}{4}$ of 60	20 da. $= \frac{1}{3}$ of 60	20 da. $= \frac{1}{3}$ of 60
45 da	6 da. $= \frac{1}{10}$ of 60	5 da. $= \frac{1}{4}$ of 20
	1 da. $= \frac{1}{6}$ of 6	85 da.
	27 da	
Subtract 15 da.	Add 20 da. 6 da. and 1 da.	Add 60 da. 20 da. and 5 da.

64. Find combinations of 60 da. and aliquot parts of 60 da. that will best amount to:

1. 55 da.	**6.** 26 da.	**11.** 84 da.	**16.** 76 da.
2. 40 da.	**7.** 66 da.	**12.** 90 da.	**17.** 96 da.
3. 33 da.	**8.** 75 da.	**13.** 11 da.	**18.** 105 da.
4. 35 da.	**9.** 72 da.	**14.** 14 da.	**19.** 110 da.
5. 25 da.	**10.** 80 da.	**15.** 16 da.	**20.** 135 da.

65. 1. Find the interest at 6 % on $ 540 for 22 da.

$ 5|40 = int. for 60 da.   22 da. is composed of the aliquot parts
$\overline{1|80}$ = int. for 20 da.   20 da. and 2 da.
$|18$ = int. for 2 da. $ 5.40 is the interest for 60 da.
$\overline{1|98}$ = int. for 22 da.   20 da. is $\frac{1}{3}$ of 60 da , hence the interest
                             for 20 da. is $\frac{1}{3}$ of $ 5.40, or $ 1.80. The re-
maining 2 da. is $\frac{1}{10}$ of 20 da., and the interest for 2 da. is $\frac{1}{10}$ of $ 1.80, or $.18.
The interest for 22 da. is $ 1.98.

2. Find the interest at 6 % on $ 1716.75 for 87 da.

$ 17|1675 = int. for 60 da.   87 da. = 60 da. + 20 da. + 6 da. + 1 da.
$5|7225$ = int. for 20 da.   Find 60 da. interest, which is $ 17.1675
$1|7167$ = int. for 6 da.   and add to it $\frac{1}{3}$ of itself (for 20 da.), $\frac{1}{10}$ of
$|2861$ = int. for 1 da.   itself (for 6 da.), and $\frac{1}{6}$ of the interest for
$\overline{$ 24|8928}$ = int. for 87 da.   6 da. (for 1 da.) which gives $ 24.89, the
or $ 24|89 = int. for 87 da.   interest for 87 da.

                             NOTES. (1) The vertical line takes the
                             place of the decimal points in the solution.
(2) The interest for fractional parts of the time should be kept correct to the
4th decimal place to insure accuracy in the cents column.

66. Find the interest at 6 % on :

1. $ 260 for 30 da.	**13.** $ 2100 for 72 da.	**25.** $ 1600 for 150 da.
2. $ 360 for 45 da.	**14.** $ 2200 for 96 da.	**26.** $ 1800 for 210 da.
3. $ 480 for 50 da.	**15.** $ 2600 for 105 da.	**27.** $ 3000 for 180 da.
4. $ 560 for 36 da.	**16.** $ 3000 for 16 da.	**28.** $ 4000 for 240 da.
5. $ 840 for 26 da.	**17.** $ 4200 for 30 da.	**29.** $ 4500 for 320 da.
6. $ 720 for 25 da.	**18.** $ 4800 for 33 da.	**30.** $ 4800 for 160 da.
7. $ 960 for 40 da.	**19.** $ 5100 for 45 da.	**31.** $ 5000 for 150 da.
8. $ 900 for 44 da.	**20.** $ 8000 for 45 da.	**32.** $ 6300 for 160 da.
9. $ 1260 for 70 da.	**21.** $ 9000 for 50 da.	**33.** $ 6600 for 140 da.
10. $ 1600 for 90 da.	**22.** $ 9600 for 90 da.	**34.** $ 7200 for 115 da.
11. $ 1800 for 80 da.	**23.** $ 1000 for 75 da.	**35.** $ 7500 for 100 da.
12. $ 2000 for 75 da.	**24.** $ 1340 for 120 da.	**36.** $ 8400 for 99 da.

67. 1. What is the interest on $100 at 6 % for 6 da.?

6 da. is what part of 60 da. ?
What is the interest on $100 at 6% for 60 da. ?
What is the interest on $100 at 6% for 6 da. ?
What is the quickest way of finding the interest on $100 at 6% for 6 da. ?
What is the quickest way of finding the interest for 6 da. at 6% on any sum of money ?

2. Find the interest on $1460 at 6 % for 6 da.

$1.460 = interest for 6 da.
Move the decimal point 3 places to the left.   The result is the interest at 6% for 6 da.

Having the interest for 6 da., how may the interest be found for

1 da. ?	4 da. ?	8 da. ?	18 da. ?
2 da. ?	5 da. ?	9 da. ?	24 da. ?
3 da. ?	7 da. ?	10 da. ?	36 da. ?

68. Find the interest at 6 % on the following :

1. $400 for 6 da.	**8.** $1560 for 36 da.	**15.** $1480 for 24 da.
2. $800 for 6 da.	**9.** $1280 for 42 da.	**16.** $1600 for 27 da.
3. $1000 for 9 da.	**10.** $2000 for 39 da.	**17.** $2100 for 7 da.
4. $1200 for 8 da.	**11.** $1250 for 48 da.	**18.** $2400 for 3 da.
5. $1400 for 18 da.	**12.** $1125 for 24 da.	**19.** $3600 for 2 da.
6. $1500 for 7 da.	**13.** $1440 for 19 da.	**20.** $4200 for 1 da.
7. $1800 for 24 da.	**14.** $1280 for 21 da.	**21.** $9000 for 4 da.

69. Find the total interest at 6 % on :

1	**2.**	**3.**
$420 for 18 da.	$630 for 11 da.	$1300 for 66 da.
$540 for 20 da.	$750 for 9 da.	$1560 for 72 da.
$640 for 21 da.	$780 for 26 da.	$1480 for 78 da.
$350 for 13 da.	$960 for 36 da.	$1260 for 84 da.
$432 for 15 da.	$1200 for 49 da.	$1450 for 90 da.
$128 for 16 da.	$840 for 50 da.	$1660 for 91 da.

4.	5.	6.
$1728 for 95 da.	$800 for 31 da.	$320 for 24 da.
$1890 for 98 da.	$900 for 41 da.	$390 for 2 da.
$1820 for 100 da.	$1000 for 48 da.	$410 for 12 da.
$1675 for 108 da.	$460 for 27 da.	$812 for 15 da.
$1440 for 110 da.	$390 for 19 da.	$915 for 20 da.
$1360 for 120 da.	$260 for 16 da.	$990 for 10 da.

7.	8.	9.
$828 for 45 da.	$736 for 105 da.	$482.75 for 12 da.
$936 for 15 da.	$888 for 110 da.	$321.50 for 18 da.
$832 for 75 da.	$960 for 115 da.	$428.48 for 15 da.
$648 for 70 da.	$822 for 150 da.	$320.10 for 20 da.
$444 for 80 da.	$440 for 130 da.	$516.74 for 24 da.
$555 for 40 da.	$375 for 88 da.	$608.31 for 30 da.

10.	11.	12.
$597.30 for 36 da.	$731.45 for 13 da.	$673.87 for 21 da.
$312.47 for 40 da.	$806.40 for 14 da.	$832.49 for 22 da.
$218.38 for 42 da.	$318.25 for 11 da.	$738.27 for 25 da.
$712.80 for 6 da.	$208.14 for 10 da.	$387.39 for 29 da.
$811.75 for 10 da.	$327.38 for 16 da.	$873.93 for 5 da.
$928.30 for 30 da.	$445.50 for 19 da.	$783.79 for 3 da.

13.	14.	15.
$720.30 for 25 da.	$475 for 26 da.	$425 for 38 da.
$628.50 for 44 da.	$938 for 33 da.	$576 for 67 da.
$731.20 for 14 da.	$840 for 17 da.	$545 for 88 da.
$906.75 for 9 da.	$729 for 19 da.	$750 for 112 da.
$887.63 for 12 da.	$325 for 39 da.	$800 for 99 da.
$674.70 for 13 da.	$480 for 47 da.	$235 for 81 da.

ADDITION

70. Addition is the process of finding one number or quantity equal in value to two or more other numbers or quantities.

71. The several numbers or quantities to be added are called addends.

72. Investigation shows that the most important arithmetical operation is simple addition. It is important that it be done rapidly and accurately.

The following suggestions will be found helpful.

73. Drill on the 45 primary combinations is the first essential. They must be perfectly memorized. They are

1	1	1	1	1	1	1	1	1
1	2	3	4	5	6	7	8	9
	2	2	2	2	2	2	2	2
	2	3	4	5	6	7	8	9
		3	3	3	3	3	3	3
		3	4	5	6	7	8	9
			4	4	4	4	4	4
			4	5	6	7	8	9
				5	5	5	5	5
				5	6	7	8	9
					6	6	6	6
					6	7	8	9
						7	7	7
						7	8	9
							8	8
							8	9
								9
								9

When the combinations on p. 31 **are** perfectly memorized, drill on the following and others like them:

11	11	11	12	13	15	**15**	16	**14**	16
2	4	6	3	4	5	7	6	8	5

16	17	15	19	17	18	19	18	19	15
7	4	8	6	7	8	8	9	9	4

Note that the units' figures present the same combinations as before. Continue the drill on such numbers as the following until the results are recognized instantly:

32	26	43	27	45	53	76	47	98	54
9	7	8	8	7	9	8	6	3	7

ORAL DRILL EXERCISES

74. 1. Count by 3's from 1 to 100; from 2 to 101.

2. Count by 4's from 1 to 101; from 2 to 102; from 3 to 103.

3. Count by 5's beginning with 0; 1; 2; 3; 4, until 100 is passed.

4. Count by 6's beginning with 0; 1; 2; 3; 4; 5, until 100 is passed.

5. Count by 7's beginning with 0; 1; 2; 3; 4; 5; 6, until 100 is passed.

6. Count by 8's beginning with 0; 1; 2; 3; 4; 5; 6; 7, until 100 is passed.

7. Count by 9's beginning with 0; 1; 2; 3; 4; 5; 6; 7; 8, until 100 is passed.

8. Count by 10's beginning with 0; 1; 2; 3; 4; 5; 6; 7; 8; 9, until 100 is passed.

9. Count by 11's beginning with 0; 1; 2; 3; 4; 5; 6; 7; 8; 9; 10, until 100 is passed.

10. Count by 12's beginning with 0; 1; 2; 3; 4; 5; 6; 7; 8; 9; 10; 11, until 100 is passed.

RAPID DRILL EXERCISES

75. Drill on the following until you can state the results immediately:

9	8	7	6	9	4	5	3	7	9
1	8	4	2	1	3	5	6	1	2
2	1	1	3	3	3	5	2	4	2
7	1	5	5	6	4	4	2	5	6

9	8	7	6	9	9	7	8	9	6
1	2	3	4	5	2	4	4	7	3
9	8	7	6	5	8	6	6	3	7

17	18	24	39	43	67	76	43	61	55
2	3	5	2	8	4	2	2	1	3
3	3	4	6	1	4	7	4	7	5
5	6	1	2	1	2	1	4	2	2

38	47	64	89	79	97	48	53	89	62
9	8	3	4	3	6	5	1	9	8
1	2	7	6	7	4	5	9	1	2

56	67	83	47	82	69	43	67	79	88
2	3	6	3	8	1	4	2	3	4
4	2	3	4	1	4	4	6	6	1
4	5	1	3	1	5	2	2	1	5

81	47	74	82	71	89	74	90	27	44
6	3	4	4	6	2	6	8	3	4
2	4	5	3	1	2	1	1	4	4
2	3	1	3	3	6	3	1	3	2

76. Two important aids to accuracy and speed in addition work are: (1) to make the figures neatly and of uniform size; and (2) to write them exactly in columns.

Add:

1.	2.	3.	4.	5.	6.	7.	8.	9.	10.
64	59	94	56	49	59	67	61	67	56
28	63	89	38	84	63	38	49	84	73
39	87	21	27	47	72	79	72	98	87

11.	12.	13.	14.	15.	16.	17.	18.	19.	20.
69	19	64	56	59	53	48	53	43	53
76	66	83	79	32	87	72	89	72	87
87	74	39	96	27	98	28	72	29	91
24	47	77	87	87	82	63	68	38	76

21.	22.	23.	24.	25.	26.	27.	28.	29.	30.
43	74	92	94	69	88	63	48	56	45
82	91	28	76	78	72	79	27	74	97
79	43	74	87	86	38	64	68	83	86
87	87	68	92	97	76	87	93	92	73
64	96	93	47	74	43	99	27	76	49
47	78	72	83	83	89	76	64	57	36
82	47	46	97	72	96	64	56	83	17
96	64	87	48	87	74	87	74	92	85

77. The rapidity of addition may be increased by adding figures by groups. Note the groups of 10 in the following. Add rapidly:

1.	2.	3.	4.	5.	6.	7.	8.	9.	10.
4	7	4	6	3	9	7	4	5	6
6	3	3	3	4	1	3	4	3	2
8	9	3	1	3	4	6	2	2	2
2	1	7	4	5	2	2	5	4	3
7	5	2	4	2	3	2	4	2	5
3	5	1	2	3	1	8	1	4	2

11.	12.	13.	14.	15.	16.	17.	18.	19.	20.
37	49	67	56	56	47	31	77	47	49
84	36	84	34	74	39	63	69	83	84
97	83	97	73	64	28	39	83	76	39
26	92	32	91	59	37	87	65	64	73
84	78	43	79	76	73	79	27	96	42
97	46	88	65	85	81	73	39	89	87
43	87	79	82	93	79	47	47	97	65

CHECKING THE WORK

78. **Checking** is a means of testing the accuracy of the work. Accuracy can be obtained by frequently checking the results.

79. The best method of checking addition is to add the columns in the reverse order. If the column is first added from the bottom up, check by adding from the top down.

80. Many bookkeepers and accountants use the method shown in the following illustrative solution:

Add:

(1st)	3967	(Check)
26	8234	27
31	7987	32
32	4372	31
27	5976	26
30536	30536	30536

By adding the units' column, a sum of 26 is obtained, which is written as shown. Next, by adding the tens' column, the sum is found to be 31, which is written as in the illustration. The sums of the hundreds' and thousands' columns are 32 and 27, respectively, which are written as here shown.

Beginning with the thousands' column and adding in the opposite direction, we find the sum to be 27, which is written as shown in the column marked "Check." The sums of the hundreds', tens', and units' columns, which are 32, 31, and 26, respectively, are written as in the illustration. By adding these several sums the result is seen to agree with the first result obtained. Hence the work is probably correct.

81. **Casting out nines.**

$$147 = 12 = 3$$
$$229 = 13 = 4$$
$$331 = 7 = 7$$
$$\overline{707 \quad 14 = 14}$$
$$5 = 5$$

The sum of the digits 1, 4, and 7 is 12, equal to one 9 and 3 over. The sum of the digits in the second addend is 13, equal to one 9 and 4 over; and the sum in the third addend is 7, equal to no 9's and 7 over. The sum of the several remainders, 3, 4, and 7, is 14, equal to one 9 and 5 over. The sum of the digits in the sum, 707, is 14, equal to one 9 and 5 over. Hence the addition is probably correct.

NOTES. (1) The digits of 147 are the separate numbers 1, 4, and 7.

(2) The remainders 3, 4, 7, and 5 are the sums of the digits in 12, 13, 7, and 14, respectively. 12 is equal to one 9 with a remainder of 3. The division is unnecessary, however, as the excess (remainder) can always be found by adding the digits.

(3) The remainders 3, 4, and 7 are generally called the "excess of nines," and the principle is stated thus: "The excess of nines in a sum is equal to the excess in the sum of the excesses."

(4) Casting out nines is not an absolute test of accuracy. Transposition of figures cannot be detected by this check, nor can the omission or addition of 9's or ciphers be discovered.

82. Casting out elevens.

$$475 = 2$$
$$382 = 8$$
$$479 = 6$$
$$\overline{1336 = 5}$$

Find the sum of the digits in the odd places (beginning at units) and from it deduct the sum of the digits in the even places. If the sum of the digits in the even places exceeds the sum of those in the odd places, add 11, or some multiple of 11, to the sum of the digits in the odd places.

Thus,

$$\left. \begin{array}{l} 5 + 4 - 7 = 2 \\ 2 + 3 + 11 - 8 = 8 \\ \underline{9 + 4 - 7 = 6} \\ \overline{6 + 3 - 3 - 1 = 5} \end{array} \right\} \begin{array}{l} 6 + 8 + 2 = 16; \\ 16 = \text{one 11 and 5 over.} \end{array}$$

NOTE. Casting out elevens will detect errors not located by the nine check, as indicated in note (4), p. 35.

83. Add the following, and check the results:

1.	2.	3.	4.	5.	6.
2874	8727	8237	9371	6738	3796
3957	3469	6489	6487	9674	3987
6498	5487	5387	5238	5976	4598
7237	3245	6959	9764	8238	3294
4596	5894	6432	5437	9783	5786
3872	7978	8643	6087	4598	5487
5295	5673	9782	9121	1359	7372
9638	9178	9176	4327	9385	9765

84. Each of the following columns should be added in fifteen seconds. Drill on them until correct results can be obtained in that time.

1.	2.	3.	4.	5.	6.
4789	7968	4894	4893	4389	1531
3847	7324	3762	3469	5767	4574
5963	8759	5437	9276	9247	8276
8769	4365	9864	1498	6389	9832
5271	9859	4767	9763	9127	4976
8728	7312	5943	7987	8784	2872
9592	8764	8976	6477	4728	3193
7352	5231	4382	9816	2854	8715
5897	9827	4767	8125	1645	2810
7684	5876	5723	7784	5196	4567

7.	8.	9.	10.	11.
32478672	42876597	45967248	12345678	24681357
45972861	27638976	53876289	98765432	91352468
56972764	45972489	93856214	87654321	57913546
36824579	72510764	72387276	34567892	94872468
59472876	63897123	10468217	45678923	74692789

12.	13.	14.	15.	16.
76945897	54763247	57689727	47897684	98753214
34872469	98674382	68972543	32076483	76543216
89778976	76948269	74202604	82478978	97531246
69877978	63897476	70876437	68764376	89765382
45987678	48769531	97687248	98765432	74974768

85. Each of the following columns should be added in one minute. Drill on them until it can be done with facility and accuracy.

	1.	2.	3.	4.	5.
1.	34568	93876	69784	38976	94768
2.	72487	47876	72864	72487	48694
3.	95438	83297	14897	92764	38972
4.	64876	45872	32571	12476	66666
5.	45987	76978	89724	53897	89764
6.	98321	32589	89724	69748	92787
7.	87643	13489	97247	48724	77777
8.	97987	87943	83971	93697	48938
9.	34786	76897	72481	54328	55555
10.	49876	37643	92482	97648	48743
11.	24689	28971	76893	63872	33333
12.	39872	59761	64762	59763	98697
13.	97246	97246	59769	97687	88888
14.	72489	38972	43872	53792	46479
15.	93247	89726	74672	47867	99999
16.	59876	97642	94897	83897	72687
17.	72489	57921	58932	46876	22222
18.	72979	38761	72468	93872	48769

SUGGESTION. The number of problems in the above columns can be increased by requiring pupils to find the sum of numbers 1 to 10, inclusive; 2 to 11; 3 to 12; 4 to 13; etc.

ADDITION

ACCURACY TEST

86. Add each column rapidly.   Check results with those given.

WORLD'S PRODUCTION OF GOLD AND SILVER.   (CALENDAR YEAR, 1907.)

COUNTRY	GOLD		SILVER		
	Ounces (fine)	Value	Ounces (fine)	Coining Value	Commercial Value
North America:					
United States . .	4,374,827	$90,435,700	56,514,700	$73,069,500	$37,299,700
Mexico	903,699	18,681,100	61,147,203	79,059,000	40,357,200
Canada . . .	405,553	8,383,500	12,779,800	16,523,400	8,434,700
Africa	7,338,468	151,699,600	790,431	1,022,000	521,700
Australasia . .	3,660,911	75,677,700	19,083,031	24,673,000	12,594,800
Europe:					
Russia	1,290,840	26,684,000	132,122	170,800	87,200
Austria-Hungary .	120,209	2,484,900	1,744,233	2,255,200	1,151,200
Germany . . .	3,220	66,600	5,088,086	6,578,500	3,358,100
Norway . . .			201,516	260,500	133,000
Sweden	903	18,700	29,761	38,500	19,600
Italy	1,914	39,600	737,843	954,000	487,000
Spain.			4,097,035	5,297,200	2,704,000
Greece			829,025	1,071,900	547,200
Turkey	216	4,500	67,351	87,100	44,500
France	24,305	502,400	719,453	930,200	474,800
Great Britain . .	1,414	29,200	137,216	177,400	90,600
Servia	2,893	59,800			
South America:					
Argentina . . .	4,985	103,000	25,178	32,600	16,600
Bolivia ⎫ . .	21,402	442,400	5,222,358	6,752,100	3,446,800
Chile ⎭					
Colombia . . .	157,491	3,255,600	1,048,719	1,355,900	692,200
Ecuador . .	12,923	267,100	2,456	3,200	1,600
Brazil . .	97,750	2,020,700			
Venezuela . . .	1,082	22,400			
Guiana:					
British . . .	63,099	1,304,400			
Dutch	30,961	640,000			
French . . .	89,923	1,858,900			
Peru	24,890	514,500	9,566,118	12,368,300	6,313,600
Uruguay . . .	2,510	51,900			
Central America .	101,965	2,107,800	1,892,896	2,447,400	1,249,300
Asia:					
Japan	134,059	2,771,200	2,835,532	3,666,100	1,871,400
China . . .	217,688	4,500,000			
Korea . . .	105,013	2,170,800			
Siam	8,038	166,200			
India (British) .	502,307	10,383,600			
East Indies:					
British . . .	75,525	1,561,300			
Dutch	79,637	1,646,200	322,560	417,000	212,900
Total . .	19,860,620	410,555,300	185,014,623	239,210,800	122,109,700

87. Find the total number of miles of railroad in continental Europe as given in the following table, for each year. (U. S. Consular and Trade Report.)

Country	1905	1906	1907	1908	1909
Germany	34,659	35,273	35,770	36,204	36,701
Austria	12,420	12,606	13,041	13,103	13,289
Hungary	10,868	11,178	11,426	11,489	11,923
Bosnia-Herzegovina.	540	540	621	621	621
Belgium.	2,795	2,795	2,857	2,857	2,857
Denmark	1,242	1,242	1,292	1,279	1,298
France	24,467	24,530	24,716	24,840	24,964
Italy	8,259	8,421	8,384	8,570	8,632
Luxemburg	233	233	233	233	233
Netherlands	1,677	1,739	1,739	1,863	1,863
Roumania	1,969	1,975	1,975	1,975	1,975
Russia	33,410	33,658	38,317	40,055	40,117
Sweden				2,416	2,627
Switzerland	2,267	2,291	2,323	2,341	2,391
Totals					

NOTE. July 1, 1909, there were 236,868 mi. of railroad in the United States.

HORIZONTAL ADDITION

88. In many instances it is desired to find the totals of numbers when written in a horizontal line. This occurs on invoices, time sheets, sales sheets, in statistical reports, etc.

89. Find, by adding from left to right, the sums of the following numbers.   Check by adding from right to left.

1. 4, 8, 7, 6, 3, 2, 7, 9.

2. 7, 9, 6, 4, 5, 8, 9, 7

3. 6, 4, 7, 9, 8, 7, 9, 5, 6.

4. 15, 18, 29, 73, 96, 47, 85, 93, 88.

5. 57, 93, 84, 96, 82, 78, 49, 56, 97.

6. 72, 89, 37, 43, 87, 92, 46, 34, 73.

7. 67, 83, 96, 38, 72, 47, 93, 89, 92.

Add the following by columns and from left to right. Check **the** result by adding the horizontal and vertical totals.

8.	9.	10.
$54 + 78 + 72 =$	$38 + 28 + 54 =$	$74 + 37 + 45 =$
$36 + 29 + 83 =$	$27 + 36 + 38 =$	$56 + 78 + 58 =$
$85 + 34 + 96 =$	$54 + 79 + 29 =$	$89 + 56 + 79 =$
$92 + 56 + 87 =$	$93 + 87 + 98 =$	$79 + 33 + 46 =$
$+ \quad + \quad =$	$+ \quad + \quad =$	$+ \quad + \quad =$

11.	12.	13.
$382 + 592 + 848 =$	$568 + 827 + 584 =$	$382 + 916 + 538 =$
$961 + 796 + 796 =$	$729 + 968 + 763 =$	$976 + 693 + 872 =$
$729 + 816 + 382 =$	$384 + 817 + 972 =$	$438 + 821 + 279 =$
$384 + 723 + 794 =$	$754 + 927 + 475 =$	$728 + 572 + 186 =$
$+ \quad + \quad =$	$+ \quad + \quad =$	$+ \quad + \quad =$

14. Find the total sales in all departments for each month, **and** the total sales in each department for the year. Check the work by adding the two sets of totals.

Summary of Monthly Sales

	Clothing	Dress Goods	Gloves	Hats	Shoes	Totals
Jan. . .	$16,427.85	$12,427.83	$3,472.50	$5,472.60	$7,532.50	
Feb. . .	15,138.20	11,428.50	2,896.75	4,854.30	6,438.90	
Mar. . .	18,278.26	12,736.25	3,124.60	4,472.60	5,372.50	
Apr. . .	17,493.87	10,964.78	2,148.75	4,134.50	5,796.25	
May . .	15,382.97	9,974.63	2,004.50	3,834.75	5,176.85	
June . .	14,582.76	9,864.59	1,995.70	3,564.75	6,487.50	
July . .	13,996.80	9,645.72	1,875.50	3,472.25	6,193.85	
Aug. . .	14,592.73	8,796.21	1,850.60	3,678.50	5,040.45	
Sept. . .	15,827.64	9,793.13	1,924.30	4,075.65	6,384.90	
Oct. . .	18,973.90	12,807.08	2,394.60	4,763.83	6,876.75	
Nov. . .	22,738.75	14,594.77	3,548.75	5,276.90	7,564.55	
Dec. . .	28,693.72	18,798.99	3,875.50	5,798.75	8,097.55	
Totals .						

15. Find (1) total tax for each purpose, (2) total tax in each county, (3) total tax for all purposes.

ABSTRACT SHOWING THE AMOUNTS OF STATE GENERAL, SCHOOL, MILITARY, AND PUBLIC HIGHWAY TAX TO BE RAISED BY EACH COUNTY IN THE STATE OF WASHINGTON FOR THE YEAR 1909

COUNTIES	AMOUNT OF STATE GENERAL TAX	AMOUNT OF STATE SCHOOL TAX	AMOUNT OF TAX FOR MILITARY FUND	AMOUNT OF TAX FOR PUBLIC HIGHWAY FUND	TOTAL AMOUNT OF TAX
1. Adams . .	$ 52,676	$ 43,897	$ 3,513	$ 17,558	
2. Asotin . . .	8,554	7,111	569	2,844	
3. Benton . .	29,846	24,873	1,989	9,949	
4. Chehalis . .	83,471	69,559	5,565	27,823	
5. Chelan . .	32,214	26,845	2,148	10,788	
6. Clallam . .	19,014	15,845	1,268	6,338	
7. Clarke . . .	33,281	27,735	2,220	11,094	
8. Columbia . .	21,303	17,752	1,420	7,101	
9. Cowlitz . .	28,447	23,706	1,896	9,482	
10. Douglas . .	21,765	18,137	1,452	7,255	
11. Ferry . . .	6,999	5,832	466	2,333	
12. Franklin . .	20,681	17,234	1,379	6,894	
13. Garfield . .	13,087	10,905	872	4,362	
14. Grant . . .	33,223	27,685	2,215	11,074	
15. Island . . .	5,039	4,199	336	1,680	
16. Jefferson . .	14,142	11,785	943	4,714	
17. King . . .	637,983	531,652	42,533	212,661	
18. Kitsap . . .	11,665	9,721	777	3,888	
19. Kittitas . .	37,148	30,956	2,476	12,382	
20. Klickitat . .	25,312	21,093	1,687	8,437	
21. Lewis . . .	55,194	45,995	3,679	18,398	
22. Lincoln . .	70,343	58,619	4,690	23,448	
23. Mason . . .	13,713	11,428	914	4,571	
24. Okanogan .	10,826	9,022	722	3,609	
25. Pacific . .	34,080	28,400	2,272	11,360	
26. Pierce . . .	245,408	204,506	16,361	81,803	
27. San Juan . .	4,848	4,040	323	1,616	
28. Skagit . . .	56,861	47,384	3,791	18,954	
29. Skamania .	12,011	10,009	800	4,004	
30. Snohomish .	99,302	82,752	6,620	33,102	
31. Spokane . .	244,941	204,118	16,329	81,647	
32. Stevens . .	29,448	24,541	1,963	9,816	
33. Thurston . .	31,629	26,358	2,108	10,543	
34. Wahkiakum .	5,710	4,759	381	1,903	
35. Walla Walla	77,102	64,251	5,140	25,700	
36. Whatcom .	72,836	60,696	4,855	24,279	
37. Whitman .	97,468	81,224	6,498	32,490	
38. Yakima . .	73,709	61,425	4,914	24,570	
Total					

SUBTRACTION

90. Subtraction is the process of finding a number or quantity equal to the difference between two other numbers or quantities.

91. In expressions involving a parenthesis ( ), as $16-(5+7)$, the operations indicated within the parenthesis must be performed before the operation indicated by the sign before the parenthesis. Thus, in the example given, 5 and 7 must be added first, and their sum, 12, taken from 16.

The vinculum ——— has the same effect as the parenthesis. Thus, $16 - \overline{5+7} = 4$.

ORAL DRILL EXERCISE

92. 1. Count backwards by 2's beginning with 100; 99.

2. Count backwards by 3's beginning with 100; 99; 98.

3. Count backwards by 4's beginning with 100; 99; 98; 97.

4. Count backwards by 5's beginning with 100; 99; 98; 97; 96.

5. Count backwards by 6's beginning with 100; 99; 98; 97; 96; 95.

6. Count backwards by 7's beginning with 100; 99; 98; 97; 96; 95; 94.

7. Count backwards by 8's beginning with 100; 99; 98; 97; 96; 95; 94; 93.

8. Count backwards by 9's beginning with 100; 99; 98; 97; 96; 95; 94; 93; 92.

9. Count backwards by 10's beginning with 100; 99; 98; 97; 96; 95; 94; 93; 92; 91.

10. Count backwards by 11's beginning with 100; 99; 98; 97; 96; 95; 94; 93; 92; 91; 90.

11. Count backwards by 12's beginning with 100; 99; 98; 97; 96; 95; 94; 93; 92; 91; 90; 89.

93. Subtract quickly:

1.	2.	3.	4.	5.	6.	7.	·8.	9.	10.
27	23	24	27	22	23	25	27	28	26
6	8	12	9	4	7	9	8	9	7

11.	12.	13.	14.	15.	16.	17.	18.	19.	20.
32	35	38	32	39	37	36	34	33	31
19	17	19	17	18	18	19	17	16	15

21.	22.	23.	24.	25.	26.	27.	28.	29.	30.
54	59	58	54	56	52	51	57	53	55
36	32	39	37	38	39	34	38	37	38

RAPID DRILL EXERCISE

94. Subtract:

1.	2.	3.	4.	5.	6.	7.
890467	740231	947204	943876	740023	630704	845700
387589	674678	678327	427628	287204	470568	687354

8.	9.	10.	11.	12.	13.	14.
472907	907240	548724	324876	597860	938701	300470
293879	689382	279876	128987	407976	764902	298784

15.	16.	17.	18.	19.	20.	21.
487246	697287	320765	596872	729684	864200	975310
398472	472349	162782	372489	537315	795301	682405

22.	23.	24.	25.	26.	27.	28.
108724	398764	587639	438076	987654	438762	487296
106897	297876	395301	278649	321076	287964	245498

29.	30.	31.	32.	33.	34.	35
568397	497872	495287	863829	397286	327693	978531
298729	387692	278469	438297	276397	278798	299729

CHECKING THE WORK

95. The best method of checking the work in subtraction is to add the remainder to the subtrahend. If the sum is equal to the minuend, the work is correct.

From 1476 take 938.

```
1476
 938
─────
 538        Check.   938 + 538 = 1476
```

Check the work on the preceding page.

THE "MAKING CHANGE" OR AUSTRIAN METHOD

96. This method of subtraction consists in adding to the amount of the purchase enough change to make the sum equal to the amount paid.

1. A man buys groceries to the value of $1.38, and gives a two-dollar bill in payment. How much change should he receive?

The cashier in making change will return to the customer 2 pennies, a ten-cent piece, and a half dollar (saying, "$1.38, 40, 50, $2.00," which means, $1.38 + $.02 =: $1.40 ; $1.40 + $.10 = $1.50 ; and $1.50 + $.50 = $2.00).

Of course, other coins than those mentioned might be returned by the cashier, but it is customary to make change in the largest coins possible.

Note. This method is used by many in ordinary subtraction problems.

2. From 1231 take 573.

```
1231      3 + 8 = 11 ; write 8.
 573      7 + 1 (carried) + 5 = 13; write 5.
─────
 658      5 + 1 (carried) + 6 = 12 ; write 6.
```

97. What change should be given for a two-dollar bill if the following purchases were made?

1.	$1.56	**8.**	$1.17	**15.**	$.72	**22.**	$1.82
2.	1.27	**9.**	.96	**16.**	.49	**23.**	1.54
3.	.87	**10.**	.88	**17.**	.27	**24.**	1.43
4.	.58	**11.**	1.49	**18.**	1.67	**25.**	.40
5.	1.33	**12.**	1.68	**19.**	1.76	**26.**	1.32
6.	1.64	**13.**	1.21	**20.**	1.87	**27.**	1.41
7.	1.29	**14.**	1.11	**21.**	1.19	**28.**	.99

This method of subtraction has an important application in all problems in which the sum of several numbers is to be taken from another number.

29. A man deposits $1625 in a bank and draws checks against it for $122, $245, and $347. Find his balance.

$1625

$7 + 5 + 2 = 14$; $14 + 1 = 15$. The units' figure of 14 is 1 less than the units' figure of the minuend; hence 1 is added to the 14 to make its units' figure equal to the units' figure of the minuend. The 1 is written as the first or units' figure of the difference. Carry 1 (the tens' figure in 15) and add the tens' column. $1 + 4 + 4 + 2 = 11$; $11 + 1 = 12$. As before, write 1 in the remainder, and carry 1. Add the hundreds' column. $1 + 3 + 2 + 1 = 7$; $7 + 9 = 16$. Write 9 in the remainder. The bank balance, then, is $911.

122
245
347
911

Find the balance of the following bank accounts:

Total Deposits	Checks Drawn
30. $ 785.00	$ 76.00, $45.00.
31. $ 427.50	$ 37.50, $ 25.00, $ 100.00.
32. $1287.70 . . .	$128.50, $ 97.60, $ 52.70.
33. $2576.28	$528.75, $ 7.50, $923.65.
34. $3789.10	$227.50, $ 1.75, $ 85.00, $ 931.80.

35. Find the balance of the following account:

Withdrawal   J. McLOUGHLIN   Investment
                                            Gain

1911				1911				
Jan.	10	3	75	Jan.	3		1	4250
					31	Loss and Gain	2	365 50

36. Find the loss or gain on merchandise:

Cost   MERCHANDISE   Returns

1911				1911				
Jan.	4	1	1764 50	Jan.	5		2	950 75
					20		8	574 50
					31	*Inventory*		*874 75*

37. Find the balance of cash :

RECEIVED			CASH				PAID OUT	
1911				1911				
Jan.	3	1	4250	Jan.	3	1	250	
	5	2	950 75		4	1	1764 50
	20	8	574 50		10	3	75
					12	4	500

38. Find the loss :⁻

COST			EXPENSE				RETURNS	
1911				1911				
Jan.	3	1	250	*Jan.*	*31*	*Inventory*		*480*
	12	4	500					

THE COMPLEMENT METHOD

98. The **complement** of a number is the difference between that number and the unit of a next higher order.

3 is the complement of 7, because 3 is the difference between 7 and 10 ; 31 is the complement of 69, because 31 is the difference between 69 and 100 ; etc.

99. The **complement method of subtraction is** based on the principle illustrated in the following examples :

1. From 23 take 8.

$$23 - 8 = (23 + 2) - (8 + 2) = 25 - 10 = 15.$$

The complement of 8 is 2.

If 2 is added to the minuend and is then deducted by adding it to the subtrahend also, it is evident that the value of the remainder will be unchanged. Hence, in the above example we think 23, 25, 15.

2. From 173 take 96.

$$173 - 96 = (173 + 4) - (96 + 4) = 177 - 100 = 77.$$

The complement of 96 is 4.

As stated in the preceding example, if 4 is added to both minuend and subtrahend, the remainder is unchanged. Think 173, 177, 77.

This method of subtraction is specially convenient when one number is to be deducted from the sum of several other numbers.

From the sum of 827, 136, and 472, take 658.

827
136
472
6̄5̄8̄
7̄7̄7̄

Arrange the numbers as shown. The complement of 658 is 342. (The complement should be carried in the mind, not written on the paper.) Add the complement to the several numbers of the minuend, and deduct 1000 (that is, 1 unit of the next higher order than the subtrahend). The result is 777.

100. By the method just explained solve the following:

1. $48 + 27 + 63 - 85$
2. $39 + 48 + 97 - 64$
3. $149 + 164 + 387 - 456$
4. $827 + 639 + 564 - 387$
5. $964 + 763 + 816 - 978$
6. $839 + 572 + 634 - 857$
7. $768 + 173 + 824 - 643$
8. $246 + 369 + 741 - 927$
9. $329 + 417 + 726 - 843$
10. $274 + 976 + 834 - 748$
11. $1238 + 1376 + 2473 - 3764$
12. $3872 + 4391 + 7684 - 8727$

13. The following statement shows the earnings and the expenses of a certain railroad company for two years. Find the total earnings and expenses; the increase or decrease of each class of earnings and expenses for 1909 as compared with 1908; and the net increase or decrease.

	1909	1908	INCREASE	DECREASE
Earnings :				
Coal traffic	$ 9,081,664.11	$ 6,665,732.09		
Merchandise	7,533,168.79	6,949,897.80		
Passengers.	2,945,408.45	2,814,081.63		
Express	198,846.64	186,722.21		
Mails	132,021.87	132,508.08		
Miscellaneous.	284,683.69	301,087.27		
Total Earnings . .				
Expenses :				
Maintenance of road . .	$ 1,836,870.86	$ 1,388,884.08		
Maintenance of equipment	2,336,272.97	2,016,385.52		
Transportation	7,227,309.90	6,504,653.43		
General.	728,626.60	760,705.39		
Total Expenses . . .				
Net Earnings				

MULTIPLICATION

101. **Multiplication** is the process of taking one number as **many** times as there are units in another.

102. The **multiplier** and the **multiplicand** are called **factors of the** product.

103. **To multiply by 10 or a multiple of 10.**

1. Multiply 4586 by 10; by 100; by 1000.

$$4586 \times \quad 10 = 45860$$
$$4586 \times \quad 100 = 458600$$
$$4586 \times 1000 = 4586000$$

Hence, to multiply by 10, 100, 1000, etc., annex as many ciphers to the multiplicand as there are ciphers in the multiplier.

2. Multiply 345 by 30; by 400; by 5000.

$$345 \times \quad 30 = 10350$$
$$345 \times \quad 400 = 138000$$
$$345 \times 5000 = 1725000$$

Hence, to multiply by 30, 400, 5000, etc., multiply the multiplicand by the 3, 4, or 5, etc., and annex as many ciphers to the product as there are ciphers in the multiplier.

3. Multiply 4300 by 6000.

$$43 \times 6 = 258; \text{ hence } 25800000.$$

Omit all ciphers on the right of the factors, and multiply the remaining numbers. To the result annex as many ciphers as were omitted from both factors.

104. Make the following multiplications mentally where possible:

1. 472 by 10   **3.** 596 by 1000   **5.** 920 by 100
2. 563 by 100   **4.** 738 by 10   **6.** 6472 by 1000

7. 5345 by 10	**12.** 93,875 by 100	**17.** 83,472 by 10
8. 824 by 300	**13.** 92,700 by 2000	**18.** 94,800 by 800
9. 9300 by 500	**14.** 4300 by 700	**19.** 7000 by 5000
10. 7250 by 6000	**15.** 89,600 by 4000	**20.** 8900 by 900
11. 82,400 by 5000	**16.** 97,620 by 100	**21.** 9800 by 3000

105. To multiply by 25, 50, 75, etc.

1. Multiply 413 by 25 ; by 50; by 75.

4)41300
10325

25 is ¼ of 100. Multiplying by 25 gives a product ¼ as great as the product obtained by multiplying by 100.

Hence, to multiply by 25, annex two ciphers to the multiplicand (that is, multiply by 100) and divide by 4.

2)41300
20650

50 is ½ of 100.

Hence, to multiply by 50, annex two ciphers to the multiplicand and divide by 2.

4)41300
10325
3
30975

75 is ¾ of 100.

Hence, to multiply by 75, annex two ciphers to the multiplicand, divide by 4, and multiply the result by 3.

2. Multiply 475 by 125; by 250.

8)475000
59375

4)475000
118750

125 is ⅛ of 1000. To multiply by 1000, annex three ciphers. The product is now 8 times as large as it should be. Therefore, divide by 8 to obtain the correct result.

In like manner, 250 is ¼ of 1000. To multiply by 250, annex three ciphers to the multiplicand and divide by 4.

106. Find the products mentally where possible:

1. 24 × 25	**6.** 87 × 50	**11.** 824 × 125	**16.** 7296 × 250
2. 64 × 25	**7.** 99 × 75	**12.** 938 × 250	**17.** 8974 × 750
3. 88 × 25	**8.** 103 × 25	**13.** 728 × 500	**18.** 4976 × 75
4. 94 × 50	**9.** 137 × 50	**14.** 1498 × 125	**19.** 8397 × 250
5. 53 × 50	**10.** 246 × 75	**15.** 2387 × 25	**20.** 17986 × 125

107. To find the product of two factors when each ends in 5.

1. Multiply 45 by 45.   45 × 45 = 2025.

To find the square of numbers ending in 5, write 25 for the right-hand two figures of the product. Next add 1 to the tens' figure (4 + 1 = 5 in this example), multiply by the other tens' figure (4 × 5 = 20), and write the product next to the 25 already written. The result is 2025.

MULTIPLICATION

101. Multiplication is the process of taking one number as many times as there are units in another.

102. The multiplier and the multiplicand are called **factors** of the product.

SHORT METHODS

103. To multiply by **10** or a multiple of **10**.

1. Multiply 4586 by 10; by 100; by 1000.

$$4586 \times 10 = 45860$$
$$4586 \times 100 = 458600$$
$$4586 \times 1000 = 4586000$$

Hence, to multiply by 10, 100, 1000, etc., annex as many ciphers to the multiplicand as there are ciphers in the multiplier.

2. Multiply 345 by 30; by 400; by 5000.

$$345 \times 30 = 10350$$
$$345 \times 400 = 138000$$
$$345 \times 5000 = 1725000$$

Hence, to multiply by 30, 400, 5000, etc., multiply the multiplicand by the **3**, **4**, or **5**, etc., and annex as many ciphers to the product as there are ciphers in the multiplier.

3. Multiply 4300 by 6000.

$$43 \times 6 = 258; \text{ hence } 25800000.$$

Omit all ciphers on the right of the factors, and multiply the remaining numbers. To the result annex as many ciphers as were omitted from both factors.

104. Make the following multiplications mentally where possible:

1. 472 by 10	**3.** 596 by 1000	**5.** 920 by 100
2. 563 by 100	**4.** 738 by 10	**6.** 6472 by 1000

48

7. 5345 by 10	**12.** 93,875 by 100	**17.** 83,472 by 10
8. 824 by 300	**13.** 92,700 by 2000	**18.** 94,800 by 800
9. 9300 by 500	**14.** 4300 by 700	**19.** 7000 by 5000
10. 7250 by 6000	**15.** 89,600 by 4000	**20.** 8900 by 900
11. 82,400 by 5000	**16.** 97,620 by 100	**21.** 9800 by 3000

105. To multiply by 25, 50, 75, etc.

1. Multiply 413 by 25; by 50; by 75.

4)41300
10325

25 is ¼ of 100. Multiplying by 25 gives a product ¼ as great as the product obtained by multiplying by 100.

Hence, to multiply by 25, annex two ciphers to the multiplicand (that is, multiply by 100) and divide by 4.

2)41300
20650

50 is ½ of 100.

Hence, to multiply by 50, annex two ciphers to the multiplicand and divide by 2.

4)41300
10325
     3
30975

75 is ¾ of 100.

Hence, to multiply by 75, annex two ciphers to the multiplicand, divide by 4, and multiply the result by 3.

2. Multiply 475 by 125; by 250.

8)475000
59375

4)475000
118750

125 is ⅛ of 1000. To multiply by 1000, annex three ciphers. The product is now 8 times as large as it should be. Therefore, divide by 8 to obtain the correct result.

In like manner, 250 is ¼ of 1000. To multiply by 250, annex three ciphers to the multiplicand and divide by 4.

106. Find the products mentally where possible:

1. 24×25	**6.** 87×50	**11.** 824×125	**16.** 7296×250
2. 64×25	**7.** 99×75	**12.** 938×250	**17.** 8974×750
3. 88×25	**8.** 103×25	**13.** 728×500	**18.** 4976×75
4. 94×50	**9.** 137×50	**14.** 1498×125	**19.** 8397×250
5. 53×50	**10.** 246×75	**15.** 2387×25	**20.** 17986×125

107. To find the product of two factors when each ends in 5.

1. Multiply 45 by 45. $45 \times 45 = 2025$.

To find the square of numbers ending in 5, write 25 for the right-hand two figures of the product. Next add 1 to the tens' figure ($4 + 1 = 5$ in this example), multiply by the other tens' figure ($4 \times 5 = 20$), and write the product next to the 25 already written. The result is 2025.

2. Multiply 35 by 55. $35 \times 55 = 1925$.

If the sum of the tens' figures is even, write 25 as the first part of the product. The sum of the tens' figures ($5 + 3 = 8$) is even. Take $\frac{1}{2}$ of the sum of the tens' figures ($\frac{1}{2}$ of $8 = 4$) and add it to the product of the tens' figures multiplied together. $5 \times 3 = 15$; $15 + 4 = 19$. Write 19 in the product, making 1925.

3. Multiply 45 by 115. $45 \times 115 = 5175$.

The sum of the tens here ($4 + 11 = 15$) is odd; hence write 75 as the first part of the product. Take $\frac{1}{2}$ the sum of the tens' figures ($\frac{1}{2}$ of $15 = 7\frac{1}{2}$), disregard the fraction, and add the half sum to the product of the tens' figures multiplied together. $(4 \times 11) + 7 = 51$. Write 51 in the product, making 5175.

108. Practice on the following until you can read the products at sight:

1. 15×15	**8.** 85×85	**15.** 95×85	**22.** 125×45
2. 25×25	**9.** 95×95	**16.** 65×55	**23.** 125×125
3. 35×35	**10.** 85×65	**17.** 45×35	**24.** 105×35
4. 45×45	**11.** 85×55	**18.** 75×85	**25.** 95×105
5. 55×55	**12.** 45×65	**19.** 95×15	**26.** 105×105
6. 65×65	**13.** 55×75	**20.** 85×45	**27.** 115×115
7. 75×75	**14.** 85×35	**21.** 65×75	**28.** 115×55

109. To square any number of two figures.

1. Multiply 53 by 53. $53 \times 53 = 2809$.

(1) $3 \times 3 = 9$; write 9. Add the units' figures ($3 + 3 = 6$). Multiply the sum by one of the tens' figures. $5 \times 6 = 30$; write 0, and carry 3. Multiply the tens' figures together, and add the 3. ($5 \times 5 + 3 = 28$); write 28, making 2809.

(2) (For students who have had algebra.)
$53 \times 53 = (50 + 3)(50 + 3) = 2500 + 2(50 \times 3) + 9 = 2809$.
Arrange the factors as the "sum of two quantities," and solve as follows: 50 squared $= 2500$; $2 \times 50 \times 3 = 300$; 3 squared $= 9$.
On adding, $2500 + 300 + 9 = 2809$.

2. Multiply 69 by 69. $69 \times 69 = 4761$.

(1) $9 \times 9 = 81$; write 1, and carry 8. Multiply the sum of the tens' figures by one of the units' figures and add 8 ($6 + 6) \times 9 + 8 = 116$; write 6, and carry 11. Multiply the tens' figures together, and add 11. $6 \times 6 + 11 = 47$. The product, then, is 4761.

(2) **By** the algebraic method this example illustrates the square of the "difference of two quantities."

$69 \times 69 = (70 - 1)(70 - 1) = 4900 - 2 (70 \times 1) + 1 = 4761.$

70 squared $= 4900$; $- 2$ times $70 \times 1 = - 140$; $- 1$ squared $= 1$.

Combining, $4900 - 140 + 1 = 4761.$

NOTE. Students should observe in the explanation of the preceding two examples that, in Ex. 1, the sum of the *units'* figures is taken in the second step, while in Ex. 2 the sum of the *tens'* figures is taken. The result is the same in problems of this nature. If the tens' figures are smaller than the units' figures, take the sum of the tens' figures, and *vice versa*. It is easier to do it as here suggested, because, as in Ex. 2, $(6 + 6) \times 9$ is more quickly solved than $(9 + 9) \times 6$, by most students. Compare with Art. 225.

110. Write the squares of all the numbers from 40 to 60.

111. To find the product of any two numbers of two figures each when the units or the tens are alike.

1. Multiply 42 by 72. $42 \times 72 = 3024.$

$2 \times 2 = 4$; write 4. Take the sum of the tens' figures, and multiply by one of the units' figures. $(7 + 4) \times 2 = 22$; write 2, and carry 2. Take the product of the tens, and add 2. $7 \times 4 + 2 = 30$; write 30, making 3024.

2. Multiply 37 by 34. $37 \times 34 = 1258.$

$4 \times 7 = 28$; write 8, carry 2. Take the sum of the units' figures, and multiply by one of the tens' figures. $(7 + 4) \times 3 = 33$; $33 + 2 = 35$; write 5, carry 3. Take the product of the tens' figures, and carry the 3 ; $3 \times 3 + 3 = 12$. Therefore, the product is 1258.

NOTE. The student should take careful notice of the fact that when the tens are alike, the units' figures are added, and *vice versa*.

112. Multiply :

1. 43 by 33	**7.** 74 by 84	**13.** 93 by 98	**19.** 67 by 63
2. 24 by 44	**8.** 56 by 86	**14.** 69 by 63	**20.** 72 by 92
3. 52 by 62	**9.** 37 by 32	**15.** 67 by 69	**21.** 86 by 46
4. 67 by 27	**10.** 48 by 45	**16.** 78 by 76	**22.** 55 by 57
5. 83 by 53	**11.** 58 by 54	**17.** 39 by 49	**23.** 39 by 38
6. 69 by 29	**12.** 88 by 87	**18.** 58 by 57	**24.** 46 by 36

113. To multiply by 11 or by a multiple of 11.

1. Multiply 287 by 11.

$287 \times 11 = 3157.$

```
  287
   11
 ----
  287
  287
 ----
 3157
```

By referring to the solution in the margin, it is seen that the first figure in the product is 7, the units' figure of the multiplicand ; the second figure of the product is the sum of the units and the tens ; the third is the sum of the tens and the hundreds, including the carrying figure, and the fourth is the hundreds' figure of the multiplicand plus the carrying figure.

The principle involved in the preceding example applies to problems in which it is desired to multiply by a multiple of 11.

2. Multiply 329 by 44. $329 \times 44 = 14476.$

44 is 4 times 11. Make the multiplication mentally by 11, and multiply the result by 4 at the same time. In multiplying by 11 the first figure of the product would be 9, which multiplied by 4 gives 36. Write 6, and carry 3.

Continue the multiplication by 11 : $9 + 2 = 11$; $4 \times 11 = 44$, which, with 3 to carry, makes 47 ; write 7, carry 4. $2 + 3 = 5$; $4 \times 5 = 20$, which, with 4 to carry, makes 24 ; write 4, carry 2. $4 \times 3 + 2 = 14$; write 14, making 14476.

114. Multiply, mentally, each of the following numbers by 11 :

1. 54	**5**. 96	**9**. 244	**13**. 396	**17**. 4872	**21**. 4507
2. 35	**6**. 58	**10**. 372	**14**. 927	**18**. 3976	**22**. 5240
3. 43	**7**. 39	**11**. 628	**15**. 489	**19**. 8498	**23**. 3007
4. 78	**8**. 76	**12**. 579	**16**. 767	**20**. 2976	**24**. 9080

115. Multiply each of the above numbers by 22; 33; 55; 77; 88.

116. To multiply by 21, 31, etc., and by 101, 201, etc.

1. Multiply 287 by 41.

$287 \times 41 = 11767.$

```
   287
    41
 -----
   287
  1148
 -----
 11767
```

The solution in the margin will help to make the explanation clear. The first figure of the product is plainly 7. Now, begin to multiply by 4 : $4 \times 7 = 28$. $28 + 8$ (the tens' figure of the multiplicand) $= 36$; write 6, carry 3. $4 \times 8 = 32$; $32 + 2$ (the hundreds' figure of the multiplicand) $+ 3$ (carried) $= 37$; write 7, carry 3. $4 \times 2 + 3 = 11$; write 11, making 11767.

2. Multiply 458 by **601.**

$458 \times 601 = 275258.$

$$
\begin{array}{r}
458 \\
601 \\
\hline
458 \\
2748 \\
\hline
275258
\end{array}
$$

The solution in the margin makes it clear that the first two figures of the product are 58. Now multiply by 6. $6 \times 8 = 48$. Add 4, the hundreds' figure of the multiplicand, which gives 52; write 2 and carry 5. Next, $6 \times 45 = 270$. To this, add the 5 to be carried, which gives 275   The entire product, then, is 275258.

117. Multiply, as directed by the teacher, the following numbers by 21, 41, 51, 81, 101, 301, 601, 901:

1. 234	**3.** 728	**5.** 371	**7.** 295	**9.** 756
2. 456	**4.** 342	**6.** 542	**8.** 468	**10.** 637

118. To multiply by a number composed of factors.

1. Multiply 369 by 497.

$$
\begin{array}{r}
369 \\
497 \\
\hline
2583 \\
18081 \\
\hline
183393
\end{array}
$$

First multiply by 7 in the usual way. The other part of the multiplier, 49, is 7 times 7; hence instead of multiplying 369 by 49, multiply 2583 by 7, placing the result as shown in the illustration, since the multiplier is 49 *tens*. Then add the partial products.

2. Multiply 657 by 321.

$$
\begin{array}{r}
657 \\
321 \\
\hline
1971 \\
13797 \\
\hline
210897
\end{array}
$$

First multiply by 3, placing the result as shown. 21 is 7 times 3. Multiply the result already found by 7, writing the product two places to the right of the first partial product, since the 3 represents *hundreds* and the 21, *units*.

119. To multiply any two numbers of two figures.

Multiply 46 by 92. $46 \times 92 = 4232.$

First, $2 \times 6 = 12$; write 2, carry 1. Next, $2 \times 4 + 1$ (carried) $= 9$; also $9 \times 6 = 54$. Add 54 and 9, giving 63; write 3, carry 6. Next, $9 \times 4 + 6$ (carried) $= 42$, which makes a completed product of 4232.

CHECKING THE WORK

120. The following checks will be helpful in proving multiplication.

121. 1. Divide the product by one of the factors: the quotient will be the other factor. If there are more than two factors, the product divided by one of them will give a quotient equal to the product of all the other factors.

(a) $18 \times 25 = 450$. PROOF. $450 \div 18 = 25$.

(b) $4 \times 8 \times 9 \times 5 = 1440$. PROOF. $1440 \div 4 = 360 = 8 \times 9 \times 5$.

122. Casting out the nines.

$$41 = 5$$
$$22 = 4$$
$$\overline{82} \quad \overline{20} = 2$$
$$82$$
$$\overline{902} = 2$$

The excess of nines in the multiplicand is 5, and in the multiplier is 4. The product of the excesses 4 and 5 is 20, and the excess of nines in 20 is 2. The excess of nines in the product, 902, is 2. Hence the multiplication is probably correct.

The excess of nines in the product is equal to the excess of nines in the product of the excesses.

123. Casting out the elevens.

$$248 = 6$$
$$37 = 4$$
$$\overline{1736} \quad \overline{24} = 2$$
$$744$$
$$\overline{9176} = 2$$

The excess of elevens in 248 is 6, $(8 - 4 + 2)$; the excess of elevens in 37 is 4, $(7 - 3)$. The product of the excesses is 4×6, or 24, and the excess of elevens in 24 is 2. The excess of elevens in 9176 is 2, $(6 + 1 + 11 - 7 - 9)$. Since the results agree, the multiplication is probably correct.

124. Multiply, and check the results. Use the check assigned by the teacher.

1. 43 by 29	10. 875 by 756	19. 64 by 36
2. 72 by 37	11. 1249 by 648	20. 93 by 67
3. 48 by 27	12. 3147 by 639	21. 94 by 69
4. 37 by 84	13. 2489 by 763	22. 81 by 74
5. 43 by 56	14. 3271 by 654	23. 82 by 29
6 91 by 28	15. 5298 by 324	24. 34 by 85
7 96 by 47	16. 6476 by 549	25. 35 by 63
8. 56 by 46	17. 8749 by 624	26. 73 by 57
9. 83 by 34	18. 9483 by 424	27. 79 by 67

DIVISION

125. **Division** is the process of finding how many times one num ber is contained in another.

126. **To divide by 10, 100, 1000, etc.**

Divide 4370 by 10.

$4370 \div 10 = 437$

$$\frac{437.0}{10)\overline{4370}}$$

The solution at the left shows that to divide by 10 it is necessary only to move the decimal point one place to the left.

In like manner, to divide by 100, 1000, etc., move the point 2 places, 3 places, etc., to the left.

127. Divide each of the following numbers by 10; 100; 1000; 10,000.

1. 130000	**6** 24750000	**11.** 3489
2. 1250000	**7.** 400000	**12.** 7248
3. 40000	**8.** 1800000	**13.** 5673
4. 750000	**9.** 2000000	**14.** 8470
5. 1160000	**10.** 14500000	**15.** 19872

128. **To divide by 25, 50, 75, etc.**

In many cases the work of division can be lessened by making the operation one of multiplication.

Divide 2800 by 25 $2800 \div 25 = 28 \times 4 = 112$

25 is $\frac{1}{4}$ of 100. Divide 2800 by 100 by dropping the two ciphers. But, in dividing by 100, you have divided by a number 4 times too large, consequently the quotient is only $\frac{1}{4}$ as large as it should be (Principle 17, page 102). To correct the error, multiply the 28 by 4.

NOTE. Further application ot the principle illustrated in the last example is made in the chapter on Aliquot Parts, page 21.

55

129. Divide:

1. 1400 by 25	**5.** 142 by 25	**9.** 27000 by 75
2. 1600 by 25	**6.** 153000 by 250	**10.** 36000 by 250
3 1800 by 50	**7.** 72000 by 500	**11.** 48000 by 500
4. 2400 by 50	**8.** 16000 by 500	**12.** 54000 by 750

130. To divide by a number composed of factors.

1. Divide 12726 by 63.

$9 \times 7 = 63.$

9)12726
7) 1414
  202

The factors of 63 are 9 and 7. Instead of making a long division, divide by the factors, 9 and 7, using short division

2 Divide 38971 by 64.

$8 \times 8 = 64.$

8)38971
8) 4871   3 rem.
  608   7 rem.
$7 \times 8 + 3$ 59 rem.

The factors of the divisor are 8 and 8. Divide by the factors as before, reserving the remainders as shown. To find the remainder, multiply the first divisor, 8, by the second remainder, 7, and add the first remainder, 3. Thus, $7 \times 8 + 3 = 59$, remainder. Hence the result is 608, with a remainder of 59.

3. Divide 174684 by 504.

$7 \times 8 \times 9 = 504.$

7)174684
8) 24954   6 rem.
9) 3119   2 rem.
  346   5 rem.

$5 \times 8 \times 7 = 280$
$2 \times 7 = 14$
$6 = 6$
  300

300, remainder.

The factors of 504 are 7, 8, and 9. Divide by the factors as before. The quotient is 346. To find the remainder, multiply the third remainder by the first and second divisors, 7, and 8, ($5 \times 8 \times 7 = 280$). Next multiply the second remainder by the first divisor, 7, ($2 \times 7 = 14$). To the sum of these two products, add the first remainder, 6, giving 300 as the remainder.

4. Divide 58750 by 2400.

$4 \times 6 \times 100 = 2400.$

4)587 . 50

6)146　3 rem.

　24　2 rem.

$2 \times 4 + 3 = 11,$

1150, remainder.

The factors are 4, 6, and 100. Divide by 100 by "cutting off" the right-hand two figures of the dividend. Then divide by the factors 4 and 6. The quotient is 24. To find the remainder, first treat the divisors and remainders as though the divisor, 100, and the remainder, 50, were not present. $2 \times 4 + 3 = 11$. To complete the remainder, annex the remainder, 50, to the 11, giving a remainder of 1150.

CHECKING THE WORK

131. The following methods will be found helpful in checking the work.

1. Multiply the divisor by the quotient and add the remainder (if any); the result should equal the dividend.

2. Cast out the nines.

$8879 \div 247 = 35$, with a remainder of 234.

Therefore, $8879 = 247 \times 35 + 234$.

The excess of 9's in 8879 is 5.

The excess of 9's in 247 is 4.

The excess of 9's in 35 is 8.

The excess of 9's in 234 is 0.

$4 \times 8 + 0 = 32.$

The excess of 9's in 32 is 5.

Since the excess of 9's in the dividend is equal to the excess of 9's in the product of the excesses in the divisor and quotient plus the excess in the remainder, the division is probably correct.

3. Cast out the elevens.

$3504 \div 54 = 64$, with a remainder of 48.

Therefore, $3504 = 54 \times 64 + 48$

The excesses of 11's in 3504 is 6.

The excess of 11's in 54 is 10.

The excess of 11's in 64 is 9.

The excess of 11's in 48 is 4.

$10 \times 9 + 4 = 94.$

The excess of 11's in 94 is 6.

Since the excesses agree, the division is probably correct.

132. Divide and check the work.   Use the check assigned by the teacher.

1.   4872 by 36
2.   7349 by 42
3.   5987 by 56
4.   7879 by 72
5.   9478 by 84
6.   9764 by 96
7. 11492 by 99

8.   24763 by 108
9.   45972 by 54
10. 79489 by 88
11. 48900 by 3500
12. 147260 by 3600
13. 248750 by 4200
14. 329764 by 4800

15. 247689 by 4900
16. 324798 by 5600
17. 498749 by 54000
18. 897200 by 63000
19. 948720 by 48000
20. 672000 by 24000
21. 486000 by 18000

DIVISION BY CONTINUED SUBTRACTION

133. This method of division is often required in civil service examinations.

134. By the method of continued subtraction divide **784** by 149, and prove the work.

```
    784
1.  149
    ───
    635

2.  149
    ───
    486

3.  149
    ───
    337

4.  149
    ───
    188

5.  149
    ───
     39
```

Subtract the divisor from the dividend, and from the successive remainders, until a remainder less than the divisor is obtained.   The number of subtractions is the quotient.

PROOF.

$5 \times 149 + 39 = 784$; or, $784 \div 149 = 5$, with a remainder of 39.

135. Divide by continued subtraction and prove:

1. 489764 by 42875
2. 447859 by 48727
3. 128976 by 54738
4. 249768 by 55987
5. 473964 by 75428
6. 963874 by 85925

7. 389726 by 98762
8. 598370 by 87820
9. 397645 by 98435
10. 947680 by 94768
11. 247682 by 31463
12. 437263 by 43721

FACTORING

136. 6 is the product of what numbers?

Because $2 \times 3 = 6$, 2 and 3 are called the **factors of 6.**

137. What are the factors of 10? 15? 18? 21?

138. A **prime** number is one not exactly divisible by any number except itself and 1.

139. Prime factors are prime numbers.

140. Factoring is the process of separating a number or quantity into its factors.

REMARK. Factoring is important, not merely for the purpose of resolving a given number into its factors, but also for its assistance in the solution of problems. In many problems in fractions, practical measurements, and percentage, in fact, in all problems in which cancellation is used, the work involved in the solution is very materially lessened by the ability to see factors quickly.

The ability quickly to factor a given number depends upon a knowledge of the tests of divisibility, with which the student should become thoroughly familiar.

141. Tests of Divisibility.

1. All even numbers are divisible by 2.

Thus, 4, 8, 18, 38, 96, 144, are divisible by 2.

2. If the sum of the digits of any number is divisible by 3, the number is divisible by 3.

Thus, 144957 is divisible by 3 because $1+4+4+9+5+7 = 30$, and 30 is divisible by 3.

3. All numbers whose right-hand two figures are ciphers, or express a number divisible by 4, are divisible by 4.

Thus, 100, 144, and 3288, are divisible by 4.

4. All numbers whose units' figure is either a cipher or 5 are divisible by 5.

Thus, 60 and 135 are divisible by 5.

5. All even numbers, the sum of whose digits is divisible by **3**, are divisible by 6.

Thus, 126, 918, and 45252 are divisible by 6.

In practice, notice first whether the number is even. If so, proceed as in test for 3.

6. 7, 11, and 13 will divide 1001 and any of its multiples; as, 3003, 8008, 12012, etc.

7. All numbers whose right-hand three digits are ciphers, or express a number divisible by 8, are divisible by 8.

Thus, 3000, 6624, and 18232 are divisible by 8.

8. All numbers, the sum of whose digits is divisible by **9**, are divisible by 9.

Thus, 45621 and 234819 are divisible by 9.

9. All numbers whose right-hand figure is 0 are divisible by **10**.

Thus, 60, 140, 1190, are divisible by 10.

10. All numbers whose right-hand two figures are ciphers, or express a number divisible by 25, are divisible by 25.

Thus, 9800, 4125, and 875 are divisible by 25.

11. All numbers whose right-hand three figures are ciphers, or express a number divisible by 125, are divisible by 125.

Thus, 3000, 4625, and 3375 are divisible by 125.

142. Applying the tests given above, find the prime factors of 12,464.

$$8)\underline{12464}$$
$$2)\underline{\ 1558}$$
$$19)\underline{\ \ \ 779}$$
$$41$$

Therefore, $2 \times 2 \times 2 \times 2 \times 19 \times 41 = 12,464$.

The "test" for 8 shows that 8 is a factor. The quotient 1558 is even; hence, 2 is a factor. By trial 19 is found to be a factor, resulting in a quotient of 41, a prime number.

8 was used as a divisor first to save time. Its prime factors are known to be $2 \times 2 \times 2$. Therefore, the prime factors of 12,464 are $2 \times 2 \times 2 \times 2 \times 19 \times 41$.

143. There is no regular method of determining large factors. Unless one has access to factor tables, nothing remains but to try the prime numbers until the right one is found. There are, however, two fundamental principles of factoring that should be borne in mind:

1. A factor of a number is also a factor of all the multiples of that number.

Thus, 7 is a factor of 21. 7 is also a factor of 42, 63, 105, 189, 420, 1071 all of which are multiples of 21.

2. A common factor of two numbers is a factor also of the sum or of the difference of those two numbers.

Thus 4 is a common factor of 20 and 36. It is a factor also of 56 (20 + 36), and of 16 (36 − 20).

144. Find the prime factors of the following numbers:

1.	48	**7.**	729	**13.**	2664	**19.**	16296	**25.**	5872
2.	64	**8.**	343	**14.**	4064	**20.**	25728	**26.**	9536
3.	80	**9.**	1125	**15.**	4512	**21.**	6875	**27.**	3856
4.	105	**10.**	3375	**16.**	3824	**22**	5375	**28.**	7328
5.	144	**11.**	3654	**17.**	9664	**23.**	4004	**29.**	15309
6.	160	**12**	3248	**18**	9750	**24.**	3224	**30.**	27783

GREATEST COMMON DIVISOR

145. The **greatest common** divisor of two or more numbers is the largest number that will exactly divide each of them.

146. The only practical application (so far as this work is concerned) of the principles involved in finding the G. C. D. is in reducing common fractions to their lowest terms

147. **1** Find the G C. D. of 177 and 295.

177—295
177—118
59—118
59 is the G. C. D.

Divide the greater number by the smaller, writing the remainder, 118, and the divisor, 177, as shown. Divide again by the smaller number (118 this time) and write results as before. Dividing again, it is seen that 59 is exactly contained in 118, hence 59 is the G. C. D.

2. Find the G. C. D. of 188, 470, 705, and 940

188—470—705—940
188— 94—141— 0
0— 94— 47— 0
47 is the G. C. D

Divide all the other numbers by the smallest number, writing the remainders as shown in the solution. The "divisor" must also be "brought down." Divide again by the smallest number. By inspection, it is now seen that 47 is the G. C. D.

---

148. Find the G. C. D. of:

1. 912 and 1121
2. 319 and 551.
3. 629 and 1147.
4. 2813 and 3589.
5. 791 and 1016.
6. 2167, 2561, and 985.
7. 3749, 8313, and 1141.
8. 6859, 11,191, and 3610.
9. 961, 1271, and 372.
10. 1681, 2501, and 410

GREATEST COMMON DIVISOR BY INSPECTION

149. 1. Find the G. C. D. of 115 and 161.

$$161 - 115 = 46. \qquad 46 = 2 \times 23.$$

The G. C. D. of two numbers is their difference, or a factor of their difference. The factors of the difference, 46, are 2 and 23. By inspection, it is seen that 2 is not a common divisor, and that 23 *is* a common divisor. Hence 23 is the G. C. D.

2. Find the G. C. D. of 57, 95, 152, and 228.

The G. C. D. of several numbers is their *smallest* difference, or a factor of their smallest difference. The smallest difference of the numbers given is 38 (95 − 57). The factors of 38 are 2 and 19. By inspection, it is seen that 2 is not a common factor, and that 19 *is* a common factor. Hence 19 is the G. C. D.

150. Find, by inspection, the G. C. D. of:

1. 145 and 203.
2. 296 and 407.
3. 387 and 559.
4. 85 and 119.
5. 52, 91, and 143
6. 68, 102, and 187
7. 48, 72, 120, and 288.
8. 90, 108, 144, and 111.

LEAST COMMON MULTIPLE

151. A multiple of any given number is a number that will contain the given number without a remainder. Thus, 24 is a multiple of 4.

152. 24 is a common multiple of 4 and 6, because it contains 4 and 6 each without a remainder.

153. 24 is the least common multiple of 4, 6, 8, and 12, because it is the smallest number that will contain them without remainders.

Finding the least common multiple has its practical application in reducing several fractions to equivalent fractions having the least common denominator.

154. Find the L. C. M. of 6, 8, 12, 15, 20, 24, 30, and 36.

$$30) \cancel{6} — \cancel{8} — \cancel{12} — \cancel{15} — 20 — 24 — 30 — 36$$
$$4) \ \underline{2— \ 4— \ 1— \ 6}$$
$$1— \ 1— \ 3$$

$30 \times 4 \times 3 = 360$, the L. C. M.

Write all the numbers in one line. Cancel all the numbers that are contained in any of the other numbers. Divide the remaining numbers by any number that will exactly divide one of them. 30 is taken as the divisor. It is contained in 30, once. Divide the other numbers (which do not exactly contain 30) by any factor that is common to the divisor, 30, and to any of the numbers. 10 is a common factor of both 30 and 20. 20 is divided by 10, which gives 2. 6 is a factor common to 30, and to both 24 and 36. Dividing 24 and 30 by 6 gives 4 and 6, respectively. A new line of numbers is now obtained. Proceed as at first, and continue the operation until the remaining numbers have no common factor except unity. The L. C. M. is the product of all the divisors and the numbers remaining in the last line.

NOTE. In case any of the given numbers do not contain the divisor, and do not have any factor of the divisor, "bring them down" into the next line.

155. Find the L. C. M. of:

1. 2, 4, 6, 12, 15.
2. 3, 6, 9, 18, 30.
3. 5, 12, 15, 30, 40.
4. 8, 12, 24, 60, 84.
5. 15, 25, 36, 75, 100.

6. 13, 15, 25, 65, 75.
7. 24, 36, 48, 42, 120.
8. 6, 20, 48, 100, 144.
9. 9, 16, 64, 80, 128.
10. 36, 45, 72, 105, 150.

CANCELLATION

156. Problems frequently arise in which the solution is best made like the following:

Divide the product of $36 \times 18 \times 21$ by the product of $9 \times 28 \times 12$.

$$\frac{\overset{3}{\cancel{36}} \times \overset{\cancel{2}}{\cancel{18}} \times \overset{3}{\cancel{21}}}{\underset{\underset{2}{4}}{\cancel{9}} \times \cancel{28} \times \cancel{12}} = \frac{9}{2} = 4\tfrac{1}{2}.$$

Write the factors of the dividend above a horizontal line, and the factors of the divisor below the line, as shown in the solution. 9, below the line, is contained in 18, above the line. Both numbers are canceled, and the quotient, 2, is written near the 18. In the same way 12 is contained in 36, 3 times. Since 28 is not divisible by 21, both numbers are divided by their common factor, 7. The numbers 21 and 28 are both canceled, and the quotients, 3 and 4, are written near the numbers, respectively. 2 of the dividend now divides the 4 in the divisor; that is, the common factor, 2, is canceled in both dividend and divisor. (See Principles of Division, No. 21, page 102.) The product of the remaining factors above the line is now divided by the product of the factors left below the line, thus $\frac{3 \times 3}{2} = \frac{9}{2} = 4\frac{1}{2}$, the desired quotient.

Using cancellation, divide:

1. $8 \times 6 \times 12 \times 15$ by $3 \times 4 \times 5 \times 18$.

2. $15 \times 12 \times 24 \times 39$ by $10 \times 13 \times 4 \times 16$.

3. $125 \times 80 \times 96 \times 216$ by $50 \times 60 \times 32 \times 540$.

4. $48 \times 56 \times 63 \times 72$ by $36 \times 78 \times 42 \times 108$.

5. $30 \times 45 \times 9 \times 12$ by $5 \times 50 \times 18 \times 60$.

6. How many tubs of butter averaging 30 lb. each, at 28¢ per pound, will pay for 12 pieces of muslin, each piece containing 49 yd. at 10¢ per yard?

The cost of the muslin is represented by the product of $12 \times 49 \times \$.10$; the value of 1 tub of butter, by the product of $30 \times \$.28$. The number of tubs of butter required is found by dividing the cost of the muslin by the value of 1 tub of butter.

Thus, $\dfrac{\overset{4}{12} \times \overset{7}{49} \times .10}{\underset{3}{30} \times \underset{7}{.28}} = 7.$ *Ans.* 7 tubs of butter.

7. How many pounds of sugar at 5¢ per pound can be had in exchange for 15 doz. eggs at 21¢ per dozen?

8. At 60¢ a bushel, how many bushels of potatoes will pay for 3 chests oi tea, weighing 36 lb. each, at 40¢ per pound?

9. How many boys earning 10¢ an hour, 8 hr. a day, in 15 da., will earn as much as 4 men at 25¢ an hour, 10 hr. a day, in 18 da.?

10. If \$560 earns \$84 interest in 2 yr. 6 mo., how much interest will \$640 earn in 3 yr. 6 mo. at the same rate?

DOMESTIC PARCEL POST

157. Domestic parcel post mail includes merchandise, farm and factory products, seeds, plants, etc., books and catalogues, miscellaneous printed matter weighing more than 4 pounds, and all other mailable matter not embraced in the first, second, or third classes.

Parcel post matter must not be of a kind likely to injure the person of any postal employee or to damage the mail equipment or other mail matter, nor must it be of a character perishable within a period reasonably required for transportation and delivery.

Rates of postage on domestic parcel post matter are as follows:

(a) Parcels weighing 4 oz. or less, except books, seeds, plants, etc., 1 ¢ for each ounce or fraction thereof, for any distance.

(b) Parcels weighing 8 oz. or less containing books, seeds, plants, etc., 1 ¢ for each 2 oz. or fraction thereof, for any distance.

(c) Parcels weighing more than 8 oz. containing books, seeds, plants, etc., parcels of miscellaneous printed matter weighing more than 4 lb., and all other parcels of domestic parcel post, or fourth class matter weighing more than 4 oz. are chargeable according to distance or zone, at the pound rates, a fraction of a pound being considered a full pound.

In addition, parcels on which the postage amounts to 25 ¢ or more are subject to a tax of 1 ¢ for each 25 ¢ or fractional part thereof. This tax must be paid by *special internal revenue stamps* affixed to the parcels.

RATES OF POSTAGE IN THE DIFFERENT ZONES

Local rate, 1st pound, 5 ¢; each additional 2 pounds, 1 ¢.

1st and 2d zones, 1 to 150 miles, 1st pound, 5 ¢; each additional pound, 1 ¢.

3d zone, 150 to 300 miles, 1st pound, 6 ¢; each additional pound, 2 ¢.

4th zone, 300 to 600 miles, 1st pound, 7 ¢; each additional pound, 4 ¢.

5th zone, 600 to 1000 miles, 1st pound, 8 ¢; each additional pound, 6 ¢.

6th zone, 1000 to 1400 miles, 1st pound, 9 ¢; each additional pound, 8 ¢.

7th zone, 1400 to 1800 miles, 1st pound, 11 ¢; each additional pound, 10 ¢.

8th zone, over 1800 miles, 1st pound, 12 ¢; each additional pound, 12 ¢.

The limit of weight for local delivery and for the first three zones is 70 pounds; for all other zones, 50 pounds. Parcel post matter may not exceed 84 inches in length and girth combined.

Insurance may be secured by a payment of 3 ¢ extra for an amount up to $ 5; 5 ¢ up to $ 25; 10 ¢ up to $ 50; 25 ¢ up to $ 100.

Parcels valued at not more than $100 may be sent "C. O. D." upon the payment of a fee of 10 cents, if the amount to be remitted does not exceed $50; or a fee of 25¢ for greater amounts up to $100. This fee includes insurance up to $50 and $100 respectively. The fee for the return, by postal money order, of the amount collected is paid by the addressee.

A parcel post guide and map are used to determine in which zone a post office is located from the office of mailing.

Determine the cost of mailing the following packages to the places mentioned :

1. A parcel of merchandise weighing 18 pounds to a post office in the fifth zone.

2. A 40-pound parcel of books to the second zone.

3. An insured parcel of dry goods weighing 14 pounds 6 ounces, valued at $32.75, to a post office in the sixth zone.

4. A "C. O. D." insured parcel weighing $12\frac{1}{2}$ pounds, valued at $18.50, to a post office in the seventh zone.

5. A 7-ounce book to a post office in the fifth zone.

6. A 10-ounce parcel of dry goods to the third zone.

7. A "C. O. D." parcel valued at $75 and weighing 20 lb. is mailed to the 8th zone. Find the cost of mailing it; also the cost to the addressee, including money order fee. (For money order rates, see p. 355.)

8. A man sends a 30-pound pail of butter to a post office in the second zone. How much is the postage?

9. A merchant in Toledo, Ohio, telegraphed a jobber in Chicago, "Send C. O. D. by parcel post 100 dozen #546 buttons." The buttons came billed at 22 cents a dozen. How much did the merchant have to pay for the buttons, including the money order fee?

10. A jeweler in Spokane, Wash., ordered from a wholesaler in San Francisco, Calif. (fifth zone), one dozen watches sent C. O. D. by parcel post. If the watches were valued at $7.50 each, and the parcel weighed $3\frac{1}{2}$ pounds, (a) how much was the total cost of mailing the watches? (b) how much did they cost the jeweler in Spokane, including the money order fee? (c) if the watches were broken en route, how much insurance could be collected?

FRACTIONS

158. What part of a dollar is 50 ¢? 25 ¢? 75 ¢? 12½ ¢? 10 ¢? 5 ¢? 1 ¢?

Each of these amounts is less than one dollar. They are fractional parts of a dollar.

159. A **fraction** is a part of a unit.

> 50 ¢ = ½ of a dollar   = .50 of a dollar.
> 25 ¢ = ¼ of a dollar   = .25 of a dollar.
> 75 ¢ = ¾ of a dollar   = .75 of a dollar.
> 5 ¢ = $\frac{1}{20}$ of a dollar   = .05 of a dollar.
> 1 ¢ = $\frac{1}{100}$ of a dollar = .01 of a dollar.

How many ways are there of writing a fraction?

> Are ½, $\frac{5}{10}$, and .5 equal in value?
> Are ¾, $\frac{75}{100}$, and .75 equal in value?
> Are $\frac{1}{20}$, $\frac{5}{100}$, and .05 equal in value?
> Are $\frac{1}{100}$ and .01 equal in value?

160. There are, then, two ways of writing a fraction. In the common fraction form any number may be used as a denominator of the fraction. In the fractions ½, ¾, $\frac{1}{20}$, and $\frac{1}{100}$, 2, 4, 20, and 100, respectively, are the denominators.

161. In the decimal fraction form, the denominator must be 10 or some power * of 10. In the fractions .5, .75, .05, and .125, the denominators are 10, 100, 100, and 1000, respectively.

162. Into how many parts is this circle divided?

What part of the circle is heavily shaded?

How much of it is lightly shaded?

How does the lightly shaded part compare in size with the heavily shaded part? ¼ = 2 × ⅛.

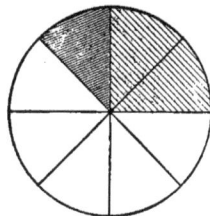

---

* By "power of 10" is meant the product resulting from using 10 a given number of times as a factor. Thus, 10 × 10, or 100, is the 2d power of 10; 10 × 10 × 10, or 1000, is the 3d power of 10, etc.

This illustration emphasizes an important principle:

The larger the denominator, the smaller the fraction. Compare with the Principles of Division (No. 17, page 102).

What part of the circle is not shaded?

Which part of the fraction, $\frac{5}{8}$, shows the number of parts into which the circle is divided?

Which part shows the number of eighths in the fraction?

In the fraction, $\frac{5}{8}$, 8 is the **denominator**, and shows that the unit (the unit, in this case, is the circle) has been divided into 8 equal parts. 5 is the **numerator**, and shows that 5 of the 8 parts have been taken to form the fraction, $\frac{5}{8}$.

Important. For all practical purposes, a common fraction represents a division. $5 \div 8 = \frac{5}{8}$. These expressions are the same except in form. Each may be read "5 divided by 8." It is clear, therefore, that 5, the numerator of the fraction, is a dividend, and that 8, the denominator, is a divisor.

163. Decimal and common fractions are so closely associated in business problems, that it has been deemed wise to treat them together.

There are three things to learn about decimals: (1) The use of decimal point. (2) Notation and numeration. (3) The common fraction equivalent, and *vice versa.*

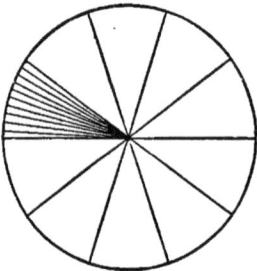

Into how many equal parts has this circle been divided?

One of these equal parts is called one tenth, written $\frac{1}{10}$ or .1.

Into how many equal parts has one of the tenths been divided?

If each of the tenths were divided into ten equal parts, into how many smaller parts would the entire circle be divided?

One of these small parts is what part of the circle? $\frac{1}{10}$ of $\frac{1}{10} = \frac{1}{100}$; or .1 of .1 = .01. (Principles 2 and 4, page 101.)

One tenth equals how many hundredths?

164. .1 and .01 are called **decimal fractions**, or, more often, **decimals**. (From the Latin word *decem*, meaning ten.) They are decimals because their denominators are 10 or some power of 10.

165. The denominator of a decimal is not written, but is indicated by the number of figures used to write the decimal.

166. In reading decimals, the denominator is named **the same as** though it were written. Thus, the decimal .015 is read "fifteen thousandths," the same as if written $\frac{15}{1000}$. The denominator is determined by the fact that the decimal contains 3 figures ("places"). For each place, one cipher is written to the right of the figure 1, making, in this case, 1000.

ORAL EXERCISE

167. Read the following decimals:

1.	.4	9.	.025	17.	.125484
2.	.14	10.	.0025	18.	236.324562
3.	.05	11.	15.0125 *	19.	.000125
4.	.125	12.	28.02345	20.	10000.0001
5.	.001	13.	75.00001	21.	.10101
6.	.101	14.	.00125	22.	.4755
7.	.427	15.	.00025	23.	125.0125
8.	.821	16.	112.01875	24.	.0000001

REDUCTION OF FRACTIONS

168. How many eighths equal $\frac{1}{4}$? How many quarters equal $\frac{1}{2}$? How many eighths equal $\frac{1}{2}$?

$$\tfrac{4}{8} = \tfrac{2}{4} = \tfrac{1}{2}, \text{ and } \tfrac{1}{2} = \tfrac{2}{4} = \tfrac{4}{8}$$

These equations illustrate what is meant by reduction of fractions. The form of the fraction, $\frac{4}{8}$, has been changed so as to read "$\frac{1}{2}$," but its value is not changed.

In the first equation, the fraction, $\frac{4}{8}$, has been reduced to its lowest terms.

In the second equation, the fraction, $\frac{1}{2}$, has been reduced to higher terms.

169. The terms of a fraction are its numerator and denominator.

170. A fraction is reduced to its lowest terms when its numerator and denominator cannot both be exactly divided by any whole number except 1.

* Read "fifteen, *and* one hundred twenty-five ten-thousandths."

171. **1.** Reduce $\frac{126}{189}$ to its lowest terms.

$$3\Big|\frac{126}{189} = 3\Big|\frac{42}{63} = 7\Big|\frac{14}{21} = \frac{2}{3}$$

Divide both terms of the fraction by any common divisor, say 3. The resulting fraction is $\frac{42}{63}$. Divide again by any common divisor, as 3, obtaining $\frac{14}{21}$. Now divide both terms by 7, and the result is $\frac{2}{3}$, the lowest terms, because the numbers 2 and 3 cannot be divided by the same number.

2. Reduce $\frac{493}{899}$ to its lowest terms.

$$29\Big|\frac{493}{899} = \frac{17}{31}$$

In this fraction the common divisor cannot be readily determined by inspection. Find the greatest common divisor by the method explained on page 63. 29 is the G. C. D. Dividing both terms of the fraction by 29 gives $\frac{17}{31}$, the lowest terms.

Compare these solutions with the principles of division (No. 21, page 102).

State the principle involved in the last two examples.

Fractions in the decimal form generally cannot be reduced to lower terms, unless they are changed to the common fraction form. Thus, .143 and .0025 as decimals cannot be reduced to lower terms, but .40 and .0030 may be reduced by omitting the ciphers at the right. .40 = .4 and .0030 = .003.

Practically the only instances in which a cipher is written at the right of a decimal is in writing 10¢, 20¢, 30¢, etc., decimally, as $.10, $.20, $.30, etc.

3. Reduce $\frac{7}{8}$ to fortieths.

$$40 \div 8 = 5$$
$$7 \times 5 = 35$$
$$8 \times 5 = 40$$

The denominator of the fraction desired is 40, which is 5 times as large as the denominator of the given fraction. If $\frac{7}{8}$ is to be changed to a fraction having a denominator 5 times as large, the numerator also must be 5 times as large. Hence, multiplying both numerator and denominator by 5 gives the desired result, $\frac{35}{40}$.

Compare with the Principles of Division (No. 21, page 102).

4. Reduce .13 to ten-thousandths.

$$10,000 \div 100 = 100$$
$$13 \times 100 = 1300. \text{ Therefore, } .1300$$

The required denominator is 100 times as large as the denominator of the given fraction. Therefore the numerator should be 100 times as large. Hence, the required decimal is .1300.

Observe that the process of reduction consists in annexing ciphers. Show the application of the principles of division.

172. Reduce the following fractions to their lowest terms (mentally, when possible):

1. $\frac{72}{80}$, $\frac{84}{120}$, $\frac{96}{116}$

2. $\frac{125}{150}$, $\frac{48}{144}$, $\frac{56}{168}$

3. $\frac{111}{117}$, $\frac{156}{207}$, $\frac{165}{220}$

4. $\frac{45}{60}$, $\frac{45}{180}$, $\frac{55}{495}$

5. $\frac{1240}{1648}$, $\frac{1640}{2050}$, $\frac{3088}{4632}$

6. $\frac{1998}{2997}$, $\frac{5922}{6782}$, $\frac{2448}{7272}$

7. $\frac{437}{551}$, $\frac{609}{783}$, $\frac{341}{465}$

8. $\frac{648}{756}$, $\frac{244}{427}$, $\frac{201}{335}$

9. $\frac{355}{568}$, $\frac{207}{552}$, $\frac{152}{171}$

10. $\frac{153}{187}$, $\frac{91}{143}$, $\frac{209}{231}$

11. $\frac{253}{299}$, $\frac{435}{464}$, $\frac{555}{740}$

12. $\frac{970}{1455}$, $\frac{410}{615}$, $\frac{684}{1026}$

Change to higher terms :

13. $\frac{3}{4}$ to 24ths.

14. $\frac{5}{8}$ to 48ths.

15. $\frac{7}{9}$ to 63ds.

16. .05 to thousandths.

17. .04 to hundred-thousandths.

18. $\frac{5}{6}$ to 36ths.

19. $\frac{8}{11}$ to 55ths.

20. $\frac{1}{2}$, $\frac{2}{3}$, $\frac{3}{4}$, $\frac{5}{6}$ to 12ths.

21. $\frac{3}{5}$, $\frac{5}{8}$, $\frac{3}{4}$, $\frac{9}{10}$ to 40ths.

22. $\frac{3}{4}$, $1\frac{3}{5}$, .5, .8 to 100ths.

23. $\frac{5}{8}$, $\frac{5}{6}$, .75, .125 to 24ths.

24. $\frac{3}{16}$, $\frac{2}{3}$, $.33\frac{1}{3}$, $.16\frac{2}{3}$ to 48ths.

173. To reduce an improper fraction to a whole or a mixed number.

How many halves are there in 1 circle ? in 2 circles? in 3 circles?
How many quarters are there in 1 circle? in 2 circles? in 3 circles?
How many eighths are there in 1 circle? in 2 circles? in 3 circles?

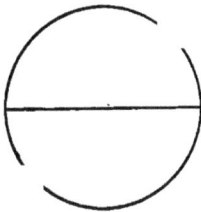

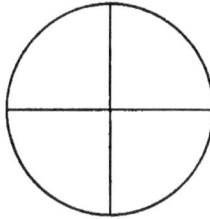

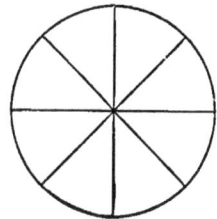

1 circle $= \frac{?}{2} = \frac{?}{4} = \frac{?}{8}$.   2 circles $= \frac{?}{2} = \frac{?}{4} = \frac{?}{8}$.   3 circles $= \frac{?}{2} = \frac{?}{4} = \frac{?}{8}$.

$\frac{3}{2}$ circles = how many circles? $\frac{9}{4}$ circles = how many circles?

$\frac{?}{?}$ circles = how many circles? $1\frac{2}{8}$ circles = how many circles?

$\frac{4}{2}$ circles = how many circles? $1\frac{6}{8}$ circles = how many circles?

$\frac{5}{4}$ circles = how many circles? $2\frac{0}{8}$ circles = how many circles?

All the fractions in this paragraph are **improper fractions**, because their numerators are equal to, or greater than, their denominators.

174. A fraction having a numerator smaller than its denominator is a **proper fraction**.

175. Improper fractions may be reduced to whole or mixed numbers.

176. A **mixed number** is the sum of a whole number and a fraction. Thus, $2\frac{1}{4}$ is a mixed number; it is the sum of 2 and $\frac{1}{4}$.

177. Reduce $\frac{37}{5}$ to a mixed number.

$$37 \div 5 = 7\frac{2}{5}.$$

There are $\frac{5}{5}$ in one unit. In $\frac{37}{5}$ there are as many units as 5 is contained times in 37, or 7 times and $\frac{2}{5}$ over.

Therefore, $\frac{37}{5} = 7\frac{2}{5}$.

EXERCISE

178. Reduce the following improper fractions to whole or mixed numbers:

1. $\frac{43}{6}$	7. $\frac{88}{9}$	13. $\frac{472}{4}$	19. $\frac{4897}{100}$	25. $\frac{675}{200}$
2. $\frac{72}{10}$	8. $\frac{124}{7}$	14. $\frac{263}{16}$	20. $\frac{2431}{1000}$	26. $\frac{990}{300}$
3. $\frac{31}{4}$	9. $\frac{263}{15}$	15. $\frac{224}{128}$	21. $\frac{6987}{100}$	27. $\frac{4897}{25}$
4. $\frac{48}{8}$	10. $\frac{328}{12}$	16. $\frac{163}{10}$	22. $\frac{3480}{100}$	28. $\frac{8000}{400}$
5. $\frac{53}{7}$	11. $\frac{264}{12}$	17. $\frac{248}{100}$	23. $\frac{5625}{1000}$	29. $\frac{4725}{25}$
6. $\frac{91}{8}$	12. $\frac{366}{15}$	18. $\frac{387}{100}$	24. $\frac{4375}{1000}$	30. $\frac{3125}{125}$

179. To reduce a mixed number to an improper fraction.

Reduce $4\frac{5}{8}$ to an improper fraction.

$4\frac{5}{8}$　　There are $\frac{8}{8}$ in one unit. In 4 units, or 4, there are 4 times $\frac{8}{8}$, or $\frac{32}{8}$;
$\frac{32}{8} + \frac{5}{8} = \frac{37}{8}$. Therefore, $4\frac{5}{8} = \frac{37}{8}$.

$\frac{32}{8}$
$\frac{5}{8}$
$\overline{\frac{37}{8}}$　　In practice, the solution is as given in the margin, viz., multiply the whole number by the denominator of the fraction, and to the product add the numerator. Under the sum write the denominator of the fraction.

EXERCISE

180. Reduce the following mixed numbers to improper fractions (mentally, when possible):

1. $3\frac{1}{2}$	4. $9\frac{3}{8}$	7. $10\frac{3}{4}$	10. $9\frac{3}{7}$	13. $72\frac{3}{4}$
2. $4\frac{2}{3}$	5. $6\frac{3}{7}$	8. $5\frac{9}{10}$	11. $17\frac{3}{4}$	14. $96\frac{5}{9}$
3. $8\frac{4}{4}$	6. $9\frac{8}{9}$	9. $6\frac{7}{8}$	12. $19\frac{3}{8}$	15. $82\frac{3}{4}$

16. $85\frac{5}{7}$	19. $38\frac{8}{11}$	22. $236\frac{7}{9}$	25. $437\frac{3}{5}$	28. $322\frac{5}{8}$
17. $45\frac{5}{8}$	20. $87\frac{1}{2}$	23. $328\frac{5}{6}$	26. $927\frac{3}{8}$	29. $972\frac{3}{16}$
18. $19\frac{9}{11}$	21. $128\frac{3}{4}$	24. $482\frac{5}{9}$	27. $726\frac{6}{7}$	30. $1846\frac{3}{7}$

In the following pages both ways of writing fractions will be used. The student must be able to handle either form alone or both together.

As illustrated in the chapter on Aliquot Parts, the common fraction is the more convenient form to use in many instances. In other cases, the decimal form is better.

INTERCHANGE OF FRACTIONAL FORMS

181. The student already knows that $\frac{1}{2}$ of a dollar = 50 cents = .50 of a dollar. If this were not known, it could be determined thus: One dollar is written \$1.00. To find $\frac{1}{2}$ of any number, divide by 2.

$$\underline{\begin{array}{r} \$.50 \\ 2)\$1.00 \end{array}}$$ This simple example illustrates the method of changing any common fraction to its equivalent decimal form, viz.,

Annex ciphers to the numerator and divide by the denominator.

EXERCISE

182. Change the following fractions to their equivalent decimal form. (Find results mentally, when possible.)

1. $\frac{3}{20}$	5. $\frac{4}{15}$ *	9. $\frac{4}{11}$	13. $\frac{1}{200}$	17. $\frac{18}{25}$
2. $\frac{5}{8}$	6. $\frac{5}{18}$	10. $\frac{3}{14}$	14. $\frac{13}{14}$	18. $\frac{21}{500}$
3. $\frac{9}{32}$	7. $\frac{9}{27}$	11. $\frac{1}{80}$	15. $\frac{15}{44}$	19. $\frac{4}{3125}$
4. $\frac{7}{25}$	8. $\frac{4}{9}$	12. $\frac{3}{40}$	16. $\frac{4}{125}$	20. $\frac{39}{75}$

* A common fraction in its lowest terms having a denominator that contains any factor other than 2 or 5 cannot be changed to a pure decimal. In such cases the division should be carried out two places and the fraction retained

Thus,   Hence, $\frac{4}{15} = .26\frac{2}{3}$.

$$\begin{array}{r} .26\frac{2}{3} \\ 15)\overline{4.00} \\ \underline{30} \\ 100 \\ \underline{90} \\ \frac{10}{15} = \frac{2}{3} \end{array}$$

If it were not known that 25 ¢ = .25 of a dollar = $\frac{1}{4}$ of a dollar, that fact could be determined by comparing 25 ¢ with 100 ¢ (100 ¢ being equal to one dollar). 25 ¢ is $\frac{25}{100}$ of 100 ¢. $5 \mid \frac{25}{100} = 5 \mid \frac{5}{20} = \frac{1}{4}$. Hence, $.25 = \frac{1}{4}$.

Observe that the decimal is written in the common fraction form by supplying its denominator, and reducing the resulting fraction to its lowest terms.

183. The denominator of the decimal is 1, with as many ciphers after it as there are places in the decimal.

Change $.46\frac{2}{3}$ to its common fraction form in its lowest terms.

$$.46\tfrac{2}{3} = \frac{46\tfrac{2}{3}}{100}$$

$$\frac{46\tfrac{2}{3}\times3}{100\times3} = 10 \left| \frac{140}{300} = 2 \right| \frac{14}{30} = \frac{7}{15}, \text{ or } \frac{.46\frac{2}{3}}{}\; \frac{140}{300} = \frac{7}{15}.$$

$.46\frac{2}{3}$ is a complex decimal. Its denominator is 100, giving a complex common fraction of $\frac{46\frac{2}{3}}{100}$. By the Principles of Division (No. 21, page 102), multiplying both dividend and divisor (numerator and denominator) by the same number does not change the value of the quotient. Hence, multiply both numerator and denominator by 3 (the denominator of the fraction $\frac{2}{3}$). The result is $\frac{140}{300}$, which equals $\frac{7}{15}$, the desired result.

EXERCISE

184. Change the following decimals to their equivalent common fraction form in their lowest terms. Determine results mentally, when possible.

1. .8	7. .032	13. $.003\frac{1}{3}$	19. .056
2. .24	8. .00125	14. .1875	20. .104
3. .56	9. .0005	15. $.09\frac{3}{8}$	21. $.0001\frac{1}{4}$
4. .625	10. $.012\frac{1}{2}$	16. $.09\frac{1}{11}$	22. $.0003\frac{1}{3}$
5. $.71\frac{3}{7}$	11. $.006\frac{1}{4}$	17. $.07\frac{1}{7}$	23. $.0142\frac{6}{7}$
6 .888	12. $.002\frac{1}{2}$	18. .0015	24. $.006\frac{2}{3}$

LEAST COMMON DENOMINATOR

185. What is the denominator of each of the following fractions: $\frac{5}{16}, \frac{11}{16}, \frac{9}{16}, \frac{13}{16}, \frac{1}{16}, \frac{3}{16}, \frac{7}{16}, \frac{15}{16}$?

Because their denominators are all alike, the fractions are said to have a **common denominator**. They are also called similar fractions.

186. Compare these denominators with each other: $\frac{2}{3}$, $\frac{3}{4}$, $\frac{5}{6}$, $\frac{4}{5}$, $\frac{9}{10}$, $\frac{11}{16}$, $\frac{8}{9}$, .18, .232.

They are unlike, or dissimilar.

What are dissimilar fractions?

187. Fractions must be made similar before they can be added. (See Principles of Addition, No. 5, page 101.)

188. What is the least common multiple of 2, 4, 8, and 16?

189. 1. Change $\frac{1}{2}$, $\frac{3}{4}$, $\frac{5}{8}$, and $\frac{9}{16}$ to similar fractions, having the smallest denominator possible.

$\frac{1}{2} = \frac{8}{16}$ If the fractions are to be similar, the denominators must be
$\frac{3}{4} = \frac{12}{16}$ alike. The least common denominator is the same as the least
$\frac{5}{8} = \frac{10}{16}$ common multiple of the denominators. The least common de-
$\frac{9}{16} = \frac{9}{16}$ nominator can be determined mentally. (See page 64 for finding L. C. M.) It is 16.

As explained on page 70, reduce each of the fractions to 16ths. (State the principle of division that applies here.)

2. Reduce $\frac{5}{8}$, $\frac{11}{24}$, .12$\frac{1}{2}$, .33$\frac{1}{3}$, and .25 to similar fractions, having their least common denominator.

$\frac{5}{8} \qquad = \frac{15}{24}$ Change the fractions in decimal form to their equivalent
$\frac{11}{24} \qquad = \frac{11}{24}$ common fraction form, in their lowest terms. By inspec-
$.12\frac{1}{2} = \frac{1}{8} = \frac{3}{24}$ tion, it is seen that 24 is the least common denominator.
$.33\frac{1}{3} = \frac{1}{3} = \frac{8}{24}$ Each fraction is now changed to a fraction having 24 for
$.25 \quad = \frac{1}{4} = \frac{6}{24}$ its denominator.

EXERCISE

190. Reduce the following fractions to similar fractions having their least common denominator. (Do the work mentally when possible, using pencil only to tabulate results.)

1. $\frac{1}{2}$, $\frac{2}{3}$, $\frac{3}{4}$ 3. $\frac{3}{4}$, $\frac{1}{8}$, .25   5. $\frac{3}{4}$, .06$\frac{1}{4}$, .12$\frac{1}{2}$ 7. $\frac{5}{8}$, $\frac{2}{3}$, $\frac{1}{12}$

2. $\frac{4}{5}$, $\frac{9}{10}$, $\frac{1}{2}$ 4. $\frac{5}{6}$, $\frac{2}{3}$, .1   6. $\frac{7}{8}$, .9, .375   8. $\frac{5}{6}$, $\frac{3}{4}$, $\frac{2}{3}$

9. $\frac{5}{16}$, $\frac{7}{24}$, .62$\frac{1}{2}$ 15. $\frac{5}{12}$, $\frac{4}{9}$, .08$\frac{1}{3}$, .33$\frac{1}{3}$

10. $\frac{8}{9}$, $\frac{2}{3}$, $\frac{5}{18}$ 16. $\frac{7}{8}$, $\frac{9}{16}$, $\frac{3}{4}$, $\frac{1}{6}$, $\frac{5}{12}$

11. $\frac{5}{9}$, $\frac{3}{4}$, $\frac{7}{18}$, $\frac{2}{3}$, $\frac{5}{36}$ 17. .25, .625, .06$\frac{1}{4}$, .5

12. $\frac{5}{16}$, .125, $\frac{3}{4}$, .18$\frac{3}{4}$ 18. $\frac{3}{7}$, $\frac{5}{21}$, $\frac{9}{28}$, $\frac{3}{4}$, $\frac{1}{2}$

13. $\frac{3}{4}$, .87$\frac{1}{2}$, $\frac{1}{2}$, .5, .25   19. $\frac{3}{8}$, .31$\frac{1}{4}$, $\frac{2}{3}$, .16$\frac{2}{3}$, $\frac{5}{12}$

14. $\frac{17}{24}$, $\frac{5}{6}$, .33$\frac{1}{3}$, .16$\frac{2}{3}$ 20. $\frac{3}{11}$, $\frac{5}{9}$, $\frac{7}{33}$, $\frac{18}{99}$, $\frac{2}{3}$

ADDITION OF FRACTIONS

191. 1 qt. + 3 qt. + 2 qt. + 3 qt. = how many quarts?

What kind of numbers can be added?

1 fourth + 3 fourths + 2 fourths + 3 fourths = how many fourths?

$\frac{1}{4} + \frac{3}{4} + \frac{2}{4} + \frac{3}{4} = ?$

What kind of fractions can be added?

192. 1. Add $\frac{5}{8}, \frac{7}{8}, \frac{3}{8}, \frac{1}{8}, \frac{2}{8}$.

$\frac{5}{8} + \frac{7}{8} + \frac{3}{8} + \frac{1}{8} + \frac{2}{8} = \frac{18}{8} = 2\frac{1}{4}$. Since these are similar fractions, they may be added by taking the sum of their numerators, which is 18, and writing it over the common denominator, 8. $\frac{18}{8}$, reduced to a mixed number, $= 2\frac{1}{4}$.

2. Find the sum of $\frac{2}{3}, \frac{3}{4}, \frac{5}{8}$, and $\frac{11}{12}$.

$\frac{2}{3} = 16$
$\frac{3}{4} = 18$
$\frac{5}{8} = 15$
$\frac{11}{12} = 22$
$\frac{71}{24} = 2\frac{23}{24}$.

Before these fractions can be added they must be changed to similar fractions. By inspection, the least common denominator is seen to be 24. Instead of writing the 24 for each fraction, write it where it is needed, under the sum of the new numerators, as shown in the solution. Now find the numerators, writing each one after the fraction it represents. This arrangement makes the addition of the numerators very easy and quick, their sum being written, as shown, over the common denominator, 24; $\frac{71}{24} = 2\frac{23}{24}$, the desired result.

3. Find the sum of $18\frac{2}{3}, 33\frac{4}{5}$, and $68\frac{1}{2}$.

$18 \ \frac{2}{3} = 20$
$33 \ \frac{4}{5} = 24$
$68 \ \frac{1}{2} = 15$
$120\frac{29}{30} \quad \frac{59}{30} = 1\frac{29}{30}$

Write the integers and fractions, slightly separated, as shown in the solution. Add the fractions first by the method already explained. Their sum is $1\frac{29}{30}$. Add the integers, carrying the 1 from the sum of the fractions. The total result is $120\frac{29}{30}$.

4. Find the sum of $13\frac{1}{3}, 18.25, 64.345$, and $56\frac{1}{7}$.

$13.333\frac{1}{3} = 7$
18.25
64.345
$56.142\frac{6}{7} = 18$
$152.071\frac{4}{21} \quad \frac{25}{21} = 1\frac{4}{21}$

These fractions are best added in their decimal form. The first thing to do is to change each common fraction to its decimal form. Next write the numbers as shown, with the decimal points under one another. In changing $\frac{1}{3}$ and $\frac{1}{7}$ to their decimal form, the student should observe that the decimals are carried out as many places as there are places in the longest decimal in any of the other numbers. Thus, the longest decimal given is .345, having 3 places. Therefore, $\frac{1}{3}$ and $\frac{1}{7}$ are changed to 3-place decimals. This being done, the remaining common fractions, $\frac{1}{3}$ and $\frac{6}{7}$, are parts of like orders and can be added. Add the common fractions as before, and carry the integral part as in the preceding example.

5. How many yards are there in 5 pieces of silk as follows: 41^2, 43^1, 44, 42^3, 47^1?

41^3
43^1
44 The small numbers written a little above and to the right of the numbers 41, 43, etc., indicate quarter yards. That is, 41^2 yd. is $41\frac{2}{4}$ yd.; 41^3 yd. is $41\frac{3}{4}$ yd., etc. This method of writing quarters is much used in the dry-goods business. These being similar fractions, the addition is made as already explained.
42^3
47^1
——
218^3

193. Add:

1. $\frac{3}{4}, \frac{1}{2}, \frac{2}{4}, \frac{5}{8}$

2. $\frac{5}{6}, \frac{1}{3}, \frac{1}{2}, 4$

3. $\frac{5}{7}, \frac{1}{2}, \frac{3}{14}, \frac{2}{3}$

4. $\frac{3}{11}, \frac{1}{2}, \frac{5}{22}, \frac{1}{3}$

5. $18\frac{1}{4}, 16\frac{2}{3}, 17\frac{3}{8}$

6. $24\frac{5}{6}, 31\frac{3}{4}, 47\frac{1}{8}$

7. $14\frac{1}{4}, 18.125, 27.5$

8. $126\frac{1}{3}, 162.5, 204.475$

9. $\frac{5}{8}, 2\frac{1}{2}, 13.324$

10. $\frac{3}{8}, .4, \frac{5}{8}, .875$

11. $\frac{4}{5}, \frac{7}{8}$

12. $\frac{5}{7}, \frac{9}{10}$

13. $18^1, 17^2, 16^3, 19^2, 22^1$

14. $48^1, 72^3, 64^2, 89^2, 94^1$

15. $75\frac{2}{3}, 81\frac{1}{2}, 63.427$

16. $41\frac{5}{8}, 72.325, 47.18\frac{2}{3}$

17. $32\frac{7}{8}, 91\frac{3}{4}, 52.135$

18. $46^1, 72^2, 84^1, 92, 87^3$

19. $19\frac{1}{2}, 18\frac{1}{4}, 82\frac{2}{3}, 76\frac{3}{4}$

20. $21\frac{2}{3}, 76\frac{4}{5}, 91\frac{2}{3}, 4.847$

MENTAL ADDITION OF FRACTIONS

194. 1. Find the sum of $\frac{1}{3}$ and $\frac{1}{4}$

$\frac{1}{3} = \frac{4}{12}$
$\frac{1}{4} = \frac{3}{12}$
$\frac{7}{12}$ Observe that, in changing the fractions to similar fractions, the numerators of the similar fractions are the same numbers as the denominators of the given fractions, and that the common denominator is the product of the same numbers, 3 and 4. Therefore, to find the sum of two fractions having 1 for a numerator, take the sum of the denominators for the numerator, and the product of the denominators for the denominator.

2. Add $\frac{1}{8}, \frac{1}{3}, \frac{5}{24}$.

$\frac{1}{8} + \frac{1}{3} = \frac{11}{24}$
$\frac{11}{24} + \frac{5}{24} = \frac{16}{24} = \frac{2}{3}$ Notice that the sum of the first two fractions, $\frac{1}{8}$ and $\frac{1}{3}$, is a fraction ($\frac{11}{24}$) similar to the third fraction, $\frac{5}{24}$, which may then be added mentally.

The student should be always on the alert, looking for short methods.

195. Add the following:

1. $\frac{1}{3}+\frac{1}{5}$	**11.** $\frac{1}{10}+\frac{1}{11}$	**21.** $\frac{1}{2}+\frac{1}{3}+\frac{5}{6}$
2. $\frac{1}{3}+\frac{1}{7}$	**12.** $\frac{1}{9}+\frac{1}{5}$	**22.** $\frac{1}{4}+\frac{1}{5}+\frac{9}{20}$
3. $\frac{1}{2}+\frac{1}{9}$	**13.** $\frac{1}{8}+\frac{1}{5}$	**23.** $\frac{1}{3}+\frac{1}{4}+\frac{5}{6}$
4. $\frac{1}{8}+\frac{1}{3}$	**14.** $\frac{1}{9}+\frac{1}{4}$	**24.** $\frac{1}{7}+\frac{1}{4}+\frac{3}{14}$
5. $\frac{1}{5}+\frac{1}{4}$	**15.** $\frac{1}{12}+\frac{1}{5}$	**25.** $\frac{1}{9}+\frac{1}{2}+\frac{2}{6}$
6. $\frac{1}{6}+\frac{1}{7}$	**16.** $\frac{1}{11}+\frac{1}{12}$	**26.** $\frac{1}{8}+\frac{1}{3}+\frac{11}{12}$
7. $\frac{1}{8}+\frac{1}{9}$	**17.** $\frac{1}{8}+\frac{1}{11}$	**27.** $\frac{1}{9}+\frac{1}{3}+\frac{5}{27}$
8. $\frac{1}{7}+\frac{1}{8}$	**18.** $\frac{1}{7}+\frac{1}{11}$	**28.** $\frac{1}{10}+\frac{1}{3}+\frac{4}{15}$
9. $\frac{1}{12}+\frac{1}{7}$	**19.** $\frac{1}{6}+\frac{1}{11}$	**29.** $\frac{1}{9}+\frac{1}{5}+\frac{8}{45}$
10. $\frac{1}{11}+\frac{1}{2}$	**20.** $\frac{1}{7}+\frac{1}{5}$	**30.** $\frac{1}{7}+\frac{1}{9}+\frac{5}{21}$

196. Find the sum of $\frac{2}{5}$ and $\frac{2}{7}$.

$$\frac{2}{5}+\frac{2}{7}=\frac{24}{35}$$
$$(7+5)\times 2 = 24$$

$\frac{2}{5}$ is 2 times $\frac{1}{5}$, and $\frac{2}{7}$ is 2 times $\frac{1}{7}$. The sum of $\frac{1}{5}$ and $\frac{1}{7}$ is $\frac{12}{35}$. Therefore, the sum of $\frac{2}{5}$ and $\frac{2}{7}$ is 2 times $\frac{12}{35}$, or $\frac{24}{35}$. That is, take 2 times the sum of the denominators for the numerator of the sum. The denominator is the product of the denominators, the same as before.

197. Add the following:

1. $\frac{2}{5}+\frac{2}{6}$	**7.** $\frac{2}{3}+\frac{2}{8}$	**13.** $\frac{4}{5}+\frac{4}{7}$	**19.** $\frac{2}{3}+\frac{2}{7}+\frac{8}{21}$
2. $\frac{2}{5}+\frac{2}{8}$	**8.** $\frac{2}{7}+\frac{2}{11}$	**14.** $\frac{4}{8}+\frac{4}{11}$	**20.** $\frac{2}{5}+\frac{2}{9}+\frac{4}{15}$
3. $\frac{2}{3}+\frac{2}{7}$	**9.** $\frac{2}{5}+\frac{2}{11}$	**15.** $\frac{5}{8}+\frac{5}{9}$	**21.** $\frac{2}{7}+\frac{2}{9}+\frac{1}{21}$
4. $\frac{2}{8}+\frac{2}{5}$	**10.** $\frac{3}{4}+\frac{3}{8}$	**16.** $\frac{5}{6}+\frac{5}{7}$	**22.** $\frac{2}{3}+\frac{2}{4}+\frac{11}{12}$
5. $\frac{2}{7}+\frac{2}{9}$	**11.** $\frac{3}{5}+\frac{3}{6}$	**17.** $\frac{6}{7}+\frac{6}{8}$	**23.** $\frac{3}{4}+\frac{3}{5}+\frac{7}{20}$
6. $\frac{2}{5}+\frac{2}{9}$	**12.** $\frac{3}{7}+\frac{3}{8}$	**18.** $\frac{6}{7}+\frac{6}{10}$	**24.** $\frac{4}{5}+\frac{4}{7}+\frac{8}{35}$

198. The student should drill carefully and thoroughly on additions like the following, till he can add simple fractions as readily as whole numbers.

Add $\frac{1}{2}$, $\frac{3}{4}$, $\frac{5}{8}$, and $\frac{9}{16}$.

To add these fractions, the student should "*think*" all operations, naming only the results. ("Think" the part in italics). *The least common denominator is* 16; *the first numerator is* 8; *the second is* 12, 20; *the third is* 10; 30, 39; $\frac{39}{16}=2\frac{7}{16}$.

199. Drill on the following:

1. $\frac{1}{3}+\frac{3}{4}+\frac{1}{6}+\frac{5}{12}$

2. $\frac{1}{2}+\frac{1}{3}+\frac{5}{6}+\frac{2}{3}$

3. $\frac{1}{4}+\frac{3}{8}+\frac{1}{2}+\frac{9}{16}$

4. $\frac{3}{5}+\frac{7}{10}+\frac{1}{2}+\frac{1}{10}$

5. $\frac{4}{5}+\frac{1}{2}+\frac{3}{4}+\frac{7}{20}$

6. $\frac{1}{2}+\frac{2}{3}+\frac{3}{4}+\frac{7}{12}$

7. $\frac{1}{3}+\frac{3}{4}+\frac{7}{12}+\frac{5}{6}+\frac{1}{2}$

8. $\frac{5}{8}+\frac{7}{16}+\frac{3}{4}+\frac{1}{2}+\frac{5}{8}$

9. $\frac{7}{9}+\frac{2}{3}+\frac{1}{2}+\frac{5}{18}+\frac{1}{6}$

10. $\frac{9}{12}+\frac{3}{4}+\frac{7}{8}+\frac{3}{6}+\frac{2}{3}$

11. $\frac{7}{12}+\frac{1}{2}+\frac{5}{6}+\frac{2}{3}+\frac{3}{4}$

12. $\frac{1}{2}+\frac{5}{16}+\frac{7}{8}+\frac{3}{4}+\frac{1}{4}$

13. $\frac{1}{3}+\frac{1}{2}+\frac{5}{6}+\frac{2}{3}+\frac{1}{6}$

14. $\frac{5}{9}+\frac{1}{2}+\frac{5}{6}+\frac{2}{3}+\frac{1}{2}$

15. $\frac{5}{8}+\frac{1}{2}+\frac{3}{4}+\frac{1}{4}+\frac{5}{8}$

16. $\frac{5}{16}+\frac{3}{8}+\frac{9}{16}+\frac{1}{2}+\frac{3}{4}$

17. $\frac{9}{10}+\frac{1}{2}+\frac{2}{3}+\frac{4}{5}+\frac{7}{10}$

18. $\frac{15}{32}+\frac{1}{16}+\frac{7}{8}+\frac{9}{32}+\frac{3}{4}$

19. $\frac{5}{18}+\frac{1}{2}+\frac{2}{3}+\frac{5}{6}+\frac{5}{12}$

20. $\frac{9}{16}+\frac{2}{3}+\frac{3}{4}+\frac{1}{2}+\frac{1}{6}$

21. $\frac{11}{15}+\frac{2}{3}+\frac{5}{6}+\frac{2}{3}+\frac{1}{6}$

22. $\frac{11}{24}+\frac{3}{4}+\frac{7}{8}+\frac{1}{2}+\frac{5}{24}$.

23. $\frac{2}{3}+\frac{3}{4}+\frac{5}{6}+\frac{11}{12}+\frac{1}{2}$

24. $\frac{5}{9}+\frac{2}{3}+\frac{3}{4}+\frac{11}{12}+\frac{1}{2}$

200. Add (mentally, when possible):

1. $\frac{2}{3}+\frac{3}{4}+\frac{5}{6}+\frac{7}{8}$

2. $5\frac{4}{5}+\frac{1}{3}+\frac{1}{2}+\frac{7}{10}$

3. $\frac{5}{8}+\frac{9}{16}+.25+.37\frac{1}{2}$

4. $16\frac{3}{4}+17\frac{2}{3}+24\frac{5}{8}+56.5$

5. $47\frac{3}{5}+96.2+143\frac{9}{10}+54.25$

6. $122.125\frac{1}{3}+63\frac{1}{2}+88.0625+15\frac{1}{8}$

7. $49\frac{3}{4}+72\frac{5}{8}+99\frac{11}{16}+\frac{1}{2}$

8. $51^1+64^2+48^3+54+50^3$

9. $32\frac{3}{16}+29\frac{3}{8}+19\frac{3}{4}+54$

10. $78.125+31.5+83.333\frac{1}{3}+.66\frac{2}{3}$

11-30. For further drill add the fractions on page 75.

SUBTRACTION OF FRACTIONS

201. 1. 5 yd. minus 3 yd. = ?

2. 5 eighths minus 3 eighths = ?

$\frac{5}{8}-\frac{3}{8}=$?

3. What kind of fractions can be subtracted?

4. From $\frac{9}{11}$ take $\frac{3}{11}$.

$\frac{9}{11}-\frac{3}{11}=\frac{6}{11}$.

Since the fractions are similar, it is necessary only to take the difference between the numerators, 9 and 3, which = 6, and under the 6 to write the common
denominator, thus, $\frac{6}{11}$.

202. Find the difference between:

1. $\frac{5}{8}$ and $\frac{3}{8}$ 5. $\frac{9}{14}$ and $\frac{7}{14}$ 9. $\frac{4}{3}$ and $\frac{2}{3}$

2. $\frac{9}{11}$ and $\frac{7}{11}$ 6. $\frac{7}{12}$ and $\frac{5}{12}$ 10. .8 and 5

3. $\frac{5}{6}$ and $\frac{1}{6}$ 7. $\frac{7}{8}$ and $\frac{5}{8}$ 11. .11 and .09

4. $\frac{3}{7}$ and $\frac{2}{7}$ 8. $\frac{9}{16}$ and $\frac{7}{16}$ 12. .50 and .25

203. 1. From $\frac{5}{8}$ take $\frac{1}{3}$.

$\frac{5}{8} = \frac{15}{}$ Reduce the fractions to similar fractions, as in addition. In-
$\frac{1}{3} = \frac{8}{}$ stead of adding the new numerators, subtract the one from the
$\frac{7}{24}$ other, which gives the result, $\frac{7}{24}$.

2. From $15\frac{1}{3}$ take $9\frac{3}{4}$.

$14 \cdot, \frac{4}{3} = \frac{16}{}$ $\frac{3}{4}$ is larger than $\frac{1}{3}$. First, therefore, "borrow" 1 from the
$9 \cdot, \frac{3}{4} = \frac{9}{}$ 15, and add it to the $\frac{1}{3}$, expressing the result as an improper
$5\frac{7}{12}$ $\frac{7}{12}$ fraction, $\frac{4}{3}$. Now proceed as before: Change to similar
fractions, and subtract, obtaining $5\frac{7}{12}$ as the result.

204. Find the value of:

1. $\frac{7}{8} - \frac{3}{4}$ 11. $14\frac{5}{9} - 7.33\frac{1}{3}$ 21. $18.127 - 12.33\frac{1}{3}$

2. $\frac{9}{10} - \frac{5}{6}$ 12. $38\frac{7}{8} - 21\frac{11}{12}$ 22. $26.872\frac{1}{3} - 5\frac{1}{6}$

3. $\frac{3}{4} - \frac{2}{3}$ 13. $56\frac{3}{4} - 29.5$ 23. $39.458\frac{1}{3} - 18.12\frac{1}{2}$

4. $\frac{9}{16} - \frac{1}{4}$ 14. $76\frac{3}{16} - 67.18\frac{3}{4}$ 24. $79\frac{3}{11} - 18\frac{3}{4}$

5. $1\frac{1}{2} - \frac{3}{4}$ 15. $98.16\frac{2}{3} - 21.135$ 25. $71\frac{1}{3} - 28\frac{7}{8}$

6. $5\frac{2}{3} - 3\frac{5}{6}$ 16. $88.124 - 57\frac{1}{3}$ 26. $82\frac{9}{10} - 43\frac{1}{4}$

7. $5\frac{1}{2} - 2.75$ 17. $74\frac{7}{16} - 28\frac{11}{12}$ 27. $67\frac{3}{16} - 51\frac{3}{4}$

8. $18\frac{2}{3} - 12.16\frac{2}{3}$ 18. $58\frac{8}{9} - 27.9$ 28. $54.128 - 27.063$

9. $24.83\frac{1}{3} - 18\frac{1}{3}$ 19. $73\frac{3}{7} - 24\frac{9}{14}$ 29. $57.45\frac{1}{3} - 33.128$

10. $63\frac{3}{8} - 12.06\frac{1}{4}$ 20. $84\frac{1}{4} - 72.13$ 30. $99\frac{5}{8} - 74.25$

MENTAL SUBTRACTION OF FRACTIONS

205. 1. From $\frac{1}{2}$ take $\frac{1}{3}$.

$\frac{1}{2} = \frac{3}{6}$ In changing the fractions to similar fractions, notice that the
$\frac{1}{3} = \frac{2}{6}$ numerators of the similar fractions are the same numbers as the
$\frac{1}{6}$ denominators of the given fractions. The difference is $\frac{1}{6}$. Therefore, take the difference between the denominators for the required
numerator, and the product of the denominators for the required denominator.

2. From $\frac{2}{3}$ take $\frac{2}{5}$.

$\frac{2}{3} = \frac{10}{15}$
$\frac{2}{5} = \frac{6}{15}$
$\frac{4}{15}$

The common denominator is 15. Observe that the new numerators are just twice the denominators of the given fractions, and that their difference, 4, is just twice the difference between the denominators 5 and 3. Therefore, take twice the difference between the denominators of the given fractions and write it over the common denominator.

206. Find the value of:

1. $\frac{1}{3} - \frac{1}{4}$ 8. $\frac{1}{4} - \frac{1}{6}$ 15. $\frac{1}{5} - \frac{1}{6}$ 22. $\frac{2}{5} - \frac{2}{7}$

2. $\frac{1}{3} - \frac{1}{5}$ 9. $\frac{1}{4} - \frac{1}{7}$ 16. $\frac{1}{5} - \frac{1}{7}$ 23. $\frac{2}{3} - \frac{2}{5}$

3. $\frac{1}{3} - \frac{1}{7}$ 10. $\frac{1}{4} - \frac{1}{8}$ 17. $\frac{1}{5} - \frac{1}{8}$ 24. $\frac{2}{3} - \frac{2}{7}$

4. $\frac{1}{3} - \frac{1}{8}$ 11. $\frac{1}{4} - \frac{1}{9}$ 18. $\frac{1}{5} - \frac{1}{9}$ 25. $\frac{2}{7} - \frac{2}{9}$

5. $\frac{1}{3} - \frac{1}{9}$ 12. $\frac{1}{4} - \frac{1}{10}$ 19. $\frac{1}{5} - \frac{1}{10}$ 26. $\frac{3}{5} - \frac{3}{7}$

6. $\frac{1}{3} - \frac{1}{10}$ 13. $\frac{1}{4} - \frac{1}{11}$ 20. $\frac{1}{5} - \frac{1}{11}$ 27. $\frac{3}{4} - \frac{3}{5}$

7. $\frac{1}{4} - \frac{1}{5}$ 14. $\frac{1}{4} - \frac{1}{12}$ 21. $\frac{1}{5} - \frac{1}{12}$ 28. $\frac{3}{7} - \frac{3}{11}$

207. Find (mentally, when possible) the value of:

1. $\frac{1}{2} + \frac{1}{3} - \frac{1}{6}$ 11. $\frac{1}{2} + \frac{2}{3} - \frac{5}{6}$ 21. $\frac{7}{8} - \frac{1}{2} + \frac{3}{4} - \frac{3}{8} + \frac{5}{16}$

2. $\frac{1}{3} + \frac{1}{4} - \frac{5}{12}$ 12. $\frac{2}{3} + \frac{1}{4} - \frac{5}{12}$ 22. $\frac{9}{16} - \frac{1}{2} + \frac{7}{8} - \frac{3}{16} + \frac{1}{4}$

3. $\frac{1}{3} + \frac{1}{5} - \frac{7}{15}$ 13. $\frac{3}{4} + \frac{1}{3} - \frac{7}{12}$ 23. $\frac{5}{6} + \frac{1}{3} - \frac{1}{2} - \frac{1}{6} + \frac{2}{3}$

4. $\frac{2}{3} + \frac{2}{7} - \frac{5}{21}$ 14. $\frac{3}{5} + \frac{1}{2} - \frac{9}{10}$ 24. $\frac{4}{9} + \frac{2}{3} - \frac{5}{6} + \frac{1}{2} - \frac{1}{3}$

5. $\frac{3}{4} + \frac{3}{5} - \frac{7}{20}$ 15. $\frac{3}{4} + \frac{3}{8} - \frac{1}{2}$ 25. $\frac{5}{8} + \frac{1}{4} + \frac{1}{2} - \frac{3}{8} - \frac{1}{2}$

6. $\frac{2}{7} + \frac{2}{9} - \frac{1}{3}$ 16. $\frac{5}{6} + \frac{1}{2} - \frac{2}{3}$ 26. $\frac{15}{16} - \frac{1}{2} - \frac{1}{4} + \frac{3}{8} + \frac{3}{4}$

7. $\frac{1}{3} + \frac{3}{8} - \frac{1}{12}$ 17. $\frac{3}{8} + \frac{2}{3} - \frac{5}{12}$ 27. $\frac{9}{10} - \frac{4}{5} + \frac{1}{2} + \frac{3}{4} - \frac{17}{20}$

8. $\frac{1}{4} + \frac{1}{5} - \frac{3}{10}$ 18. $\frac{4}{5} + \frac{1}{3} - \frac{11}{15}$ 28. $\frac{8}{9} - \frac{1}{6} - \frac{1}{3} + \frac{2}{18} - \frac{1}{4}$

9. $\frac{1}{8} + \frac{1}{9} - \frac{5}{24}$ 19. $\frac{5}{8} + \frac{3}{4} - \frac{7}{8}$ 29. $\frac{11}{12} - \frac{3}{4} + \frac{1}{6} + \frac{1}{3} - \frac{1}{2}$

10. $\frac{1}{4} + \frac{1}{7} - \frac{5}{14}$ 20. $\frac{3}{5} + \frac{3}{7} - \frac{9}{35}$ 30. $\frac{11}{16} + \frac{1}{4} - \frac{5}{8} + \frac{1}{2} - \frac{3}{4}$

MULTIPLICATION OF FRACTIONS

208. To multiply a whole number by a fraction or a fraction by a whole number.

1. 4 times 3 gallons = how many gallons?

·2. 4 times 3 fifths = how many fifths? $4 \times \frac{3}{5} = ?$

3. Does $4 \times \frac{3}{5} = \frac{3}{5} \times 4$?   (Principle 13, page 102.)

4. Compare these two solutions.

$$5 \times \tfrac{3}{7} = \tfrac{15}{7} = 2\tfrac{1}{7}. \qquad \tfrac{3}{7} \times 5 = \tfrac{15}{7} = 2\tfrac{1}{7}.$$

($\times$ may be read "times," "multiplied by," or "of." The first of these two equations may be read "5 times $\tfrac{3}{7} = \tfrac{15}{7}$," or "5 multiplied by $\tfrac{3}{7} = \tfrac{15}{7}$"; the second may be read "$\tfrac{3}{7}$ of $5 = \tfrac{15}{7}$," or "$\tfrac{3}{7}$ multiplied by $5 = \tfrac{15}{7}$." It is not proper to say "5 of $\tfrac{3}{7}$," or "$\tfrac{3}{7}$ times 5.")

State the principle of division that applies in these equations. See page 102.

5. Compare these two solutions:

$$7 \times \tfrac{5}{14} = \tfrac{35}{14} = \tfrac{5}{2} = 2\tfrac{1}{2}.$$

$$7 \times \frac{5}{\underset{2}{\cancel{14}}} = \frac{5}{2} = 2\tfrac{1}{2}.$$

Observe that, in the first solution, the numerator of the fraction is multiplied by 7, and that the resulting fraction is reduced to its lowest terms by *dividing both* terms of the fraction by 7.

In the second solution, notice that by dividing the denominator of the fraction, 14, by the multiplier, 7, the result, $\tfrac{5}{2}$, is obtained at once, without the intermediate step of multiplying by 7 (Principles 18, 19, page 102).

To THE STUDENT. Study each problem carefully and determine the shortest solution. As far as possible, make all computations mentally, using pen or pencil to tabulate results.

209. Find the value of: ·

1. $5 \times \tfrac{7}{8}$	**10.** $5 \times \tfrac{9}{10}$	**19.** $\tfrac{11}{12} \times 11$	**28.** $\tfrac{5}{16}$ of 38
2. $8 \times \tfrac{3}{4}$	**11.** $6 \times \tfrac{5}{6}$	**20.** $\tfrac{5}{9} \times 15$	**29.** $\tfrac{11}{12}$ of 66
3. $9 \times \tfrac{7}{9}$	**12.** $8 \times \tfrac{9}{16}$	**21.** $\tfrac{8}{9} \times 24$	**30.** $\tfrac{7}{8}$ of 84
4. $6 \times \tfrac{3}{5}$	**13.** $\tfrac{5}{8} \times 16$	**22.** $\tfrac{9}{8} \times 24$	**31.** $\tfrac{15}{16}$ of 5
5. $5 \times \tfrac{4}{7}$	**14.** $\tfrac{7}{10} \times 15$	**23.** $\tfrac{7}{15} \times 60$	**32.** $\tfrac{19}{20}$ of 50
6. $9 \times \tfrac{8}{11}$	**15.** $\tfrac{7}{9} \times 27$	**24.** $\tfrac{8}{15} \times 40$	**33.** $\tfrac{21}{22}$ of 66
7. $3 \times \tfrac{2}{7}$	**16.** $\tfrac{15}{16} \times 48$	**25.** $\tfrac{5}{9}$ of 148	**34.** $\tfrac{12}{25}$ of 75
8. $4 \times \tfrac{8}{15}$	**17.** $\tfrac{3}{4} \times 64$	**26.** $\tfrac{3}{4}$ of 123	**35.** $\tfrac{13}{20}$ of 48
9. $7 \times \tfrac{6}{11}$	**18.** $\tfrac{8}{11} \times 55$	**27.** $\tfrac{8}{9}$ of 72	**36.** $\tfrac{15}{19}$ of 57

210. In all cases in which the fraction is in its decimal form, proceed in one of two ways: (1) If the decimal is one of the familiar aliquot or aliquant parts, use it in its common fraction form as illustrated in the chapter on Aliquot Parts. (2) In all other cases solve as follows:

Multiply 425 by .217.

```
   425
   .217
   2975
   8925
 92.225
```

Perform the multiplication in the same manner as for simple numbers, and point off as many places in the product as there are places in either or both the multiplier and multiplicand.

211. Multiply:

1.	48 by .16⅔	**9.**	368 by .649	**17.**	59 by .101
2.	93 by .123	**10.**	487 by .1245	**18.**	84 by .0125
3.	133 by .416	**11**	696 by .875	**19.**	96 by .125
4.	216 by .75	**12.**	389 by .0625	**20.**	88 by .09
5.	874 by .613	**13.**	496 by .083	**21.**	138 by .0001
6.	838 by .456	**14.**	979 by .005	**22.**	4585 by .01
7.	971 by .375	**15.**	845 by .001	**23.**	7345 by .003
8.	1240 by .625	**16.**	859 by .0101	**24.**	250 by .004

The student should observe carefully that on multiplying by such decimals as .1, .01, .001, .0001, and so on, the product contains the same figures as the multiplicand, with as many places pointed off as there are places in the decimal. Thus, multiply 4789 by .001. The product is 4.789, the same order of digits, with 3 places pointed off.

Write the answers to the following:

25.	8432 × .0001	**30.**	64873 × .00001	**35.**	34876 × .0001
26.	9678 × .1	**31.**	69478 × .000001	**36.**	48749 × .01
27.	8439 × .001	**32.**	83902 × .01	**37.**	87634 × .001
28.	9637 × .00001	**33.**	84800 × .001	**38.**	85634 × .1
29.	45000 × .01	**34.**	12489 × .01	**39**	7964 × .0001

212. To find the product of an integer and a mixed number.

1. At $½ a yard, how much will 4 yd. of cloth cost?

2. At $1 a yard, how much will 4 yd. of cloth cost?

Compare the cost of 4 yd. at $½ a yard and 4 yd. at $1 a yard, with the cost of 4 yd. at $1½ a yard.

3. At $3 a yard, how much will $\frac{1}{2}$ a yard of silk cost? How much will 6 yd. cost? How much will $6\frac{1}{2}$ yd. cost?

4. How many are $6 \times 4\frac{1}{2}$? $8 \times 3\frac{1}{4}$? $9 \times 5\frac{1}{3}$? $3\frac{1}{2} \times 8$? $7\frac{1}{5} \times 10$? $9\frac{1}{4} \times 4$? $6\frac{1}{3} \times 6$?

213. 1. Multiply 864 by $13\frac{3}{8}$.

$$\begin{array}{r} 864 \\ 13\frac{3}{8} \\ \hline 324 \\ 2592 \\ 864 \\ \hline 11556 \end{array}$$

First multiply mentally by the fraction $\frac{3}{8}$; thus, $\frac{1}{8}$ of 864 is 108; $3 \times 108 = 324$. Next, multiply by the whole number, 13, and add the results.

2. Multiply $231\frac{2}{3}$ by 43.

$$\begin{array}{r} 231\frac{2}{3} \\ 43 \\ \hline 3)86 \\ 28\frac{2}{3} \\ 693 \\ 924 \\ \hline 9961\frac{2}{3} \end{array}$$

First, multiply the fraction $\frac{2}{3}$ by 43. If the integer, 43, is not exactly divisible by the denominator of the fraction, multiply the numerator of the fraction by 43; $43 \times 2 = 86$. Divide 86 by 3, the denominator of the fraction, obtaining $28\frac{2}{3}$. Then multiply 231 by 43, and add results.

3. Multiply 583 by $31.64\frac{3}{4}$.

$$\begin{array}{r} 31.64\frac{3}{4} \\ 583 \\ \hline 4)1749 \\ 437\frac{1}{4} \\ 9492 \\ 25312 \\ 15820 \\ \hline 18450.49\frac{1}{4} \end{array}$$

Note that the multiplicand, 583, is used as the multiplier for the purpose of shortening the work (Prin. 13, page 102). First multiply 583 by the numerator of the fraction, and divide the product by the denominator, as in the preceding example. Multiply 31.64 by 583 the same as though 31.64 were a whole number, and add the results. In the product point off 2 places, for the 2 places in the number $31.64\frac{3}{4}$.

214. Find the value of:

1. $10 \times 8\frac{1}{2}$	**5.** $33 \times 11\frac{5}{6}$	**9.** $49 \times 88\frac{3}{8}$
2. $12 \times 4\frac{2}{3}$	**6.** $76 \times 51\frac{3}{5}$	**10.** $83 \times 31\frac{2}{3}$
3. $16 \times 8\frac{3}{4}$	**7.** $21 \times 7\frac{3}{7}$	**11.** $59 \times 83\frac{3}{5}$
4. $24 \times 5\frac{3}{8}$	**8.** $67 \times 18\frac{4}{9}$	**12.** $38 \times 17\frac{5}{7}$

13. $16\frac{3}{5} \times 28$

14. $18\frac{3}{4} \times 115$

15. $14\frac{5}{12} \times 93$

16. $15\frac{6}{7} \times 99$

17. $24\frac{3}{14} \times 50$

18. $89\frac{11}{15} \times 160$

19. $39\frac{3}{11} \times 138$

20. $58\frac{4}{7} \times 44$

21. $91\frac{15}{16} \times 110$

22. $38\frac{3}{16} \times 92$

23. $47\frac{7}{9} \times 86$

24. $69\frac{3}{17} \times 85$

25. $67 \times 4.13\frac{3}{8}$

26. $87 \times .312\frac{1}{3}$

27. $34 \times 4.37\frac{5}{6}$

28. $39 \times 3.82\frac{3}{4}$

29. $98 \times 74.66\frac{2}{3}$

30. $38.13\frac{3}{7} \times 45$

31. $56.42\frac{3}{11} \times 66$

32. 3.123×86

33. $71 \times 16.10\frac{2}{3}$

34. $5.91\frac{2}{3} \times 84$

35. $8.31\frac{1}{4} \times 34$

36. $19.36\frac{3}{16} \times 59$

215. **To multiply a fraction by a fraction.**

1. How much is $\frac{1}{2}$ of $\frac{1}{2}$? $\frac{1}{2}$ of $\frac{1}{4}$? $\frac{1}{2}$ of $\frac{1}{8}$?

2. If a yard of cloth costs $\frac{2}{3}$ of a dollar, how much will $\frac{1}{2}$ of a yard cost?

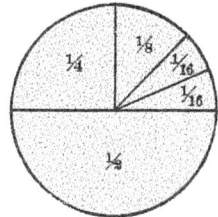

216. 1. What is $\frac{3}{4}$ of $\frac{2}{3}$?

Let the circle represent a unit divided into thirds. If each third is divided into four equal parts, the circle will be divided into 12 equal parts, or twelfths. $\frac{2}{3} = \frac{8}{12}$. $\frac{3}{4}$ of $\frac{2}{3}$, then, is the same as $\frac{3}{4}$ of $\frac{8}{12}$, which is equivalent to $\frac{3}{4}$ of 8 parts. $\frac{1}{4}$ of 8 parts is 2 parts; 3×2 parts is 6 parts, or $\frac{6}{12} = \frac{1}{2}$. Therefore, $\frac{3}{4}$ of $\frac{2}{3} = \frac{1}{2}$.

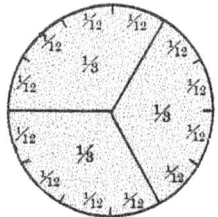

Instead of all the work indicated in this explanation, which has been given to make clear the principles involved, the solution may be made very short. Observe the following:

$$\frac{3}{4} \text{ of } \frac{2}{3} = \frac{3 \times 2}{4 \times 3} = \frac{1}{2}.$$

By applying cancellation, the work of the solution is reduced to a minimum.

Some problems do not admit of cancellation. For instance, in finding the value of $\frac{5}{6}$ of $\frac{7}{8}$, cancellation cannot be applied. The solution consists in multiplying together the numerators for the numerator of the result, and multiplying together the denominators for the denominator of the result.

The expression is equivalent to $\frac{5 \times 7}{6 \times 8} = \frac{35}{48}$.

2. Multiply .43 by .032.

$$\begin{array}{r} .43 \\ .032 \\ \hline 86 \\ 129 \\ \hline .01376 \end{array}$$

Multiply as in whole numbers, pointing off as many places in the product as there are places in both multiplicand and multiplier.

217. Find the products:

1. $\frac{2}{3}$ of $\frac{9}{10}$	**10.** $\frac{3}{4}$ of $\frac{7}{8}$ of $\frac{2}{3}$	**19.** $.4123 \times .814$
2. $\frac{5}{8}$ of $\frac{4}{5}$	**11.** $\frac{9}{10}$ of $\frac{5}{6}$ of $\frac{6}{7}$	**20.** $.963 \times .0001$
3. $\frac{7}{9}$ of $\frac{9}{14}$	**12.** $\frac{7}{8}$ of $\frac{8}{9}$ of $\frac{9}{14}$	**21.** $8.24 \times .001$
4. $\frac{5}{6}$ of $\frac{11}{15}$	**13.** $\frac{15}{16}$ of $\frac{32}{45}$ of $\frac{90}{128}$	**22.** $.4237 \times 1.001$
5. $\frac{3}{4}$ of $\frac{7}{8}$	**14.** $\frac{11}{12}$ of $\frac{6}{7}$ of $\frac{1}{2}$	**23.** $.954 \times 10.11$
6. $\frac{11}{12}$ of $\frac{3}{4}$	**15.** $\frac{3}{5}$ of $\frac{10}{21}$ of $\frac{11}{15}$	**24.** 76.98×11.11
7. $\frac{7}{15}$ of $\frac{9}{11}$	**16.** $\frac{5}{6}$ of $\frac{3}{8}$ of $\frac{9}{16}$	**25.** 8.329×8.96
8. $\frac{8}{9}$ of $\frac{11}{12}$	**17.** $\frac{9}{10}$ of $\frac{8}{15}$ of $\frac{45}{56}$	**26.** 4.73×100.10
9. $\frac{7}{10}$ of $\frac{15}{28}$	**18.** $\frac{7}{12}$ of $\frac{24}{35}$ of $\frac{15}{16}$	**27.** 89.48×3.824

218. To multiply a mixed number by a fraction.

1. What is $\frac{1}{2}$ of $\frac{1}{2}$? $\frac{1}{2}$ of 4? Add the two answers.

2. What is $\frac{1}{2}$ of $4\frac{1}{2}$? $\frac{1}{3}$ of $6\frac{1}{2}$? $\frac{1}{4}$ of $8\frac{1}{4}$?

3. What is $\frac{2}{3}$ of $6\frac{1}{2}$? $\frac{3}{4}$ of $12\frac{1}{2}$? $\frac{4}{5}$ of $20\frac{1}{2}$?

219. Find the value of $\frac{7}{8}$ of $313\frac{4}{5}$.

$$\frac{7}{\underset{2}{8}} \text{ of } \frac{\overset{4}{5}}{5} = \frac{7}{10}$$

First, multiply the fractions together, obtaining $\frac{7}{10}$.

$$\frac{7}{8} \text{ of } 313 = \frac{2191}{8} = 273\frac{7}{8}$$

Next, take $\frac{7}{8}$ of 313, which gives $273\frac{7}{8}$. (For explanation, see page 84.)

$$273\ \tfrac{7}{8} = {}^{3\,5}$$
$$\tfrac{7}{10} = {}^{2\,8}$$
$$\overline{274\tfrac{23}{40}} \qquad \tfrac{63}{40} = 1\tfrac{23}{40}$$

Now add $\frac{7}{10}$ and $273\frac{7}{8}$, and the desired result is obtained.

220. Find the products:

1. $\frac{3}{5} \times 18\frac{8}{9}$	**5.** $\frac{5}{6} \times 91\frac{3}{7}$	**9.** $\frac{6}{7} \times 148\frac{7}{8}$
2. $\frac{4}{7} \times 33\frac{3}{8}$	**6.** $\frac{7}{8} \times 88\frac{4}{5}$	**10.** $\frac{3}{4} \times 131\frac{3}{5}$
3. $\frac{9}{11} \times 16\frac{4}{7}$	**7.** $\frac{9}{14} \times 84\frac{7}{10}$	**11.** $48\frac{7}{8} \times \frac{9}{10}$
4. $\frac{8}{9} \times 39\frac{4}{5}$	**8.** $\frac{7}{12} \times 127\frac{5}{8}$	**12.** $68\frac{7}{9} \times \frac{3}{4}$

13. $137\frac{3}{4} \times \frac{6}{7}$ **19.** $591\frac{1}{4} \times \frac{4}{5}$ **25.** $\frac{3}{4} \times 414\frac{5}{8}$

14. $246\frac{7}{8} \times \frac{5}{6}$ **20.** $563\frac{2}{3} \times \frac{3}{4}$ **26.** $\frac{7}{8} \times 241\frac{2}{3}$

15. $337\frac{4}{5} \times \frac{3}{4}$ **21.** $\frac{3}{16} \times 416\frac{3}{8}$ **27.** $\frac{3}{8} \times 328\frac{2}{5}$

16. $498\frac{3}{7} \times \frac{2}{3}$ **22.** $\frac{5}{9} \times 382\frac{1}{4}$ **28.** $\frac{4}{5} \times 564\frac{1}{5}$

17. $468\frac{15}{16} \times \frac{3}{4}$ **23.** $\frac{7}{11} \times 438\frac{2}{3}$ **29.** $\frac{7}{12} \times 366\frac{6}{7}$

18. $483\frac{3}{11} \times \frac{4}{5}$ **24.** $\frac{5}{6} \times 938\frac{3}{4}$ **30.** $\frac{9}{10} \times 455\frac{5}{9}$

221. To multiply a mixed number by a mixed number.

Find the value of $31\frac{3}{4} \times 17\frac{5}{8}$.

$31\frac{3}{4}$
$17\frac{5}{8}$
$\frac{15}{32} = {}^{15}$
$19\frac{3}{8} = {}^{12}$
$12\frac{3}{4} = {}^{24}$
217
31
$559\frac{19}{32}$ $\frac{51}{32} = 1\frac{19}{32}$

1. $\frac{5}{8}$ of $\frac{3}{4} = \frac{15}{32}$.
2. $\frac{5}{8}$ of $31 = \frac{155}{8} = 19\frac{3}{8}$.
3. $17 \times \frac{3}{4} = \frac{51}{4} = 12\frac{3}{4}$.
4. $17 \times 31 = 217$
   $\underline{31}$
   527

This solution is what is known as the "four-step process" of multiplying mixed numbers. To make the work clearer, the four steps are written out in full. The student should observe these four steps very carefully : (1) $\frac{5}{8}$ of $\frac{3}{4} = \frac{15}{32}$, which is written as a part of the product. (2) $\frac{5}{8}$ of $31 = 19\frac{3}{8}$, which is written as in the solution. (3) $17 \times \frac{3}{4} = 12\frac{3}{4}$, which is written under the $19\frac{3}{8}$. (4) $31 \times 17 = 217 + 310$. Add the several results. The final result is $559\frac{19}{32}$.

In adding the fractions of the several partial products, the student will find abundant opportunity to make use of the "short cuts" in addition of fractions. He should use them wherever possible.

222. Multiply :

1. $15\frac{3}{8} \times 16\frac{1}{4}$ **8.** $76\frac{3}{4} \times 12\frac{1}{2}$ **15.** $74\frac{3}{4} \times 19\frac{1}{5}$

2. $18\frac{3}{4} \times 26\frac{1}{2}$ **9.** $89\frac{4}{5} \times 17\frac{1}{2}$ **16.** $67\frac{7}{8} \times 18\frac{1}{4}$

3. $27\frac{5}{6} \times 34\frac{3}{5}$ **10.** $93\frac{3}{5} \times 27\frac{3}{4}$ **17.** $47\frac{3}{5} \times 21\frac{4}{5}$

4. $37\frac{3}{7} \times 29\frac{4}{7}$ **11.** $87\frac{3}{8} \times 29\frac{7}{9}$ **18.** $78\frac{2}{3} \times 37\frac{3}{4}$

5. $38\frac{3}{4} \times 34\frac{1}{4}$ **12.** $91\frac{1}{2} \times 41\frac{1}{3}$ **19.** $98\frac{3}{7} \times 42\frac{5}{9}$

6. $56\frac{4}{5} \times 9\frac{3}{4}$ **13.** $47\frac{2}{3} \times 38\frac{5}{6}$ **20.** $87\frac{1}{4} \times 56\frac{3}{8}$

7. $4\frac{1}{2} \times 9\frac{3}{8}$ **14.** $93\frac{3}{4} \times 86\frac{2}{3}$ **21.** $91\frac{3}{4} \times 36\frac{4}{13}$

22. $88\frac{4}{5} \times 55\frac{5}{8}$ **26.** $.246\frac{4}{9} \times .96\frac{3}{4}$ **30.** $5.98\frac{3}{4} \times 1.22\frac{3}{4}$

23. $1.34\frac{3}{8} \times .19\frac{4}{7}$ **27.** $3.24\frac{5}{8} \times .56\frac{1}{4}$ **31.** $8.46\frac{5}{8} \times 1.38\frac{5}{8}$

24. $2.28\frac{4}{5} \times .67\frac{3}{4}$ **28.** $6.32\frac{2}{3} \times .16\frac{2}{3}$ **32.** $9.66\frac{2}{3} \times 1.44\frac{4}{7}$

25. $2.14\frac{5}{16} \times .48\frac{1}{4}$ **29.** $4.56\frac{3}{5} \times .59\frac{3}{5}$ **33.** $8.62\frac{1}{2} \times 2.33\frac{1}{8}$

Find the total value of each of the following:

Fractions, when not expressed in full, are to be considered as fourths. Thus, $\$.07^2 = \$ 07\frac{2}{4}$.

34.
27³ yd. at $.07².
31¹ yd. at .08³.
45² yd. at .09¹.
56³ yd. at .07³.
38² yd. at .06¹.
51² yd. at .07³.

35.
35¹ yd. at $.07¹.
43³ yd. at .08¹.
72¹ yd. at .05².
53³ yd. at .06².
58² yd. at .04³.
63¹ yd. at .07¹.

36.
64¹ yd. at $.08².
76² yd. at .07³.
48³ yd. at .08².
39² yd. at .04³.
17² yd. at .12³.
19³ yd. at .13³.

37.
72³ yd. at $.10¹.
42² yd. at .09².
63¹ yd. at .07².
16³ yd. at .15².
22² yd. at .18³.
31³ yd. at .23².

38.
21³ yd. at $.07².
34³ yd. at .06³.
56² yd. at .08².
47¹ yd. at .09¹.
53³ yd. at .07³.
29¹ yd. at .07¹.

39.
44¹ yd. at $.08¹.
29³ yd. at .07¹.
64² yd. at .07³.
29³ yd. at .05².
12² yd. at .26¹.
14³ yd. at .35².

40.
59² yd. at $.05².
39³ yd. at .09³.
45¹ yd. at .06².
73² yd. at .08².
67³ yd. at .07².
83¹ yd. at .09¹.

41.
26¹ yd. at $.07².
32³ yd. at .07¹.
49² yd. at .08³.
38² yd. at .22².
59³ yd. at .33².
47² yd. at .17².

42.
32² yd. at $.06².
54¹ yd. at .07¹.
69³ yd. at .09³.
57¹ yd. at .07¹.
48² yd. at .06².
27³ yd. at .05³.

MENTAL MULTIPLICATION OF MIXED NUMBERS

223. When each fraction is $\frac{1}{2}$.

1. Multiply $6\frac{1}{2}$ by $6\frac{1}{2}$.

(a) $6\frac{1}{2}$ (b) $6\frac{1}{2}$

$\frac{6\frac{1}{2}}{3\frac{1}{4}}$
$\frac{3}{36}$
$\overline{42\frac{1}{4}}$

$\frac{6\frac{1}{2}}{42\frac{1}{4}}$

In (a) the complete "four-step" solution (page 87) is given. Note that the sum of the two cross products (3 and 3) is 6. Hence, instead of taking 6 × 6, and adding 6, add 1 to the multiplicand, and take 6 × 7, as in (b).

2. Multiply 9.5 by 9.5.

$\frac{9.5}{9.5}$
$\overline{90.25}$

As in (b), problem 1, .5 × .5 = .25. Add 1 to one of the 9's and multiply by the other 9, and obtain 90. Hence, 9.5 × 9.5 = 90.25.

224. Multiply (mentally):

1. $8\frac{1}{2}$ by $8\frac{1}{2}$	**6.** $1\frac{1}{2}$ by $1\frac{1}{2}$	**11.** $5\frac{1}{2}$ by $5\frac{1}{2}$
2. $7\frac{1}{2}$ by $7\frac{1}{2}$	**7.** $10\frac{1}{2}$ by $10\frac{1}{2}$	**12.** 6.5 by 6.5
3. 3.5 by 3.5	**8.** 11.5 by 11.5	**13.** $9\frac{1}{2}$ by $9\frac{1}{2}$
4. 4.5 by 4.5	**9.** $13\frac{1}{2}$ by $13\frac{1}{2}$	**14.** $15\frac{1}{2}$ by $15\frac{1}{2}$
5. $2\frac{1}{2}$ by $2\frac{1}{2}$	**10.** 12.5 by 12.5	**15.** $20\frac{1}{2}$ by $20\frac{1}{2}$

225. 1. Multiply $7\frac{1}{2}$ by $5\frac{1}{2}$.

$\frac{7\frac{1}{2}}{5\frac{1}{2}}$
$\overline{41\frac{1}{4}}$

$\frac{1}{2}$ of 7 + $\frac{1}{2}$ of 5 = $\frac{1}{2}$ of (7 + 5) = 6. Take 5 × 7, and add the 6, which gives $41\frac{1}{4}$ as the product.

2. Multiply $9\frac{1}{2}$ by $4\frac{1}{2}$.

$\frac{9\frac{1}{2}}{4\frac{1}{2}}$
$\overline{42\frac{3}{4}}$

The sum of the integers (4 + 9 = 13) being an odd number, $\frac{1}{2}$ of the sum gives a mixed number, $6\frac{1}{2}$. $\frac{1}{2}$ of $\frac{1}{2}$ = $\frac{1}{4}$, to which is added the $\frac{1}{2}$ of the mixed number ($6\frac{1}{2}$), giving $\frac{3}{4}$, the fractional part of the product. Now take 4 times 9, and add the 6, obtaining $42\frac{3}{4}$, the desired result.

In problems like the last two examples, observe first whether the sum of the integers is even or odd. If even, write as the first part of the product $\frac{1}{4}$, .25, or 25, as the case may require; if odd, write $\frac{3}{4}$, .75, or 75. For the remainder of the product multiply the integers together, and add the integral part of half their sum. (Compare with Art. 107.)

When the fractions are alike.

3. Multiply $8\frac{2}{3}$ by $5\frac{2}{3}$.

$8\frac{2}{3}$
$5\frac{2}{3}$
$\frac{4}{9}$
$8\frac{2}{3}$
40
$\overline{49\frac{1}{9}}$

First multiply $\frac{2}{3}$ by $\frac{2}{3}$. Next find $\frac{2}{3}$ of 13, $(8+5)$, which is $8\frac{2}{3}$. Then find 5×8, which is 40. It should be observed that although $\frac{2}{3}$ of 13 gives a mixed number, $8\frac{2}{3}$, the solution is shorter than by the other method. Now there are only two fractions to be added, instead of three, as there would be if 8 and 5 were multiplied separately.

The student will be surprised at the rapidity with which he will be able to make multiplications like the above examples, after a little practice.

226. Multiply (mentally):

1. $6\frac{1}{2} \times 10\frac{1}{2}$	**8.** $7\frac{2}{3} \times 8\frac{2}{3}$	**15.** $15\frac{3}{7} \times 6\frac{3}{7}$
2. $8\frac{1}{3} \times 7\frac{1}{3}$	**9.** $10\frac{1}{5} \times 15\frac{1}{5}$	**16.** $18\frac{2}{5} \times 7\frac{2}{5}$
3. $5\frac{3}{4} \times 11\frac{3}{4}$	**10.** 11.5×8.5	**17.** $16\frac{2}{3} \times 8\frac{2}{3}$
4. 8.25×4.25	**11.** $14\frac{1}{4} \times 8\frac{1}{4}$	**18.** $11\frac{3}{4} \times 7\frac{3}{4}$
5. $6\frac{1}{3} \times 9\frac{1}{3}$	**12.** $5\frac{3}{5} \times 7\frac{3}{5}$	**19.** $13\frac{2}{3} \times 2\frac{2}{3}$
6. $7\frac{1}{2} \times 9\frac{1}{2}$	**13.** $12\frac{3}{4} \times 9\frac{3}{4}$	**20.** 15.75×1.75
7. 11.25×9.25	**14.** $14\frac{1}{3} \times 7\frac{1}{3}$	**21.** $17\frac{4}{7} \times 4\frac{4}{7}$

227. Find the total value of each of the following groups:

1.	**2.**	**3.**
65 crates at $ 6.50.	$10\frac{1}{2}$ lb. at $.75.	$12\frac{1}{2}$ A. at $ 125.
85 crates at 7.50.	$5\frac{1}{2}$ lb. at .65.	$15\frac{1}{4}$ A. at 155.
$4\frac{1}{2}$ bu. at .75.	$7\frac{1}{2}$ lb. at $.07\frac{1}{2}$.	$16\frac{1}{4}$ A. at 85.
$9\frac{1}{2}$ bu. at .85.	$3\frac{1}{2}$ lb. at $.15\frac{1}{2}$.	$20\frac{1}{2}$ A. at 75.

4	**5.**	**6.**
$5\frac{1}{4}$ yd. at $.07\frac{1}{4}$.	$8\frac{3}{4}$ lb. at $ .08\frac{3}{4}$.	$15\frac{1}{2}$ bu. at $ 1.50.
$9\frac{1}{4}$ yd. at .18$\frac{1}{4}$.	$9\frac{3}{4}$ lb. at .11$\frac{3}{4}$.	350 lb. at 1.50 per cwt.
$7\frac{1}{4}$ yd. at .09$\frac{1}{4}$.	$5\frac{3}{4}$ lb. at 3.75.	2500 ft. at 18.50 per M.
$13\frac{1}{4}$ yd. at .07$\frac{1}{4}$.	$12\frac{3}{4}$ lb. at 4.75.	4500 ft. at 6.50 per M.

228. The following is a form of pay roll generally used by time clerks. In addition to the time, rate, and amount due for each clerk, it also shows the number of both bills and minor coins necessary to pay each and all of the clerks. The total of all the different denominations must agree with the total of the amount column.

No.*	M.	T.	W.	T.	F.	S.	Total Time	Rate	Am't	$10	$5	$2	$1	¢50	¢25	¢10	¢5	¢1
1.	8	7	7½	6	9	8	45½	.30	$13.65	1		1	1	1		1	1	
2.	7	8½	6	9	9	7½	47	.33⅓	15.67	1	1			1		1	1	2
3.	9	10	7	6	7½	6½	46	.27½	12.65	1		1		1		1	1	
4.	8	8	7	8	8	6	45	.22	9.90		1	2		1	1	1	1	
5.	9	7	6	4½	8½	9½	44½	.17½	7.79		1	1		1	1			4
6.	10	8	7	9	8	8½	50½	.35	17.68	1	1	1		1		1	1	3
7.	9	7	8	6	4½	9½	44	.37½	16.50	1	1		1	1				
8.	8½	7	6	9½	8½	7½	47	.30	14.10	1		2				1		
									$107.94	6	5	8	2	7	2	6	5	9
										60	25	16	2	3.50	.50	.60	.25	.09

At the right is a currency memorandum and on p. 92 will be found a pay roll check. The memorandum accompanies the check drawn for the total amount of the pay roll to show the paying teller at the bank just what denominations and how many of each are required.

The Corn Exchange Bank
CURRENCY MEMORANDUM
New York, *March 18* 19 *11*
Depositor *Robert Levitt*.

		DOLLARS	CENTS
BILLS	$1		2
"	2		16
"	5		25
"	10		60
"	20		
"	50		
"	100		
COIN:	PENNIES		09
"	NICKELS		25
"	DIMES		60
"	QUARTERS		50
"	HALVES	3	50
"	DOLLARS		
"	GOLD		
	TOTAL	107	94

* Frequently the names as well as the numbers of the clerks are given, as shown in the form on page 92.

PAY ROLL CHECK

229. **1.** In the following time sheet the rate is for a day of 8 hr., with ¼ more for overtime.

Prepare a pay roll, a currency memorandum, and a check like the above.

TIME SHEET

WEEK ENDING, JAN. 21, 1911

CLERK Nos.	NAMES OF EMPLOYEES	M.	T.	W.	T.	F.	S.	TOTAL TIME	RATE
1.	Addison, James	8	10	9	11	8	9		4.50
2.	Buell, Henry	9	8	9	8½	9	8		4.00
3.	Crane, John	10	9	8	9½	8	8		4.00
4.	Crawford, Wm.	9	9½	8	9	10½	12		3.50
5.	Jones, Albert	10	8	9	9	10	11		3.25
6.	Landis, Frank	8	8	9	8	10	9		3.00
7.	Money, Theo.	9	9	8	8½	9½	10		3.00
8.	Newton, James	8	10½	9½	8	9	8		3.00
9.	Orton, George	9	8	9	9	10	8		2.75
10.	Rooney, Thos.	9	8	8	10	10	11		2.50

2. The following time card is used by a large drug corporation for recording the time of beginning and ending work each day. The first two columns are for the forenoon; the second two for the afternoon, and the third two for the evening.

Find the total time and the total wages.

In	Out	In	Out	In	Out
7.50	11.30	12.20	5.30		
		12.30	5.30	6.45	11.30
7.45	11.30	12.45	5.45		
7.55	12.15			6.10	11.15
		12.00	5.40	6.30	11.20
8.00	12.20	12.50	6.10		
8.30	12.15	1.30	6.10		

Form No. 3

No.......... *18*.........

Name...... *William MacCormack*...........................

Total Time_____Hr.   Rate, 27½¢.

Total Wages_____$......

Week Ending...... *January 7*_____19....

DIVISION OF FRACTIONS

230. **To divide a fraction by an integer.**

If each half of the circle is divided into 4 equal parts, into how many parts will the entire circle be divided? Into how many parts is $\frac{1}{2}$ of the circle divided? If $\frac{1}{2}$ of the circle is divided into 4 equal parts, one of those parts is equal to what part of the whole circle? $\frac{1}{2} \div 4 = \frac{1}{8}$.

How does the number of eighths in the circle compare with the number of halves?

How do $\frac{1}{8}$ and $\frac{1}{2}$ compare in size?

How does the denominator of the fraction $\frac{1}{8}$ compare in size with the denominator of the fraction $\frac{1}{2}$?

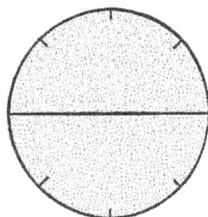

The fraction $\frac{1}{8}$ is only one fourth as great as the fraction $\frac{1}{2}$, because its denominator is 4 times as large as the denominator of $\frac{1}{2}$.

How, then, may a fraction be divided by an integer?

In the same way divide $\frac{1}{3}$ by 2; $\frac{1}{4}$ by 3; $\frac{1}{2}$ by 5; $\frac{1}{5}$ by 3, $\frac{2}{3}$ by 4; $\frac{3}{4}$ by 5.

What effect does multiplying the denominator of a fraction by an integer have on the value of a fraction? (Principle 17, page 102.)

231. Into how many parts is this circle divided? What part of the circle is marked off by the curved line? If these $\frac{9}{16}$ are divided into 3 equal parts, how many sixteenths will there be in each part? $\frac{9}{16} \div 3 = \frac{3}{16}.$

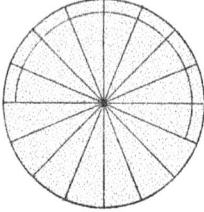

How does the fraction $\frac{3}{16}$ compare in value with the fraction $\frac{9}{16}$? Formulate a second way of dividing a fraction by an integer. (Prin. 20, page 102.)

232. 1. Divide $\frac{5}{8}$ by 3. $\frac{5}{8} \div 3 = \frac{5}{24}.$

Multiply the denominator of the fraction by the integer, 3, and write the numerator, 5, over the product. This divides the fraction, because multiplying the divisor (denominator) divides the quotient (which, in this case is the fraction). (Prin. 17, page 102.)

2. Divide $\frac{6}{11}$ by 2. $\frac{6}{11} \div 2 = \frac{3}{11}.$

Divide the numerator of the fraction by the integer, 2, and write the result over the denominator, 11. This divides the fraction, because dividing the dividend (numerator) divides the quotient (the fraction). (Prin. 20, page 102.)

This solution should always be used when the integer is an exact divisor of the numerator of the fraction.

3. Find the value of .0344 ÷ 8.

```
   .0043
8).0344
```

The first thing to do when the divisor is a whole number, as in this case, is to place the point for the quotient. Place it directly above the point in the dividend, as shown in the example. By dividing thus, "8 is contained in 0, 0 times; 8 in 3, 0 times; 8 in 34, 4 times, etc."; placing the figures of the quotient above the figures of the dividend, as here shown, the position of the decimal point in the quotient is always right.

The important part of division of fractions in the decimal form is locating the point in the quotient. The student should give the most careful attention to these solutions and explanations.

233. Find the value of:

1. $\frac{6}{7} \div 3$	**5.** $.0125 \div 5$	**9.** $\frac{11}{15} \div 3$	**13.** $\frac{7}{20} \div 5$
2. $\frac{8}{9} \div 4$	**6.** $464 \div 8$	**10.** $.968 \div 4$	**14.** $\frac{19}{17} \div 8$
3. $\frac{5}{8} \div 3$	**7.** $\frac{7}{8} \div 3.$	**11.** $.8464 \div 16$	**15.** $\frac{11}{18} \div 9$
4. $\frac{9}{10} \div 6$	**8.** $\frac{9}{16} \div 3$	**12.** $.0035 \div 5$	**16.** $\frac{24}{31} \div 6$

17 $.3125 \div 25$	**23.** $\frac{11}{4} \div 5$	**29.** $.0125 \div 500$	**35.** $1\frac{144}{12} \div 104$
18. $.738 \div 9$	**24.** $\frac{25}{6} \div 5$	**30.** $.8448 \div 8$	**36.** $.00126 \div 63$
19. $.001 \div 100$	**25.** $\frac{19}{8} \div 3$	**31.** $\frac{159}{100} \div 3$	**37.** $.216 \div 24$
20. $.00128 \div 16$	**26.** $\frac{31}{32} \div 6$	**32.** $\frac{249}{328} \div 83$	**38.** $.4888 \div 8$
21. $\frac{19}{21} \div 4$	**27.** $\frac{43}{8} \div 7$	**33.** $\frac{896}{1024} \div 128$	**39.** $.03036 \div 11$
22. $\frac{15}{26} \div 5$	**28.** $.125 \div 50$	**34.** $\frac{1089}{10} \div 121$	**40.** $.3069 \div 11$

234. To divide a mixed number by an integer.

If $\$1\frac{1}{2}$ is divided equally among 3 boys, how much will each receive? If 9 lb. of coffee cost $\$2\frac{1}{4}$, what part of a dollar does 1 lb. cost?

$$1\frac{1}{2} \div 3 = ? \qquad 2\frac{1}{4} \div 9 = ? \qquad 2\frac{2}{3} \div 8 = ? \qquad 1\frac{3}{5} \div 4 = ?$$

235. 1. Divide $5\frac{5}{8}$ by 7.

$5\frac{5}{8} = \frac{45}{8}$ When the mixed number is small, change it to an improper
$\frac{45}{8} \div 7 = \frac{45}{56}$ fraction, and divide as in dividing a fraction by an integer.

2. Divide $3478\frac{2}{9}$ by 8

$434\frac{7}{9}$ In problems like this, in which the mixed number is much
$8\overline{)3478\frac{2}{9}}$ larger than the divisor, *do not* change the mixed number
$6\frac{2}{9} = \frac{56}{9}$ to an improper fraction. Divide the integral part of the
$\frac{56}{9} \div 8 = \frac{7}{9}$ dividend by the divisor as if there were no fraction in the
dividend. 8 is contained in 3478, 434 times, with a remainder of
6, which, with the fraction $\frac{2}{9}$, makes a total remainder of $6\frac{2}{9}$. Dividing $6\frac{2}{9}$ by
8 as explained in the preceding example gives the completed quotient, $434\frac{7}{9}$

The student should divide $6\frac{2}{9}$ by 8, and all similar expressions, mentally. It is written out in full here for the sake of clearness.

3. Divide 42.3 by 25

```
      1.692
25)42.300
     25
     ___
     17 3
     15 0
     ____
      2 30
      2 25
      ____
        50
        50
        __
```

Place the point for the quotient directly above the point in the dividend. The important part, now, is to write the first figure of the quotient in the right place. Observe that 1, the first figure of the quotient, is written over the 2, the figure farthest to the right in the dividend, that is used to contain the divisor, 25. Continue the division, annexing ciphers in the dividend as needed, and placing the figures of the quotient over each succeeding figure in the dividend, respectively.

NOTE. If the figures of the quotient are placed correctly, the last, or right-hand figure, of the quotient will stand directly above the last figure of the dividend.

236. Find the value of:

1. $3\frac{3}{4} \div 5$	**16.** $3924\frac{5}{8} \div 12$	**31.** $478.336 \div 16$
2. $6\frac{2}{3} \div 10$	**17.** $4964\frac{3}{4} \div 11$	**32.** $394.128 \div 8$
3. $8\frac{3}{4} \div 7$	**18.** $6847\frac{5}{8} \div 96$	**33.** $48.3744 \div 24$
4. $3\frac{3}{8} \div 7$	**19.** $3478\frac{2}{3} \div 84$	**34.** $9674.88 \div 48$
5. $9\frac{4}{5} \div 14$	**20.** $9674\frac{5}{9} \div 99$	**35.** $364.64 \div 160$
6. $18\frac{3}{4} \div 8$	**21.** $8496\frac{7}{11} \div 64$	**36.** $511.375 \div 125$
7. $26\frac{7}{8} \div 4$	**22.** $4589\frac{7}{10} \div 5$	**37.** $895.4 \div 110$
8. $47\frac{5}{9} \div 2$	**23.** $896\frac{3}{11} \div 17$	**38.** $6.4872 \div 4$
9. $167\frac{8}{9} \div 21$	**24.** $974\frac{3}{16} \div 11$	**39.** $9.674\frac{2}{3} \div 8$
10. $568\frac{3}{7} \div 43$	**25.** $897\frac{1}{5} \div 15$	**40.** $4.592 \div 16$
11. $828\frac{3}{11} \div 69$	**26.** $969\frac{1}{14} \div 7$	**41.** $9.34\frac{1}{2} \div 12$
12. $939\frac{4}{15} \div 22$	**27.** $896\frac{3}{20} \div 8$	**42.** $6.43\frac{2}{3} \div 5$
13. $896\frac{4}{5} \div 15$	**28.** $487\frac{5}{7} \div 10$	**43.** $6.38\frac{7}{8} \div 9$
14. $948\frac{5}{8} \div 12$	**29.** $389\frac{7}{12} \div 6$	**44.** $9.72\frac{5}{9} \div 11$
15. $776\frac{7}{9} \div 9$	**30.** $967\frac{3}{4} \div 5$	**45.** $106.88\frac{8}{9} \div 15$

237. To divide an integer, a fraction, or a mixed number by a fraction.

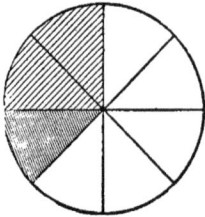

How many times does the whole circle contain $\frac{1}{2}$ of the circle?

How many times does it contain $\frac{1}{4}$ of the circle? $\frac{1}{8}$ of the circle?

$$1 \div \tfrac{1}{2} = ? \quad 1 \div \tfrac{1}{4} = ? \quad 1 \div \tfrac{1}{8} = ?$$

If $\frac{1}{2}$ is contained in 1, 2 times, how many times is $\frac{1}{2}$ contained in 2? in 3? in 4? in 6? in 10?

$$1 \div \tfrac{1}{2} = 2 = \tfrac{2}{1}$$
$$3 \div \tfrac{1}{2} = 6 = \tfrac{6}{1}$$

Observe, in the first equation, that the answer, 2, or its equivalent, $\frac{2}{1}$, is the fraction $\frac{1}{2}$ inverted; that is, the denominator is written for the numerator, and the numerator for the denominator.

In the second equation, notice that the answer 6, or $\frac{6}{1}$, is 3 times as large as the answer in the first equation. That is, to divide by $\frac{1}{2}$ is to multiply by its reciprocal, 2.

In the same circle count off 3 eighths. How many times will these $\frac{3}{8}$ be contained in the circle?

Call the " eighths " " parts," and divide 8 parts (that is, the circle) by 3 parts. 8 parts $\div$ 3 parts $= 2\frac{2}{3}$ times. That is, $1 \div \frac{3}{8} = 2\frac{2}{3}$. But $2\frac{2}{3} = \frac{8}{3}$, and $\frac{8}{3}$ is the fraction $\frac{3}{8}$ inverted.

The result of 1 divided by any fraction is that fraction inverted.

238. The reciprocal of a fraction is the fraction inverted.

239. The reciprocal of any number is 1 divided by that number. The reciprocal of 4 is $\frac{1}{4}$; of 8 is $\frac{1}{8}$; etc.

What is the reciprocal of $\frac{3}{4}$? of $\frac{5}{8}$? of $\frac{2}{3}$? of 10? of 12?

1. $1 \div \frac{3}{4} = \frac{4}{3}$.

2. $3 \div \frac{3}{4} = 3 \times \frac{4}{3} = \frac{12}{3} = 4$.

How does the dividend of the second equation compare with the dividend of the first equation?

What effect does multiplying the dividend by 3 have on the quotient? (See Principles of Division, No. 19, page 102.)

How may an integer be divided by a fraction?

To divide by a fraction, multiply by its reciprocal.

240. 1. Divide $\frac{5}{8}$ by $\frac{2}{5}$.

$\frac{5}{8} \div \frac{2}{5} = \frac{5}{8} \times \frac{5}{2} = \frac{25}{16} = 1\frac{9}{16}$.

To divide by $\frac{2}{5}$ is to multiply by its reciprocal, $\frac{5}{2}$. $\frac{5}{8}$ multiplied by $\frac{5}{2}$ gives $\frac{25}{16} = 1\frac{9}{16}$.

2. Divide $6\frac{2}{3}$ by $\frac{4}{5}$.

$6\frac{2}{3} \div \frac{4}{5} = 6\frac{2}{3} \times \frac{5}{4} = \frac{\overset{5}{\cancel{20}}}{3} \times \frac{5}{\cancel{4}} = \frac{25}{3} = 8\frac{1}{3}$.

To divide by $\frac{4}{5}$ is to multiply by its reciprocal, $\frac{5}{4}$. $6\frac{2}{3}$ multiplied by $\frac{5}{4}$ gives $\frac{25}{3} = 8\frac{1}{3}$.

3. Divide $\frac{4}{5}$ of $\frac{5}{8}$ of $\frac{2}{3}$ by $\frac{3}{4}$ of $\frac{2}{3}$ of $\frac{9}{10}$.

$\frac{4}{5}$ of $\frac{5}{8}$ of $\frac{2}{3} \div \frac{3}{4}$ of $\frac{2}{3}$ of $\frac{9}{10} =$

$\frac{4}{5} \times \frac{5}{8} \times \frac{2}{3} \times \frac{4}{3} \times \frac{3}{2} \times \frac{10}{9} = \frac{20}{27}$.

In dividing by a compound fraction, invert all the separate fractions composing the divisor, and treat the problem as an exercise in cancellation.

NOTE. A compound fraction is a fraction of a fraction: thus $\frac{2}{3}$ of $\frac{3}{4}$; $\frac{4}{5}$ of $\frac{5}{8}$ of $\frac{7}{8}$, etc.

THEORY OF INVERTING THE DIVISOR

Let it be required to divide 24 by $\frac{3}{5}$.

Dividing 24 by 3 (the numerator of the divisor) gives $\frac{24}{3}$, or 8.

Since 3 is 5 times as large as $\frac{3}{5}$, dividing by 3 is dividing by a divisor 5 times too large, and gives a quotient only $\frac{1}{5}$ as large as it should be. To correct the error, multiply the quotient by 5 (the denominator of the divisor).

$$24 \div 3 = \tfrac{24}{3}, \text{ or } 8.$$

$$\frac{24}{3} \times 5 = \frac{24 \times 5}{3} = 24 \times \frac{5}{3} = 40.$$

(See Prin. 17 and 18, page 102.)

4. Divide 4.8325 by .025.

$$\begin{array}{r} 193.3 \\ \times 025.)\overline{4\times832.5} \\ \underline{25} \\ 233 \\ \underline{225} \\ 82 \\ \underline{75} \\ 75 \\ \underline{75} \\ 75 \end{array}$$

When the divisor is a decimal, change it to an integer by moving its point to the right-hand side of the decimal. In this example the point is moved three places, which is equivalent to multiplying by 1000. Since the divisor is multiplied by 1000, the dividend must likewise be multiplied by 1000 (Prin. 21, page 102). Therefore, the point in the dividend is moved 3 places to the right. Before dividing, place the point for the quotient directly above the point in the dividend after it is moved.

The division is now performed in the same manner as in dividing by integers.

5. Divide 5000 by .0005.

$$\begin{array}{r} 1000\ 0000 \\ \times 0005.)\overline{5000\times0000.} \end{array}$$

As in the preceding example, change the divisor to an integer by moving the point to the right-hand side of the decimal, that is, by multiplying by 10,000. The dividend also must be multiplied by 10,000; that is, the point must be moved 4 places to the right. In this example it is necessary to make the places by annexing ciphers.

Remember, *the point in the dividend* MUST *be moved in the same direction as many places as the point in the divisor is moved to make the divisor a whole number. Annex ciphers when necessary.*

Remember, also, to place the point for the quotient *before* dividing.

The process of division is then the same as in simple numbers.

241. Find the value of:

1. $\frac{2}{3} \div \frac{3}{4}$		**4.** $\frac{6}{7} \div \frac{5}{9}$		**7.** $9\frac{1}{3} \div \frac{7}{8}$
2. $\frac{4}{5} \div \frac{9}{10}$		**5.** $1\frac{1}{2} \div \frac{2}{3}$		**8.** $15\frac{1}{4} \div \frac{3}{4}$
3. $\frac{5}{11} \div \frac{7}{8}$		**6.** $2\frac{1}{4} \div \frac{4}{9}$		**9.** $28\frac{2}{3} \div \frac{4}{15}$

10. $18\frac{3}{4} \div \frac{5}{8}$

11 $\frac{2}{3}$ of $\frac{4}{5} \div \frac{5}{8}$ of $\frac{6}{7}$

12. $\frac{5}{8}$ of $\frac{3}{4} \div \frac{9}{10}$ of $\frac{5}{4}$

13. $\frac{8}{9}$ of $\frac{3}{4} \div \frac{11}{15}$ of $\frac{15}{22}$

14. $28\frac{1}{2} \times \frac{3}{19} \div \frac{5}{16}$ of $\frac{4}{5}$

15. $\frac{7}{15}$ of $22\frac{1}{2} \div \frac{3}{13}$ of $\frac{26}{33}$

16. $2862 \div .009$

17. $34.74 \div .018$

18 $49.52 \div .16$

19 $004 \div .0002$

20. $.400 \div .004$

21. $.800 \div .0001$

22 $008 \div .0001$

23. $.12\frac{1}{2} \div .62\frac{1}{2}$

24 $4.875 \div .0625$

25 $8.5 \div 3125$

26. $40,000 \div .002$

27 $404.04 \div .20$

28. $96.848 \div 1.60$

29. $74.64 \div 1.25$

30. $395.20 \div .0125$

31 $16 \div \frac{5}{8}$

32. $28 \div \frac{4}{7}$

33. $36 \div \frac{12}{17}$

34. $89 \div \frac{2}{3}$

35. $56 \div \frac{6}{7}$

36. $38 \div \frac{19}{20}$

37. $42 \div \frac{6}{11}$

38. $32 \div \frac{16}{17}$

39 $48 \div \frac{9}{10}$

242. To divide an integer or a mixed number by a mixed number.

1. Divide 487 by $6\frac{3}{4}$.

$6\frac{3}{4}$) 487
4   4
$72\frac{4}{27}$
27) 1948
189
58
54
$\frac{4}{27}$

The simplest and most compact form of solution for problems of this nature is the solution shown here. Write the dividend and divisor as for division of simple numbers. The divisor being a mixed number, it is first necessary to "clear of fractions." This is done by multiplying the divisor by 4, the denominator of the fraction in the divisor. Multiplying the divisor by 4 gives 27. If the divisor is multiplied by 4, the dividend must also be multiplied by 4 (why?), which gives 1948. The divisor and the dividend are now both integral, and the division is easily completed.

2. Divide $683\frac{2}{3}$ by $44\frac{3}{5}$.

9   10
$44\frac{3}{5}$) $683\frac{2}{3}$
15   15
$15\frac{220}{669}$
669)10255
669
3565
3345
220
669

Here both divisor and dividend are mixed numbers, their fractions having different denominators. The divisor can be cleared of fractions by multiplying it by 5, and the dividend, by 3. But to retain the right value for the quotient, both divisor and dividend must be multiplied by the *same* number. The simplest number to multiply by is the least common denominator of the fractions $\frac{3}{5}$ and $\frac{2}{3}$, which is 15. Observe that the numerators of the fractions are changed to correspond to the common denominator, 15. Multiplying both dividend and divisor by 15, the new dividend and divisor, 10,255 and 669, are obtained. The division results in a quotient of $15\frac{220}{669}$.

3. Divide $76\,87\frac{5}{6}$ by $3.18\frac{2}{3}$.

$$4$$
$$3.18\tfrac{2}{3})76.87\tfrac{5}{6}$$
$$6\qquad 6$$

$$\overline{24.125}$$
$$19 \times 12\,)\overline{461 \times 27.000}$$

$$382\ 4$$
$$\overline{78\ 87}$$
$$76\ 48$$
$$\overline{2\ 39\ 0}$$
$$1\ 91\ 2$$
$$\overline{47\ 80}$$
$$38\ 24$$
$$\overline{9\ 560}$$
$$9\ 560$$

Multiply both dividend and divisor by 6, the least common multiple of the denominators of the fractions $\frac{2}{3}$ and $\frac{1}{6}$, retaining the decimal points in the products. Move the points each two places to the right, and divide as already explained. The quotient is 24.125.

243. Find the value of:

1. $472\frac{3}{4} \div 24\frac{2}{3}$	16. $978 \div 85\frac{1}{8}$	31. $963\frac{5}{9} \div 8\frac{2}{3}$
2. $867 \div 15\frac{1}{6}$	17. $125 \div 1\frac{2}{3}$	32. $458\frac{3}{4} \div 19\frac{1}{2}$
3. $978\frac{5}{8} \div 19\frac{1}{2}$	18. $250 \div 1.66\frac{2}{3}$	33. $732\frac{2}{3} \div 19\frac{3}{4}$
4. $5.26\frac{2}{3} \div .46\frac{2}{3}$	19. $375 \div 3.33\frac{1}{3}$	34. $864\frac{5}{8} \div 21\frac{1}{2}$
5. $8.49\frac{1}{2} \div 2.12\frac{2}{3}$	20. $964\frac{5}{8} \div 16\frac{1}{4}$	35. $438\frac{6}{7} \div 219\frac{3}{7}$
6. $349\frac{3}{5} \div 16\frac{1}{4}$	21. $493\frac{2}{3} \div 42\frac{3}{4}$	36. $644\frac{3}{8} \div 16\frac{1}{2}$
7. $8749 \div 22\frac{1}{4}$	22. $967\frac{5}{6} \div 91\frac{1}{3}$	37. $849\frac{5}{9} \div 13\frac{2}{3}$
8. $8396 \div 78\frac{2}{3}$	23. $83.92 \div 5.125$	38. $674\frac{5}{9} \div 11\frac{1}{3}$
9. $7469\frac{3}{5} \div 152\frac{1}{2}$	24. $49.87\frac{1}{2} \div 2.49\frac{3}{8}$	39. $84.96\frac{1}{2} \div 15\frac{1}{4}$
10. $9119\frac{2}{3} \div 428\frac{1}{2}$	25. $39.37\frac{1}{2} \div 1.31\frac{1}{4}$	40. $60.83\frac{1}{3} \div 12\frac{1}{6}$
11. $16.31\frac{1}{2} \div 1.21\frac{1}{4}$	26. $478\frac{3}{7} \div 21\frac{2}{3}$	41. $.0482 \div .0002$
12. $329\frac{2}{5} \div 26\frac{1}{4}$	27. $674\frac{5}{6} \div 91\frac{1}{2}$	42. $59.68\frac{1}{2} \div 0005$
13. $897\frac{2}{3} \div 31\frac{1}{4}$	28. $845\frac{3}{8} \div 14\frac{1}{4}$	43. $11.25 \div .003\frac{3}{4}$
14. $598\frac{4}{7} \div 24\frac{1}{2}$	29. $7689 \div 18\frac{5}{6}$	44. $131 \div .008\frac{1}{3}$
15. $939\frac{3}{8} \div 42\frac{2}{3}$	30. $598\frac{3}{5} \div 14\frac{1}{2}$	45. $45.75 \div .001\frac{2}{3}$

SOME FUNDAMENTAL PRINCIPLES OF ARITHMETIC

244. Every problem in arithmetic is based on one or more fundamental principles or laws. A knowledge of these principles and of their application to the solution of problems is essential to success in arithmetical calculations.

Instead of stating the principles of notation, addition, subtraction, etc., in connection with these topics, it has been deemed advisable to group them in one chapter, in order that the proper emphasis may be placed upon them.

It is suggested that special attention be given to these principles, as frequent reference will be made to them.

PRINCIPLES OF NOTATION AND NUMERATION

245. 1. *Each removal of any significant figure* one place to the left toward or from the decimal point multiplies its value by ten.*

2. *Each removal of any significant figure one place to the right toward or from the decimal point divides its value by ten.*

3. *Each cipher annexed to an integer multiplies its value by ten.*

4. *Each cipher prefixed to a decimal divides its value by ten.*

PRINCIPLES OF ADDITION

5. *Only like numbers or parts of like numbers can be added.*

6. *The sum is composed of the same kind of units as are the addends.*

PRINCIPLES OF SUBTRACTION

. *Only like numbers or parts of like numbers can be subtracted.*

8. *The remainder is composed of the same kind of units as are the minuend and subtrahend.*

9. *The minuend is equal to the sum of the subtrahend and the remainder.*

---

* A significant figure is a figure that possesses a value of its own. Thus, all the figures, 1, 2, 3, etc., except 0, are significant figures. 0 has no value of its own.

PRINCIPLES OF MULTIPLICATION

10. *The multiplicand may be either an abstract or a concrete number.*

11. *The multiplier is an abstract number.*

12. *The multiplicand and the product are like numbers.*

13. *The multiplier and the multiplicand may be interchanged without affecting the value of the product. (This principle is called the commutative law of multiplication.)*

14. *The product of two factors divided by either of them will give the other one. (If there are more than two factors in any product, the product divided by any one of the factors will give a quotient equal to the product of all the other factors.)*

PRINCIPLES OF DIVISION

15. *The dividend is equal to the product of the divisor and the quotient (plus the remainder, if any.)*

16. *The dividend and the remainder are like numbers.*

17. *Multiplying the divisor divides the quotient by the same number.*

18. *Dividing the divisor multiplies the quotient by the same number.*

19. *Multiplying the dividend multiplies the quotient by the same number.*

20. *Dividing the dividend divides the quotient by the same number.*

21. *Multiplying or dividing both divisor and dividend by the same number does not change the value of the quotient.*

REMARK. Some of the above principles are already familiar to the student and are applied unconsciously to the solution of problems. Others (Nos. 11, 12, 14, 15, 16, 17, 18, 19, 20, and 21) are not so familiar, and need emphasis. In the number of its practical applications, Principle No. 14 is the most important.

APPLICATION OF THE FUNDAMENTAL PRINCIPLES OF ARITHMETIC TO THE SOLUTION OF PROBLEMS

246. The solution of problems requires a knowledge of three things: (1) The student must know how to add, subtract, multiply, and divide; (2) he must know in what order these operations are to be performed; and (3) he must be able to make the right combinations of numbers in any given problem.

The order of performing the operations and the proper combining of numbers depend upon a knowledge of the fundamental principles of arithmetic.

The following illustrations show the practical application of the most important principles to the solution of problems.

Very many problems are reducible to one or more of three type forms. As type problems, the following may be taken as examples:

Type 1. $\frac{3}{4}$ of $ 48 = ?

Type 2. $\frac{5}{8}$ of what number = $ 45 ?

Type 3. What part of $ 90 = $ 75 ?

In the first example given, the operation is obviously one of multiplication, and the product is $ 36 (Prin. 12).

In the second example one factor and a product are given. The product is $ 45, and the given factor is $\frac{5}{8}$. The missing factor is indicated by the words "what number." By Principle 14, if $ 45 is divided by $\frac{5}{8}$, the other factor will be obtained. $ 45 ÷ $\frac{5}{8}$ = $ 72

Many problems are reducible to this type form. It is worthy of the careful attention of the student.

In the third example there are also given a product, $ 75, and a factor, $ 90. The product divided by the given factor gives the other factor. $ 75 ÷ $ 90 = $\frac{75}{90}$ = $\frac{5}{6}$. $\frac{5}{6}$ is the missing factor.

In the second example the missing factor was the multiplicand (Prin. 12). In the third example it was the multiplier (Prin. 11).

Practice in reducing problems to their appropriate type form will be found helpful. Following are a few illustrations and explanations:

1. A farmer had 48 cd. of wood and sold $\frac{3}{4}$ of it. How many cords did he sell ?

First, after reading the problem, note the question, "How many cords did he sell ?" The problem answers this question by saying that he "sold $\frac{3}{4}$ of it." What is "it" ? "It" is 48 cd. Now put the number in the place of "it," and you have $\frac{3}{4}$ of 48 cd. = number of cords sold. The problem is now reduced to the form of the first type problem (Prin. 12).

2. A street is paved for a distance of 960 yd., which is $\frac{4}{5}$ of its length. What is the length of the street ?

Read the problem. Note the question to be answered. Is the street all paved ? What part of it is paved ? How many yards are paved ? How do

"⅘ of the length of the street" and "960 yd." compare ?   (They are alike.)
Therefore write,

• ⅘ of the length of the street = 960 yd. or
⅘ of ? yards = 960 yd.
960 yd. is the product obtained by multiplying some number by ⅘. Therefore,
960 yd. ÷ ⅘ = 1200 yd., length of street.   (Prin. 14, Type 2.)

3. A man's annual income is $ 2500, and his expenses for the year are $ 1500.   What part of his income does he save ?

Read the problem till you know what it means.   Note the question carefully
The question is, "What part of his income did he *save ?*"   He saved the difference between $ 2500 and $ 1500, which is $ 1000.   He saved $ 1000.   His income was $ 2500.

Now write the question in the shortest way, using numbers instead of words where possible.
? part of $ 2500 = $ 1000 ?   (Prin. 14, Type 3.)

Reduce each of the following problems to its appropriate type form.

1. A merchant bought 480 yd. of muslin and sold ⅝ of it.   How many yards did he sell ?

2. A man's total expenses for one year were $ 1280, of which amount $ 400 was spent for board.   His board bill was what part of all his expenses ?

3. A manufacturing concern pays profits amounting to $\frac{3}{16}$ of its capital.   If the profits are $ 15,000, what is the capital ?

4. A man sold ⅜ of his farm for $ 4612½.   At the same rate how much was the whole farm worth ?

5. A dealer bought goods for $ 12½, and sold them for $ 15.   The gain was what part of the cost?   The gain was what part of the selling price ?

6. If a hatter buys a hat for $ 6, and has to sell it for $ 4½, he loses what part of the cost ? of the selling price ?

7. The first of these lines is 1½   1909._____
in. long, the second 1¼ in. long.   1910._____
If the first line represents a merchant's sales in 1909, and the second one, his sales in 1910, the amount of sales in 1910 was what part of the sales in 1909 ?   The sales in 1910 were how much (fractional part) less than in 1909 ?   The sales in 1909 were how much greater than in 1910 ?

8. If an acre of land yields 96 bu. of corn, and another acre produces 36 bu. of oats, the oat crop is what part of the corn crop? If oats are worth 40 ¢ a bushel, and corn is worth 60 ¢ a bushel, the value of the oat crop is what part of the value of the corn crop?

9. A man buys a house and lot for $4800. He receives $640 for rent, and pays for expenses $160. The expenses are what part of the total income? of the cost of the house? The total income is what part of the cost of the house? The net income is what part of the cost of the house?

10. A fruit dealer buys 15 doz. oranges for $3. 2 doz. oranges spoil. If he sells the remainder of them at 30 ¢ a dozen, what part of the cost does he gain?

247. In problems like the following, the student must give careful attention to the reading of the problem, and make the solution in exact accordance with its meaning.

1. At the end of the first year's service, a young man's salary was increased $\frac{1}{5}$, and at the end of the second year it was increased $\frac{1}{6}$. If his third year's salary was $1728, what was his salary the first year?

About which year's salary is nothing said in the problem? With which year's salary are the other years' salaries compared? the first year. Let the first year's salary stand for unity, and represent unity by $\frac{25}{25}$.

Then $\frac{25}{25}$ of first year's salary = first year's **salary**,

and $\frac{30}{25}$ of first year's salary = second year's salary,

and $\frac{36}{25}$ of first year's salary = third year's salary.

The increase at the end of the first year was $\frac{1}{5}$ of $\frac{25}{25}$, or $\frac{5}{25}$, which, added to the $\frac{25}{25}$ for the first year, gives $\frac{30}{25}$ for the second year. At the end of the second year there was an increase of $\frac{1}{5}$ of $\frac{30}{25} = \frac{6}{25}$, which, added to the $\frac{30}{25}$, gives $\frac{36}{25}$ for the third year. Therefore, $\frac{36}{25}$ of the first year's salary = $1728. Hence $1728 \div \frac{36}{25}$ = $1200, the first year's salary. (State the principles that apply in this example.)

248. Reduce each of the following problems to one of the type forms given on page 103:

1. A merchant's sales in 1909 amounted to $\frac{1}{4}$ more than they did in 1908, and to $\frac{1}{5}$ more in 1910 than in 1909. If the sales in 1910 amounted to $30,000, find the amount of sales in 1908; in 1909.

2. If a dealer's profits are $\frac{1}{3}$ greater the second year than the first, $\frac{1}{4}$ greater the third year than the second, $\frac{1}{3}$ greater the fourth year than the third, and $\frac{1}{5}$ less the fifth year than the fourth, and if the profits for the five years are $\$21,900$, what were the profits for the first year?

3. In 1909, the receipts of a certain railroad company increased $\frac{1}{4}$ over the receipts of 1908. In 1910 there was a decrease of $\frac{1}{6}$ of the receipts of 1909. If the receipts for 1910 were $\$3,768,053.10$, what were they in 1908?

4. A merchant's sales in 1906 amounted to $\$426,000$; in 1907, to $\$532,500$; and in 1908, to $\$639,000$. Find the rate of increase each year.

5. If goods are bought for $\$145$ and sold for $\$174$, what part of the cost is gained? what part of the selling price?

The most important practical applications of the principles of division are noted in reduction and division of fractions.

REVIEW PROBLEMS IN FRACTIONS

MENTAL

249. Find the results as indicated:

1. $\frac{1}{2}+\frac{1}{3}$; $\frac{1}{2}-\frac{1}{3}$; $\frac{1}{2}\times\frac{1}{3}$; $\frac{1}{2}\div\frac{1}{3}$.

2. $\frac{4}{5}+\frac{1}{2}$; $\frac{4}{5}-\frac{1}{2}$; $\frac{4}{5}\times\frac{1}{2}$; $\frac{4}{5}\div\frac{1}{2}$.

3. $\frac{3}{7}+\frac{3}{8}$; $\frac{3}{7}-\frac{3}{8}$; $\frac{3}{7}\times\frac{3}{8}$; $\frac{3}{7}\div\frac{3}{8}$.

4. $4\frac{1}{2}+2\frac{1}{4}$; $4\frac{1}{2}-2\frac{1}{4}$; $4\frac{1}{2}\times2\frac{1}{4}$; $4\frac{1}{2}\div2\frac{1}{4}$.

5. $7\frac{1}{3}+3\frac{2}{3}$; $7\frac{1}{3}-3\frac{2}{3}$; $7\frac{1}{3}\times3\frac{2}{3}$; $7\frac{1}{3}\div3\frac{2}{3}$.

6. If $\frac{3}{4}$ is subtracted from a given fraction, the remainder will be $\frac{1}{6}$. What is the fraction?

7. $\frac{5}{8}$ added to a certain fraction gives $1\frac{7}{24}$. What is the fraction?

8. If a man spends, on the average, $\$\frac{5}{8}$ a day for cigars and liquor, in what time will he spend $\$100$?

9. A boy bought a bicycle for $\$45$, and sold it for $\frac{8}{9}$ of what he gave for it. How much less did he receive than he paid?

10. Two men bought a motor boat for $\$450$, each paying half. Later each sold one third of his share at cost. How much had each invested then, and what part of the boat did each own?

11. I bought $11\frac{3}{8}$ lb. of tea at $\$\frac{1}{2}$ a pound and 6 lb. of coffee at $\$\frac{3}{8}$ a pound. How much change should I receive from a ten-dollar bill?

12. A young lady bought $7\frac{1}{2}$ yd. muslin at $8\frac{1}{2}\cancel{c}$ a yard. What change ought she to receive from a half dollar and a quarter?

13. A woman bought $10\frac{1}{2}$ yd. of silk at $\$2\frac{1}{2}$ a yard, and $3\frac{1}{2}$ yd. lining at $\$.25$ a yard. She gave in payment three ten-dollar bills. What change should she receive?

14. I sold a horse for $75, which was $\frac{5}{6}$ of what it cost. How much did I lose?

15. A crate containing 10 doz. oranges cost $1.75. If the oranges sell at the rate of 2 for $5\cancel{c}$, what is the gain?

16. A bunch of 8 doz. bananas cost $1. If one dozen spoil and the rest are sold at the rate of 3 for $10\cancel{c}$, what is the gain?

17. Two railroads carry merchandise 450 mi., for $3.60. If the first road carries it 200 mi., how much should each receive?

18. Two men engage to do a piece of work for $157.50. The first works 18 da. and the other works 27 da. How shall the money be divided?

19. $3\frac{1}{2}$ doz. eggs at $\$\frac{1}{4}$ a dozen will pay for how many pounds of rice at $8\frac{3}{4}\cancel{c}$ a pound?

20. I sold $\frac{3}{8}$ of my apples for $90. At that rate, how much were they all worth?

21. A merchant by selling cloth at $42\cancel{c}$ a yard gained $\frac{1}{6}$ of the cost. Find the cost.

22. I bought a quantity of cloth for $180. By selling it at $2 a yard, I gained $\frac{1}{3}$ of the cost. How many yards were there?

23. A buying agent charges $\frac{1}{16}$ of the net cost of goods as his commission. If his commission for one month amounts to $240, what is the value of the goods bought?

24. I sold $\frac{3}{7}$ of my land for $1500. If I sold 75 A., how much land have I left, and how much is it worth at the same rate?

25. A man picked $\frac{3}{4}$ of a bushel of pears and sold $\frac{1}{2}$ of them for $.45. At that rate, how much would a bushel cost?

26. A man had $37\frac{1}{2}$, and spent $12\frac{1}{2}$. What part of his money did he spend?

27. A clerk earns $87½ a month, and spends $30 for board and $20 for other expenses. What part of his money does he save?

28. In a certain factory 56 men were employed. During a dull season 24 of them were laid off. What part of the men continued working?

29. In an orchard there are 120 trees; 96 of them are apple trees, 16 are pear trees, and the remainder are cherry trees. What part of all is each kind of trees?

30. Three railroads carry a piece of freight 540 mi. The first carries it 180 mi., the second 300 mi., and the third the remainder of the distance. What fractional part of the distance does each carry it?

PROBLEMS

(Solve mentally when possible.)

250. 1. ⅜ of a pole 32 ft. long is decayed. How many feet are good?

2. A tree 76 ft. high breaks 19 ft. from the ground. What part is broken off?

3. A man has ½ of his money in one bank, ⅓ of it in another, and the remainder in a third bank. If the amount in the first bank is $660, how much is there in each of the other banks?

4. A man sold a horse for $150, which was ¾ of what he paid for it. Find the loss.

5. The wholesale price of Rio coffee at New York increased from 7¼¢ in August, 1909, to 8¾¢ in August, 1910. What fractional part of the price in 1909 was the increase?

6. A farmer sold ⅔ of his sheep for $360. At the same rate what is the value of all his sheep? What is the value of those he did not sell?

7. A boy spent ⅔ of his money for a bicycle, and had $24 left. How much money had he at first?

8. A man had ½ of his sheep in one pasture, ¼ of them in another pasture, and the rest of them, which was 33 sheep, in a third pasture. How many sheep had he?

9. A gentleman's estate is so divided that $\frac{2}{5}$ of it is woodland, $\frac{1}{4}$ of it is water, and the remaining 600 A. are under cultivation. How many acres are there in the estate?

10. A house and lot are worth $ 6300. If the lot is worth $\frac{4}{5}$ as much as the house, how much is each worth?

11. A gentleman left his son $ 6500, which amount was $2\frac{1}{6}$ times as much as he left to his daughter. How much did the daughter receive?

12. A owned $\frac{4}{7}$ of a farm and B the remainder. If A had $24\frac{3}{4}$ A. more than B, how many acres had each?

13. A house and a barn are valued at $ 14,805. Find the value of each if the barn is worth $\frac{2}{5}$ as much as the house.

14. Find the cost of each, if a house and lot together cost $ 14,000, the lot costing 4 as much as the house.

15. A gentleman left his estate to his three sons. The eldest received .40 of it, the next son .35, and the youngest, $ 7500. Find the value of the estate.

16. The cost of furnishing two rooms is $ 480. If the cost of one is $\frac{7}{9}$ as much as the cost of the other, what is the cost of each?

17 The income from a certain business for two years was $ 27,500. Find the income for each year if the second year's income was .20 greater than the first.

18 What number increased by $\frac{3}{4}$ of itself is 651?

19 If goods are sold for $ 90, and a profit of $\frac{1}{5}$ of the cost is made, find the cost.

20. If the profit on eggs is $3\mathcal{c}$ per dozen, and the eggs sell for $30\mathcal{c}$ a dozen, what part of the cost is gained?

21. The profit on a pair of shoes is $ 75. If the profit is .20 of the cost, for how much per pair do the shoes sell?

22. What number decreased by $\frac{3}{7}$ of itself is equal to 56?

23. Find the cost of an article sold for $ 36, if $\frac{1}{7}$ of the cost was lost.

24. A hatter marked all his straw hats down $\frac{1}{4}$. What was the marked price of a hat that he now sells for $ 4.50? If he made a profit of $\frac{1}{2}$ of the cost *before* marking the hat down, what part of the cost does he make *after* marking it down?

25. If a man buys $868\frac{1}{2}$ yd. of cloth and sells $\frac{2}{3}$ of it at $14\frac{3}{4}$ ¢ a yard, and the remainder at $12\frac{3}{4}$ ¢ a yard, how much does he receive for all of it?

26. A field of $9\frac{1}{2}$ A. was bought for $380. The first year's wheat crop averaged 20 bu. to the acre. If wheat was worth 90 ¢ per bushel, the value of the crop was what part of the purchase price?

27. Find the total value of the following articles: $1\frac{3}{4}$ doz. buttons at 15 ¢; $2\frac{3}{4}$ yd. ribbon at $9\frac{1}{2}$ ¢; $3\frac{1}{2}$ yd. lace at 65 ¢; $\frac{3}{4}$ yd. velvet at 79 ¢.

28. I bought 360 sheep for $1200. I sold $\frac{2}{3}$ of them at $4 per head, and $\frac{1}{4}$ of the remainder died. At what price per head must the rest of them be sold to gain $\frac{1}{5}$ of the cost of all?

29. A fruit dealer buys 10 doz. oranges for $2.30 If 2 doz. spoil, at what price per dozen must he sell the good ones to gain $\frac{1}{4}$ of the cost?

30. From the sum of $36\frac{3}{4}+47\frac{1}{3}+18\frac{1}{8}$ take the sum of $31\frac{2}{3}$ and $24\frac{3}{4}$.

31. Multiply the sum of $27\frac{1}{4}$ and $34\frac{2}{3}$ by the difference between $18\frac{2}{3}$ and $16\frac{1}{2}$.

32. Divide the product of $83\frac{13}{16}\times79\frac{1}{12}$ by the product of $18\frac{1}{4}\times9\frac{5}{16}$.

33. A young man buys $13\frac{3}{4}$ A. of land at $36.50 per acre. If he works for $27\frac{1}{2}$ ¢ an hour, 8 hr. a day, and spends $6.60 a week for board and other expenses, how long will it take him to pay for the land?

34. If eggs are bought at 28 ¢ a dozen and sold at the rate of 10 eggs for 25 ¢, find the gain on 500 doz eggs.

35. On the first Saturday of each month of the fiscal year ending June 30, 1910, the wholesale weekly price of rubber per pound at New York City was $1.45, $1.87, $1.92, $2.03, $1.84, $1.75, $1.76, $1.84, $2.06, $2.80, $2.47, $2.28, respectively. Find (1) the average price for the year; (2) what fractional part the average price is greater than the first price given.

NOTE. The average of several numbers is found by dividing their sum by the number of numbers.

36. Make a pay roll from the following data: The rate is for a week of 48 hr., with $\frac{1}{4}$ extra for overtime.   Clerk #1 worked 9, $8\frac{1}{2}$, 10, $10\frac{1}{2}$, 8, and $9\frac{1}{4}$ hr. at $16.50; #2 worked 8, 8, $11\frac{1}{2}$, 9, $7\frac{1}{2}$, and 10 hr. at $15; #3 worked $9\frac{1}{4}$, $8\frac{1}{2}$, 8, $8\frac{3}{4}$, 8, and 8 hr. at $20; #4 worked $8\frac{1}{2}$, 8, 8, 9, 10, and $9\frac{1}{4}$ hr. at $18; #5 worked 8, 8, 8, 9, 8, and 8 hr. at $14.50; #6 worked 8, 8, 9, 9, 8, and 10 hr. at $16; #7 worked 8, 9, 8, 10, $8\frac{1}{2}$, and 10 hr. at $17; and #8 worked $8\frac{1}{2}$, 9, 9, 8, 9, and $8\frac{1}{2}$ hr at $19.50.

The following facts are taken from the weekly reports of a cheese factory as given to a dairyman who supplies part of the milk. Sales are made weekly, and the proceeds returned to the milk producers.   Sale #7 is worked out in detail.   Find the missing items in the other sales.

	37.	**38.**	**39.**	**40.**	**41.**
	SALE #7	SALE #8	SALE #9	SALE #10	SALE #11	SALE #12
From . . .	May 25	June 1	June 8	June 15	June 22	June 29
To	May 31	June 7	June 14	June 21	June 28	July 5
Total Milk . .	59572 lb.	53781 lb.	63114 lb.	63639 lb.	63256 lb.	61247 lb.
Total Cheese .	5767 lb.	5070 lb.	6119 lb.	6084 lb.	5997 lb.	5847 lb.
	@ $11\frac{1}{2}$¢	@ $11\frac{1}{3}$¢	@ $11\frac{3}{4}$¢	@ $11\frac{1}{4}$¢	@ $10\frac{3}{4}$¢	@ $11\frac{3}{8}$¢
Total Money .	$663.20					
Total Expense	$74.59	$67.13	$78.99	$78.55	$77.46	$75.59
Pounds Milk to						
1 lb. Cheese .	10.3298					
Ratio	$.98806					
Your Milk . .	1209 lb.	1018 lb.	1165 lb.	1141 lb.	1150 lb.	1074 lb.
	$11.95					
Less Cheese . .	4 lb. .46				$3\frac{1}{2}$ lb.	
Your Money .	$11.49					

The total money is the product of the total number of pounds of cheese by the price per pound.   The number of pounds of milk to a pound of cheese is found by dividing the total amount of milk by the total amount of cheese.   The ratio is the net return per 100 lb. of milk, found by dividing the net proceeds of the sale by the total milk in hundredweight.   Thus, $663.20 − $74.59 = $588.61, net proceeds. $588.61 ÷ 595.72 = $.98806, the ratio.   1209 lb. multiplied by the ratio per 100 gives $11.95, from which is deducted the value of the cheese used by the dairyman, leaving $11.49.

RATIO AND PROPORTION

RATIO

251. Ratio is the relation existing between two quantities of the same kind expressed as the quotient of the first divided by the second.

252. The sign of ratio is the colon (:).

The ratio of 8 to 4 is written 8 : 4. It equals $\frac{8}{4}$, or 8 ÷ 4, or 2. The ratio of 4 to 7 is written 4 : 7, and equals $\frac{4}{7}$.

253. The terms of a ratio are the numbers compared.

The first term is the **antecedent**; the second term is the **consequent**.

254. The foregoing statements may be represented graphically thus:

$$\text{The ratio of 5 to 8} = \frac{5}{8} = \frac{\text{antecedent}}{\text{consequent}} = \frac{\text{dividend}}{\text{divisor}} = \frac{\text{numerator}}{\text{denominator}}.$$

255. A **direct ratio** is the quotient of the antecedent divided by the consequent.

256. An **inverse ratio** is the quotient of the consequent divided by the antecedent.

257. Since a ratio may be expressed in the form of a common fraction, all the principles of fractions apply to ratio.

258. Ratio may be **simple** or **compound**. 7 : 9 is a simple ratio $\begin{matrix} 4 : 7 \\ 9 : 10 \end{matrix}$ is a compound ratio.

259. Find the ratio of :

1. 4 to 8; 5 to 6; 7 to 9; 8 to 6.

2. 6 to 18; 8 to 32; 45 to 15; 56 to 7.

3. 9 to 30; 30 to 9; 128 to 4; 5 to 125.

112

4. $\frac{4}{5}$ to $\frac{2}{3}$.

$\frac{4}{5} = \frac{12}{15}$

$\frac{2}{3} = \frac{10}{15}$

$\frac{12}{15} : \frac{10}{15} = 12 : 10 = \frac{12}{10} = 1\frac{1}{5}$. Or, $\frac{4}{5} \div \frac{2}{3} = \frac{4}{5} \times \frac{3}{2} = \frac{6}{5} = 1\frac{1}{5}$.

Fractions having a common denominator are in the ratio of their numerators. Fractions *not* having a common denominator should be reduced to equivalent fractions having a common denominator, then they will be in the ratio of their numerators. Or the antecedent should be divided by the consequent.

5. $\frac{5}{8}$ to $\frac{7}{9}$; $\frac{3}{4}$ to $\frac{7}{8}$; $\frac{5}{6}$ to $\frac{9}{10}$; $\frac{6}{7}$ to $\frac{5}{16}$.

6. $\frac{3}{8}$ to $\frac{1}{9}$; $\frac{9}{10}$ to $\frac{1}{3}$; $\frac{3}{11}$ to $\frac{11}{16}$; $\frac{5}{24}$ to $\frac{3}{8}$.

7. $4 : 7$

$8 : 12$

$\frac{4 \times 8}{7 \times 12} = \frac{8}{21}$ Arrange all the antecedents as a dividend and all the consequents as a divisor, and apply cancellation.

8. $\frac{3:6}{9:12}$; $\frac{5:8}{7:14}$; $\frac{9:4}{12:2}$; $\frac{7:14}{12:2}$.

PROPORTION

260. Name two numbers that have the same ratio as 4 to 8; as 6 to 10; as 9 to 6; as 12 to 4

261. Proportion is an equality of ratios

262. The sign of proportion is the double colon (: :). Thus, the expression $4 : 7 :: 8 : 14$ is a proportion, and is read 4 is to 7 as 8 is to 14.

The sign of proportion is sometimes written as a sign of equality (=), from which it is derived. In the proportion given, the ratio of 4 to 7 ($\frac{4}{7}$) = the ratio of 8 to 14. ($\frac{4}{7} = \frac{8}{14}$.)

263. The first and third terms of a proportion are the **antecedents**, and the second and fourth terms are the **consequents**. In the above proportion 4 and 8 are the antecedents, and 7 and 14 are the consequents.

264. The first and fourth terms are the **extremes**; the second and third terms are the **means**. Thus, 4 and 14 are the extremes; 7 and 8 are the means.

VAN TUYL'S BUS. ARITH. — 8

265. The solution of problems in proportion depends upon the following principles:

1. *The product of the means equals the product of the extremes.*

2. *The product of the means divided by either extreme gives the other extreme.*

3. *The product of the extremes divided by either mean gives the other mean.*

266. Find the missing term in each of the following:

1. $14 : 25 :: 21 : ?$

$\dfrac{25 \times 21}{14} = 37\frac{1}{2}.$ To find the missing extreme, divide the product of the means by the given extreme, 14. The other extreme is $37\frac{1}{2}$.

2. $? : 54 :: 28 : 42.$ **5.** $75 : ? :: 25 : 40.$

3. $32 : 50 :: ? : 25.$ **6.** $16\frac{1}{2} : 24\frac{3}{4} :: 30 : ?$

4. $55 : 60 :: 11 : ?$ **7.** $33\frac{1}{3} : 50 :: ? : 200.$

267. Proportion may be simple, compound, or partitive.

SIMPLE PROPORTION

268. Simple proportion is an equality of simple ratios. All the proportions given above are simple.

Cause and Effect

269. The simplest method of making a proportion statement from a given problem is by the method of "Cause and Effect." In every problem there are certain "causes" the operation of which results in certain "effects." Every effect, or result, has a cause, and every cause produces an effect. In any given problem, the effects are in the same ratio as the causes.

270. If 11 bbl. of flour cost $ 60.50, how much will 21 bbl. cost at the same rate?

$$\overset{\substack{\text{1st} \\ \text{cause}}}{11 \text{ bbl.}} : \overset{\substack{\text{2d} \\ \text{cause}}}{21 \text{ bbl.}} :: \overset{\substack{\text{1st} \\ \text{effect}}}{\$ 60.50} : \overset{\substack{\text{2d} \\ \text{effect}}}{\$?}$$

$$\dfrac{21 \times \$ \overset{5.50}{\cancel{60.50}}}{\cancel{11}} = \$ 115.50.$$

The 11 bbl. is the 1st cause, the effect of which is an expenditure of $60.50. The 2d cause is the 21 bbl., the effect, or cost, of which is to be found. Dividing the product of the means by the given extreme, the cost of 21 bbl. is found to be $115.50.

NOTE. Observe that the two ratios are each composed of *like* quantities. If this fact is kept in mind, it will aid the student in making statements of proportion.

PROBLEMS

271. 1. If 17 A. of land produce 442 bu. of grain, how many bu. will 51 A. produce?

2. 18 horses eat $23\frac{5}{8}$ bu. of grain a week. How much will 27 horses eat in 8 wk.?

3. If 24 yd. of cloth cost $2.64, how much will 33 yd. of the same kind of cloth cost?

4. At 45 ¢ a dozen how much will 16 oranges cost?

5. The shadow of a post 9 ft. high is 10 ft. long. How high is a tree that casts a shadow 90 ft. long at the same time?

6. For how long a time should $500 be loaned to balance a loan of $800 for 10 mo.?

7. The capacity of a shoe factory is 2580 pairs per week of 6 da. How long will it take to fill an order for 34,400 pairs?

8. If 12 men can do a piece of work in 9 da., how long will it take 18 men to do it?

SUGGESTION. The time required is in inverse ratio to the number of men employed.

9. It is estimated that 45 men can do a piece of work in 12 da. How many additional men should be employed to perform the work in 9 da.?

COMPOUND PROPORTION

272. Compound proportion is an equality of compound ratios. Thus,   4 : 8
7 : 21 : : 15 : 90
9 : 18   4 : 8 is a compound proportion.

273. The principles underlying solutions in compound proportion are the same as those applied in simple proportion.

274. If an army of 2500 men in 15 da. requires 46,875 lb. of food, how many pounds of provisions will 3500 men need in 25 da.?

1st cause   2d cause   1st effect   2d effect

2500 men : 3500 men : : 46875 lb. : ? lb.

15 da. : 25 da.

$$\frac{3500 \times \overset{3125}{\cancel{25}} \times \cancel{46875} \text{ lb.}}{\cancel{2500} \times \cancel{15}} = 109,375 \text{ lb.}$$

The 1st cause is the 2500 men requiring food for 15 da., as a result of which 46,875 lb. were required. The 2d cause is similar to the 1st cause, but is greater, that is, 3500 men for 25 da. The second effect is to be determined. Dividing the product of the means by the given extreme gives the required amount of food as 109,375 lb

PROBLEMS

275. 1. 75 sheets of paper $36'' \times 54''$ weigh 24 lb. How much do 5000 sheets $17'' \times 22''$ of the same grade of paper weigh?

2. If 3 doz. sheets of drawing paper $24'' \times 36''$ weigh 2 lb., find the weight of 6 reams (500 sheets $= 1$ ream) $8'' \times 12''$ of the same grade of paper.

PARTITIVE PROPORTION

276. 1. Three partners, A, B, and C, agree that the profits of their business shall be divided into 9 equal parts or shares, of which A is to have 4, B, 3, and C, 2 shares. If the profits amount to $6300, how much does each partner receive?

A has $\frac{4}{9}$ of $6300, or $2800.   Since there are 9 parts or shares, and A
B has $\frac{3}{9}$ of $6300, or $2100.   has 4 of them, he has $\frac{4}{9}$ of all the shares,
C has $\frac{2}{9}$ of $6300, or $1400.   or $\frac{4}{9}$ of the profits. For the same reason B
has $\frac{3}{9}$, and C, $\frac{2}{9}$, of the profits.

In the above problem, the profits are divided among A, B, and C in the proportion of 4, 3, and 2.

The process of dividing a number into parts proportional to other given numbers may be called **partitive proportion**.

2. Two men do a piece of work for $6. The first man works $\frac{1}{2}$ of a day, and the second $\frac{3}{4}$ of a day. How shall they divide the $6?

Only similar fractions can be compared. $\frac{1}{2} = \frac{2}{4}$, $\frac{3}{4} = \frac{3}{4}$. Similar fractions are in proportion to their numerators. Hence, $\frac{2}{4}$ and $\frac{3}{4}$ are in proportion to 2 and 3. That is, the $6 is to be divided into parts proportional to 2 and 3. 2 parts + 3 parts = 5 parts.

The first man has 2 parts, or $\frac{2}{5}$ of $6, or $2.40.

The second man has 3 parts, or $\frac{3}{5}$ of $6, or $3.60.

PROBLEMS

277. 1. A worked 3 da. and B 4 da. for $14. How much should each receive?

2. Two men rent a pasture for $50. A puts in 4 horses, and B 6 horses. How much should each pay?

3. Three men performed a piece of work for $288. The first man worked 24 da., the second 40 da., and the third 32 da. How much should each receive?

4. Divide $540 among three families so that the first shall receive $2 and the second $3, as often as the third receives $4.

5. Four partners' investment are in proportion to 6, 7, 8, and 9. They gain in one year $10,800. How much is each partner's share?

6. Divide 25 into parts proportional to $\frac{1}{2}$ and $\frac{1}{3}$.

7. Divide 35 into parts proportional to $\frac{1}{3}$ and $\frac{1}{4}$.

8. The sides of a triangle are in proportion to 4, 5, and 6. Its perimeter is 360 ft. Find the sides.

9. Four men and eight boys earn $440. If one man earns as much as two boys, how much does each earn?

10. Two building lots cost $14,000. One cost $\frac{1}{3}$ more than the other. How much did each cost?

11. A field rents for $36. B puts in 12 sheep for 15 wk. and C 20 sheep for 18 wk. How much should each pay?

12. Divide 180 into parts proportional to 3, $1\frac{1}{2}$, $\frac{3}{4}$, and $\frac{3}{8}$.

13. Three partners agree to divide their profits in proportion to $\frac{2}{3}$, $1\frac{1}{4}$, and 2. If their profits are $7050, what is the share of each?

14. An estate of $50,000 is divided among four sons in proportion to their ages, which are 20 yr., 17 yr., 14 yr., and $11\frac{1}{2}$ yr., respectively. Find the share of each.

15. A and B are partners. A invests $5000 for 8 mo. and B, $7500 for 6 mo. Their gains amount to $3400. Find the share of each.

16. C, D, and E are partners. C invests $8000 for a year. D invests $10,000 for 10 mo., and E invests $6000 for 8 mo. Divide a profit of $9760 among them.

EXAMINATIONS

278. Minimum time, twenty minutes; maximum time, one hour. Suggestion for marking: Allow 100 credits if test is completed correctly in minimum time. Deduct one credit for each minute required beyond minimum time.

1. Add the following:

241693798	$ 693.05
473185459	7.847
391533768	78.98
427936857	1489.86
819348673	.75
473925165	974.98
274639827	6.
315987352	87.87
675431298	9748.098
897316984	7.96
931258369	75.899
687316984	17.

2.

From	Take	Difference
752361	582834	
1021314	987476	
7934.19	986.74	
784	95.9876	

3. Find the total value of the following:

222^1 yd. @ 3^2 ¢
166^2 yd. @ 4^1 ¢
265^1 yd. @ 7^1 ¢
163^2 yd. @ 9^2 ¢
254^3 yd. @ 8^1 ¢
212 yd. @ 4^3 ¢

5. Find the total value of the following:

3290 lb. @ $ 2.75 per hundredweight.
4332 lb. @ 2.75 per hundredweight.
3570 lb. @ 1.87½ per hundredweight.
5792 lb. @ 22.50 per 1000 lb.
2396 lb. @ 16.50 per 1000 lb.
32690 lb. @ 12.80 per 1000 lb

4. Complete the following bill:

Chicago, Ill., July 1, 1910.		
H. P. Jones		
Bought of		
D. C. Roy.		
3 bbl. Sugar,	640# @ 6¼ ¢	
5 chests Tea,	680# @ 75 ¢	
3 chests Tea,	498# @ 66⅔ ¢	
6 bbl. Flour,	1176# @ 4 ¢	
Total		

118

6. Add 873 to 428 continuously till the sum equals or exceeds **10,000.** (Show all the work.)

7. Divide 25,473 by 2237 by continued subtraction. (Show all the work.)

8. Reduce to equivalent fractions having the least common denominator:

(a) $\frac{1}{2}$, $\frac{2}{3}$, $\frac{5}{12}$, $\frac{7}{8}$.   (c) $\frac{9}{16}$, $\frac{5}{8}$, $\frac{5}{24}$, $\frac{3}{32}$.

(b) .75, $\frac{5}{6}$, $\frac{7}{9}$, $\frac{5}{18}$.   (d) .80, $\frac{2}{3}$, $\frac{4}{15}$, .3.

9. Reduce to common fractions in their lowest terms:

.065   .0625   225   .00125

10. Perform the operation indicated:

$15\frac{3}{5} \times 17\frac{1}{3}$ $426\frac{3}{4} \div 21\frac{1}{2}$ $48.72 \times .00125$ $54.7325 \div .001$

SPEED TEST

279. 1. Add horizontally and vertically. Check by adding the totals:

$36,968	$ 7,634	$14,262	$———
22,635	28,329	37,257	———
9,564	82,378	6,295	———
49,877	69,764	34,472	———

Grand total . . . . $———

2. Find the gain on each of the following items and the total gain:

SELLING PRICE	COST	GAIN
$40,000.00	$37,624.37	$———
9,437.83	5,819.24	———
4,341.85	2,838.96	———
57,640.33	48,924.29	———
89,754.00	68,947.25	———
Total		$———

Perform the following indicated operations:

3. $425 \div \frac{5}{6} =$

$234 \div .33\frac{1}{3} =$

$.25 \times \frac{8}{9} =$

$9\frac{1}{8} - 4\frac{3}{16} =$

$\frac{1}{8} + \frac{1}{6} + \frac{1}{9} =$

4. $9 \times 12 \times 14 \div 42 =$

$95 \times 8 \times 3 \div 19 =$

$52 \times 8 \times 7 \div 91 =$

$20 \times 21 \times 22 \div 154 =$

$(48 \times 36 \times 24) \div (16 \times 12) =$

5. Find the value of each of the following:

$$\frac{\frac{3}{4} + \frac{5}{6} - \frac{1}{3}}{\frac{3}{4}} \times 12\frac{1}{2} = \qquad \frac{\frac{13}{15}}{\frac{7}{5}} \times \frac{15}{16} \times \frac{14}{65} =$$

6. Reduce each of the following groups to equivalent fractions having the least common denominators:

(a) $\frac{9}{24}$ $\frac{13}{18}$ $\frac{7}{12}$ $\frac{11}{36}$ =   (c) $\frac{4}{15}$ $\frac{5}{21}$ $\frac{4}{27}$ $\frac{2}{63}$ =

(b) $\frac{3}{5}$ $\frac{4}{7}$ $\frac{11}{14}$ $\frac{9}{10}$ =   (d) $\frac{5}{12}$ $\frac{8}{21}$ $\frac{11}{28}$ $\frac{13}{42}$ =

7. Multiply and prove:

76897 × 328
45723 × 126

8. Divide and prove:

$6806\frac{1}{4} \div 412\frac{1}{2}$

9. Find the total cost of the following:

250 bu @	$ .75	= $ ——
340 bu @	.80	= ——
1360 bu. @	1.12½	= ——
1840 bu. @	1.25	= ——
1480 bu. @	.62½	= ——
Total	. .	$ ——

10. Find the total interest at 6 % on the following:

$ 500.00 for 90 da.	= $ ——
750.00 for 80 da.	= ——
1000.00 for 30 da.	= ——
1200.00 for 45 da.	= ——
895.50 for 20 da.	= ——
Total . .	$ ——

WRITTEN TEST

280. 1. A company employs 17 laborers at $1.50 per day; 28 mechanics at $3.60 per day; 9 teamsters at $2.25 per day; 2 book-keepers at $15 a week; and a superintendent at $1560 a year. What is the weekly pay roll of the company?

2. A man sold a farm of 248½ A. at $45¾ per acre, and with the proceeds bought another farm of 125 A. What price per acre did he pay for the new farm?

3. A person owned $\frac{5}{8}$ of a mine and sold $\frac{3}{4}$ of his share for $1710. What was the value of the mine?

4. Find the net cost of the following bill of hardware: 12 kegs cut nails, 8d., 1200 lb. @ 3½ ¢ per pound; 5 doz. handsaws, 26 in., @ $15 per dozen; 4 doz. carpet stretchers at $2.87½ per dozen; 20 doz. garden rakes @ $3.25 per dozen; and 12 doz. try squares @ $5.40 per dozen. Less 20 % on the entire bill.

5. Find the total cost of the following: 8 pieces broadcloth, 56^1, 58^3, 57, 59^2, 55^1, 60^1, 61^2, 58^3 yd. at \$ $3.12\frac{1}{2}$; 5 pieces corduroy, 47^2, 44^2, 43^3, 46^2, 45^1 yd. at \$ $0.57\frac{1}{2}$; and 7 pieces storm serge, 43^1, 47^2, 48^3, 44, 45^2, 48^2, 47^3 yd. at \$ $0.62\frac{1}{2}$.

6. A man had on hand in the morning cash in the safe amounting to \$ 206.20, and in the bank \$ 1379.30. During the day he received cash in currency amounting to \$ 909.24 and checks amounting to \$489.36. Cash deposited in the bank, \$ 604.37. Checks drawn on the bank account, \$ 3.97, \$ 47.86, \$ 396.25, \$ 49.83, \$ 246.97. Cash paid out in bills and coin, \$ 49.86. Show the condition of the cash and bank accounts in the evening.

7. Find the total cost of 64 bags of sugar, each containing $12\frac{1}{2}$ lb. at $5\frac{1}{2}$ ⊄ a pound; 19 bbl. of flour at \$ 5.50 a barrel; 1740 eggs at 33 ⊄ a dozen; 375 pineapples at \$12.75 per hundred; 42 qt. of walnuts at $8\frac{1}{2}$ ⊄ a quart.

8. A stenographer gets \$ $83\frac{1}{3}$ per month. How long will it take him to pay for a house and lot costing \$ 2300, if his yearly expenses average \$ 625.60? Reduce time to years, months, and days.

9. Find the cost of the following: 348 eggs at 14 ⊄ per dozen; 643 lb. sugar at $6\frac{1}{2}$ ⊄ per pound; and 7750 lb. coal at \$ 5.87 per ton.

10. In a hospital having 1048 inmates there were consumed the following articles: Meat, fresh, 287,138 lb., average price, \$.082; meat, smoked, 7614 lb., average price, \$0.12; meat, canned, 7239 lb., average price, \$.066; poultry, 4128 lb., average price, \$.167. Find the total cost and the cost per capita.

DENOMINATE NUMBERS

281. A denominate number is a concrete number whose unit of value has been fixed by law or custom. $5; 2 qt.; 3 rd.; are denominate numbers.

282. A simple denominate number is one whose value is expressed in one denomination. 4 qt.; 7 in.; 8 gal.; are simple denominate numbers.

283. A compound denominate number is one whose value is expressed in two or more denominations. 5 sq. yd., 2 sq. ft., 18 sq. in.; 8 bu., 2 pk., 6 qt., are compound denominate numbers.

284. A measure is a standard unit used to determine quantity.

285. Quantity is anything that can be measured, as length, area, volume, capacity, time, value, etc.

286. A standard unit of measure is a unit established by law or custom, by which other units are to be adjusted.

The gallon of 231 cu. in. is the established unit of measure for liquids; from it the other units, gill, pint, quart, and barrel are derived.

The yard is the standard unit of measure for length; the inch, foot, rod, and mile are derived from the yard, etc.

287. A quantity is measured by determining how many times it contains its appropriate unit of measure.

Linear Measure

288. Linear measure is used for measuring distance.

TABLE

12 inches (in.)	= 1 foot (ft.)
3 feet	= 1 yard (yd.)
5½ yards	= 1 rod (rd.)
320 rods	= 1 mile (mi.)

1 mi. = 320 rd. = 1760 yd. = 5280 ft. = 63,360 in.

The unit of length is the **yard**.

Other Linear Measures

1 size, $\frac{1}{3}$ **in.** Used by shoemakers.
1 hand, 4 in. Used in measuring the height of horses.
1 fathom, 6 ft. Used in measuring depths at sea.
1 knot, nautical or geographical mile — 1.152$\frac{2}{3}$ mi. or 6086 ft.
The knot is used in measuring distances at sea. It is equivalent to **1 min.** of longitude at the equator.

Square Measure

289. Square measure is used to measure the areas of surfaces.

TABLE

144 square inches (sq. in.)	= 1 square foot (sq. ft.)
9 square feet	= 1 square yard (sq. yd.)
30$\frac{1}{4}$ square yards	= 1 square rod (sq. rd.)
160 square rods	= 1 acre (A.)
640 acres	= 1 square mile (sq. mi.)

1 sq. mi. = 640 A. = 102,400 sq. rd. = 3,097,600 sq. yd. = 27,878,400 sq. ft. = 4,014,489,600 sq. in.
A **square** is 100 sq. ft.
In measuring land, other than city lots, the acre is the unit.
In measuring other surfaces, the **square yard** is the unit (except in roofing, where the square is used).

Cubic Measure

290. Cubic measure is used to measure the volume of solids and the contents or capacity of hollow bodies.

TABLE

1728 cubic inches (cu. in.)	= 1 cubic foot (cu. ft.).
27 cubic feet	= 1 cubic yard (cu. yd.).
24$\frac{3}{4}$ cubic feet	= 1 perch (P.).
128 cubic feet	= 1 cord (cd.).
1 cubic yard (of earth)	= 1 load.

A **perch** of stone or masonry is 16$\frac{1}{2}$ ft. (1 rd.) long, 1$\frac{1}{2}$ ft. wide, and **1 ft.** high.
A **cord** of wood is a pile 8 ft. long, 4 ft. wide, and 4 ft. high.
A **cubic foot** of water weighs 62$\frac{1}{2}$ lb. (Avoir.).

APPROXIMATE MEASURES

A ton of timothy hay in a well-settled mow is about 450 cu. ft.
A ton of clover hay in a well-settled mow is about 550 cu. ft.
A ton of stove coal is about 34$\frac{1}{2}$ cu. ft.

NOTES. Tables for surveyors' linear and square measure are omitted because they are going out of use. Surveyors now use, instead of the Gunter's chain, a steel tape 100 ft. long, divided into feet and tenths of a foot (sometimes into inches). Measurements are more nearly accurate by the tape because there are no joints to wear as in the chain. Areas are measured by finding the number of square feet and dividing by 43,560, the number of square feet in an acre.

MEASURES OF CAPACITY

Liquid Measure

291. Liquid measure is used for measuring liquids, and in estimating the capacity of cisterns, tanks, reservoirs, etc.

TABLE

$$4 \text{ gills} \quad (\text{gi.}) = 1 \text{ pint (pt.)}$$
$$2 \text{ pints} \qquad = 1 \text{ quart (qt.)}$$
$$4 \text{ quarts} \qquad = 1 \text{ gallon (gal.)}$$
$$1 \text{ gal.} = 4 \text{ qt.} = 8 \text{ pt.} = 32 \text{ gi.}$$

Technically, the barrel contains $31\frac{1}{2}$ gal. In practice, barrels are of various sizes, as are also hogsheads, pipes, butts, etc. The capacity of each is marked upon it.

The unit of liquid measure is the wine gallon of 231 cu. in. A gallon of water weighs about $8\frac{1}{3}$ lb. (Avoir.).

The imperial gallon of England contains 277.274 cu. in. and is equal, very nearly, to $1\frac{1}{5}$ wine gallons. It contains 10 lb. of water.

Apothecaries' Fluid Measure

292. Apothecaries' fluid measure is used by druggists in prescribing and compounding liquid medicines.

TABLE

$$60 \text{ minims (m.)} = 1 \text{ fluid drachm (f}\!\!\!\!\!\!\!\text{3)}$$
$$8 \text{ fluid drachms} = 1 \text{ fluid ounce (f}\!\!\!\!\!\!\!\text{℥)}$$
$$16 \text{ fluid ounces} = 1 \text{ pint (O)}$$
$$8 \text{ pints} \qquad = 1 \text{ gallon (Cong.)}$$
$$1 \text{ Cong.} = 8 \text{ O} = 128 \text{ f℥} = 1024 \text{ f3} = 61{,}440 \text{ m.}$$

The gallon of this measure is the same as the wine gallon.

Dry Measure

293. Dry measure is used for measuring grain, fruit, vegetables etc.

TABLE

```
2 pints   (pt.) = 1 quart (qt.)
8 quarts        = 1 peck (pk.)
4 pecks         = 1 bushel (bu.)
1 bu. = 4 pk.   = 32 qt. = 64 pt.
```

The unit of dry measure is the **Winchester** bushel, which contains 2150.42 cu. in. It is a hollow cylinder 18½ in. in diameter and 8 in. deep.

The Winchester bushel is used in measuring grain, sand, etc.

The heaped bushel of 2747.71 cu. in. is used for measuring fruits, vegetables, and other coarse articles.

The imperial bushel of England contains 2218.192 cu. in. It is exactly 8 times the imperial gallon.·

Liquid and Dry Measures Compared

	GALLON	QUART	PINT
Liquid . . .	231 cu. in.	57¾ cu. in.	28⅞ cu. in.
Dry	268⅘ cu. in. (½ peck)	67¼ cu. in.	33¾ cu. in.

Weight

294. Weight is the measure of the earth's attraction of matter on its surface.

295. There are four kinds of weight in use in the United States: Troy, Apothecaries', Avoirdupois, and Metric.*

296. The **Troy pound** is the standard unit of weight in the United States.

Troy Weight

297. Troy weight is used in weighing gold, silver, diamonds, and other precious minerals. It is used by the government in weighing coins at the mint, by the jewelry trade and manufactures, and by importers and exporters of gold and silver. In the jewelry trade and manufacturing industries where gold, silver, and platinum are employed, a thorough knowledge of the Troy units and their comparisons is important.

* Metric tables are given on pages 132–135.

TABLE

24 grains (gr.)   = 1 pennyweight (pwt.)
20 pennyweights = 1 ounce (oz.)
12 ounces   = 1 pound (lb.)
   1 lb. = 12 oz. = 240 pwt. = 5760 gr.

The term "carat" (or karat) has two meanings. In weighing dia-
monds, it is a denomination of weight, and is equal to 3.168 gr.
Its second use is to denote the fineness of gold, and means $\frac{1}{24}$ part.
Gold marked 18K (18 carats) is $\frac{18}{24}$, by weight, pure gold and $\frac{6}{24}$ alloy.

Apothecaries' Weight

298. Apothecaries' weight is used by druggists and physicians
in compounding and prescribing medicines.   .

TABLE

20 grains (gr.) = 1 scruple (sc. or ℈)
3 scruples   = 1 dram (dr. or ℨ)
8 drams   = 1 ounce (oz. or ℥)
12 ounces   = 1 pound (lb. or ℔)
1 lb. = 12 ℥ = 96 ℨ = 288 ℈ = 5760 gr.

Drugs and chemicals are bought and sold wholesale by avoirdupois
weight.

Avoirdupois Weight

299. Avoirdupois weight is used in weighing all sorts of coarse,
heavy articles.

TABLE

16 ounces (oz.)   = 1 pound (lb.)
100 pounds   = 1 hundredweight (cwt.)
20 hundredweight = 1 ton (T.)
   1 T. = 20 cwt. = 2000 lb. = 32,000 oz.

Long Ton

300. The long ton (or gross ton) contains 2240 lb. It is used in
the United States customhouse in determining the duty on merchandise taxed by the ton. Coal and iron are sold wholesale at the mines
by the long ton.

TABLE

16 ounces (oz.)   = 1 pound (lb.)
28 pounds   = 1 quarter (qr.)
4 quarters   = 1 hundredweight (cwt.)
20 hundredweight = 1 ton (T.)
1 T. = 20 cwt. = 80 qr. = 2240 lb. = 35,840 oz.

Comparison of Weights

	POUND	OUNCE	GRAIN
Troy	5760 gr.	480 gr.	1 gr.
Apothecaries'	5760 gr.	480 gr.	1 gr.
Avoirdupois	7000 gr.	437½ gr.	1 gr.

Note that the pound and the ounce of Troy and apothecaries' weights are alike, and that the grain is the same in all three kinds of weight.

301. TABLE SHOWING WEIGHTS IN POUNDS PER BUSHEL LEGALLY ESTABLISHED FOR CERTAIN PRODUCTS BY THE SEVERAL STATES AND (FOR CUSTOMS PURPOSES) BY CONGRESS

	U.S.	STATES	EXCEPTIONS
Barley	48	48	Ala., Ga., Ky., Pa., 47 ; Ariz., 45 ; Calif., 50.
Beans	60	60	Ariz., 55 ; N. H., Vt., 62.
Buckwheat	48	52	Calif., 40 ; Conn., Maine, Mass., Mich., Miss., N. Y., Pa., R. I., Vt., 48 ; Idaho, N. Dak., Okla., Oregon, S. Dak., Tex., Wash., 42 ; Ind., Kan., Minn., N. J., N. C., Ohio, Tenn., Wis., 50.
Clover seed	60	60	N. J., 64.
Corn (in ear) . . .	70	70	Miss., 72 ; Ohio, 68.
Corn (shelled) . . .	56	56	Mass., 50.
Corn meal	48	50	Ark., Fla., Ga., Ill., Miss., S. C., 48.
Oats	32	32	N. J., Va., 30.
Onions	57	57	Conn., Maine, Mass., Minn., N. Dak., Okla., S. Dak., Vt., 52 ; Fla., Tenn., 56 ; Ind., 48 ; Mich., 54 ; Ohio, 55 ; Pa., R. I., 50.
Peas	60	60	
Potatoes	60	60	Md., Pa., 56.
Rye	56	56	Calif., 54 ; Maine, 50.
Timothy seed . . .	45	45	Ark., 60 ; Okla., S. Dak., 42.
Wheat	60	60	

Other Common Measures

			POUNDS
Beef,	Barrel	. . . .	200
Butter,	Firkin	. . . .	56
Flour,	Barrel	. . . .	196
Nails,	Keg	. . . .	100
Pork,	Barrel	. . . .	200
Salt,	Barrel	. . . .	280

Circular or Angular Measure

302. Circular or angular measure is used in measuring angles or arcs of circles as applied to surveying, civil engineering, astronomical calculations, latitude, longitude, etc.

TABLE

60 seconds ($''$) = 1 minute ($'$)
60 minutes   = 1 degree ($°$)
360 degrees   = 1 circle (cir.)
1 cir. = $360° = 21,600' = 1,296,000''.$

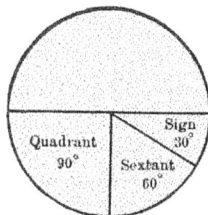

In astronomical calculations the circle is sometimes divided into signs, sextants, and quadrants.

In a circle are :   12 signs of 30° each.
6 sextants of 60° each.
4 quadrants of 90° each.

303. The unit of circular measure is the degree, and is equal to $\frac{1}{360}$ of the circumference of a circle.

304. On the earth's surface at the equator 1° of distance is equal to $69\frac{1}{6}$ statute miles, or 60 geographical miles or knots. Hence, 1 min. of distance equals 1 knot.

305. In measuring circles the length of a degree is dependent upon the size of the circle, but in measuring angles, the degree has a fixed value.

Time

306. Time is the measure of duration.

TABLE

60 seconds (sec.) = 1 minute (min.)	52 weeks = 1 year (yr.)
60 minutes = 1 hour (hr.)	12 months = 1 year
24 hours = 1 day (da.)	365 days = 1 common year
7 days = 1 week (wk.)	366 days = 1 leap year
30 days = 1 month (mo.)	100 years = 1 century

307. The months of the year.

January (Jan.)	31 days	July	31 days
February (Feb.)	28–9 days	August (Aug.)	31 days
March (Mar.)	31 days	September (Sept.)	30 days
April (Apr.)	30 days	October (Oct.)	31 days
May	31 days	November (Nov.)	30 days
June	30 days	December (Dec.)	31 days

The length of the year is determined by the time required for the earth to make one complete revolution around the sun. The exact time is 365 da. 5 hr., 48 min. 46 sec. (very nearly 365¼ da.).

It is because of this extra ¼ da. each year that we have a leap year once in 4 yr. But by allowing ¼ of a day each year in making a leap year each 4 yr., a little too much time is allowed, as the true year lacks a little of being 365¼ da. To correct the error, all centennial years (1800, 1900, 2100) not divisible by 400 are not leap years. All other years divisible by 4 are leap years.

The extra day of leap year is added to the month of February.

Standard Time

308. Since the earth makes a complete rotation of 360° in 24 hr., any point on the earth's surface passes through 15° of space in 1 hr. The earth rotates from west to east, and makes the sun appear to "rise" in the east and "move" toward the west. From New York to San Francisco is a distance of about 50°, or a difference in time of about 3⅓ hr. Hence the sun rises in New York 3⅓ hr. before it does in San Francisco. A person traveling from New York to San Francisco would find that his watch was too fast. It has seemed to gain an hour for every 15° of distance.

To remedy this condition, the principal railroads of the United States in 1883 agreed upon a system of reckoning time since known as " Standard Time." The territory of the United States was divided into four north and south belts. These belts lie approximately half and half on each side of the 75th, the 90th, the 105th, and the 120th meridans of longitude, respectively. It was agreed that the time for all places approximately 7½° either east or west of any one of these meridians should have the same time as that of the chief meridian for that belt, and that there should be a difference of 1 hr. in the time between any place in one belt and any place in an adjoining belt.

The time for all places in the belt having the meridian of 75° longitude west from Greenwich as the chief meridian is called Eastern Time ; for all places in the belt having the meridian of 90° as the chief meridian, as Central Time ; for the belt having the meridian of 105° as the chief meridian, as Mountain Time ; and for the belt in which the meridian of 120° is the chief meridian, as Western or Pacific Time.

The dividing line between any two adjoining time belts is not a straight north and south line, but is very irregular, so that the change in the time can be made at the points most convenient to the different railroads.

What is the difference in time between Boston and Chicago? Boston and Denver? New York and San Francisco? Charleston and Santa Barbara? Cincinnati and Denver?

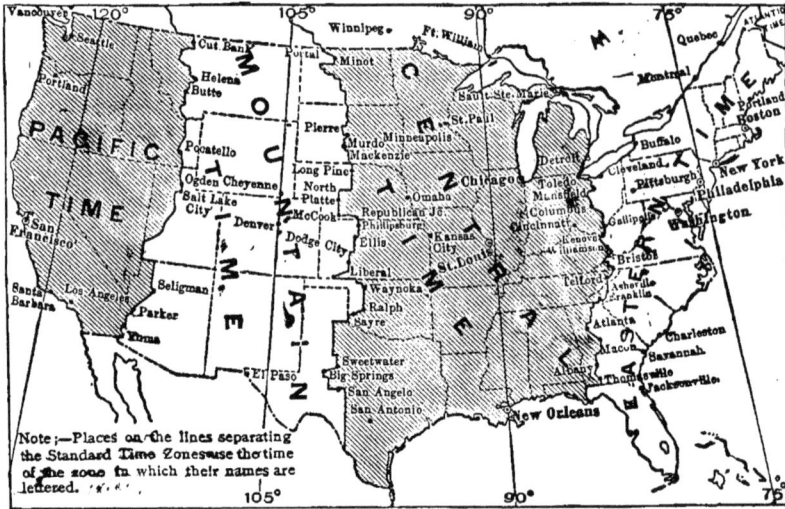

MAP OF STANDARD TIME BELTS

At what city in the United States is there no "Mountain Time"? If it is noon at San Antonio, what time is it at El Paso?

English Money

309. English money is the legal currency of Great Britain. There are gold, silver, and copper coins, and bills.

TABLE

4 farthings (far.)	= 1 penny (*d.*)
12 pence	= 1 shilling (*s.*)
20 shillings	= 1 pound, or sovereign (£)

The unit of English monetary value is the **sovereign**. It contains 113.0016 gr. of pure gold. Its value in United States money is, therefore, as many dollars as 23.22 gr. is contained times in 113.0016 gr., or $4.8665. (Called par of exchange or par value. See page 361.)

The gold coins are the sovereign, and the half sovereign. They are 22K (or $\frac{11}{12}$) fine.

The silver coins are crown (5s.), half crown, florin (2s.), sixpence, and three-pence. They are .925 fine.

The copper coins are the penny, halfpenny, and the farthing.

French Money

310. French money is the legal currency of France. It is a decimal currency.

TABLE

10 millimes (m.)	= 1 centime (c.)	
10 centimes	= 1 decime (dc.)	
10 decimes	= 1 franc (F.)	

The unit of French money is the **franc**, and is equal to **$.193** of United States money. The franc contains 4.4803 gr. pure gold.

Gold and silver coins of France are $\frac{9}{10}$ fine.

Gold coins are 5, 10, 50, and 100 franc pieces.

Silver coins are 25 and 50 centime pieces, and the 1, 2, and 5 franc pieces.

Bronze coins are 1, 2, 5, and 10 centime pieces.

F. 5.27 is read " 5 francs 27 centimes."

German Money

311. German money is the legal currency of the German Empire.

TABLE

100 pfennig (pf.) = 1 mark (M.)

The unit is the **mark**. Its value in United States money is $ .238.

There are gold, silver, nickel, and copper coins. The gold and silver coins are $\frac{9}{10}$ fine. The mark contains 5.5313 gr. of pure gold.

The gold coins are 5, 10, and 20 mark pieces.

The silver coins are 20 and 50 pfennig pieces and 1 and 2 mark pieces.

The nickel coins are 5 and 10 pfennig pieces.

The copper coins are 1 and 2 pfennig pieces

Miscellaneous Measures

312. Paper.

TABLE

24 sheets	= 1 quire (qr.)	
20 quires	= 1 ream (rm.)	
2 reams	= 1 bundle (bdl.)	
5 bundles	= 1 bale (bl.)	

313. Counting.

TABLE

20 units = 1 score
12 units = 1 dozen (doz.)
12 dozen = 1 gross (gro.)
12 gross = 1 great gross (gr **gro.**)

THE METRIC SYSTEM

314. The metric system is a decimal system of weights and measures. Nearly all the civilized nations of the world except the United States and England use the system.

315. The metric system was legalized in the United States in 1866, but has not been generally adopted except in scientific work.

316. The fundamental unit of the system is the meter, because every other unit of measure or weight is based on it.

317. The length of the meter was determined by taking one ten-millionth of the distance from the equator to the pole. Its length is 39.37 in.

318. The advantages of the metric system are:

(1) "The decimal relation between the units;
(2) "The extremely simple relation of the units of length, area, volume, and weight to one another;
(3) "The uniform and self-defining names of units."

319. The primary units are the following:

For length, meter = 39.37 inches.
For capacity, liter = .908 dry quarts; 1.0567 liquid quarts.
For weight, gram = 15.432 grains.

320. There are three Latin, and four Greek, prefixes, as follows: The Latin prefixes are:

$$\text{milli} = \tfrac{1}{1000} \ (\text{millimeter} = \tfrac{1}{1000} \text{ meter})$$
$$\text{centi} = \tfrac{1}{100} \ \ (\text{centimeter} = \tfrac{1}{100} \text{ meter})$$
$$\text{deci} = \tfrac{1}{10} \ \ \ (\text{decimeter} = \tfrac{1}{10} \text{ meter})$$

The Greek prefixes are:

deca = 10	(decagram	= 10 grams)
hecto = 100	(hectogram	= 100 grams)
kilo = 1000	(kilogram	= 1000 grams)
myria = 10000	(myriagram	= 10000 grams)

321. The United States government requires the use of the metric system of measures in all medical work of the navy and the war departments, and in the public health and marine hospital service. Its use is obligatory in Porto Rico.

322. For postal purposes " fifteen grams . . . shall be the equivalent . . of one half ounce avoirdupois, and so on in progression."

At the mint $12\frac{1}{2}$ g. is the weight of a half-dollar. The quarter-dollar and the dime in proportion. (See Weights of Coins, page 11.)

Linear Measure

323. The unit of linear measure is the **meter.**

TABLE

10 millimeters (mm.)	= 1 centimeter (cm.)
10 centimeters	= 1 decimeter (dm.)
10 decimeter	= 1 meter (m.)
10 meters	= 1 decameter (Dm.)
10 decameters	= 1 hectometer (Hm.)
10 hectometers	= 1 kilometer (Km.)
10 kilometers	= 1 myriameter (Mm.)

NOTES. (1) The most commonly used denominations in the above and in the following tables are indicated by the heavy-faced type.

(2) It should be observed that the abbreviations of the Latin prefixes begin with small letters, and those of the Greek prefixes begin with capital letters.

Square Measure

324. The unit of square measure is the **square meter** for small areas, and the **are** of 100 sq. m. for land areas.

TABLE

100 square millimeters (sq. mm.)	= 1 square centimeter (sq. cm.)
100 square centimeters	= 1 square decimeter (sq. dm.)
100 square decimeters	= 1 square meter (sq. m.)
100 square meters	= 1 square decameter (sq. Dm.)
100 square decameters	= 1 square hectometer (sq. Hm.)
100 square hectometers	= 1 square kilometer (sq. Km.)

Land Measure

325. The unit of land measure is the **are.**

TABLE

100 centares (ca.)	= 1 are (a.)	= 100 sq. **m.**
100 ares	= 1 hectare (Ha.)	= 10000 sq. **m.**

Cubic Measure

326. The unit of volume is the cubic **meter.**

TABLE

1000 cubic millimeters (cu. mm.) = 1 cubic centimeter (cu. cm.)
1000 cubic centimeters = 1 cubic decimeter (cu. dm.)
1000 cubic decimeters = 1 cubic meter (cu. m.)

Table of Wood Measure

327. The unit of wood measure is the stere.

10 decisteres (ds.) = 1 stere   (s.) = 1 cu. m.
10 steres   = 1 decastere (Ds.) = 10 cu. m.

Measure of Capacity

328. The unit of capacity for either solids or liquids is the liter, which is equal in volume to 1 cu. dm.

TABLE

10 milliliters (ml.) = 1 centiliter (cl.)   10 liters   = 1 decaliter (Dl.)
10 centiliters   = 1 deciliter (dl.)   10 decaliters = 1 hectoliter (Hl.)
10 deciliters   = 1 liter (l.)   10 hectoliters = 1 kiloliter (Kl.)

For measuring liquids in ordinary quantities the liter is used. For minute quantities the centiliter or the milliliter is used.

For measuring grain, vegetables, etc., the hectoliter is used.

Measure of Weight

329. The unit of weight is the **gram,** which is the weight of 1 cu. cm. of distilled water in a vacuum, at its greatest density (39.2° F.).   It weighs 15.4324 gr.

TABLE

10 milligrams (m.g.) = 1 centigram (c.g.)   10 hectograms = 1 kilogram (Kg.)
10 centigrams   = 1 decigram (d.g.)   10 kilograms = 1 myriagram (Mg.)
10 decigrams   = 1 gram (g.)   10 myriagrams = 1 quintal (Q.)
10 grams   = 1 decagram (Dg.)   10 quintals   = 1 tonneau (T.)
10 decagrams   = 1 hectogram (Hg.)

NOTES.   (1) The gram is used in weighing precious metals, drugs, letters, etc.
(2) The kilogram is used for weighing ordinary merchandise, except in large quantities, for which the quintal or tonneau is used.

330. The legal equivalent values of the units of both the English and the metric systems are given in the following:

TABLES OF EQUIVALENTS
Linear Measure

1 inch = 2.54 centimeters	1 centimeter = .3937 of an inch
1 foot = .3048 of a meter	1 decimeter = .328 of a foot
1 yard = .9144 of a meter	1 meter = 1.0936 yards
1 rod = 5.029 meters	1 dekameter = 1.9884 rods
1 mile = 1.6093 kilometers	1 kilometer = .62137 of a mile

Surface Measure

1 sq. inch = 6.452 sq. centimeters	1 sq. centimeter = .155 of a sq. inch
1 sq. foot = .0929 of a sq. meter	1 sq. decimeter = 1076 of a sq. foot
1 sq. yard = .8361 of a sq. meter	1 sq. meter = 1.196 sq. yards
1 sq. rd. = 25.293 sq. meters	1 are = 3.954 sq. rods
1 acre = 40.47 ares	1 hectare = 2.471 acres
1 sq. mile = 259 hectares	1 sq. kilometer = .3861 of a sq. mile

Cubic Measure

1 cu. inch = 16.387 cu. centimeters	1 cu. centimeter = .061 of a cu. inch
1 cu. foot = 28.317 cu. decimeters	1 cu. decimeter = .0353 of a cu. foot
1 cu. yard = .7646 of a cu. meter	1 cu. meter = 1.308 cu. yards
1 cord = 3.624 steres	1 stere = .2759 of a cord

Measures of Capacity

1 dry quart = 1.101 liters	1 liter = .908 of a dry quart
1 liquid quart = .9463 of a liter	1 liter = 1.0567 liquid qt.
1 liquid gallon = .3785 of a decaliter	1 decaliter = 2.6417 liquid gal.
1 peck = .881 of a decaliter	1 decaliter = 1.135 pecks
1 bushel = .3524 of a hectoliter	1 hectoliter = 2.8377 bushels

Measures of Weight

1 grain, Troy = .0648 of a gram	1 gram = 15.432 grains, Troy
1 ounce, Troy = 31.104 grams	1 gram = .03215 of an oz. Troy
1 ounce, avoir. = 28.35 grams	1 gram = .03527 of an oz. avoir
1 lb., Troy = .3732 of a kilogram	1 kilogram = 2.679 lb. Troy
1 lb., avoir. = .4536 of a kilogram	1 kilogram = 2.2046 lb. avoir.
1 ton (short) = .9072 of a tonneau or ton	1 tonneau = 1.1023 tons (short)

Convenient Equivalent Values

1 cu. cm. of water = 1 ml. of water, and weighs 1 gram	= 15.432 gr.
1 cu. dm. of water = 1 l. of water, and weighs 1 Kg.	= 2.2046 lb.
1 cu. m. of water = 1 Kl. of water, and weighs 1 tonneau	= 2204.6 lb.

REDUCTION OF DENOMINATE NUMBERS

331. Reduction of denominate numbers consists in changing the form or denomination of a number without changing its value.

332. Reduction descending consists in changing denominate numbers from higher to lower denominations.

333. Reduction ascending consists in changing denominate numbers from lower to higher denominations.

REDUCTION DESCENDING

334. Reduce 4 lb. 8 oz. 16 pwt. 14 gr. to grains.

4×12 oz. $= 48$ oz.

48 oz. $+ 8$ oz. $= 56$ oz.

56×20 pwt. $= 1120$ pwt.

1120 pwt. $+ 16$ pwt. $= 1136$ pwt.

1136×24 gr. $= 27,264$ gr.

$27,264$ gr. $+ 14$ gr. $= 27,278$ gr.

There are 12 oz. in 1 lb., and in 4 lb. there are 4 times 12 oz., or 48 oz., to which are added the 8 oz., making 56 oz.

In each ounce are 20 pwt., and in 56 oz. there are 56 times 20 pwt., or 1120 pwt. Adding the 16 pwt. makes 1136 pwt.

In each pennyweight are 24 gr., and in 1136 pwt. there are 1136 times 24 gr., or 27,264 gr. Adding the 14 gr. gives 27,278 gr., the desired result.

PROBLEMS

335. Reduce to lower denominations:

1. 4 bu. 2 pk. 6 qt. 1 pt.
2. 8 rd. 4 yd. 2 ft. 8 in.
3. 16 T. 1425 lb.
4. 6 da. 14 hr. 27 min. 38 sec.
5. 15 cu. yd. 21 cu. ft. 1276 cu. in.
6. 4 Dm. 8 m. 7 dm. 5 cm.
7. 5 Kg. 7 Hg. 4 Dg. 9 g.
8. 7 Hl. 5 Dl. 3 l.
9. 15 sq. rd. 26 sq. yd. 7 sq. ft. 86 sq. in.

10. A man buys 2 bu. 2 pk. of chestnuts at $1.50 per bushel and sells them at 5 ¢ a pint. How much does he gain?

11. A cask of molasses containing 48 gal. cost 32 ¢ a gallon and sold for 15 ¢ a quart. Find the gain.

12. If a drug costs $1.50 per kilogram and sells for 5 ¢ a decagram, what is the profit?

13. If a kilogram of quinine costs $8' and is put up into powders of 1 decigram each, and sold at the rate of a dozen powders for 10 ¢, find the profit.

14. Reduce $\frac{3}{5}$ of a day to hours, minutes, etc.

$\frac{3}{5} \times 24$ hr. $= \frac{72}{5}$ hr. $= 14\frac{2}{5}$ hr.

$\frac{2}{5} \times 60$ min. $= 24$ min.

14 hr. 24 min.

In $\frac{3}{5}$ of a day there are $\frac{3}{5}$ of 24 hr., or 14$\frac{2}{5}$ hr. In $\frac{2}{5}$ of an hour there are $\frac{2}{5}$ of 60 min., or 24 min. Hence, $\frac{3}{5}$ of a day $=14$ hr. 24 min.

15. Reduce $\frac{5}{8}$ of a bushel to lower denominations.

16. Reduce $\frac{3}{7}$ of a pound Troy to lower denominations.

17. Reduce $\frac{2}{3}$ of an acre to lower denominations.

18. A man bought $\frac{5}{8}$ of an acre of land and sold it in building lots of 20 sq. rd. each. How many lots did he sell?

19. A fence is to be $\frac{3}{8}$ of a mile long. How many boards 12 ft. long will be required if the fence is 4 boards high?

REDUCTION ASCENDING

336. Reduce 113,628 sq. in. to higher denominations.

144)113628

    9)789, number of square feet —12 sq. **in.**

   30$\frac{1}{4}$)87, number of square yards —6 sq. ft.

      2, number of square rods —26$\frac{1}{2}$ sq. yd.

2 sq. rd. 26($\frac{1}{2}$) sq. yd. 6 sq. ft. 12 sq. in.

                 $\frac{1}{2}$ sq. yd. $= 4$ sq. ft. 72 sq. in.

2 sq. rd. 27   sq. yd. 1 sq. ft. 84 sq. in.

First, divide by 144, the number of square inches in a square foot; the result is 789 sq. ft. and 12 sq. in. over. Next divide 789 by 9, the number of square feet in a square yard; the result is 87 sq. yd., and 6 sq. ft. over; then divide 87 by 30$\frac{1}{4}$, the number of square yards in a square rod; the quotient is 2 sq. rd., with a remainder of 26$\frac{1}{2}$ sq. yd.

Arrange the last quotient and the several remainders as shown in the illustration. To eliminate the $\frac{1}{2}$ sq. yd. change it to square feet and square inches, and add it to the 6 sq. ft. 12 sq. in., as illustrated in the solution. The final result is 2 sq. rd. 27 sq. yd. 1 sq. ft. 84 sq. in.

PROBLEMS

337. Reduce to higher denominations.

1.   2489 gills.

2. 15,729 grains (Troy).

3. 198,726 seconds.

4. 48,729 square inches.

5. 139,287 inches.

6.   4372 minutes.

7. 38,476 seconds (of longitude).

8. 48,792 centigrams.

9. 53,986 deciliters.

10. 58,371 millimeters.

11. Reduce 2 sq. ft. 36 sq. in. to the fraction of a square yard.

144)36.00 number of sq. in.   Divide the 36 sq. in. by 144 (number
  9) 2.25 number of sq. ft.   of square inches in a square foot), which
     .25 number of sq. yd.   gives .25 of a square foot. Add the 2 sq.
                                 ft. and divide by 9 (the number of square
feet in a square yard). The result is .25 of a square yard.

12. Reduce 2 ft. 8 in. to the fraction of a yard.

13. Reduce 5 oz. 8 pwt. 12 gr. to the fraction of a pound.

14. Reduce 5 hr. 48 min. to the fraction of a day.

15. Reduce 4 cwt. 50 lb. to the fraction of a ton.

16. Reduce 48 sq. rd. 21 sq. yd. 7 sq. ft. to the fraction of an acre.

17. Reduce 4 hr. 30 min. to the fraction of a day of 8 hr.

18. A laborer worked 6 hr. 45 min. If a working day is 9 hr., what part of the day did he work?

19. A garden contains 64 sq. rd 4 sq. ft 72 sq. in. What part of an acre is it?

COMPARISON OF MEASURES

Weight

338. For comparative tables of weights, see pages 127 and 135.

339. 1. Which is heavier, a pound avoirdupois or a pound Troy? an ounce avoirdupois or an ounce Troy? Explain.

2. How do the pound and ounce of Troy and apothecaries' weights compare?

340. How many pounds Troy are equal to 12 lb. avoirdupois?

$$\frac{12 \times 7000}{5760} = \frac{175}{12} = 14\tfrac{7}{12} \text{ lb. Troy.}$$

In one pound avoirdupois there are 7000 gr.; in 12 lb. there are 12 times 7000 gr. Dividing by 5760 gr. in one pound Troy gives the number of Troy pounds, which is $14\tfrac{7}{12}$.

The cancellation form of solution is best adapted to these problems.

PROBLEMS

341. 1. Change 16 lb. avoirdupois to Troy pounds.

2. Change 25 lb. Troy to avoirdupois pounds.

3. Reduce 22 lb. avoirdupois to apothecaries' pounds.

4. How much is gained per pound by buying a drug at $3.60 per avoirdupois pound, and selling it at 10¢ per dram ?

5. A druggist buys medicine at $1.80 per pound avoirdupois, and sells it at 20¢ an ounce apothecaries' weight. Find his gain on 5 lb.

6. A jeweler buys 3 lb. 8 oz. of gold and manufactures it into 64 watch chains. What is the average weight of the chains ?

7. How many pounds are there in 26 Kg. ?

26 × 2.2046 lb. = 57.3196 lb.   A kilogram weighs 2.2046 lb., and 26 Kg. weigh 26 times 2.2046 lb., or 57.3196 lb.

8. Reduce 20 Kg. to pounds.

9. Reduce 45 Kg. 5 Hg. to pounds.

10. Reduce 27 lb. to kilograms.

11. Reduce 48 lb. 8 oz. to kilograms.

12. An importation of sugar weighed **12,425 Kg.** What was its value at $4\frac{1}{2}$ ¢ a pound ?

13. How many rings of 4 pwt. each can be made from 4 Kg. of gold ?.

14. How many grams of pure gold will be required to make 24 gold rings 18 karats fine each weighing 5 pwt. 8 gr. ?

15. .Find the gain on a carload of coal weighing 24 long tons bought at $4.75 per ton, and sold at $6 per short ton.

MONEY

342. In dealing with other nations it is necessary to reduce foreign money to United States money, and *vice versa.*

343. 1. Change £ 228 16s. 8d. to United States money.

16s. ÷ 20s. = £ .8.

228.8 × $4.8665 = $1113.46.

8d. = 16 ¢.

$ 1113.46 + 16 ¢ = $1113.62.

First reduce the 16s. to the decimal of a pound by dividing by 20, and add the result, £ .8, to the £ 228, making £ 228.8. Multiply the value of £ 1, $4.8665, by 228.8, and the result, $ 1113.46, is the value in United States money of £ 228 16s. Each penny of English money is worth two cents of our money; 8d. is worth 16 ¢. Adding $ 1113.46 and $ 16 gives $1113.62, the value of £ 228 16s. 8d. in United States money.

2. Change $683.62 to English money.

$683.62 ÷ $4.8665 = £140.4748.

.4748 × 20s. = 9.496s.

.496 × 12d. = 5.952d. = 6d.

Therefore £140 9s. 6d.

Divide the $683.62 by the value of £1, $4.8665. The result, £140.4748, is the value in pounds and decimal of a pound. To change the decimal part of a pound to shillings, multiply by 20, the number of shillings in a pound. The result is 9.496s. To change the decimal part of a shilling to pence, multiply by 12, the number of pence in a shilling. The result is £140 9s. 6d.

3. Change F. 7284 to United States money.

$1.00 ÷ $.193 = 5.18⅛

7284 ÷ 5.18⅛ = $1405.83

In changing francs to dollars, bankers reckon on the number of francs equal to one dollar. By dividing $1 by $.193, the value of F. 1, the number of francs equal to $1 is found to be 5.18⅛ (very nearly). If it takes F. 5.18⅛ to equal $1, F. 7284 will equal as many dollars as 5.18⅛ is contained times in 7284 or $1405.83.

4. Change $575 to francs if exchange is at 5.18¾.

575 × 5.18¾ = F. 2982.81

The number of francs equal to $1 varies with market conditions. The expression "at 5.18¾" means the number of francs to the dollar. If $1 is equal to F. 5.18¾, $575 is equal to 575 times F. 5.18¾, or F. 2982.81.

5. Change M. 760 to dollars.

190

$$\frac{760 \times \$.952}{4} = \$180.88.$$

In changing marks to dollars, or dollars to marks, the price of 4 marks is stated instead of the price of 1 mark. If no price is stated, the par value is meant. Since 1 mark is worth $.238, 4 marks are worth 4 × $.238, or $.952. To find the value of 760 marks, multiply by $.952 and divide by 4. (Divide by 4 because $.952 is the price of 4 marks.)

6. Change $875 to marks if exchange is at 96¼.

7

35

$$\frac{875 \times 4}{.9625} = 3636.36,$$ num-

.9625

.0385

.0077

.0011

ber of marks.

In changing dollars to marks the preceding operation has to be reversed. $.96¼ is the price of 4 marks. Dividing 875 by .96¼ would give only ¼ as many marks as is desired. Hence, multiply by 4 to obtain the correct number. (Prin. 17 and 19, page 102.)

NOTE. The price of foreign money, i.e. the rate of exchange, varies in accordance with the laws of supply and demand, just as the price of wheat, potatoes, or clothing varies.   When the demand for, and the supply of, foreign money are equal, the price is said to be at par.

The prices, or rates of exchange, at par are for

English money . . . . $4.8665 = £1
French money . . . . $1.00 = F. 5.18⅛
German money . . . . $ .952 = M. 4

In the following problems the **par** value is to be used unless a different value is stated.

PROBLEMS

344. **1.** Reduce £ 147 7*s.* 5*d.* to dollars.

  2. Reduce £ 325 18*s.* 7*d.* to dollars.

  3. Reduce £ 458 15*s.* 9*d.* to dollars.

  4. Reduce $ 685 to English money.

  5. Reduce $ 796 to English money.

  6. Reduce $ 1200 to English money

  7. Reduce $ 1825.50 to English money at 4.84.

  8. Reduce £1650 12*s.* 6*d.* to dollars at 4.86.

  9. Reduce £1764 13*s.* 4*d.* to dollars at 4.885.

  10. Reduce $ 3487.64 to English money at 4.8625.

  11. Reduce F. 598 to dollars.

  12. Reduce F. 1285 to dollars.

  13. Reduce F. 3484 to dollars.

  14. Reduce $ 1593 to francs at 5.18⅛.

  15. Reduce $ 3327.75 to francs at 5.19⅜.

  16. Reduce $ 5897.50 to francs at 5.17½.

  17. Reduce F. 7500 to dollars at 5.18¾.

  18. Reduce F. 6488.5 to dollars at 5.19⅜.

  19. Reduce $ 3971.54 to francs at 5.18⅛.

  20. Reduce $ 4782.35 to francs at 5.20.

  21. Reduce M. 1372 to dollars.

  22. Reduce M. 2468 to dollars at **.96**.

  23. Reduce M. 3896 to dollars.

  24. Reduce $ 4385 to marks at .955.

  25. Reduce $ 3681.50 to marks.

  26. Reduce $ 1782.75 to marks.

  27. Reduce M. 2342.50 to dollars at **.95**.

  28. Reduce M. 3289.75 to dollars at .96¼.

29. Reduce $4987.75 to marks at 94⅞.

30. Reduce $5983.40 to marks at 95.

Find the missing values in the following table:

	VALUE IN UNITED STATES MONEY	VALUE IN ENGLISH MONEY	VALUE IN FRENCH MONEY	VALUE IN GERMAN MONEY
31.	$1699.50			
32.		£ 965 18s. 10d.		
33.			F. 14852	
34.				M. 9875
35.			F. 8964	
36.				M. 12978
37.	$13878.75			
38.		£ 1264 5s. 1d.		
39.			F. 12379	
40.		£ 1585 9s. 8d.		

MISCELLANEOUS MEASURES

345. For comparison of measures see pages 123 to 125.

346. Find the weight of a barrel of water.

$$\frac{\overset{3.5}{\cancel{31.5}} \times \overset{77}{\cancel{231}} \times 62.5 \text{ lb.}}{\underset{\underset{64}{576}}{\cancel{1728}}} = \frac{16843.75 \text{ lb.}}{64} = 263.18 \text{ lb.}$$

A barrel is 31½ gal. A gallon contains 231 cu. in. 31½ times 231 gives the number of cubic inches in a barrel. Dividing by 1728 cu. in. to the cubic foot gives the number of cubic feet in a barrel. Multiplying by 62½ lb. to the cubic foot gives the weight of a barrel of water as 263.18 lb.

PROBLEMS

347. 1. Find the weight of 25 gal. of water.

2. How much does a 10-qt. pail of water weigh?

3. A can of water weighs 45 lb. 10 oz. If the can weighs 8 lb., find the quantity of water in gallons.

4. How many liquid gallons are equal in volume to 24 dry gallons?

5. If a pail contains 10 qt. of berries, how many quarts of milk will it hold?

6. A merchant bought cloth at 22 ¢ per meter and sold it at 30 ¢ per yard. Find his gain on 160 m.

$$\frac{\overset{40}{\cancel{160}} \times 39.37}{\underset{9}{\cancel{36}}} = \frac{1574.8}{9} = 174.97\tfrac{7}{9}, \text{ no. yd.}$$

160 × $.22 = $ 35.20, cost.

174.97$\tfrac{7}{9}$ × $.30 = $ 52.49, selling price.

$ 52.49 − $ 35.20 = $ 17.29, gain.

Change the 160 m. to yards by multiplying 160 by 39.37 in. in a meter, and dividing by 36 in. in a yard. The number of yards is 174.97$\tfrac{7}{9}$.

The gain is found by taking the difference between the cost and the selling price.

7. Change 48 m. to yards.

8. Change 120 m. to yards.

9. Change 45 Km. to miles, rods, etc.

10. Change 12 mi. to kilometers.

11. An express train in France runs at a speed of 76 Km. per hour. In the United States an express train runs 50 mi. per hour. Which train makes the better time, and how many miles per hour?

12. If cloth costs 40 ¢ per meter and sells for 50 ¢ a yard, find the gain on 348 m.

13. Find the difference between 384 m. and 406 yd.

14. How many square yards are equal to 450 sq. m. ?

15. Change 12 gal. to liters.

12 × 4 qt. = 48 qt.

48 ÷ 1.0567 = 45.42, no. liters.

12 gal. = 48 qt. 1 l. = 1.0567 qt. In 48 qt. there are as many liters as 1.0567 is contained times in 48, or 45.42 l.

16. How many decaliters are there in a barrel of vinegar?

17. A farmer is offered $ .90 a bushel, or $ 2.50 a hectoliter, for his wheat. Does he gain or lose, and how much, on 480 bu. by accepting $ .90 a bushel?

18. How many kilograms are there in a barrel of flour (196 lb.) ?

19. One wholesaler offers sugar at 5 ¢ per pound, and another at 11 ¢ per kilogram Which offer is the better, and how much, for the purchaser, on 20 bbl. averaging 280 lb. each?

20. Reduce 4 A. 28 sq. rd. to ares.

21. In excavating for a basement, 430 cu. yd. of earth were removed. Find the cost at $ 2.50 per cubic meter.

22. How many bushels per acre are equivalent to 40 Hl. of wheat to the hectare ?

ADDITION OF DENOMINATE NUMBERS

348. Add: 3 A. 57 sq. rd. 22 sq. yd. 7 sq. ft. 68 sq. in., 5 A. 100 sq. rd. 17 sq. yd. 5 sq. ft. 120 sq. in., 7 A. 72 sq. rd. 28 sq. yd. 8 sq. ft. 64 sq. in.

A.	sq. rd.	sq. yd.	sq. ft.	sq. in.
3	57	22	7	68
5	100	17	5	120
7	72	28	8	64
16	231	69	21	252
16	71	8 ($\frac{1}{2}$)	3	108
		$\frac{1}{2}$ =	4	72
			8	· 180
16	71	8	8	36

Arrange the addends so that like units are in the same column. Add the lowest denomination first, and write the result, 252, as illustrated. 252 sq. in. equal 1 sq. ft. and 108 sq. in. as a remainder. Write the 108 beneath the 252 and carry the 1 sq. ft. to the next column and add, obtaining 21, which write as shown. Reducing the 21 sq. ft. to square yards gives 2 sq. yd. with a remainder of 3 sq. ft. Write 3 and carry the 2 to the next column. The sum is 69 sq. yd., which, divided by 30¼, the number of square yards in a square rod, gives 2 sq. rd., and 8½ sq. yd. remaining. Write 8½ and carry 2 to the square rod column. The sum is 231. Dividing 231 sq. rd. by 160, the number of square rods in an acre, gives 1 A. and 71 sq. rd. over. Write 71 and carry 1 to the next column, obtaining 16 as the sum.

It is desired to have the answer free from fractional units except in the lowest denomination. To eliminate the ½ sq. yd. reduce it to square feet and square inches, writing the result, 4 sq. ft. 72 sq. in., as illustrated. Now add, obtaining 180 sq. in., which is equal to 1 sq. ft. and 36 sq. in. Write the remainder (36 sq. in.) as before, and carry the 1 sq. ft., making 8 sq. ft. "Bring down" the other results. The completed answer is 16 A. 71 sq. rd. 8 sq. yd. 8 sq. ft. 36 sq. in.

PROBLEMS

349. 1. Add: 5 rd. 4 yd. 2 ft. 6 in., 11 rd. 2 yd. 1 ft. 8 in., 7 rd. 3 yd. 10 in., 12 rd. 1 yd. 2 ft.

2. Add: 5 lb. 4 oz. 16 pwt. 12 gr., 2 lb. 9 oz. 10 pwt. 14 gr., 9 lb. 8 oz. 18 pwt. 16 gr., 7 lb. 5 oz. 12 gr.

3. Add: 5 A. 44 sq. rd. 15 sq. yd. 7 sq. ft. 98 sq. in., 13 A. 58 sq. rd. 18 sq. yd. 7 sq. ft. 56 sq. in., 9 A. 68 sq. rd. 24 sq. yd. 5 sq. ft., 17 A. 120 sq. rd. 22 sq. yd. 106 sq. in.

4. A broker sold the following bills of exchange on London: £ 138 16s. 8d., £ 424 7s. 9d., £ 227 15s. 10d., £ 650 12s. 7d.   Find the total of his sales.

5. A rectangular field is 26 rd. 2 yd. 1 ft. long, and 18 rd. 4 yd. 2 ft. wide.   What length of wire will be required for a fence five wires high ?

6. A farmer sold 4 loads of potatoes containing 48 bu. 24 lb., 51 bu. 26 lb., 53 bu. 44 lb., and 43 bu. 50 lb. respectively.   What was the total value at 75 ¢ a bushel?

7. A milk dealer bought 6 gal. 2 qt. 1 pt. of milk from one man, 8 gal. 3 qt. from another, and 12 gal. 2 qt. from a third.   At 12 ¢ a gallon, what did it cost him ?

8. Add: 8 Km. 7 Hm. 5 m. 6 dm., 5 Km. 4 Dm. 4 m., 3 Km. 7 m. 9 dm.

9. Find the sum of 7 Kg. 5 Hg. 4 Dg. 7 g., 9 Kg. 8 Dg. 4 g. 5 dg., 7 Kg. 8 Hg. 3 g. 8 dg., 2 Hg. 5 Dg. 6 g. 8 dg.

10. Three casks of molasses contain 1 Hl. 2 Dl. 5 l., 9 Dl. 8 l. 7 dl., and 1 Hl. 3 Dl. 1 l. 3 dl. respectively.   Find the total amount in the three casks.

11. A room is 4 m. 6 dm. 5 cm. long, and 3 m. 4 dm. 8 cm. wide. How long a border is required to reach around it ?

12. Four fields contain 8 Ha. 54 a. 63 ca., 16 Ha. 27 a. 85 ca., 24 Ha. 49 a. 50 ca. and 36 Ha. 72 a. 65 ca. respectively.   Find the total area.

13. Add $\frac{1}{4}$ A. $16\frac{1}{2}$ sq. rd. $4\frac{1}{4}$ sq. yd. and $\frac{1}{4}$ sq. ft.

$\frac{1}{4}$ A. $= 40$ sq. rd.			
$16\frac{1}{2}$ sq. rd. $= 16$ sq. rd.	15 sq. yd.	1 sq. ft.	18 sq. in.
$4\frac{1}{4}$ sq. yd. $=$ -	4 sq. yd.	2 sq. ft.	36 sq. in.
$\frac{1}{4}$ sq. ft. $=$			72 sq. in.
· 56 sq. rd.	19 sq. yd.	3 sq. ft.	126 sq. in.

Reduce each of the fractional units to lower denominations (Art. 335). Add the results as already explained.

14. Add $\frac{5}{8}$ bu. $2\frac{1}{2}$ pk. $6\frac{1}{2}$ qt.   **17.** Add .65 lb. .45 oz. .25 dr. .5 sc.

15. Add $\frac{3}{4}$ rd. $\frac{1}{2}$ yd. $1\frac{1}{2}$ ft.   **18.** Add .76 m. .38 dm. .56 cm.

16. Add $\frac{5}{8}$ lb. $\frac{3}{4}$ oz $\frac{4}{5}$ pwt.   **19.** Add 9 Hl. .7 Dl. 6 l. .15 dl.

SUBTRACTION OF DENOMINATE NUMBERS

350. From 16 rd. 4 yd. 2 ft. 4 in. take 12 rd. 5 yd. 2 ft. 8 in.

rd.	yd.	ft.	in.
16	4	2	4
12	5	2	8
3	$3(\frac{1}{2})$	2	8
	$\frac{1}{2} = 1$		6
3	4	1	2

Arrange the numbers with like denominations in the same column. Since 8 in. cannot be taken from 4 in., "borrow" 1 ft., reduce it to inches, and add to the 4 in. $(12 + 4 = 16)$. Now take 8 in. from 16 in. leaving 8 in. Write 8 in the remainder. 1 ft. was "borrowed," leaving 1 ft. 2 ft. cannot be taken from 1 ft. "Borrow" 1 yd., reduce to feet and add to the 1 ft. $(3 + 1 = 4)$. 2 from 4 leaves 2 for the remainder. Next borrow 1 rd., reduce and add to the yards $(5\frac{1}{2} + 3 = 8\frac{1}{2})$. 5 from $8\frac{1}{2}$ leaves $3\frac{1}{2}$ as a remainder. 12 from 15 leaves 3. To eliminate the $\frac{1}{2}$ yd. in the remainder, reduce it to feet and inches, $(\frac{1}{2}$ yd. $= 1$ ft. 6 in.), and add as illustrated.

PROBLEMS

351. Subtract:

da.	hr.	min.	sec.
1. 5	10	26	32
3	12	48	16

bu.	pk.	qt.	pt.
2. 16	3	4	0
8	3	6	1

£	s.	d.	far.
3. 28	11	6	2
14	14	8	3

cd.	cu. ft.
4. 28	48
14	72

Km.	Hm	Dm.	m.
5. 8	4	3	5
6	7	5	9

cu. m.	cu dm.	cu. cm.
6. 716	127	243
285	458	759

7. From a cask containing 56 gal. 2 qt. 1 pt. of vinegar, 27 gal. 3 qt. were drawn out. How much vinegar remained?

8. A field contains 38 A. 116 sq. rd. 18 sq. yd. 9 A. 28 sq. rd. are planted to corn. 6 A. 96 sq. rd. 20 sq. yd. are planted to pota- toes, and the remainder of the field is meadow. Find the area of the meadow.

9. A vat contains 116 Hl. 8 Dl. 5 l. of water. A tank contain- ing 34 Hl. 9 Dl. 7 l. is filled from the vat. What quantity of water remains in the vat?

10 From $\frac{3}{4}$ sq. yd. take $2\frac{1}{2}$ sq ft.

$\frac{3}{4}$ sq. yd. = 6 sq. ft	108 sq..in.
$2\frac{1}{2}$ sq. ft. = 2 sq. ft	72 sq. in.
4 sq. ft.	36 sq. in.

Reduce the fractional units to lower denominations (Art. 335), and subtract as before.

11. From $\frac{3}{4}$ bu. take $2\frac{1}{2}$ pk.   **13.** From .7 rd. take .45 yd.

12. From $\frac{5}{6}$ A. take $54\frac{1}{2}$ sq. rd.   **14.** From $\frac{1}{2}$ Kg. take $\frac{2}{5}$ Hg.

15. From $\frac{3}{4}$ sq. m. take $44\frac{1}{2}$ sq. dm.

16. From .76 cu. m. take 347.5 cu. cm.

17. From $\frac{3}{4}$ rd. take the sum of 2 yd. 1 ft. 8 in. and $\frac{3}{4}$ yd.

18. From a field containing $\frac{5}{6}$ of an acre a lot having an area of $10\frac{1}{2}$ sq. rd. was sold. Find the area of the part remaining.

19. From a quantity of sugar weighing 324 Mg. 8 Kg. 6 Hg. there was sold 128 Mg. 9 Kg. 7 Hg. How much remained? What was it worth at 11 ¢ per kilogram?

MULTIPLICATION OF DENOMINATE NUMBERS

352. Multiply 5 lb. 7 oz. 8 pwt. 15 gr. by 6.

lb.	oz.	pwt.	gr.
5	7	8	15
			6
33	44	51	90
33	8	11	18

Arrange the numbers as shown in the illustration. Begin with the lowest denomination. The first product is 90 gr. Dividing by 24 gr. (in a pwt.) gives 3 pwt. 18 gr. Write 18 gr. and carry the 3 pwt. The next product is 51 pwt. which is equal to 2 oz. and 11 pwt. Write 11 and carry 2. Continue this process until each denomination in the multiplicand has been multiplied.

PROBLEMS

353. 1. Multiply 5 gal. 3 qt. 1 pt. 2 gi. by 7.

2. Multiply 6 da. 5 hr. 14 min. 17 sec. by 9.

3. Multiply 64 sq. rd. 13 sq. yd. 7 sq. ft. 54 sq. in. by 8.

4. Multiply **4 mi. 124 rd. 3 yd. 2 ft. 9 in. by 15.**

5. Multiply **7 A. 57 sq. rd. 5 sq. ft. by 6.**

6. Multiply **2 lb. 5 oz. 6 dr. 2 sc. 8 gr. by 9.**

7. A druggist bought 7 bottles of sulphuric acid, each containing 8 lb. 6 oz. (Avoir.).   Find the total weight.

8. If 1 bag holds 2 bu. 1 pk. 4 qt. 1 pt. of grain, how much will 36 such bags hold?

9. A jeweler bought a dozen silver spoons, each weighing 2 oz. 8 pwt. 16 gr.   At 5 ¢ a pennyweight, what did they cost?

10. If 1 cask of molasses contains 43 gal. 3 qt. 1 pt., find the cost of 12 casks at 30 ¢ per gallon.

11. Multiply **4 Kg. 7 Hg. 8 Dg. 5 g. by 7.**

12. Multiply 8 Hl. 6 Dl. 9 l. 4 dl. by 8.

13. Multiply 24 sq. m. 36 sq. dm. 47 sq. cm. by **15.**

14. Multiply 427 cu. m. 356 cu. dm. 860 cu. cm. by **9.**

15. A load of wood measures 5 s. 7 ds.   Find the cost of 10 such loads at $ .75 a stere.

16. At $750 per are, what will be the cost of **14** lots each containing 1 Ha. 56 a. 32 ca. ?

17. A block of marble contains 3 cu. m. 155 cu. dm. 625 cu. cm. Find the weight of 8 similar blocks, if 1 cu. dm. of marble weighs 2.7 Kg.

18. An importation of glass consisted of 225 boxes, each box containing 24 plates.   If each plate measures 55 sq. dm. 25 sq. cm., find the cost at $1.10 per square meter.

19. The drive wheel of an engine is 5 yd. 2 ft. 4 in. in circumference, and makes 4 revolutions per second.   What fraction of a mile does the engine travel per minute ?   What is the speed per hour ?

DIVISION OF DENOMINATE NUMBERS

354. Divide 48 sq. rd. 22 sq. yd. 7 sq. ft. 108 sq. in. **by 9.**

sq. rd.	sq. yd.	sq. ft.	sq. in.
9)48	22	7	108
5	12	5	88

$3 \times 30\frac{1}{4} = 90\frac{3}{4}$

$90\frac{3}{4} + 22 = 112\frac{3}{4}$

$112\frac{3}{4} \div 9 = 12 \quad 4\frac{3}{4}$

$4\frac{3}{4} \times 9 = 42\frac{3}{4}$

$42\frac{3}{4} + 7 = 49\frac{3}{4}$

$49\frac{3}{4} \div 9 = 5 \quad 4\frac{3}{4}$

$4\frac{3}{4} \times 144 = 684$

$684 + 108 = 792$

$792 \div 9 = 88$

Arrange the dividend and divisor as illustrated. Divide as in simple numbers. The first partial quotient is 5, with a remainder of 3 sq. rd. As shown in the margin, reduce the 3 sq. rd. to square yards. and to the result, 90¾, add the 22 sq. yd. of the dividend, making 112¾ sq. yd. 112¾ sq. yd. divided by 9 gives 12, with a remainder of 4¾ sq. yd. Write the 12 as a part of the quotient, reduce the 4¾ sq. yd. to square feet, and add the 7 sq. ft. of the dividend. The result is 49¾ sq. ft., which, divided by 9, gives 5, with a remainder of 4¾ sq. ft. Write the 5 in the quotient and continue the reduction and division until the lowest denomination is reached.

PROBLEMS

355. 1. Divide 4 da. 5 hr. 27 min. 18 sec. **by 6.**

2. Divide 15 lb. 8 oz. 5 dr. 2 sc. by 4.

3. Divide 2 bbl. 17 gal. 3 qt. 1 pt. 2 gi. by **7.**

4. Divide 17 cu. yd. 1264 cu. in. by 16.

5. A field containing 12 A. 106 sq. rd. 14 sq. yd. was cut up into 48 city lots after a deduction of 2 A. 120 sq. rd. 20 sq. yd. 6 sq. ft. was made for the streets. What was the area of each lot?

6. If 18 horses in 60 da. eat 624 bu. 1 pk. 4 qt. **of oats,** how much does each horse eat per day?

7. Divide 479 cu. m. 827 cu. dm. 448 cu. cm. by **16.**

8. Divide 8 Kl. 7 Hl. 5 Dl. 7 l. 6 dl. 4 cl. 5 ml. by **5.**

9. A bin contains 25 Kl. 5 Hl. of wheat. If all the wheat is put into 125 sacks of equal size, what is the capacity of each sack?

10. Of a field containing 13 A. 29 sq. rd. 4 sq. yd., ⅕ was sold and the balance divided into 60 plots of equal size. How large was each plot?

INVOLUTION AND EVOLUTION

INVOLUTION

356. A power of a number is the product obtained by multiplying the number by itself. 16 is a power of 4, because $4 \times 4 = 16$.

357. Powers are named from the number of times a given factor is used to produce the power.

Thus, 36 is the *second* power (called " square ") of 6, because 6 is used twice as a factor to produce 36.

27 is the *third* power (called " cube ") of 3, because 3 is used three times as a factor to produce 27.

256 is the *fourth* power of 4, because 4 is used four times as a factor to produce 256, etc.

358. The number of times a given factor is to be used is shown by a small figure, called an **exponent**, written to the right of the factor.

Thus in 5^2, 2 is the exponent and shows that 5 is to be used twice as a factor ;

$$5^2 = 5 \times 5 = 25.$$
In the same manner $\quad 4^3 = 4 \times 4 \times 4 = 64.$
$$5^4 = 5 \times 5 \times 5 \times 5 = 625, \text{ etc.}$$

359. Involution is the process of finding the power of a number.

PROBLEMS

360. 1–20. Find the square, cube, and the fourth power of all the numbers from 1 to 20 inclusive.

21–40. Find the square, cube, and the fourth power of all the two-place decimals .01 to .20 inclusive.

Find the square, cube, and the fourth power of:

41. $\frac{1}{2}$	**45.** $\frac{3}{4}$	**49.** $\frac{4}{5}$	**53.** $\frac{9}{16}$	**57.** $\frac{8}{9}$
42. $\frac{1}{3}$	**46.** $\frac{1}{5}$	**50.** $\frac{5}{6}$	**54.** $\frac{5}{11}$	**58.** $\frac{7}{12}$
43. $\frac{2}{3}$	**47.** $\frac{2}{5}$	**51.** $\frac{7}{8}$	**55.** $\frac{4}{7}$	**59.** $\frac{9}{10}$
44. $\frac{1}{4}$	**48.** $\frac{3}{5}$	**52.** $\frac{5}{9}$	**56.** $\frac{3}{8}$	**60.** $\frac{10}{17}$

EVOLUTION

361. 1. What number multiplied by itself will produce 9? 16? 25? 36? 49? 64? 81?

2. What number used three times as a factor will produce 8? 27? 64? 125?

362. The root of a number is one of the equal factors of that number. Thus, 4 is a root of 16; 5 is a root of 125; etc.

363. The square root of a number is one of the *two* equal factors which produce the number. 5 is the square root of 25.

364. The cube root of a number is one of the *three* equal factors which produce the number. 6 is the cube root of 216.

365. Evolution is the process of finding the root of a number.

366. The radical sign ($\sqrt{\ }$) denotes that a root of a number is to be found.

$\sqrt{625}$ indicates that the square root is to be found.

$\sqrt[3]{343}$ indicates that the cube root is to be found.

367. The following diagram will illustrate the principle involved in finding the square root of a number.

The square is 15 ft. on a side, and is divided into four parts as follows.

1 square 10 ft. on a side

$= (10 \text{ ft.})^2 = 100$ sq. ft.

2 rectangles 10 ft. long and 5 ft. wide

$= 2 \times (10 \times 5)$ sq. ft. $= 100$ sq. ft.

· **1** square 5 ft. on a side

$= (5 \text{ ft.})^2 = 25$ sq. ft.

Adding the several results gives

$$[10^2 + 2 \times (10 \times 5) + 5^2] \text{ sq. ft.} = 225 \text{ sq. ft.}$$

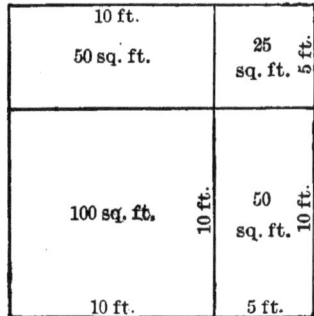

This result may be expressed as follows:

The square of a number of two figures is equal to the square of the tens plus twice the product of the tens by the units plus the square of the units.

By careful inspection and application of this principle, the square root of any number may be found.

368. 1. Find the square root of 225.

$$\sqrt{2'25} = 10 + 5 = 15.$$

$2 \times 10 = 20$, trial divisor.   1,00

$20 + 5 = 25$, complete divisor. $\overline{1,25}$

                             1,25

Beginning at units, divide the number into periods of two figures each. The largest perfect square in the left-hand period is 1. The square root of 1 is 1. Write, as part of the square root, 1 with a cipher to represent the remaining period in the number 225. The square of 10 is 100, which subtract from 225, leaving 125 as the new dividend.

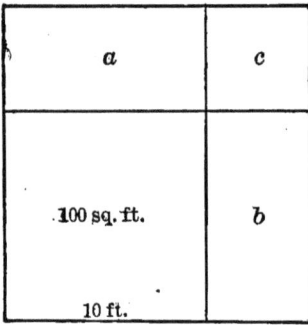

The accompanying diagram illustrates the condition of the problem at this point.

The remaining 125 sq. ft. is represented by the two rectangles a and b, and by the square c. Each rectangle is 10 ft. long. Hence, both rectangles are 2×10 ft., or 20 ft. long. 20 ft. is the trial divisor as shown in the solution. $125 \div 20 = 6$, but 6 is too large because, when it is added to the 20 to complete the divisor, it makes the divisor too large. Try 5 as the quotient figure. Complete the divisor by adding the 5 to the 20, making 25. The 5 is the width of the rectangles a and b, and is the length of one side of the square c. The complete divisor represents the total length of both rectangles a and b, and of the square c.

Now multiply the total length, 25, by the width, 5, and the result, 125, is just equal to the remaining area of the large square. Hence the square root of 225 is 15.

NOTE. The student should observe that

$$2 \times (10 \times 5) + 5^2 = [(2 \times 10) + 5] \times 5.$$

2. Find the square root of 1489.96.

$$\sqrt{14'89'.96} = 38.6$$

                            9

$2 \times 3 = 6^0$ $\overline{5\,89}$

$60 + 8 = 68$ 5 44

   $\overline{2 \times 38 = 76^0}$ $\overline{45\,96}$

$760 + 6 = 766$ $\underline{45\,96}$

The preceding solution and explanation were made in full that the principle involved might be thoroughly understood. The actual work of finding the square root is reduced to the minimum in this solution.

In dividing a number involving a decimal, into periods of two figures each, begin at the decimal point and point off both to the right and to the left.

The largest perfect square in the left-hand period is 9, and its square root is 3. Write 3 in the root and its square, 9, under the 14, and subtract. Bring down the next period. The new dividend is 589. Take twice the root already found, and annex one cipher, giving 60 as a trial divisor. Divide and obtain 8 (9 is too large). Complete the divisor by adding 8 to the 60. Multiplying 68 by 8 gives 544 to be subtracted from 589, leaving 45. Bring the next period down. Since this period is decimal, place a point in the root after the 8 to separate the integral and decimal parts of the root.

Now proceed exactly as before. Take two times the root already found, and annex one cipher. The trial divisor is 760. On dividing, the next figure in the root is found to be 6. Add the 6 to the trial divisor, and multiply the sum by the same 6. There is no remainder; hence, the square root of 1489.96 is 38.6.

NOTE. The number of figures in the root is equal to the number of periods in the power.

The following solutions are given to make clear several minor points not touched upon in the preceding solutions.

3. Find the square root of 5.247 to three places of decimals.

$$\sqrt{5.24'70'00'00} = 2.2906 = 2.291$$

$$
\begin{array}{rl}
2 \times 2 = & 4^0 \\
40 + 2 = & 42 \\
2 \times 22 = & 44^0 \\
440 + 9 = & 449 \\
2 \times 229 = & 458^{00} \\
45800 + 6 = & 45806
\end{array}
\qquad
\begin{array}{l}
4 \\
\overline{124} \\
84. \\
\overline{40\ 70} \\
40\ 41 \\
\overline{29\ 00\ 00} \\
27\ 48\ 36
\end{array}
$$

Annex a sufficient number of ciphers to make four full periods to the right of the decimal point. Proceed with the solution as already explained. Observe that the third decimal figure in the root is 0. In such cases bring down another period, and annex another 0 to the trial divisor as illustrated. Find the root to the fourth place of decimals and then take the nearest third place as the root.

4. Find the square root of .000729.

$$\sqrt{.00'07'29} = .027$$

$$
\begin{array}{rl}
2 \times 2 = & 4^0 \\
40 + 7 = & 47
\end{array}
\qquad
\begin{array}{l}
4 \\
\overline{3\ 29} \\
3\ 29
\end{array}
$$

As the left-hand period is composed of ciphers, write a cipher in the root to the right of the decimal point. The next period is 07. Treat it as though the 7 were an integer, and proceed as before.

5. Find the square root of $\frac{169}{256}$.

$\sqrt{169} = 13$

$\sqrt{256} = 16$

$\sqrt{\frac{169}{256}} = \frac{13}{16}$.

To find the square root of a common fraction, find the square root of the numerator for the numerator of the root, and the square root of the denominator for the denominator of the root.

6. Find the square root of $\frac{3}{8}$.

$\frac{3}{8} = .375$

$\sqrt{.37'50'00} = .612$

$2 \times 6 = 12^0$	36
$120 + 1 = 121$	$\overline{1\ 50}$
	$1\ 21$
$2 \times 61 = 122^0$	$\overline{29\ 00}$
$1220 + 2 = 122\overset{.}{2}$	$24\ 44$

When the numerator and denominator of the fraction are not perfect squares, reduce the fraction to its decimal form and find the square root of the resulting decimal. The degree of accuracy of the result is determined by the number of places the result is carried out.

PROBLEMS

369. Find the square root of the following:

1.	289	**7.**	15.452*	**13.**	.144	**19.**	8.
2.	961	**8.**	18.215	**14.**	.576	**20.**	$\frac{576}{961}$
3.	1764	**9.**	18.0625	**15.**	.8925	**21.**	$\frac{529}{625}$
4.	2809	**10**	54.76	**16.**	.78095	**22.**	$\frac{15}{16}$
5.	4489	**11.**	156.25	**17.**	.0000138	**23.**	$\frac{3}{4}$
6.	11449	**12.**	139.052	**18.**	2.	**24.**	$\frac{1}{4}$

The problems requiring the application of square root in their solution will be found in Mensuration, pages 161 to 170.

* Find the root to the nearest third place of decimals in each imperfect square.

MENSURATION

LINES AND ANGLES

370. A **line** has only one dimension — length.

Straight Line

371. **Parallel lines** are lines that are the same distance apart throughout their length.

Parallel Lines

372. A **horizontal line** is a line parallel to the horizon.

373. An **angle** is the opening between **two** lines that meet.

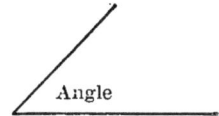

Angle

374. A **right angle** is one of the equal angles formed when two straight lines cross, making four *equal* angles.

Right Angle

375. A **perpendicular line** is one that forms one or more right angles with another line.

376. A **vertical line** is one that makes a right angle with a horizontal line.

Perpendicular Lines

SURFACES OR AREAS

377. Surfaces have two dimensions — length and breadth.

378. A **quadrilateral** is a plane surface having four sides and four angles.

Quadrilateral

379. A **parallelogram** is a quadrilateral having its opposite sides parallel.

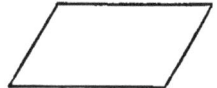

Parallelogram

155

380. A **rectangle** is a parallelogram having right angles.

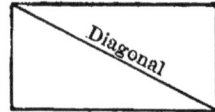

Rectangle

381. A **square** is a rectangle having four equal sides

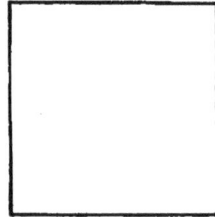

Square

382. A **diagonal** is a straight line between opposite angles of a quadrilateral.

383. A triangle is a plane surface having three sides and three angles

Triangle

384. A **right-angled triangle** is one that has one right angle.

NOTE. No triangle can have more than one right angle. The three angles of any triangle are equal to two right angles.

Right-angled Triangle

385. The **hypotenuse** of a right-angled triangle is the side opposite the right angle

386. An isosceles triangle is one that **has two** sides and two angles equal.

Isosceles Triangle

387. An **equilateral triangle is** one that **has all** sides and all angles equal.

Equilateral Triangle

388. The base of any plane figure is the side on which it is as-sumed to stand.

389. The altitude of any figure is the perpendicular distance from the base to the highest point opposite.

390. A polygon is a plane surface bounded by straight lines. Polygons derive their names from the number of their sides; as, pentagon, hexagon, heptagon, octagon, etc.

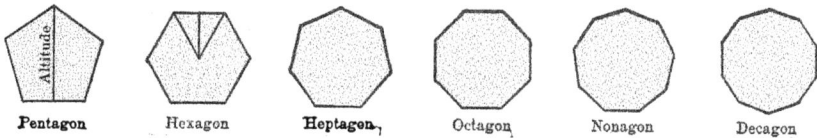

Pentagon   Hexagon   Heptagon   Octagon   Nonagon   Decagon

391. A circle is a plane surface bounded by a curved line, every point of which is equally distant from the center of the circle.

392. The circumference of a circle is its boundary line.

393. The diameter of a circle is a straight line through its center terminating at each end in the circumference.

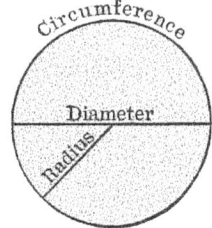

394. The radius of a circle is one-half the diameter.

395. The perimeter of a plane figure is the distance around it.

396. The area of a plane figure is the number of square units within its boundary line.

397. To find the area of any parallelogram.

1. Find the area of a rectangle 7 rd. long and 5 rd. wide.

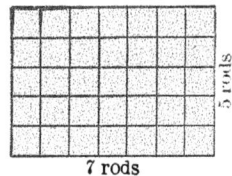

7 rods

By drawing lines as shown in the diagram the surface of the rectangle is divided into square rods.

Since the rectangle is 7 rd. long, there are 7 sq. rd. in the lower row of squares. There are 5 rows of squares, because the rectangle is 5 rd. wide. If in one row there are 7 sq. rd., in 5 rows there are 5 times 7 sq. rd., or 35 sq. rd., the required area.

2. Find the area of a parallelogram the base of which is 9 rd. and the altitude, 6 rd.

Represent the parallelogram, as in the diagram, by *abcd*. It is easily proved by Geometry that if that part of the parallelogram indicated by the triangle *dec* were cut off and placed in the position of the triangle *afb*, the resulting figure would be the rectangle *afed*, having a base *fe* equal to the base *bc* of the parallelogram.

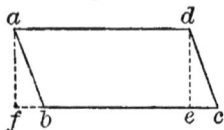

The area of the rectangle *afed* is 6 times 9 sq. rd., or 54 sq. rd., as explained in the preceding example. But the areas of the rectangle and of the parallelogram are equal. Hence the area of the parallelogram is equal to 6 times 9 sq. rd., or 54 sq. rd.

The area of any parallelogram is equal to the product of its base by its altitude.

NOTE. In this and similar rules the product of the dimensions means the product of the *numbers* that represent them, when they are expressed in *like units.*

PROBLEMS

398. Find the perimeter and the area of the following rectangles:

(Make a diagram of each.)

1. 40 ft. by 36 ft.
2. 15 yd. by 14 yd.
3. 18 yd. square.
4. 20 rd. by 18 rd.
5. 56 rd. by 48 rd.

6. 7 rd. by 14 yd.
7. 9 yd. by 8 ft. 6 in.
8. 128.5 rd. by 83.75 rd.
9. $\frac{1}{2}$ mi. by 76 rd.
10. 8 Hm. by 5 Hm. 6 Dm.

11. How many acres are there in a rectangular field $\frac{1}{2}$ mi. long and 66 rd. wide?

12. A field in the form of a rectangle contains 1 A. What length of fence is required to inclose it if the perpendicular distance between its sides is 10 rd.?

NOTE. Since the product of the base by the altitude is equal to the area, the base may be found by dividing the area by the altitude. What principle applies? Before dividing, reduce the area (1 A.) to square rods.

13. Find the perimeter of a square field containing 10 A.

How do the base and altitude of a square compare in length?
By what process may one of two equal factors of a product be found?

14. A rectangular field contains 36 A. Find the cost at 75¢ a rod of building a fence around it if the width is $\frac{5}{8}$ of the length.

15. Find the cost of a cement sidewalk 80 ft. long and 4 ft. 6 in. wide, at $1.25 per square yard.

16. How many paving blocks 1 ft. long and 5 in. wide will be required for a street 1 mi. long and 40 ft. wide ?

17. At 15¢ per square yard, find the cost of painting the four walls of a room 12 ft. long, 9 ft. 6 in. wide, and 8 ft. high.

18. The Hudson Terminal Building in New York, N. Y., contains 4000 offices, and has a floor space of 27 A. Find the average floor space per office.

19. A walk 6 ft. wide is built along two sides of a corner lot 80 ft. square. Find the area of the walk : (*a*) if it is laid on the lot; (*b*) if it is laid on the outside of the lot.

399. To find the area of any triangle.

1. Find the area of a triangle having a base of 12 ft. and an altitude of 8 ft.

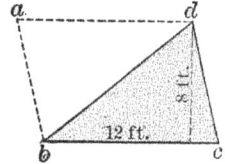

How does the triangle

$$\frac{12 \times 8}{2} = 48, \text{ no. sq. ft., area.}$$

abd compare in size with the triangle *bcd* ? What name is given to the figure *abcd* ? Note that the diagonal *bd* divides the parallelogram *abcd* into two equal parts. Since the area of the parallelogram is equal to the product of its base by its altitude, the area of the triangle *bcd* is equal to *one half* the product of the base by the altitude.

The area of any triangle is equal to one half the product of its base by its altitude.

To find the area of a triangle when the three sides are given.

It sometimes occurs that the altitude of a triangle is not known, but the length of each side is given. In such cases find the area as in the following example.

2. Find the area of a triangle the sides of which are 18 ft., 25 ft., and 31 ft.

$$18 + 25 + 31 = 74 \qquad 37 \times 19 \times 12 \times 6 = 50616.$$
$$\tfrac{1}{2} \text{ of } 74 = 37 \qquad \sqrt{50616} = 224.98, \text{ no. sq. ft.}$$
$$37 - 18 = 19$$
$$37 - 25 = 12$$
$$37 - 31 = 6$$

Find one half the sum of the three sides, which is 87 ft. From 37 ft. subtract each side of the triangle, leaving 19, 12, and 6 ft., respectively. Multiply the half sum, 37, and all the remainders, 19, 12, and 6, together, obtaining a product of 50,616. Extract the square root of 50,616, and the result, 224.98, is the number of square feet in the area.

NOTE. Any triangle having sides in the ratio of 3, 4, and 5 is a right-angled triangle.

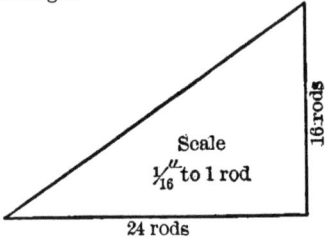

3. Make a diagram of a right-angled triangle having a base of 24 rd., and an altitude of 16 rd., to the scale of $\frac{1}{16}''$ to a rod.

Since the diagram is to be on the scale of $\frac{1}{16}''$ to a rod, the base, 24 rd,. will be represented by a line $\frac{24}{16}$ of an inch, or $1\frac{1}{2}$ in. long, and the altitude by a line 1 in. long. The triangle is completed by drawing the hypotenuse.

Scale $\frac{1}{16}''$ to 1 rod

16 rods

24 rods

PROBLEMS.

400. Make a diagram, and find the area, of triangles having dimensions as follows:

1. Base 56 ft., altitude 32 ft.   (Scale $\frac{1}{8}''$ to a foot.)

2. Base 48 yd., altitude 29 yd.   (Scale $\frac{1}{8}''$ to a yard.)

3. Base 64 rd., altitude 38 rd.   (Scale $\frac{1}{8}''$ to a rod.)

4. Base 88 rd., altitude 75 rd.   (Scale $\frac{1}{16}''$ to a rod.)

5. Base 19 in., altitude 13 in.

6. With sides 18 ft., 21 ft., and 25 ft.   (Scale $\frac{1}{4}''$ to a foot.)

7. With sides 27 ft., 35 ft., and 40 ft.   (Scale $\frac{1}{8}''$ to a foot.)

8. With sides 72 rd., 85 rd., and 91 rd.   (Scale $\frac{1}{16}''$ to a rod.)

9. With sides 65 rd., 73 rd., and 85 rd.   (Scale $\frac{1}{8}''$ to a rod.)

10. With sides 5 in., 6 in., and 7 in.

11. Find the value of a triangular field having a base of 64 rd. and an altitude of 48 rd. at $115 per acre.

12. The gables of a house are 32 ft. wide, and the ridge of the roof rises 20 ft. above the foot of the rafters. How many square feet of boards are required to cover both gables?

13. The sides of a triangular garden measure 56 ft., 49 ft., and 45 ft., respectively. How many square yards are there in its area?

14. The sides of an equilateral triangular field are 100 rd. each. How many acres are there in the field?

15. A house 28. ft. wide has rafters on one side of the roof 19 ft. long, and on the other side 17 ft. long. If the foot of the rafters is 18 ft. above the cellar wall, find the area of both gable ends of the house.

Find the altitude of the following triangles:

16. Area 180 sq. ft., base 20 ft.

Compare with note, page 158.

17. Area 270 sq. yd., base 45 yd.

18. Area 6 A., base 60 rd.

19. Area 15 A., base 100 rd.

Find the base of the following triangles:

20. Area 120 sq. in., altitude 15 in.

21. Area 360 sq. yd., altitude 120 ft.

22. Area 4 A., altitude 176 yd.

23. Area 20 A., altitude 80 rd.

24. The area of a triangle is 600 sq. ft. The differences between one half the sum of all the sides and each side are 10 ft., 20 ft., and 30 ft., respectively. Find the three sides of the triangle.

401. To find any side of a right-angled triangle having two sides given.

a. To find the hypotenuse.

1. Find the hypotenuse of a right-angled triangle whose base and altitude are 8 ft. and 6 ft., respectively.

$$8^2 = 64$$
$$6^2 = 36$$
$$8^2 + 6^2 = 64 + 36 = 100$$
$$\sqrt{100} = 10, \text{ no. ft., hypotenuse.}$$

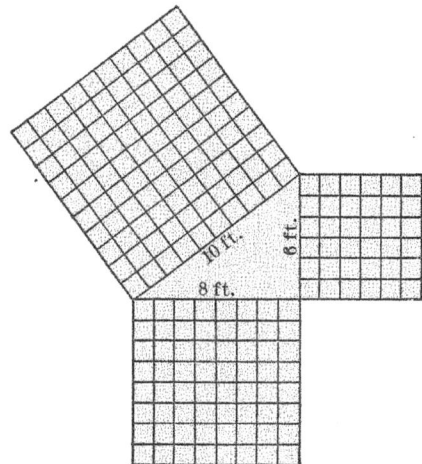

In the diagram it is seen that the square having the base of the triangle as the length of its sides contains 8 × 8, or 64, sq. ft. Likewise, the square having the

VAN TUYL'S BUS. ARITH. — 11

altitude of the triangle as the length of its sides contains 36 sq. ft. By adding these two areas, 64 sq. ft. and 36 sq. ft., the result, 100 sq. ft., is found to be equal to the area of the square having the hypotenuse of the triangle as the length of its sides. To find one side of a square, its area being known, extract the square root of its area. The square root of 100 is 10. That is, the length of the hypotenuse is 10 ft.

To find the hypotenuse of a right-angled triangle, extract the square root of the sum of the squares of the base and the perpendicular.

b. **To find the base or the perpendicular.**

1. Find the perpendicular of a right-angled triangle whose hypotenuse is 15 yd. and base 12 yd.

$$15^2 = 225$$
$$12^2 = 144$$
$$15^2 - 12^2 = 225 - 144 = 81$$
$$\sqrt{81} = 9, \text{ no. ft., perpendicular}$$

Since the sum of the squares of the base and the perpendicular equals the square of the hypotenuse (see illustration in the preceding example), it follows that the difference between the square of the hypotenuse and the square of one of the sides is equal to the square of the other side. $15^2 - 12^2 = 81$. Since 81 is the square of the perpendicular, the perpendicular is equal to the square root of 81, or 9. Therefore the perpendicular is 9 ft.

To find either side of a right-angled triangle, having the hypotenuse and one side given, extract the square root of the difference of the squares of the hypotenuse and of the given side.

PROBLEMS

402. Make a diagram of, and solve, triangles measuring as follows :

1. Base 24 ft., perpendicular 18 ft.   Find hypotenuse and area.

2. Base 30 ft., perpendicular $22\frac{1}{2}$ ft.   Find hypotenuse and area.

3. Perpendicular 18 ft., base 13 ft. 6 in.   Find hypotenuse and area.

4. Hypotenuse 45 yd., perpendicular 27 yd.   Find base and area.

5. Hypotenuse 13 ft., base 12 ft.   Find perpendicular and area.

6. Hypotenuse 25 rd., perpendicular 7 rd.   Find base and area.

7. How long a ladder placed 9 ft. from the side of a building will reach a window 40 ft. high?

8. A guy rope 60 ft. long is attached to the top of a pole 48 ft. high. How far from the foot of the pole can the rope be fastened? (No allowance for sag.)

9. The base of a square pyramid is 14 ft. on a side. The vertical height of the pyramid is 24 ft. Find the area of its four sides.

10. The rafters on a house are 25 ft. long. If the ridge of the roof is 16 ft. above the foot of the rafters, how wide is the house?

11. A room is 14 ft. long, 11 ft. wide, and 8 ft. high. Find the length of a diagonal from one of the upper corners to the opposite lower corner.

12. A ladder 61 ft. long stands close against a wall. If the bottom of the ladder is drawn out 11 ft., how far will the top of the ladder be lowered?

13. A square park 16 rd. on a side has a gravel walk across it on both diagonals. If the walk is 8 ft. wide, find its area.

CIRCLES

403. **To find the diameter or circumference of a** circle.

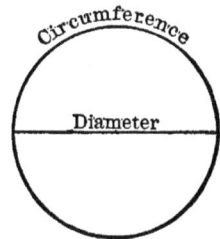

1. Find the circumference of a circle whose diameter is 18 ft.

3.1416 × 18 ft. = 56.5488 ft., circumference.

It has been found by accurate measurement that the circumference of a circle is 3.1416 times the length of the diameter.

To find the circumference of a circle, multiply the diameter by **3.1416.**

2. Find the diameter of a circle whose circumference is **78.54 ft.**

78.54 ÷ 3.1416 = 25 ft., diameter.

The circumference is 3.1416 times the diameter.

The circumference divided by 3.1416 will give the diameter. (**What** principle applies?)

NOTE. For approximate measurements, 3⅐ may be used instead of 3.1416.

164 MENSURATION

PROBLEMS

404. Find the approximate and the accurate circumferences of circles having diameters as follows:

1. 14 ft.	**5.** 45 ft. 6 in.	**9.** 6 ft. 5 in.
2. 28 yd.	**6.** 56 yd.	**10.** 1 rd. 2 yd. 2 ft. 7 in.
3. 35 in.	**7** 4 ft. 8 in.	**11.** 112 rd.
4. 42 ft	**8.** 5 ft. 3 in.	**12.** 100 rd.

Find the approximate and the accurate diameters of circles having the following circumferences:

13. 11 ft.	**17.** 110 yd.	**21.** 51 yd. 1 ft
14. 66 ft	**18.** 140 yd.	**22.** 53 ft. 2 in.
15. 5 ft. 6 in.	**19.** 55 ft.	**23.** 1 mi.
16. 33 rd.	**20.** 7 ft. 4 in.	**24.** 1 rd.

25. A bicycle wheel is 2 ft. 4 in. in diameter. Over what distance will the rider have passed when the wheel has made 5000 rotations?

26. A horse is tied to a post with a rope 30 ft. long. What is the circumference of the largest circle over which he can graze?

405. To find the area of a circle.

By examining the diagram in the margin it is seen that a circle is composed of a great number of small triangles whose bases form the circumference of the circle and whose vertices meet at the center of the circle.

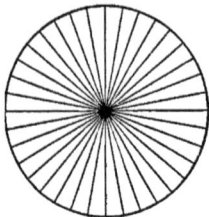

NOTE. While the bases of the triangles shown in the diagram are not straight, but curved, lines, yet the truth of the above statement is readily proved by Geometry.

How may the area of a triangle be found?

What is the shortest way to find the total area of three triangles having equal altitudes?

What is the altitude of the triangles in the diagram?

The sum of all their bases is equal to what line?

How may the area of a circle be found?

406. Find the area of a circle 18 ft. in diameter.

FIRST SOLUTION

3.1416 × 18 ft. = 56.5488 ft., circumference.

18 ft. ÷ 4 = 4½ ft., one half the radius.

4½ × 56.5488 = 254.4696, no. sq. ft., area.

· Since the diameter of a circle is given, find the circumference by multiplying the diameter by 3.1416. The circumference is the sum of the bases of all the triangles in the circle. Find half the radius of the circle by dividing the diameter by 4. One half the radius is equal to one half the altitudes of the triangles. Multiply the circumference, 56.5488, by the half radius, $4\frac{1}{2}$, and the result, 254.4696, is the number of square feet in the area.

The labor of computing the area of a circle may be lessened by writing the solution in the cancellation form, and observing the resulting combinations of numbers. Thus, in the problem given above, we have:

SECOND SOLUTION

$$\frac{\overset{.7854}{\cancel{3.1416}} \times 18 \times 18}{\cancel{4}} = .7854 \times 18^2 = .7854 \times D^2.$$

$$18^2 = 324.$$

$$.7854 \times 324 = 254.4696, \text{ no. sq. ft., area.}$$

Having the numbers written as illustrated, note that the numerator consists of 3.1416, which is the same for all circles, and of the diameter written twice. The denominator consists of 4, because half the radius is one fourth of the diameter, which is true for all circles.

Now by canceling the 4 into 3.1416, the decimal .7854 is obtained.

Therefore, by memorizing the decimal .7854, the area of a circle is readily found.

To find the area of a circle multiply the square of the diameter (D^2) *by .7854.*

In some problems the work may be still further abbreviated. The same problem will serve as an illustration.

THIRD SOLUTION

$$\frac{3.1416 \times \overset{9}{\cancel{18}} \times \overset{9}{\cancel{18}}}{\underset{2}{\cancel{4}}} = 3.1416 \times 9^2 = 3.1416 \times R^2.$$

$$9^2 = 81; \quad 3.1416 \times 81 = 254.4696, \text{ no. sq. ft., area.}$$

Instead of canceling the 4 into 3.1416, divide the 18's each by 2 (2 and 2 being the factors of 4). The result is two 9's. But 9 is the radius of the circle. Therefore, the area of a circle is equal to the square of the radius multiplied by 3.1416.

The area of a circle is equal to :

(1) *The circumference multiplied by one half the radius, or,*

(2) *The square of the diameter multiplied by .7854 ($\frac{1}{4}$ of 3.1416), or*

(3) *The square of the radius multiplied by 3.1416.*   .

Practice alone can tell the student which rule is the best for the individual problem   In general, however, it may be said that, if both diameter and circumference are known, the first rule is good.

If the diameter is a small number and can be squared mentally, the second rule is good.

If the diameter is a number that is not readily squared mentally, and the radius can be so squared, the third rule is a good one.

PROBLEMS

407. 1–26. Find the area of all the circles in the problems on page 164.

For approximate areas, $3\frac{1}{7}$ may be used instead of 3.1416.

408. To find the diameter or the circumference of a circle when the area is known.

Find the diameter of a circle whose area is 201.0624 sq. rd.

201.0624 ÷ .7854 = 256, the square of the diameter.

$\sqrt{256}$ = 16, no. rd., diameter.

Since the square of the diameter multiplied by .7854 gives the area of the circle, the area divided by .7854 gives the square of the. diameter, or 256. Extracting the square root of 256 gives 16, the number of rods in the diameter. (Prin. 14, page 102.)

To find the diameter of a circle whose area is given, divide the area by .7854 and extract the square root of the quotient.   To find the circumference, multiply the diameter by 3.1416.

PROBLEMS

409. Find the diameter and the circumference of circles whose areas are :

1. 12.5664 sq. ft.	**4.** 78.54 sq. yd.
2. 28.2744 sq. rd.	**5.** 113.0976 sq. yd.
3. 50.2656 sq. ft.	**6.** 153.9384 sq. rd.

7. 452.3904 sq. rd.   **9.** 400 sq. rd.   **11.** 15 acres.

8. 100 sq. rd.   **10.** 1 acre.   **12.** 502.656 acres.

13. 125.664 acres.   **14.** 1000 acres.

REVIEW OF PLANE FIGURES

PROBLEMS

410. 1. How much more will it cost at $1.15 per rod to build a fence around 40 A. in the form of a square than in the form of a circle?

2. Find the area of a circle that can be described with a radius of 85 ft.

3. A rectangular field containing 32 A is $\frac{4}{5}$ as wide as it is long. What are its dimensions?

4. An isosceles triangle has a base 10 ft. long, and an altitude of 12 ft. Draw a diagram and find the other sides of the triangle.

5. An equilateral triangle is 12 ft. on a side. Find its altitude.

6. A series of triangles having a common altitude of 9 ft. have bases of 4 ft., 6 ft., 3 ft. 6 in., 7 ft. 4 in., and 8 ft. 2 in., respectively. Find their total area.

7. A hexagon is 6 ft. on a side. Find its area.

8. If a walk around a court 200 ft. square occupies one fourth the area of the court, find the width of the walk.

9. If a triangular area having sides 36 ft., 45 ft., and 60 ft. long, respectively, is paved at a cost of $2.60 per square yard, how much does it cost?

10. A square park has a walk diagonally through it. If the length of the walk is 50 rd., find the area of the park.

SOLIDS

411. A solid is a figure having three **dimensions** — length, breadth, and thickness.

412. A prism is a solid whose bases or ends are any similar, equal, and parallel plane figures, and whose lateral faces are parallelograms.

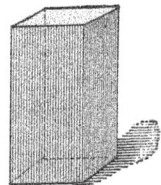

Note. Prisms are named from the number of their sides — three sides, *triangular;* four sides, *quadrangular* or *square;* etc.

Square Prism

413. A cube is a prism having six equal square faces.

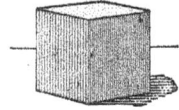

Cube

414. A cylinder is a solid whose bases are equal, parallel circles, and whose lateral surface is a uniform curve.

Note. In this book the term "cylinder" is used for "circular cylinder."

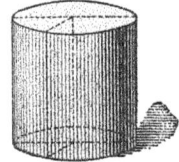

Cylinder

415. The altitude of a solid is the perpendicular distance between its bases.

416. A pyramid is a solid having a polygon for a base and triangles for its sides. The vertices of the triangles form the vertex of the pyramid.

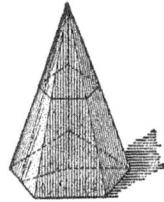

Pyramid

417. A cone is a solid having a circular base, and a lateral surface that tapers uniformly to a point.

Note. In this book the term "cone" is used for "circular cone."

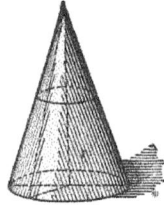

Cone

418. The slant height of a pyramid is the perpendicular distance from one side of the base to the vertex.

419. The slant height of a cone is the shortest distance from the circumference of the base to the vertex.

420. To find the volume of a prism or a cylinder.

1. Find the volume of a prism 8 ft. square and 10 ft. long.

$$8 \times 8 \times 1 \text{ cu. ft.} = 64 \text{ cu. ft.}$$
$$10 \times 64 \text{ cu. ft.} = 640 \text{ cu. ft., volume.}$$

What is the area of a square 8 ft. on a side?
What is the volume of a cube 1 ft. on each side?
By consulting the diagram it is seen that one foot
of the length of the prism contains 8 × 8, or 64,
cubes, one foot on a side, that is, 64 cu. ft. If 1 ft.
of the length contains 64 cu. ft., the whole length
will contain 10 times 64 eu. ft., or 640 cu. ft.

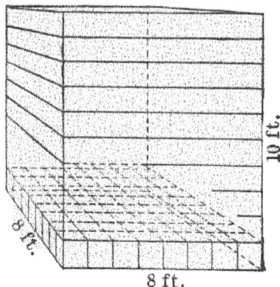

2. Find the volume of a cylinder 4 ft. in
diameter and 8 ft. long.

$$3.1416 \times 2 \times 2 \times 1 \text{ cu. ft.} = 12.5664 \text{ cu. ft.}$$
$$8 \times 12.5664 \text{ cu. ft.} = 100.5312 \text{ cu. ft., volume.}$$

As in the preceding example, one foot of the length of the cylinder contains
as many cubic feet as the number of square feet in the area of the base. The
area of the base is the area of a circle 4 ft. in diameter, which is equal to 3.1416
times the square of the radius, or 12.5664. The entire cylinder contains 8 times
12.5664 cu. ft., or 100.5312 cu. ft.

*To find the volume of a prism or of a cylinder, multiply the area of
the base by the altitude and express the product in cubic units.*

PROBLEMS

421. Find the volumes of the following solids:

1. Square prism, base 6 ft. by 6 ft., altitude 7 ft.

2. Rectangular prism, base 3 ft. by 4 ft., altitude 6 ft.

3. Triangular prism, sides of base 4 ft., altitude 8 ft.

4. Cylinder, diameter 5 ft., altitude 9 ft.

5. Cylinder, diameter 5 ft. 6 in., altitude 10 ft.

6. Cylinder, diameter 3 in., altitude 5 in.

7. Hexagonal prism, side of base 3 ft., altitude 12 ft.

8. Triangular prism, sides of base 3 ft., 4 ft., and 5 ft., altitude 9 ft.

9. Triangular prism, sides of base 6 ft., 5 ft., and 5 ft., altitude 8 ft.

10. Cylinder, diameter 16 ft., altitude 12 ft.

11. A cylindrical water tank 25 ft. in diameter, and 30 ft. deep is
full of water. How many tons of water are there?

12. A block of marble is 6 ft. long, 3 ft. 6 in. wide, and 2 ft. 6 in.
thick. What is its weight if marble is 2.7 times as heavy as water?

13. A rectangular tank is 12 ft. long, 4 ft. 9 in. wide, and 2 ft. 3 in. deep. How long will it take to fill it at the rate of a gallon a minute?

14. Find the dimensions of the base of a square prism whose altitude is 15 in., and whose volume is 540 cu. in.

540 cu. in. ÷ 15 = 36 cu. in., volume of base of the prism one inch high. $\sqrt{36} = 6$, no. in., one side of base.

Since the area of the base multiplied by the altitude equals the volume, the volume divided by the altitude equals the area of the base, and since the base is square, its dimensions are found by extracting the square root of the area.

15. The volume of a prism is 360 cu. ft. and its base is 4 ft. by 5 ft. Find its altitude.

16. A cylinder is 10 ft. long and its volume is 785.4 cu. ft. Find its diameter.

17. The diameter of a cylinder is 20 inches, and its volume is 1570.8 cu. in. Find its length.

422. **To find the lateral surface and the entire surface of a prism or of a cylinder**

1. Find the lateral-surface and the entire surface of a triangular prism, if its base is 4 ft. on a side, and its altitude is 8 ft.

3 × 4 ft. = 12 ft., perimeter of the base.

8 × 12 sq ft = 96 sq. ft., area of lateral surface.

(4 + 4 + 4) ÷ 2 = 6.

6 − 4 = 2

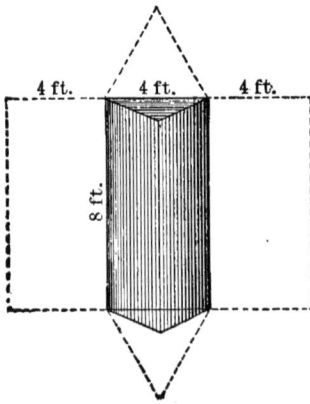

6 × 2 × 2 × 2 = 48.

$\sqrt{48} = 6.928$, no. sq. ft., area of 1 base.

2 × 6.928 sq. ft. = 13.856 sq. ft., area of both bases.

96 sq. ft. + 13.856 sq ft. = 109.856 sq. ft., area of entire surface.

If a sheet of paper were cut so as just to cover all the faces of the prism, its form would be that shown by the dotted lines in the diagram. That part of the diagram which would cover the three lateral faces of the prism is a rectangle whose length is equal to the perim-

eter of the base of the prism, or 12 ft. The width of the rectangle is the altitude
of the prism. The area of the rectangle is 8 × 12 sq. ft., or 96 sq. ft.

To find the area of the entire surface the area of the two bases must be included. The bases are equilateral triangles 4 ft. on a side, whose areas are each
6.928 sq. ft. (See page 159.) The entire area, therefore, is the sum of 96 sq.
ft. and 13.856 sq. ft. (both bases), or 109.856 sq. ft.

2. Find the area of the entire surface of a cylinder 5 ft. in diameter and 10 ft. long.

3.1416 × 5 ft. = 15.708 ft., circumference of base.

10 × 15.708 sq. ft. = 157.08 sq. ft., lateral area.

(5^2) sq. ft. × .7854 = 19.635 sq. ft., area of 1 base.

2 × 19.635 sq. ft. = 39.27 sq. ft., area of
both bases.

157.08 sq. ft. + 39.27 sq. ft. = 196.35 sq.
ft., area of entire surface.

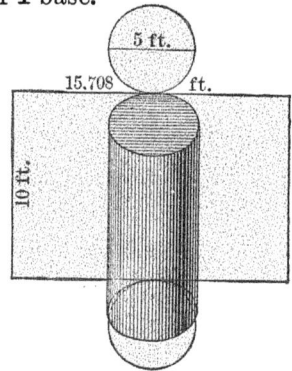

As in the preceding example, a sheet of paper
just large enough to cover the lateral surface of
the cylinder would be in the form of a rectangle
whose length is equal to the circumference of the
cylinder and whose width is the altitude of the
cylinder. The area of the lateral surface is found
by multiplying the circumference (3.1416 × 5 ft.)
by the altitude of the cylinder. The area of the
bases is the area of the two circles whose diameters are 5 ft. (For explanation
see pages 164–5.) The lateral area, 157.08 sq. ft., plus the area of the two bases,
39.27 sq. ft., equals 196.35 sq. ft., the area of the entire surface.

*To find the lateral surface of a prism or of a cylinder, multiply the
perimeter of the base by the altitude.*

*To find the entire surface, add to the area of the lateral surface the
area of the two bases.*

PROBLEMS

423. 1–10. Find the lateral, and the entire surface, of the cylin-
ders and prisms mentioned in problems **1–10** on page 169.

11. How many square yards of sheet iron would be required for
a cylindrical reservoir 40 ft. in diameter and 40 ft. deep, if the top
were open?

12. Find the cost of painting the top, sides, and ends of a box
8 ft. 4 in. long, 6 ft. 6 in. wide, and 4 ft. deep, at 20¢ a square yard.

13. How many square yards of t will be required to make 8 dozen pails, without covers, 9 in. in di eter and 7½ in. deep, allowing 3 sq. ft. for seams and waste on ea dozen pails ?

14. A wagon box is painted b< inside and outside except the under side of the bottom. If it 2 ft. long, 3 ft. 3 in. wide, and 18 in. deep, find the area painted. Make no allowance for corners.)

15. A cylinder 6 ft. in diameter as 245.0448 sq. ft. in its entire surface. Find its altitude.

(2 × 6 × 6) sq. ft. × .7854 = 56.⸱ 8 sq. ft., area of both ends.
245.0448 sq. ft. − 56.5488 sq. ft 188.496 sq. ft., lateral area.
6 ft. × 3.1416 = 18.8496 ft., circ ference of cylinder.
188.496 ÷ 18.8496 = 10, no. ft., tude.

From the entire area deduct the area (oth ends ; the remainder is the lateral area. Since the lateral area is the prod of the circumference and the altitude, the area divided by the circumference v give the altitude (Prin. 14, page 102).

16. Find the altitude of a recta ılar prism if its base is 5 ft. by 8 ft. and its entire surface is 470 ft.

17. Find the diameter of a cyl er whose altitude is 11 ft. and whose lateral surface is 190.0668 ft.

18. A box 10 ft. square with cover is made from 500 sq. ft. of u. lumber. How deep is it? (Ma no allowance for corners.)

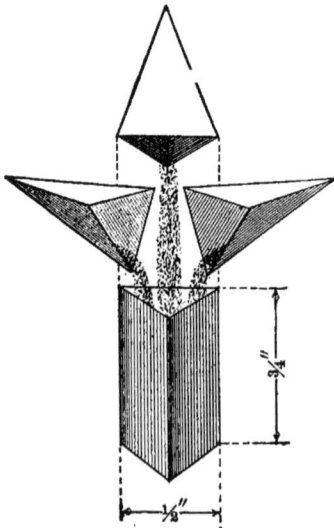

4 To find the volume of a pyramid or o cone.

S ly the accompanying diagrams. Not the dimensions of the several for The contents of how many pyr ids are being poured into one pris ? What does that suggest as to t volume of a pyramid compared wit ı prism of the same dimensions ?

I w many cones full are equal to one ylinder full ? The volume of a con s what part of the volume of a cyl ler?

Find the volume of a pyramid 4 ft. sq. and 6 ft. high.

4×4 sq. ft. $= 16$ sq. ft., area of the base.

$\frac{1}{3}$ of $6 = 2$.

2×16 cu. ft. $= 32$ cu. ft., volume of the pyramid.

Since, as illustrated above, the volume of a pyramid one third the volume of a prism of like dimensions, the volume of the pyramid is found by multiplying the area of the base by one third the altitude.

2. Find the volume of a cone if its base is 6 ft. in diameter and its altitude is 8 ft.

$(3.1416 \times 3 \times 3)$ sq. ft. $= 28.2744$ sq. ft., area of the base.

$$\frac{8 \times 28.2744 \text{ cu. ft.}}{3} = 75.3984 \text{ cu. ft.,}$$

volume of the cone.

The volume of a cone is equal to one third the volume of a cylinder of like dimensions, and is found by multiplying the area of the base by one third the altitude.

To find the volume of a pyramid or of a cone, multiply the area of the base by one third of the altitude and express the result as cubic units.

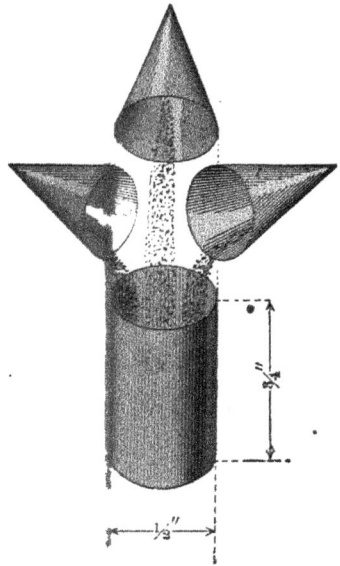

PROBLEMS

425. Find the volumes of the following pyramids and cones:

PYRAMIDS

1. Square, base 6 ft. on a side, altitude 9 ft.

2. Triangular, base 5 ft on a side, altitude 16 t.

3. Rectangular, b . ft . ., altitude 16 t.

ES

4. Diameter of ba e 9 ft.

5. Circumfer ce 2 dm., altitude 5 m.

6. Circumference altitude 100 f

13. How many square yards of tin will be required to make 8 dozen pails, without covers, 9 in. in diameter and $7\frac{1}{2}$ in. deep, allowing 3 sq. ft. for seams and waste on each dozen pails?

14. A wagon box is painted both inside and outside except the under side of the bottom. If it is 12 ft. long, 3 ft. 3 in. wide, and 18 in. deep, find the area painted. (Make no allowance for corners.)

15. A cylinder 6 ft. in diameter has 245.0448 sq. ft. in its entire surface. Find its altitude.

$(2 \times 6 \times 6)$ sq. ft. $\times .7854 = 56.5488$ sq. ft., area of both ends.
245.0448 sq. ft. $- 56.5488$ sq. ft. $= 188.496$ sq. ft., lateral area.
6 ft. $\times 3.1416 = 18.8496$ ft., circumference of cylinder.
$188.496 \div 18.8496 = 10$, no. ft., altitude.

From the entire area deduct the area of both ends; the remainder is the lateral area. Since the lateral area is the product of the circumference and the altitude, the area divided by the circumference will give the altitude (Prin. 14, page 102).

16. Find the altitude of a rectangular prism if its base is 5 ft. by 8 ft. and its entire surface is 470 sq. ft.

17. Find the diameter of a cylinder whose altitude is 11 ft. and whose lateral surface is 190.0668 sq. ft.

18. A box 10 ft. square with a cover is made from 500 sq. ft. of 1 in. lumber. How deep is it? (Make no allowance for corners.)

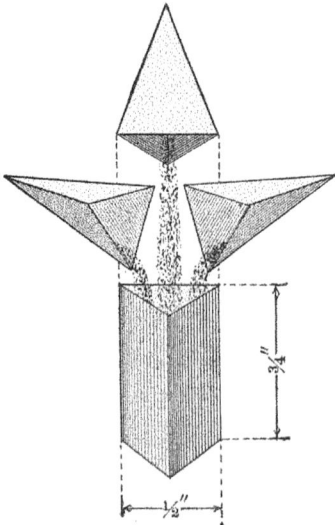

424. To find the volume of a pyramid or of a cone.

Study the accompanying diagrams. Note the dimensions of the several forms. The contents of how many pyramids are being poured into one prism? What does that suggest as to the volume of a pyramid compared with a prism of the same dimensions?

How many cones full are equal to one cylinder full? The volume of a cone is what part of the volume of a cylinder?

Find the volume of a pyramid 4 ft. sq. and 6 ft. high.

4×4 sq. ft. $= 16$ sq. ft., area of the base.

$\frac{1}{3}$ of $6 = 2$.

2×16 cu. ft. $= 32$ cu. ft., volume of the pyramid.

Since, as illustrated above, the volume of a pyramid is one third the volume of a prism of like dimensions, the volume of the pyramid is found by multiplying the area of the base by one third the altitude.

2. Find the volume of a cone if its base is 6 ft. in diameter and its altitude is 8 ft.

$(3.1416 \times 3 \times 3)$ sq. ft. $= 28.2744$ sq. ft., area of the base.

$$\frac{8 \times 28.2744 \text{ cu. ft.}}{3} = 75.3984 \text{ cu. ft.,}$$

volume of the cone.

The volume of a cone is equal to one third the volume of a cylinder of like dimensions, and is found by multiplying the area of the base by one third the altitude.

To find the volume of a pyramid or of a cone, multiply the area of the base by one third of the altitude and express the result as cubic units.

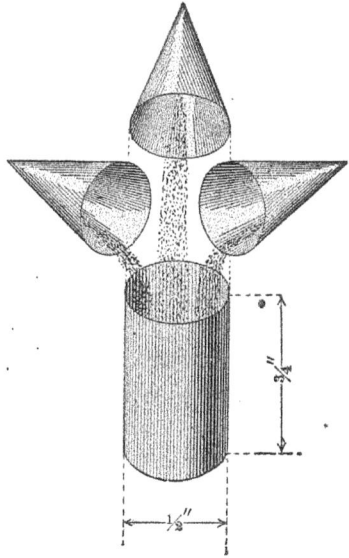

PROBLEMS

425. Find the volumes of the following pyramids and cones:

PYRAMIDS

1. Square, base 6 ft. on a side, altitude 9 ft.

2. Triangular, base 5 ft. on a side, altitude 10 ft.

3. Rectangular, base 6 ft. by 8 ft., altitude 16 ft.

CONES

4. Diameter of base 5 ft., altitude 9 ft.

5. Circumference of base 37.6992 dm., altitude 5 m.

6. Circumference of base 24 ft., altitude 100 ft.

7. A square pyramid, the base being 8 ft. on a side, and the altitude 24 ft., has its top cut off 12 ft. from the apex, parallel to the base. The top part is what fractional part of the original pyramid?

8. Find the diameter of the base of a cone 15 ft. high whose volume is 251.328 cu. ft.

$$251.328 \div 5 = 50.2656, \text{ no. sq. ft., area of base.}$$
$$50.2656 \div .7854 = 64.$$
$$\sqrt{64} = 8, \text{ no. ft., diameter.}$$

Since the volume of a cone is equal to the area of the base multiplied by one third the altitude, the volume divided by one third the altitude must give the area of the base. Having the area of a circle, its diameter is found as explained on page 166.

What principle applies?

Will the same principle apply in the case of a pyramid?

9. Find the dimensions of the base of a square pyramid if its volume is 128 cu. ft. and its altitude is 24 ft.

10. A cone 42 in. high contains 1583.3664 cu. in. Find the diameter of the base.

426. To find the lateral or convex surface of a pyramid or of a cone.

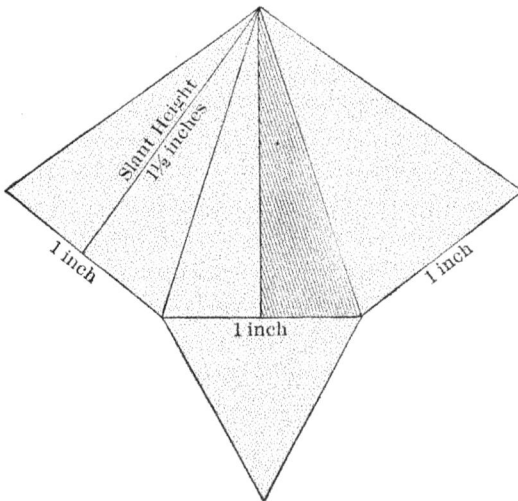

By inspecting the diagram in the margin, it is seen that the lateral surface of a triangular pyramid consists of three triangles whose altitudes are the slant height of the pyramid and whose bases are equal to the sides of the base of the pyramid. How may the total area of several triangles having a common altitude be found?

How may the lateral surface of a pyramid be found? Can the lateral surface of a cone be found in the same way?

Find the lateral surface of a pyramid whose base is 4 ft. square and whose slant height is 8 ft.

$$4 \times 4 \text{ ft.} = 16 \text{ ft., perimeter of the base.}$$

$$\frac{8 \times 16}{2} \text{ sq. ft.} = 64 \text{ sq. ft., area of lateral surface.}$$

The lateral surface of a square pyramid consists of four triangles having a common altitude, 8 ft. The sum of their bases is the perimeter of the base of the pyramid, or 16 ft. The area is found by multiplying the 16 ft. by 4 (half the slant height).

To find the lateral surface of a pyramid or of a cone, multiply the perimeter of the base by one half the slant height.

To find the entire surface add to the lateral surface the area of the base.

PROBLEMS

427. Find the lateral surface of pyramids and cones having dimensions as follows:

PYRAMIDS

1. Square, side of base 5 ft., slant height 10 ft.*

2. Square, side of base 4 ft., slant height 12 ft.

3. Square, side of base 12 ft., slant height 27 ft.

4. Square, side of base 3 m., slant height 1 Dm.

CONES

5. Diameter 5 ft., slant height 12 ft.

6. Circumference 21.992 m., slant height 20 m.

7. Circumference 31.416 ft., slant height 10 ft.

8. Diameter 2 m. 4 dm., slant height 40 dm.

9. A church spire is in the form of an octagonal pyramid, having a base 8 ft. on a side and a slant height of 75 ft. Find the cost of roofing at 60 ¢ a square yard.

10. The Pyramid of Cheops in Egypt is 746 ft. square, and 480 ft. high. Find the area of its sides in acres, square rods, etc., and its volume in cubic yards. How many acres does its base cover?

---

* Note the difference between "slant height" and "altitude."

SIMILAR SURFACES

428. How do the following pairs of plane figures differ?

How does the diameter of the small circle compare with the diameter of the larger one?

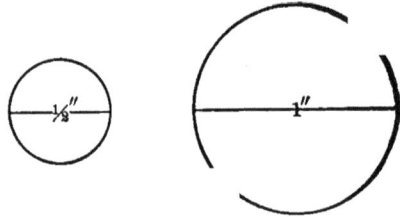

Compare the base of the larger square with the base of the smaller one.   Compare their areas.

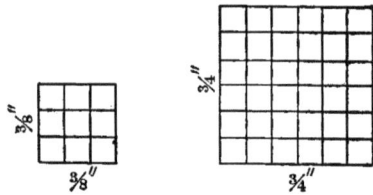

Compare the bases and the areas of the two rectangles.

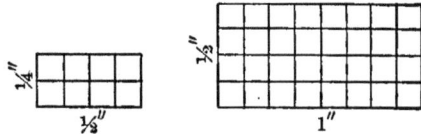

429. Surfaces that have exactly the same shape, though they differ in size, are similar surfaces.

430. Corresponding dimensions of similar surfaces are proportional.

431. PRINCIPLES: **1.** *The areas of similar surfaces are to each other as the squares of their like dimensions. And, conversely,*

2. *The like dimensions of similar surfaces are to each other as the square roots of their areas.*

432. 1. A rectangle has a base of 6 ft. and an altitude of 5 ft. Find the base of a similar rectangle whose altitude is 15 ft.

$$6 \text{ ft.} : x \text{ ft.} : : 5 \text{ ft.} : 15 \text{ ft.}$$

$$\frac{6 \times 15}{5} = 18, \text{ no. ft., base of similar rectangle.}$$

Since the like dimensions of similar rectangles are proportional, the base of the one rectangle bears the same ratio to the base of the other rectangle as the altitude of the one bears to the altitude of the other. By solving the resulting proportion, the base of the second rectangle is found to be 18 ft.

2. A circle 20 rd. in diameter has an area of 314.16 sq. rd. Find the area of a circle 50 rd. in diameter.

$$20^2 : 50^2 :: 314.16 \text{ sq. rd.} : x \text{ sq. rd.}$$

$$\frac{50 \times 50 \times \overset{78.54}{314.16}}{20 \times 20} \text{ sq. rd.} = 25 \times 78.54 \text{ sq. rd.} = 1963\tfrac{1}{2} \text{ sq. rd.}$$

The areas are proportional to the squares of the diameters. Hence the above proportion, the solution of which gives the area of the second circle as 1963¼ sq. rd.

3. The areas of two similar triangles are 81 sq. in. and 169 sq. in. If the altitude of the smaller is 9 in., find the altitude of the larger.

$$\sqrt{81} : \sqrt{169} :: 9 \text{ in.} : x \text{ in.}$$
$$9 : 13 :: 9 \text{ in.} : x \text{ in.}$$
$$\frac{9 \text{ in.} \times 13}{9} = 13 \text{ in., altitude}$$

of the larger triangle.

The like dimensions of the triangles are to each other as the square roots of the areas. Hence, the proportion as stated above. On solving, the altitude of the larger triangle is found to be 13 in.

PROBLEMS

433. 1. A rectangular field is 60 rd. long and 49 rd. wide. Find the width of a similar field 75 rd. long.

2. A field 80 rd. long contains 30 A. How many acres are there in a similar field 120 rd. long?

3. A farm one half a mile long contains 66 A. How long is a similar farm of 264 A.?

4. The area of a certain circle is 40 sq. ft. Find the area of another circle whose diameter is three times as great.

5. A man desires to find the height of a certain tree. To find it he measures its shadow, and at the same time measures the shadow of a post 7 ft. high. If the shadow of the post is 10 ft. 6 in. long and the shadow of the tree 120 ft. long, what is the height of the tree?

6. The areas of two circles are in the ratio of 16 to 25. If the diameter of the greater is 100 rd., what is the diameter of the lesser?

7. If it costs $74.50 to build a fence around a square field of 4 A., how much will it cost to build the same kind of a fence around a similar field containing 36 A.?

8. One sphere has twice the diameter of another sphere. What is the ratio of their surfaces?

9. A given cube has 384 sq. ft. of surface. What is the surface of a cube whose edge is three times that of the given cube?

SIMILAR SOLIDS

434. How do the lengths, altitudes, and widths of these solids compare?

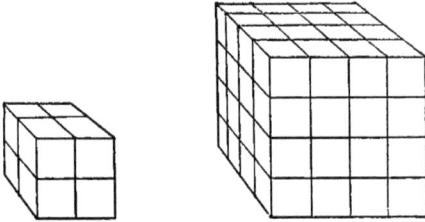

They are similar solids because they have the same *form*, though they differ in volume.

Compare their volumes. The larger solid is how many times the smaller?

435. Like dimensions of similar solids are proportional.

436. PRINCIPLES: **1.** *The volumes of similar solids are to each other as the cubes of their like dimensions.*

2. *The like dimensions of similar solids are to each other as the cube roots of their volumes.*

437. 1. A solid is 8 ft. long, 6 ft. high, and 3 ft. wide. Find the height of a similar solid 10 ft. long.

8 ft. : 10 ft. : : 6 ft. : x ft.

$\dfrac{10 \times 6}{8}$ ft. $= 7\frac{1}{2}$ ft., height of similar solid.

the height is found to be $7\frac{1}{2}$ ft.

Since the like dimensions are proportional, the lengths of the two solids are in the same ratio as the heights. Hence, the proportion is as stated. On solving,

2. The volume of a given solid 6 ft. long is 120 cu. ft. What is the volume of a similar solid twice as long?

$6^3 : 12^3 : : 120$ cu. ft. : x cu. ft.

$\dfrac{12 \times 12 \times 12 \times 120}{6 \times 6 \times 6}$ cu. ft. $= 960$ cu. ft., volume of second solid.

Since one dimension of each solid and the volume of one of the solids are given, the cubes of the given dimensions are in proportion to the volumes. On solving the resulting proportion, the volume of the second solid is found.

3. The diameter of a sphere is 3 in. What is the diameter of a sphere that is 27 times as large ? ·

$$\sqrt[3]{1} : \sqrt[3]{27} :: 3 : x.$$
$$1 : \quad 3 :: 3 : x.$$
$$\frac{3 \times 3}{1} = 9, \text{ no. in., diameter of second sphere.}$$

Since the volumes are known, the ratio of their cube roots is equal to the ratio of the diameters. The cube root of 1 is 1, and of 27 is 3. Hence, the ratio of the diameters is as 1 : 3. Therefore, the diameter of the second sphere is 9 in.

PROBLEMS

438. 1. If a bin 10 ft. long holds 160 bu. of wheat, how many bushels will a similar bin hold that is 20 ft. long ?

2. The volume of a sphere 1 in. in diameter is .5236 cu. in. Find the volume of a sphere 6 in. in diameter.

3. The city water reservoir of a small village is 40 ft. in diameter, and 40 ft. deep A new reservoir is necessary, and to provide for the future growth of the city, it is decided to erect a reservoir that will hold eight times as much water as the old one. If it is similar in form, what are its dimensions ?

4. A sphere having a 3-in. radius weighs 2 lb. 13 oz. What is the weight of a sphere of the same material whose radius is 12 in. ?

SPECIFIC GRAVITY

439. Why does a piece of wood float in water ? Why does a piece of iron sink?

Name five substances lighter than water; five heavier than water.

440. The ratio of the weight of a substance to the weight of an equal volume of water is the specific gravity of that substance. Thus, a cubic foot of cork weighs 15 lb., and a cubic foot of water weighs $62\frac{1}{2}$ lb. The ratio of 15 lb. to $62\frac{1}{2}$ lb. is expressed by the fraction $\frac{15}{62\frac{1}{2}} = 15 \div 62\frac{1}{2} = .24$ Therefore, the specific gravity of cork is .24.

441. In like manner find the specific gravity of each of the fol
lowing :

Article	Weight of Cu. Ft.		Article	Weight of Cu. Ft.
1. Gold . . .	1203⅝ lb	**8.**	Sugar	100 lb.
2. Silver . . .	656¼ lb.	**9.**	Honey . . .	91 lb
3. Copper . .	549¼ lb.	**10.**	Milk	64½ lb.
4. Cast iron .	450 lb	**11.**	Butter . . .	58⅞ lb.
5. Plate glass . .	172½ lb	**12.**	Petroleum . .	55 lb.
6. Salt . .	133⅛ lb	**13.**	Hickory, dry . .	52½ lb.
7. Brick . . .	125 lb	**14.**	Alcohol . . .	52½ lb.

Find the weight of a cubic foot of each of the following :

Article	Specific Gravity		Article	Specific Gravity
15 Platinum . . .	21.5	**20**	Beeswax	.96
16. Mercury . . .	14	**21.**	Lard	.947
17. Brass	8	**22.**	Ice	.92
18. Tin . . .	7.19	**23.**	Turpentine . . .	.87
19. Marble . .	2.72	**24.**	White pine, dry . .	.40

442.   Table showing Specific Gravities

Iridium . . .	23.000	Diamond . . .	3.550	Beeswax . . .	.960
Platinum . .	21.500	Plate glass . .	2.760	Lard	.947
Gold, pure . .	19.258	Marble . . .	2.720	Butter	.942
Mercury, pure .	14.000	Salt	2.130	Ice	.920
Silver . . .	10.500	Brick	2.000	Petroleum . . .	.880
Copper . . .	8.788	Ivory	1.870	Turpentine . .	.870
Brass	8.000	Sugar . . .	1.600	Hickory, dry . .	.840
Steel	7.840	Honey	1.456	Alcohol . . .	.840
Tin	7.290	Milk	1.032	White pine, dry .	.400
Cast iron . .	7.200	Water	1.000	Cork	.240

PROBLEMS

443. 1. A block of marble measures 4' by 2' by 1'6". How much
does it weigh ?

2. Find the weight of a plate glass window 10' by 12' by ¼".

3. A bar of cast iron is 21" long, 18" wide, and 1' thick How
much does it weigh ?

4. A brick of pure gold is $8'' \times 4'' \times 2''$. Find its weight.

5. A load of ice is 12 ft. long, 3 ft. 3 in. wide, and 2 ft. high. Allowing 8 cu. ft. for waste space, find its weight.

6. Find the weight of 1000 brick $8'' \times 4'' \times 2''$ (to the nearest pound).

Any object floating in water displaces a weight of water equal to its own weight.

Any object that sinks in water displaces a volume of water equal to its own volume.

The weight of an object heavier than water is diminished by the weight of the water it displaces when weighed under water.

7. What weight of water is displaced by a cubic foot of ice?

8. A copper rod having a volume of $1\frac{1}{2}$ cu. ft. is under water. A pull of how many pounds is necessary to lift it?

9. A dry white pine plank 12 ft. long, 1 ft. wide, and 3 in. thick will support how many pounds without sinking in water?

10. A vessel displaces 80,000 cu. ft. of water. What is its displacement? (Displacement means weight.)

11. What is the specific gravity of a substance that floats with $\frac{3}{4}$ of its volume under water?

12. A block of ice 1 ft. thick, 8 ft. long, and 4 ft. wide will support how many pounds in water?

13. A can contains 10 gallons of milk. How much does the milk weigh?

14. Into a jar full of water is placed a piece of metal weighing 120 grams. The water that overflows weighs 15 grams. What is the specific gravity of the metal?

PRACTICAL MEASUREMENTS

TIME

444. The time between two dates may be expressed in three ways; viz. compound time, exact time, and bankers' time.

445. Compound time is expressed in years, months, and days. It is determined by the method of compound subtraction. The time from Dec. 14, 1907 to Oct. 11, 1910 is reckoned thus:

1910	10	11
1907	12	14
2	9	27

The compound time is 2 yr. 9 mo. 27 da. Compound time is used in the majority of the ordinary business transactions involving time, especially if the period is more than a year.

446. The method is based on a 360-da. year — 12 mo. of 30 da each. "One half a month" is fifteen days; one quarter of a year is three calendar months; and one half of a year is six calendar months

447. A month from a given day in any month is the same day in the next month except in those months in which there is no day to correspond with the given day in the month from which the time is reckoned. For instance, 1 mo. from Jan. 14 is Feb. 14, and 5 mo. from Feb. 1 is July 1. But 1 mo. from Jan. 29, 30, or 31 is Feb. 28 (29th in leap year), and 3 mo from Aug. 31 is Nov. 30, etc.

448. There is a discrepancy of 5 da. per annum between compound time and exact time, but business men generally are willing to forego the difference on account of the facility of reckoning compound time.

449. Exact time is expressed in days, or years and days It is determined by counting the exact number of days between two dates The exact time from May 3 to Sept 28 is reckoned thus:

May, 28 days remaining.
June, 30 days.
July, 31 days.
Aug., 31 days.
Sept., 28 days.
148 days.

The first date, May 3, is omitted ; the last date, Sept. 28, is included. For exceptions, see Art. 695, note 1.

182

450. Exact time is used chiefly for periods of time less than one year, by the government in making interest calculations, by bankers in reckoning bank discount, and by some business men

451. Bankers' time is expressed in days, or months and days. It is reckoned by counting months for the whole months and exact days for any remaining part of a month. To illustrate: From Jan. 20 to June 15, is 4 mo. and 26 da. 4 mo. from Jan. 20 is May 20, and from May 20 to June 15 is 26 da.

This method is used by bankers in reckoning interest on bonds, mortgages, notes, etc., for fractional periods of time.

PROBLEMS

452. Find the compound time between:

1. May 15, 1907 and Aug. 12, 1910.
2. Apr. 4, 1903 and Jan. 3, 1909.
3. Dec. 26, 1906 and Nov. 13, 1910.
4. Oct. 13, 1907 and June 1, 1910.
5. Sept. 1, 1905 and July 30, 1910.
6. Sept. 28, 1906 and Feb. 14, 1909.
7. Nov. 20, 1904 and Jan. 1, 1911.
8. Dec. 31, 1908 and May 1, 1910.
9. June 30, 1906 and May 15, 1909.
10. July 4, 1905 and Dec. 25, 1909.

Find the exact time and the bankers' time between:

11. May 4 and Oct. 9.		16. Feb. 18 and June 12
12. Apr. 24 and Dec. 1.		17. May 13 and Dec. 10.
13. Feb. 15 and Nov. 10.		18. June 28 and Nov. 4.
14. June 30 and Nov. 20.		19. Apr. 16 and Sept. 5.
15. Jan. 15 and July 1.		20. Mar. 30 and Sept. 8.

Find the compound time and the exact time between:

21. Apr. 1, 1907 and Jan. 30, 1909.
22. Mar. 30, 1908 and Feb. 28, 1909.
23. Dec. 26, 1906 and Jan. 1, 1911.
24. Nov. 20, 1908 and July 21, 1910.
25. Oct. 18, 1907 and July 8, 1910.
26. Sept. 17, 1908 and Aug. 12, 1909.

27. July 16, 1907 and June 1, 1910.

28. June 14, 1908 and May 5, 1910.

453. This is a condensed time-table of the Atchison, Topeka and Santa Fe Railway, extending from Chicago to San Francisco.

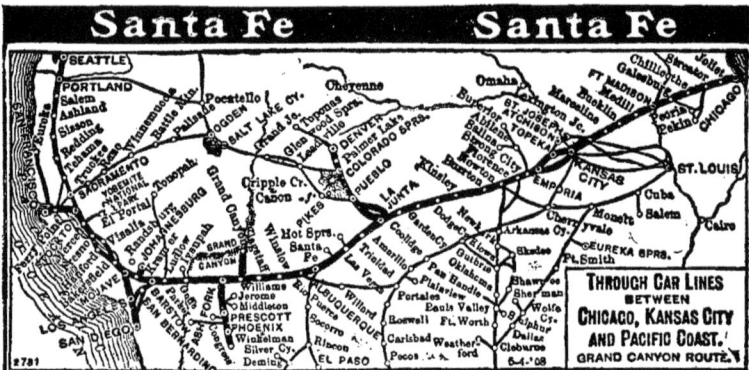

Chicago, Kansas City and Pacific Coast.
(Grand Canyon Line.)   WEST.

Mls.	STATIONS		California Limited	By Days						
0	Lv Chicago............Santa Fe		8.00 PM	Sun.	Mon.	Tues.	Wed.	Thu.	Fri.	Sat.
458	Ar Kansas City .... "		9.00 AM	Mon.	Tues.	Wed.	Thu.	Fri.	Sat.	Sun.
....	Lv St. Louis......Burlington		11.28 PM	Sun.	Mon.	Tues.	Wed.	Thu.	Fri.	Sat.
....	" St. Louis......C. & A.		11.28 PM	"	"	"	"	"	"	"
....	" St. Louis......Mo. Pacific		11.30 PM	"	"	"	"	"	"	"
....	" St. Louis......Rock Island		10.02 PM	"	"	"	"	"	"	"
....	Lv St. Louis..........Wabash		11.31 PM	"	"	"	"	"	"	"
458	Lv Kansas City ....Santa Fe		9.10 AM	Mon.	Tues.	Wed.	Thu.	Fri.	Sat	Sun.
586	" Emporia......... "		12.30 PM	"	"	"	"	"	"	"
660	" Newton.........: "		2.25 PM	"	"	"	"	"	"	"
827	" Dodge City..... "		5.35 PM	"	"	"	"	"	"	"
1029	" La Junta .... "		11.20 PM	"	"	"	"	"	"	"
1245	" Las Vegas ..... "		7.00 AM	Tues.	Wed.	Thu.	Fri.	Sat.	Sun.	Mon.
1377	Lv Albuquerque .. "		1.00 PM	"	"	"	"	"	"	"
1663	Ar Winslow......... "		9.00 PM	"	"	"	"	"	"	"
1755	Ar Williams........ "		12.50 AM	Wed.	Thu.	Fri.	Sat.	Sun.	Mon.	Tues.
1819	Ar Grand Canyon..G. C. Ry.		9.00 AM	"	"	"	"	"	"	"
1819	Lv Grand Canyon.. "		8.30 PM	Tues.	Wed.	Thu.	Fri.	Sat.	Sun.	Mon.
1778	Ar Ash Fork ......Santa Fe		1.52 AM	Wed.	Thu.	Fri.	Sat.	Sun.	Mon.	Tues.
1835	Ar Prescott ...... S.F.P.&P.		10.12 AM	Wed.	Thu.	Fri.	Sat,	Sun.	Mon.	Tues.
1972	Ar Phœnix .... "		4.40 PM	"	"	"	"	"	"	"
1955	Ar Needles ....Santa Fe		6.35 AM	"	"	"	"	"	"	"
2124	Ar Barstow .... "		12.30 PM	"	"	"	"	"	"	"
2205	Ar San Bernardino. "		3.40 PM	"	"	"	"	"	"	"
2215	" Riverside....... "		4.00 PM	"	"	"	"	"	"	"
2214	" Redlands .... "		4.37 PM	"	"	"	"	"	"	"
2255	" Pasadena....... "		5.31 PM	"	"	"	"	"	"	"
2265	" Los Angeles .... "		6.00 PM	"	"	"	"	"	"	"
2347	Ar San Diego ...... "		7.00 AM	Thu.	Fri.	Sat.	Sun.	Mon.	Tues.	Wed.
2265	Lv Los Angeles ..So. Pacific		7.30 PM	Wed.	Thu.	Fri.	Sat.	Sun.	Mon.	Tues.
2369	Ar Santa Barbara "		*11.15 PM	"	"	"	"	"	"	"
2124	Lv Barstow ....Santa Fe		1.15 PM	"	"	"	"	"	"	"
2195	Lv Mojave........ "		3.10 PM	"	"	"	"	"	"	"
2265	Ar Bakersfield ..... "		7.00 PM	"	"	"	"	"	"	"
2375	" Fresno.......... "		11.50 PM	"	"	"	"	"	"	"
2498	" Stockton .... "		4.10 AM	Thu.	Fri.	Sat.	Sun.	Mon.	Tues.	Wed.
2577	" Oakland .... "		7.08 AM	"	"	"	"	"	"	"
2576	Ar San Francisco .. "		7.35 AM	"	"	"	"	"	"	"

For equipment, see table A, page 13. * Passengers may remain in Pullman until 7.00AM

454. Use Central Time from Chicago to Dodge City; Mountain Time from Dodge City to Grand Canyon, and Pacific Time from Needles to San Francisco.

1. Find the average speed of the "California Limited" from Chicago to Kansas City; from Kansas City to Albuquerque; from Chicago to Grand Canyon; from Kansas City to Los Angeles; from Albuquerque to Needles; from Needles to San Francisco; from Chicago to San Francisco.

2. Find the running time from Chicago to Dodge City; from Chicago to La Junta; from Chicago to San Diego; from Chicago to Oakland; from Kansas City to Ash Fork; from Needles to Los Angeles; from Kansas City to San Francisco.

455. Counting forward or backward from a given date.

1. Count forward 76 da. from June 23.

76 da
7 da. remaining in June.
69
31 da in July
38
31 da. in August
7 September.
Therefore, September 7.

There are 7 da. left in June after June 23. Subtract 7 da. from 76 da., obtaining 69 da. to count forward after June 30. Subtract 31 da. for July, which leaves 38 da. to count forward after July 31. Subtract 31 da. for August, which leaves 7 da. to count into September, or September 7.

2. Count back 83 da. from May 10.

83 da.
10 da. back in May.
73
30 da. in April
43
31 da. in March
12 da. back into Feb.
Feb. 28 − 12 da. =
Feb 16

In counting backward from May 10, note that 10 da. are subtracted for May, leaving 73 da. to count back from Apr. 30. Subtract 30 da. for April, and 31 da. for March. The 12 da. remaining are to be deducted from the 28 da. in February, which leaves February 16.

NOTE. Fractional parts of a day are not recognized in law. Hence, as soon as it is past midnight on May 9, it is May 10, and the whole day is considered, in law, to be gone. Therefore, in counting back from the 10th, there are 10 days to count back through to count back out of May.

PROBLEMS

456. Count both forward and backward:

1. 95 da. from July 10.
2. 110 da. from June 30.
3. 57 da. from Apr. 15.
4. 132 da. from June 1.
5. 160 da. from June 30.
6. 173 da. from Oct. 15, 1909.
7. 208 da. from Sept. 1, 1909.
8. 142 da. from Dec 30, 1908.
9. 111 da. from Jan 10, 1909.
10. 131 da. from Feb. 1, 1909.

11. 168 da. from Mar. 31, 1909.
12. 174 da. from Feb. 28, 1908.
13. 60 da. from July 1, 1909.
14. 90 da. from May 1, 1909.
15. 89 da. from May 31, 1909.
16. 101 da. from Nov. 1, 1909.
17. 201 da. from July 31, 1909.
18. 179 da. from Aug. 19, 1909.
19. 120 da from Apr 30, 1909
20. 229 da. from Dec. 1, 1909.

PAINTING, PLASTERING, PAPERING, CARPETING

457. Painting and plastering are estimated by the square yard.

There is no definite rule regarding allowances for openings. It is a matter for special agreement, and should be mentioned in the contract when made.

458. Papering is estimated by the roll.

459. Most American wall papers are 18 in. wide, a single roll being 24 ft. long, and a double roll, 48 ft. long. Imported papers vary in width and length.

460. As there is always more or less waste in cutting and matching wall paper, it is better to use double rolls. The exact number of rolls cannot always be determined in advance.

461. In estimating the number of rolls for a room, paper hangers generally deduct the total width of openings from the perimeter of the room. Then they reckon the number of strips required to cover the remaining surface. The number of rolls is found by dividing the total number of strips required by the number of whole strips that can be cut from 1 roll.

The spaces over and under the windows, over the doors, etc., are covered with the parts of strips left from each roll.

462. Carpet is estimated and sold by the linear yard. Linoleum is generally sold by the square yard.

463. Carpets vary in width, the more common widths being 1 yd. for Ingrain, and $\frac{3}{4}$ yd. for Axminster, Brussels, Moquette, Velvet, and Wilton.

464. Carpet may be bought in any *length* desired, but only in *whole* strips; hence, in order to determine the number of yards for a floor, the number of strips must be known.* The number of yards is then found by multiplying the yards in 1 strip by the number of strips.

465. A room is 16 ft. long, 14 ft. wide, and $8\frac{1}{2}$ ft. high. It has 3 windows each 3 ft. by 6 ft. and 1 door 3 ft. by 7 ft. Find the total cost of plastering the walls and ceiling at 50 ¢ per square yard, allowing for the openings; of carpeting the floor with Brussels carpet at $1.75 per yard; and of papering the walls at $1.60 per double roll.

PLASTERING

$2 \times (16 \text{ ft.} + 14 \text{ ft.}) = 60 \text{ ft.}$, perimeter of room.

$8\frac{1}{2} \times 60$ sq. ft. $= 510$ sq. ft. in the walls.

16×14 sq. ft. $= 224$ sq. ft. in the ceiling.

510 sq. ft. $+ 224$ sq. ft. $= 734$ sq. ft. in ceiling and walls.

$3 \times 3 \times 6$ sq. ft. $= 54$ sq. ft. in the windows.

3×7 sq. ft. $= 21$ sq. ft. in the door.

54 sq. ft. $+ 21$ sq. ft. $= 75$ sq. ft. in all openings.

734 sq. ft. $- 75$ sq. ft. $= 659$ sq. ft., net area of ceilings and walls.

$$\frac{659 \times \$.50}{9} = \$36.61, \text{ cost of plastering.}$$

CARPETING

$14 \text{ ft.} \div 2\frac{1}{4} \text{ ft.} = 6\frac{2}{9} = 7$, no. strips.

$$\frac{7 \times 16}{3} = 37\frac{1}{3}, \text{ no. yd. of carpet.}$$

$37\frac{1}{3} \times \$1.75 = \$65.33$, cost of carpeting.

---

*If any whole number of strips exactly covered a floor, and there were no waste in matching the pattern, the number of yards for a floor could be found by dividing the area of the floor by the area of a yard of carpet, but these conditions occur so rarely that the statement above may be taken as correct.

PAPERING

$2 \times (16 \text{ ft.} + 14 \text{ ft.}) = 60 \text{ ft.}$, perimeter of room.

$4 \times 3 \text{ ft.} = 12 \text{ ft.}$, total width of openings.

$60 \text{ ft.} - 12 \text{ ft.} = 48 \text{ ft.}$, net perimeter.

$48 \text{ ft.} \div 1\frac{1}{2} \text{ ft.} = 32$ strips required.

$48 \text{ ft.} \div 8\frac{1}{2} \text{ ft.} = 5$ whole strips from 1 roll.

$32 \text{ strips} \div 5 \text{ strips} = 6\frac{2}{5} = 7$, no. rolls.

$7 \times \$1.60 = \$11.20$, cost of papering.

$\$36.61 + \$65.33 + \$11.20 = \$113.14$, total cost.

Plastering. The area of the walls is the perimeter of the room multiplied by the height. The ceiling is a rectangle whose area is 224 sq. ft., making the total area of the walls and ceiling equal to the sum of 510 sq. ft. and 224 sq. ft., or 734 sq. ft. The windows and the door are rectangles whose total area is 75 sq. ft. Since allowance is made for the openings, the net area to be plastered is found by deducting the 75 sq. ft. from 734 sq. ft., leaving 659 sq. ft. Dividing by 9 sq. ft. to the square yard and multiplying $.50 by the quotient, gives the cost of plastering as $36.61.

Carpeting. Unless otherwise specified, carpet is generally considered to be laid with strips running lengthwise of the room. Hence, to find the number of strips, divide the width of the floor by the width of the carpet. $14 \text{ ft.} \div 2\frac{1}{4} = 6\frac{2}{9}$, the number of strips, but as carpet has to be bought in whole strips, 7 strips are necessary. Each strip is 16 ft. long. The number of yards, therefore, is $\frac{1}{3}$ of 7 times 16, or $37\frac{1}{3}$. At $1.75 per yard, the cost is $65.33.

Papering. From the perimeter of the room deduct the total width of the windows and the door, leaving 48 ft. as the net perimeter to be papered. There are as many strips of paper required as $1\frac{1}{2}$ ft., the width of one strip, is contained times in 48 ft., or 32. The number of strips from one roll is found by dividing the length of a roll, 48 ft., by the length of one strip, $8\frac{1}{2}$ ft., which gives 5 strips from a roll. Dividing 32 strips by 5 strips gives $6\frac{2}{5}$, or 7 rolls, which, at $1.60 a roll costs $11.20. The total cost is found by adding the costs of the several kinds of work.

466. In writing the dimensions of rooms it is customary to write the length first, then the width, then the height. It is also usual to write 10 ft. thus, 10′, and 8 in., 8″. Hence a room 15 ft. long, 13 ft. wide, and 8 ft. 4 in. high, may be written "a room $15' \times 13' \times 8' 4''$."

PROBLEMS

467. Five rooms have dimensions and openings as follows:

1. $18' \times 16' \times 9'$; 2 windows $3' \times 6'\ 6''$; 2 doors $3'\ 6'' \times 7'$.
2. $16' \times 12' \times 8'\ 6''$; 3 windows $4' \times 5'\ 6''$; 1 door $3' \times 6'\ 6''$.
3. $26' \times 16' \times 10'\ 6''$; 4 windows $4' \times 7'\ 6''$; 3 doors $4' \times 8'$.
4. $22'\ 6'' \times 18' \times 9'\ 6''$; 3 windows $4' \times 7'$; 2 doors $3'\ 6'' \times 8'$.
5. $19'\ 6'' \times 15' \times 9'$; 3 windows $5' \times 6'$; 2 doors $3' \times 7'$.

For the first room find the total cost of plastering walls and ceiling at $60\cent$ per square yard, allowing for one half the openings, of papering the walls at $1.20 per double roll, and of carpeting with Wilton carpet at $1.40 per yard.

For the second room find the total cost of plastering the walls and ceiling (no allowance for openings) at $70\cent$ per square yard, of papering the walls at $90\cent$ a double roll, and of painting and varnishing the floor at $75\cent$ per square yard

For the third room find the total cost of kalsomining the ceiling at $80\cent$ per square yard, of papering the walls at $2.50 per double roll, and of carpeting the floor with velvet carpet at $2.25 per yard.

For the fourth room find the total cost of plastering walls and ceiling at $65\cent$ per square yard (no allowance), of papering walls and ceiling at $1.75 per double roll, and of carpeting the floor with Moquette carpet at $1.80 per yard.

For the fifth room find the total cost of plastering the walls and ceiling at $55\cent$ a square yard (100 sq. ft. for openings), of papering the walls at $1.30 per double roll, and of carpeting the floor with Ingrain carpet at $1.10 per yard.

The diagram on the following page is the first-floor plan of a modern farmhouse.

The rooms have a uniform height of $8'\ 6''$. The dining room and kitchen each have a wainscot $3'$ high. All others have a base board $9''$ wide. All windows are $5'\ 9'' \times 3'$. Outside doors are $7' \times 3'\ 4''$; inside doors are $7' \times 3'\ 2''$. The double doors between the sitting room and the parlor are $7' \times 5'\ 8''$. (Measurements of windows and doors include casings.)

6. Find the number of square yards of plaster required for the walls of the sitting room and parlor Make allowance for openings and baseboard.

7. How many square yards of plaster are there in the walls and ceilings of the three chambers ?   Make all allowances.

8. How many square yards of linoleum will cover the floors of the kitchen and the pantry?

9. Find the cost of ingrain carpet at $.90 a yard for the dining room (carpet to be laid most economical way).

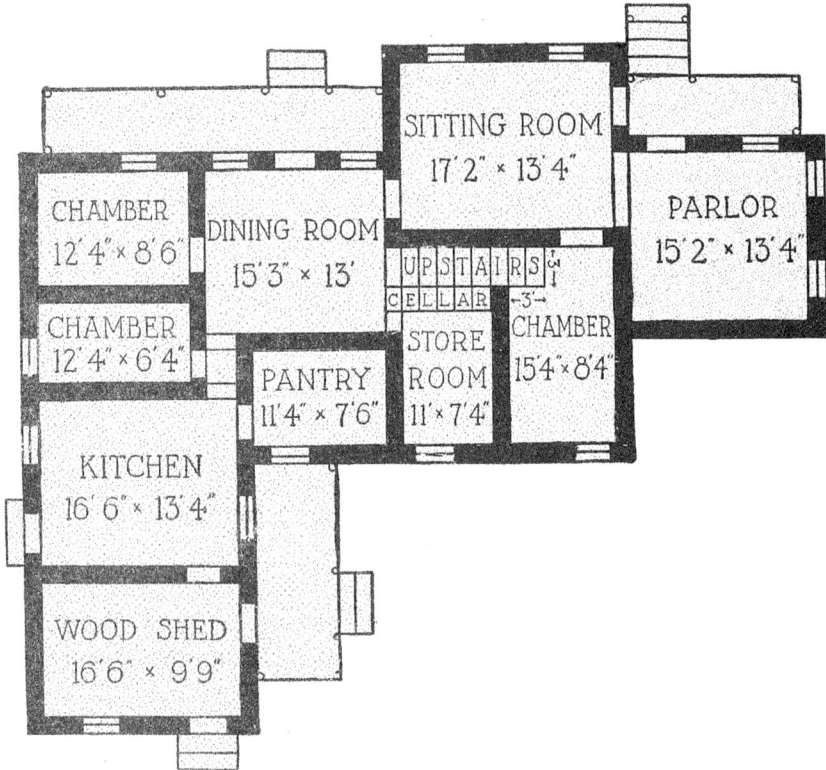

10. How much will it cost to paper the dining room at 60 ¢ a double roll?

11. If the kitchen walls are painted above the wainscot, find the the cost at 10 ¢ per square yard.

12. Find the cost of papering the walls and the ceiling of the parlor at $1.25 per double roll.

13. How many double rolls of paper are required for the three chambers, allowing for one foot of waste for each strip in matching the pattern ?

14. How many yards of Brussels carpet will cover the floor of the sitting room ?

15. At $1.35 per yard, find the cost of Wilton carpet for the parlor.

FLOORING

468. Flooring is estimated by the square or by the thousand board feet.

469. When lumber is "tongued" and "grooved," or matched, as it is called, there is some waste, as lumber dealers always measure the lumber at its full width before it is matched. The amount of waste depends upon the width of the boards. Carpenters generally allow one fifth for waste. That is, for 1000 sq. ft. of floor space, 1200 sq. ft. (board feet) of flooring is needed.

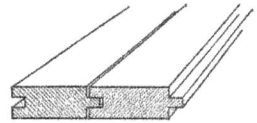

Matched Lumber

470. How many square feet of flooring are required for the parlor represented on page 190 ?

$$15' \ 2'' = 15\tfrac{1}{6}'; \ 13' \ 4'' = 13\tfrac{1}{3}';$$

$$\frac{\overset{8}{\cancel{91}}}{\cancel{6}} \times \frac{\overset{40}{\cancel{40}}}{3} \times \frac{\cancel{6}}{\cancel{5}} = \frac{728}{3} = 242\tfrac{2}{3},$$

Ans. 243 sq. ft.

Reduce feet and inches to feet and multiply the length of the room by the width. The allowance for waste is best reckoned by multiplying by $\tfrac{6}{5}$. Using the cancellation form of solution, the result is readily obtained.

PROBLEMS

471. 1. A floor is 18' 6" × 14' 4". How many square feet of lumber are required to build it ?

2. Find the total amount of lumber required for all the floors in the diagram on page 190, except the woodshed. How much will it cost at $55 per 1000 ?

3. Find the cost of a mosaic floor 60' × 32' if the tile cost $1.25 per square foot, and the cost of laying is $8 per square.

ROOFING

472. Roofing is generally estimated by the square of 100 square feet.

473 The most common roofing materials are shingles, slate, and tin.

474. Shingles are estimated as having an average width of 4 in., and are laid 4 in., $4\frac{1}{2}$ in., 5 in., or $5\frac{1}{2}$ in. to the weather, depending on the pitch of the roof. The steeper the pitch the greater the length exposed to the weather. The usual estimate per square is shown in the following table:

TABLE

LENGTH OF SHINGLE EXPOSED TO THE WEATHER	NUMBER OF SHINGLES PER SQUARE
4 inches	1000
$4\frac{1}{2}$ inches	900
5 inches	800
$5\frac{1}{4}$ inches	700

475. A bunch contains 250 shingles. A part of a bunch is not sold. Estimates, therefore, have to be made in whole bunches.

476. The pitch of a roof is numerically expressed by dividing the height by the span.

When the height of the ridgepole is $\frac{1}{4}$ of the span of the roof above the building, the pitch of the roof is $\frac{1}{4}$.

477. The accompanying diagram shows how the carpenter determines, or "lays off," some of the more common pitches.

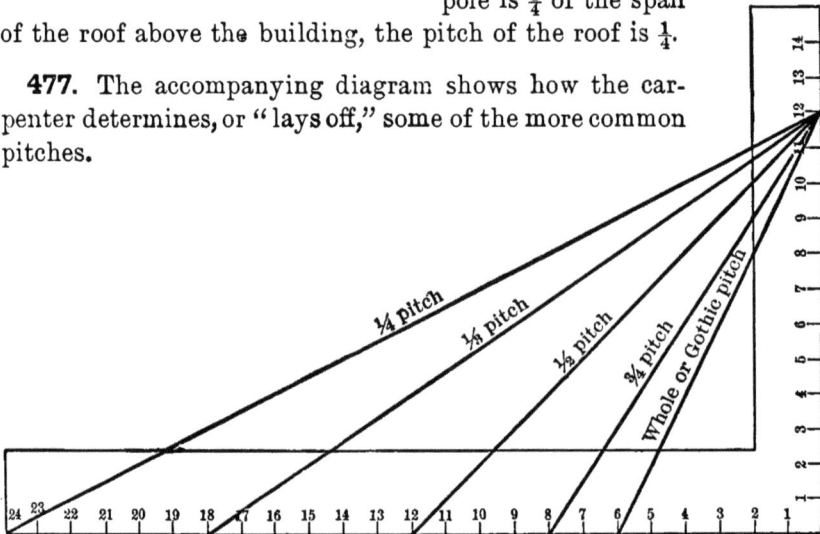

PROBLEMS

478. **1.** Using a span of 36 ft., make a diagram illustrating one third pitch; one fourth pitch; five twelfths pitch; five eighths pitch; equilateral or true pitch (60°).

2. If the foot of the rafters projects 14″ over the side of the building, find the length of the rafters for each of the pitches in 1.

The rafter is the hypotenuse of a right triangle.

A roof is 60 ft. long and 30 ft. wide on each side. How many shingles laid 5 in. to the weather must be bought to cover it?

$$\frac{2 \times 60 \times 30}{100} = 36 \text{ squares.}$$

36×800 shingles $= 28,800$ shingles $= 29,000$ shingles.

Each side of the roof is a rectangle 60 ft. by 30 ft. The area of both sides divided by 100 sq. ft. in a square, gives 36 squares. It takes 800 shingles for 1 square, and for 36 squares it takes 36 times 800 shingles or 28,800 shingles, making it necessary to buy 29,000 shingles.

4. A building is 36 ft. wide. The pitch of the roof is one half and the rafters project 15 in. How many shingles laid $4\frac{1}{2}$ in. to the weather are required if the ridgepole is 50 ft. long?

5. How many slates 6 in. wide, 5 in. to the weather, are required for a building 24 ft. wide, the pitch of the roof being five eighths, the ridgepole 36 ft. long, and the rafters projecting 15 in.?

6. A roof 44 ft. long and 18 ft. wide on each side is covered with tin. If each sheet is $18'' \times 27''$, how many sheets are there?

7. How many slates are needed to cover a roof $32' \times 16'$ on a side if the slates are $6''$ wide and are exposed $8''$ to the weather?

8. The diagram shows part of the end of a barn having a gambrel roof. If the projection of the rafters is $18''$ and the ridgepole is 59 ft. long, how many shingles are required for the roof if they are laid $5\frac{1}{2}$ in. to the weather on the lower half of the roof and 4 in. to the weather on the upper half?

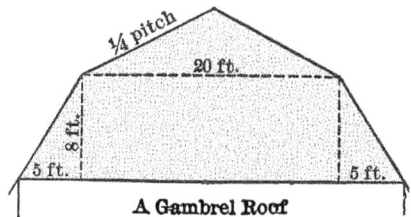

A Gambrel Roof

PAVING

479. Paving is estimated by the square foot or by the square yard.

PROBLEMS

480. 1. Find the cost of paving a street 8 rd. long and 30 ft. wide, at $1.75 per square foot.

2 How many paving blocks $4'' \times 8''$ are required for a section of street 80 rd. long and 36 ft. wide? Find the cost of paving at $14 per square yard

3. If paving stones are $5'' \times 9''$, how many will it take to pave a street 1 mi. long and 40 ft. wide?

4. A man who owns a corner lot with a frontage of 75 ft. and a depth of 100 ft., builds a cement walk 6 ft. wide on the side and front. If he puts the walk on his lot, find the cost at $3.75 per square yard.

PAPER AND BOOKS

481. Paper is of various kinds, and has many uses, some of which are:

1. Wrapping—made of straw, manila hemp, or wood pulp. It is used for wrapping bundles, making paper bags, etc. It is one of the cheapest kinds of paper.

2. News—made of wood pulp. It is of a higher grade than wrapping paper, and is used in making newspapers, and cheap books and magazines. It is sold chiefly in large rolls.

3. Book—made of wood pulp and cotton cloth (called "paper rags"). It is better than news and is used in making the better class of books and magazines. It is sold chiefly in large unfolded sheets; also in large rolls like news.

4. Flat—made of wood pulp and cotton or linen cloth, mixed with an animal size, or glue. The sizing produces a firm smooth surface and prevents the ink from spreading in the paper when written upon with a pen. It is used for correspondence, and for records and documents in which permanency is desired. It is sold wholesale in large unfolded sheets. For retail purposes it is cut into various-sized sheets dependent upon the use to which it is to be put.

482. Paper is sold in large quantities by weight; in small quantities, by the quire or ream. In some cases a ream consists of 500 sheets instead of 480.

483.　　　　　Common Sizes of Paper

Book	Flat	
$25'' \times 38''$	Cap	$14'' \times 17''$
$28'' \times 42''$	Demy	$16'' \times 21''$
$30\frac{1}{2}'' \times 41''$	Folio	$17'' \times 22''$
$32'' \times 44''$	Medium	$18'' \times 23''$
$35'' \times 46''$	Super Royal . . .	$20'' \times 28''$

The $25'' \times 38''$ size is the basis from which the weights of other sizes of book paper are reckoned. A ream $25'' \times 38''$ weighs 40 lb., 50 lb., 60 lb., etc., according to the quality of the paper.

484. Formerly the terms folio, quarto, 8vo, 12mo, etc., indicated the number of times a sheet was folded in making a book. For instance, folio was the name given to a book made of sheets which had been folded but once, thus making two leaves or four pages from each sheet; a quarto indicated a book made of sheets folded twice, making four leaves or eight pages; etc.

At the present time the terms folio, quarto, octavo, or 8vo, have reference to the size of the page, regardless of the number of times a sheet is folded.

485. The following sizes are the commercial standard board (cover) measurements of various kinds of books:

Quarto	$12\frac{3}{8}'' \times 8\frac{1}{4}''$	16mo	$6\frac{7}{8}'' \times 4\frac{1}{4}''$
8vo	$9\frac{3}{4}'' \times 6''$	18mo	$6\frac{1}{4}'' \times 3\frac{7}{8}''$
12mo	$7\frac{5}{8}'' \times 4\frac{3}{4}''$	32mo	$4\frac{5}{8}'' \times 3\frac{1}{8}''$

In practice any of the above sizes may run large or small; hence it is impossible to tell the exact size of a book by the terms folio, quarto, octavo, etc.

486. Most books, except very large ones, are printed in 16's or 32's; that is, each sheet has 16 or 32 pages on each side. The sheets are then folded and cut into two or four sections called "signatures." Each signature contains 16 pages. Hence a sheet is folded to make either 32 or 64 pages.

487. 1. If a sheet is folded into 4 signatures, each page $7\frac{1}{2}'' \times 4\frac{7}{8}''$, which of the above sizes of paper is the most economical to use?

4 signatures = 64 pages = 32 pages on one side of a sheet.

32 pages = 8 pages by 4 pages.

$4 \times 7\frac{1}{2}'' = 30''$, width of sheet.

$8 \times 4\frac{7}{8}'' = 39''$, length of sheet.

Hence the size $30\frac{1}{2}'' \times 41''$ is best.

2. A book containing 448 pages is printed in 32's. If the pages are $7\frac{3}{4}'' \times 5\frac{1}{4}''$, what size paper would be used? How many reams (500 sheets) would be required to print 5000 copies of the book?

32's means 32 pages on one side of a sheet.

32 pages = 8 pages by 4 pages.

$4 \times 7\frac{3}{4}'' = 31''$, width of sheet.

$8 \times 5\frac{1}{4}'' = 42''$, length of sheet.

Hence, use sheets $32'' \times 44''$.

448 pages ÷ 64 pages (on 1 sheet) = 7, no. sheets in 1 book.

$$\frac{5000 \times 7}{500} = 70, \text{ no. reams.}$$

3. If 70 sheets of strawboard $26'' \times 38''$ weigh 50 lb., what will be the cost of 1000 sheets of the same grade of strawboard $8\frac{1}{2}'' \times 17''$ at $28 per ton ?

First cause : Second cause : : First effect : Second effect.

70 sheets　　: 1000 sheets
26 in.　　　: $8\frac{1}{2}$ in.　　　∴ 50 lb.　　: ? lb.
38 in.　　　: 17 in.

$$\frac{1000 \times 8\frac{1}{2} \times 17 \times 50}{70 \times 26 \times 38} \text{ lb.} = 104\frac{1}{2} \text{ lb.}$$

$28 \div 2000 = \$.014$ per pound.
$104\frac{1}{2} \times \$.014 = \$1.46$, cost.

By the "cause and effect" method of stating the proportion, it is readily seen that the first cause is the 70 sheets $26'' \times 38''$, and that the corresponding effect is 50 lb. The second cause is the 1000 sheets $8\frac{1}{2}'' \times 17''$, leaving the second effect to be found. Dividing the product of the means by the given extreme gives the other extreme, or $104\frac{1}{2}$ lb., the weight of 1000 sheets $8\frac{1}{2}'' \times 17''$. Multiplying by the price per pound gives $1.46, the required cost.

With a little practice the student should be able to write the above statement directly in the form for solution, thus :

$$\frac{50 \times 8\frac{1}{2} \times 17 \times 1000 \times \$28}{70 \times 26 \times 38 \times 2000} = \$1.46$$

reasoning in this way : If 70 sheets $26'' \times 38''$ weigh 50 lb., one sheet will weigh $\frac{1}{70}$ of 50 lb., and 1 sq. in. of 1 sheet will weigh as much as 26×38 is contained times in the weight of one sheet. Thus far the statement would be $\dfrac{50}{70 \times 26 \times 38} =$ weight of 1 sq. in. If 1 sq. in. of 1 sheet weighs that much, one sheet $8\frac{1}{2}'' \times 17''$ will weigh $8\frac{1}{2} \times 17$ times that, and 1000 sheets will weigh 1000 times that result. To this point the statement is $\dfrac{50 \times 8\frac{1}{2} \times 17 \times 1000}{70 \times 26 \times 38}$. Having the weight, divide by 2000 lb. to a ton, and multiply by the price per ton. The complete statement is as given above.

PROBLEMS

488. 1. If two signatures are printed from one sheet, what size paper is best for a book whose pages are $11\frac{1}{4}'' \times 8\frac{1}{2}''$? .

2. A 24-signature book is printed in 32's. If the pages are $6\frac{3}{4}'' \times 5''$, what size of sheet would be best to use ? How many volumes could be printed from 100 reams (480 sheets) of paper ?

3. If 1 ream (500 sheets) of book paper $26'' \times 38''$ weighs 50 lb., find the cost of 70 reams of the same grade $28'' \times 42''$ at $85 per ton.

4. A ton of flat paper $17'' \times 22''$ costs \$.15 a pound and is made into 10 ¢ notebooks $5\frac{1}{2}'' \times 8\frac{1}{2}''$, of 64 pages each. If 12 sheets of paper weigh 1 lb., and the cost of manufacturing the books including the covers is \$1 per hundred, find the profit.

5. If a ream of Demy weighs 50 lb., and a firm is charged for 1075 lb. at 22 ¢ per pound for 80 reams $8'' \times 10\frac{1}{2}''$, how much too great is the charge? (A discrepancy of 1 lb. in 20 lb., or 5%, is not considered an overcharge in the paper trade.)

The regulation size of letter paper is $8'' \times 10''$; of note paper is $5'' \times 8''$. (Commercial note is $5\frac{1}{2}'' \times 8\frac{1}{2}''$.)

6. How many reams of paper, letter size, can be cut from 2000 sheets $16'' \times 21''$?

7. If 24 reams $17'' \times 22''$ (500 sheets to a ream) are cut into commercial note size and sold at \$1.10 per ream (480 sheets to a ream), what is the amount received?

8. How many pads, 80 sheets each, $5\frac{3}{4}'' \times 9''$, can be made from 4000 sheets $18'' \times 23''$?

9. How many reams (500 sheets to a ream) Super Royal are required for 500 ledgers of 320 pages each, if the page is $10'' \times 14''$?

10. A publisher has an order for 10,000 copies of a book containing 512 pages, each page $8\frac{1}{2}'' \times 5\frac{1}{4}''$. He prints in 32's, and the paper costs \$100 per ton. If the same grade of paper $25'' \times 38''$ weighs 50 lb. to the ream, find the cost of the paper for the order.

LUMBER

489. The unit of lumber measure is the **board foot**. A board foot is a square foot of board one inch (or less) thick. Each of these illustrations represents a board foot of lumber.

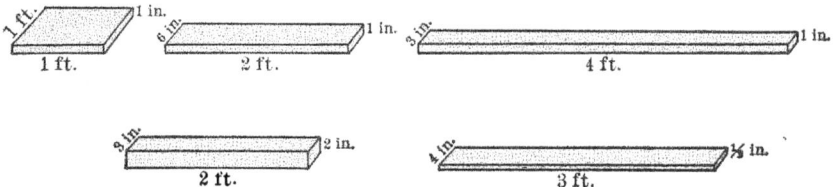

The volume of a board foot of lumber is thus seen to be **144 cu. in.**, except when the lumber is less than 1 in. thick, in which case a board foot is equal to a square foot of surface of the board.

490. **1.** How many feet of lumber are there in a board $16' \times 9''$?

$$9 \text{ in.} = \tfrac{3}{4} \text{ of a foot.}$$
$$\tfrac{3}{4} \text{ of } 16 = 12 \text{ board feet; or}$$
$$16 - (\tfrac{1}{4} \text{ of } 16) = 12 \text{ board feet.}$$

Reduce the width to feet. 9 in. $= \tfrac{3}{4}$ of a foot. The number of board feet is equal to $\tfrac{3}{4}$ of the length. $\tfrac{3}{4}$ of $16 = 12$ board feet; or since $9''$ is $\tfrac{1}{4}$ less than a foot, deduct $\tfrac{1}{4}$ of 16 from itself.

2. How many board feet are there in 48 scantlings $2'' \times 4'' \times 16'$ (2 in. $\times 4$ in. $\times 16$ ft. long)?

$$\frac{\overset{4}{\cancel{48}} \times 16 \times 2 \times 4}{\cancel{12}} = 512, \text{ no. of board feet in 48 scantlings.}$$

When there is more than one piece of lumber, the simplest method is to find the total length of all the pieces, and then multiply by the width and by the thickness, in inches, and divide the product by 12, using cancellation, as shown in the solution.

LUMBER YARD PRACTICE

491. To find the number of board feet in one piece of lumber

4 inches wide, take $\tfrac{1}{3}$ of the length.
6 inches wide, take $\tfrac{1}{2}$ of the length.
8 inches wide, take $\tfrac{1}{3}$ less than the length.
9 inches wide, take $\tfrac{1}{4}$ less than the length.
10 inches wide, take $\tfrac{1}{6}$ less than the length.
12 inches wide, take the length.
14 inches wide, add $\tfrac{1}{6}$ to the length.
15 inches wide, add $\tfrac{1}{4}$ to the length.
2 in. by 4 in. ($2 \times 4 = 8$) take $\tfrac{1}{3}$ from the length.
2 in. by 8 in. ($2 \times 8 = 16$) add $\tfrac{1}{3}$ to the length.
8 in. by 8 in. ($8 \times 8 = 64$) take $5\tfrac{1}{3}$ times the length, etc.

NOTE   When no thickness is mentioned, lumber is understood to be one inch thick.

PROBLEMS

492. Find, mentally when possible, the number of board feet in the following:

1. 20 pcs. $3'' \times 8'' \times 14'$ **4.** 90 pcs. $2'' \times 4'' \times 16'$

2. 44 pcs. $4'' \times 4'' \times 18'$ **5.** 160 pcs. $6'' \times 8'' \times 12'$

3. 60 pcs. $12'' \times 1'' \times 16'$ **6.** 16 pcs. $10'' \times 12'' \times 18'$

7.	120 pcs.	$9'' \times 11'' \times 16'$	**13.** 28 pcs. $6'' \times 12'' \times 18'$
8.	75 pcs.	$4'' \times 6'' \times 14'$	**14.** 38 pcs. $12'' \times 12'' \times 16'$
9.	49 pcs.	$8'' \times 8'' \times 22'$	**15.** 560 pcs. $10'' \times 1'' \times 16'$
10.	200 pcs.	$4'' \times 8'' \times 10'$	**16.** 960 pcs. $8'' \times 1'' \times 14'$
11.	14 pcs.	$3'' \times 5'' \times 14'$	**17.** 760 pcs. $6'' \times 1'' \times 18'$
12.	18 pcs.	$4'' \times 9'' \times 16'$	**18.** 1000 pcs. $5'' \times 1'' \times 16'$

493. This diagram shows the framework of the ground floor plan of a house $34' \times 51'$. The sills are $6'' \times 8''$. The side sills are spliced 5 ft. in the middle. The joists are $3'' \times 8''$, and are $18''$ apart from center to center. Find the cost of the lumber at $18 per M.

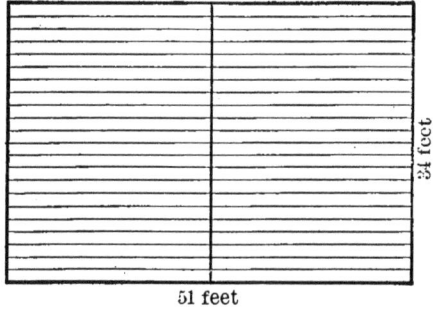

34 feet

51 feet

- 3 sills $6'' \times 8'' \times 34' = $ 408 ft.
- 4 sills $6'' \times 8'' \times 28' = $ 448 ft.
- 42 joists $3'' \times 8'' \times 26' = $ 2184 ft.

Total,   3040 ft.

$3.040 \times \$18 = \$54.72$ cost.

($18 per M $= \$.02 - \$.002$ per foot.)

The three cross sills are each 34 ft. long, making 408 board feet. The side dimensions are so long as to make splicing necessary. The length of each piece of timber is found by adding the length of the splice to the length of the sill and dividing into two equal parts. $(5 + 51) \div 2 = 28$, no. ft. The four pieces contain 448 board feet.

Since the sills are $6''$ thick, the width inside the sills is $33'$. To find the number of joists divide $33'$ by $18''$, and subtract 1 from the quotient. $33' \div 18'' = 22$; $22 - 1 = 21$. (The number of spaces is 22, but the number of joists is one less. Prove it by counting spaces and joists in the diagram.) As one joist reaches only to the middle sill, there are 2 times 21, or 42, joists in all. They are mortised into the sills $2''$ at each end. The total width of the three cross sills is $18''$, which, deducted from $51'$, leaves $49' 6''$, net length of house exclusive of the sills. One

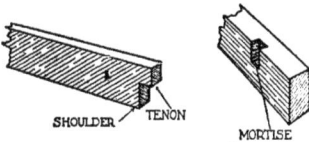

SHOULDER   TENON   MORTISE

half of $49' 6''$ equals $24' 9''$, length of joist exclusive of the tenon, $— 2''$ at each end, — which makes total length of a joist $25' 1''$.

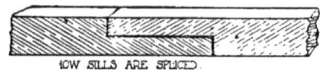

OW SILLS ARE SPLICED

$(24' 9'' + 2'' + 2''.)$ But as lumber is sawed in even foot lengths, a joist $26'$ long must be bought. 42 joists $3'' \times 8'' \times 26' = 2148$ board feet.

The total amount of lumber is 3040 ft., which at $18 per M is worth $54.72.

NOTE. Lumber is practically never cut in an odd number of feet lengths, as 13', 15', 19', 25', but in even numbers, as 10', 12', 16', 22', etc. Hence, when the actual length required in building is an odd or fractional number, the nearest length in *even* number of feet greater is purchased.

PROBLEMS

494. 1. Estimate the cost of the lumber at $22 per M for a similar plan 25' 4" × 40'. The sills are 8" × 8", those on the sides being spliced 4 ft. The joists are 3" × 8" and are 16" from center to center.

2. A house is 25 ft. by 30 ft. The sills are 6" × 6". A sill extends across the middle of the plan as in the preceding diagram. The joists are 3" × 6" and are 16" from center to center. How many feet of lumber are required for the sills and joists?

3. Make a diagram of an original plan similar to those of the preceding problems, and estimate the amount of lumber necessary to construct.

WOOD

495. Wood is measured by the cord of 128 cu. ft. A pile of wood 8 ft. long, 4 ft. wide, and 4 ft. high is a cord. **1 ft.** of the length of such a pile is sometimes called a **cord foot.**

PROBLEMS

496. 1. A pile of wood is 44 ft. long, 24 ft. wide, and 12 ft. high. How many cords are there? How many cords of wood 1 ft. long will it make?

2. A woodshed is 16 ft. long, 12 ft. wide, and 12 ft. high. How many cords of wood can be piled in it?

3. A dealer bought a pile of 4-ft. wood 120 ft. long, 6 ft. high, and 35 ft. wide at $3.75 a cord. He hired it sawed into stove lengths (1 ft.) at $1 per cord (128 cu. ft.). He sold it at $1.75 a cord, paying 25¢ a cord for delivering. Find his gain.

4. A wagon rack is 18 ft. long, 3 ft. 3 in. wide. How high must the wood be piled to make a load of 2 cd.?

5. How many steres of wood are there in a pile 48 ft. long, 24 ft. wide, and 9 ft. high? (1 stere = 1 cu. m.)

CAPACITY

BINS AND GRAIN ELEVATORS

497. A bin is a box or inclosed space for holding grain, coal, fruit, vegetables, etc.

498. A grain elevator is a building for the storage of large quantities of grain.

499. Grain and vegetables (especially potatoes) are sold by the bushel. The number of bushels is determined by weight rather than by volume. Estimates of the number of bushels a box or bin will hold are made in bushels of capacity.

500. Estimates of grain are made by the *stricken bushel* of 2150.42 cu. in., fruits and vegetables by the heaped bushel, 2747.71 cu. in.

Fruits are sold in barrels, boxes, and baskets of various sizes.

501. Coal is sold by weight. It is sold wholesale at the mines by the long ton; it is usually retailed by the short ton. In estimating the capacity of a bin, $34\frac{1}{2}$ cu. ft. of space are reckoned to contain 1 T. of hard stove coal.

APPROXIMATE MEASURES

502. Since a bushel of grain measures 2150.42 cu. in., and there are 1728 cu. in. in a cubic foot, a cubic foot of space contains $\frac{1728}{2150.42}$, or $.8 +$ bu. Hence,

To find the number of bushels of grain a bin will hold, multiply its capacity measured in cubic feet by .8.

This rule is sufficiently accurate for all practical purposes, the error being less than 4 bu. per thousand cubic feet.

In like manner, a cubic foot of space contains $\frac{1728}{2747.71}$, or .63 bu., of potatoes, apples, etc., and to find the number of bushels that can be put into a box or bin, multiply its capacity in cubic feet by .63. The error is less than $1\frac{1}{2}$ bu. per thousand cubic feet.

Use approximate measures, unless otherwise specified.

503. How many bushels of wheat can be put into a bin $20' \times 8' \times 6'$?

$$20 \times 8 \times 6 \times .8 \text{ bu.} = 768 \text{ bu.}$$

Since a cubic foot of space contains .8 of a bushel of wheat, a bin $20' \times 8' \times 6'$ contains $20 \times 8 \times 6 \times .8$ bu., or 768 bu.

PROBLEMS

504. 1. Estimate the quantity of grain that can be stored in a grain elevator $30' \times 40' \times 28'$.

Estimate the quantity of (*a*) grain, (*b*) potatoes, that can be put into each of the following bins:

2.	$8' \times 6' \times 4'$		9.	$4' 6'' \times 5' \quad \times 8'$	
3.	$9' \times 8' \times 10'$		10.	$5' 4'' \times 7' \quad \times 9'$	
4.	$4' \times 5' \times 20'$		11.	$3' 6'' \times 8' \quad \times 10' 6''$	
5.	$12' \times 8' \times 36'$		12.	$4' 4'' \times 8' 3'' \times 12'$	
6.	$15' \times 9' \times 40'$		13.	$8' 6'' \times 9' \quad \times 10'$	
7.	$10' \times 5' \times 4'$		14.	$7' 6'' \times 9' \quad \times 12'$	
8.	$9' \times 10' \times 12'$		15.	$2' 6'' \times 3' 3'' \times 4'$	

16. How long a bin 6 ft. wide, 4 ft. deep, is necessary to hold 480 bu. of oats?

17. A bin is 36 ft. long and 5 ft. wide. To what depth will 756 bu. of potatoes fill it?

18. If a bin holds 1000 bu. of grain, how many bushels of potatoes will it hold?

19. How many tons of stove coal can be put into a bin $8' \times 8' \times 6'$?

20-24. Find the exact capacities for both grain and potatoes of the bins mentioned in problems 3–7.

25. A farmer sells 8 loads of potatoes whose gross weights are 4229 lb., 4301 lb., 4408 lb., 4027 lb., 4187 lb., 4360 lb., 4110 lb., and 4568 lb. If the wagon weighs 1073 lb., how much does he receive for his potatoes at $.65 per bushel?

26. A carload of wheat weighs 45,000 lb. If the car is 33 ft. long, 7 ft. wide, and is filled to a depth of 4 ft., how much does the measure by weight differ from the measure by capacity?

27-33. How many hectoliters of grain will each of the bins in problems 3–9 hold?

CISTERNS AND RESERVOIRS

505. Cisterns and reservoirs are storage places for water, oil, etc. The unit of measure is the gallon of 231 cu. in. One cubic foot of space contains $\frac{1728}{231}$, or $7.48 +$ gal. For mental work, $7\frac{1}{2}$ gal. are reckoned for each cubic foot.

At 7.48 gal. to the cubic foot, the result varies from the result by exact measurement by less than 1 gal. for each thousand cubic feet of space.

To find the number of gallons in a cistern or reservoir, multiply the capacity in cubic feet by 7.48.

506. A reservoir is $44' \times 35' \times 12'$. Find (a) the exact number of gallons, and (b) the approximate number of gallons it will hold.

$$(a) \quad \frac{44 \times 35 \times 12 \times .1728}{2^3 1} = 138240, \text{ no. gal.}$$

$$(b) \quad 44 \times 35 \times 12 \times 7.48 = 138230.4, \text{ no. gal.}$$

(a) To find the exact number of gallons, reduce the cubic feet to cubic inches and divide by 231 cu. in. in a gallon. The result is 138,240 gal.

(b) To find the approximate capacity, multiply the number of cubic feet by 7.48. The result is 138,230.4 gal.

PROBLEMS

507. Find (a) the exact, and (b) the approximate, number of gallons in each of the following cisterns and reservoirs:

1.	$8' \times 8' \times 8'$	**6.**	$5' \times 5' \times 10'$
2.	$9' \times 6' \times 7'$	**7.**	$15' \times 12' \times 6' \, 6''$
3.	$46' \times 25' \times 10'$	**8.**	$8' \, 3'' \times 10' \times 12' \, 6''$
4.	$55' \times 49' \times 18'$	**9.**	$9' \, 6'' \times 12' \times 15'$
5.	$6' \times 7' \times 6'$	**10.**	$12' \times 12' \times 12'$

11–16. Find approximately (a) the capacity in liters, (b) the weight in kilograms, of pure water in the cisterns and reservoirs in problems 1–6, supposing each is full of water.

17. How many gallons will a cylindrical tank 24 ft. in diameter and 16 ft. deep hold?

$$24 \times 24 \times .7854 \times 16 \times 7.48 = \text{no. of gallons.}$$
$$24 \times 24 \times 16 \times 5\tfrac{7}{8} = 54144 \text{ gal.}$$

The volume of a cylinder is found by multiplying the square of the diameter by .7854 and multiplying that product by the altitude. To find the number of gallons, multiply the result thus found by 7.48.

This process results in a long and tedious multiplication. Note that the second statement in the solution differs from the first in that the numbers .7854 and 7.48 are omitted, and $5\tfrac{7}{8}$ is used in their stead. $5\tfrac{7}{8}$ is the product of $7.48 \times .7854$, very nearly. Therefore the capacity of a cylindrical cistern or reservoir may be found (very nearly exactly) by multiplying the square of the diameter times the depth by $5\tfrac{7}{8}$.

18. A cylindrical cistern is 10′ in diameter and **10′ deep. Find** by
the method just explained how many gallons it will hold. How
much does the answer vary from the exact capacity?

Find the approximate capacity in gallons of tanks having measurements as follows:

19. Diameter 4′, depth 8′.   **22.** Diameter 18′, depth 20′.

20 Diameter 5′, depth 10′   **23.** Diameter 44′, depth 42′.

21. Diameter 55′, depth 56′.   **24.** Diameter 16′, depth 10′.

25. If a standpipe 235 ft. high holds 22,090 gal. of water, what is
its diameter?

26. A cylindrical reservoir 48 ft. in diameter holds 649,728 gal.
What is its depth?

LAND

508. Land areas are estimated by the acre of 160 sq. rd.

509. In laying out public lands, surveyors select a north and south
line as a **principal meridian,** and an east and west line as a **base line.**

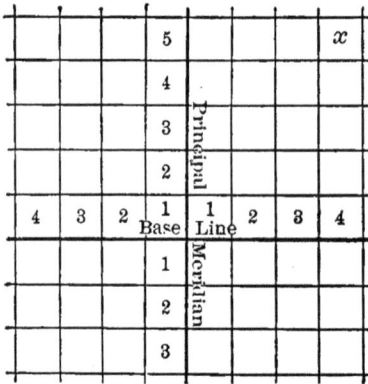

Townships, 6 miles square

A Township divided into Sections

Other lines 6 mi. apart are run parallel to the principal meridian and
to the base line, thus dividing the land into **townships,** as shown in
the diagram.

Rows of townships north and south are called **ranges,** and are
numbered both east and west from the principal meridian. The
townships are also numbered north and south of the base line.
Thus, Township x is 5 north, in the 4th range east of the principal
meridian.

Each township is again divided into squares by parallel lines **1 mi.** apart. Each square is called **a section, and** contains 640 A.

The sections in a township **are always** numbered as shown in the diagram. If **the diagram** of a township on p. 204 represents Township *x* of the preceding diagram, section **17** would be designated as section **17** of township **5** north, in the 4th range east of the principal meridian.

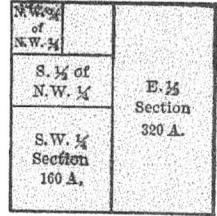

A Section

Sections are divided into halves, quarters, half-quarters, **quarter-** quarters, etc.

PROBLEMS

510. 1. A man owned the southeast quarter **of a** section of land. He bought the south half of the northeast quarter, and the northeast quarter of the northeast quarter of the same section. Make a diagram of his original farm, adding thereto the areas bought, showing number of acres in each parcel.

2. A man owning the north half of a section bought the east half of the southeast quarter, and the west half of the southwest quarter of the same section. How many acres of land has he? Draw a diagram of his farm.

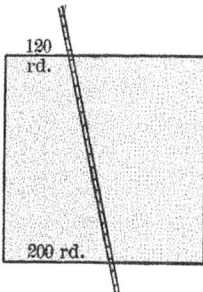

3. A railroad extends through a section of land as shown in the diagram. It occupies a strip 4 rd. wide. Find the land damage **for the** right of way at $40 an acre.

4. Find the cost at $.65 a rod of building a fence around **a section** of land; a quarter section of land.

5. Your deed shows that **you own the** southerly half of section 8 in Clark Co., S. Dak. Draw a township plan, and locate your land. Locate a house in the southeastern corner of your land. Locate a haystack in the northwestern corner of your land. Find distance in a direct line from the house to the haystack.

6. How many rods of fence **are** required to inclose each of the three parcels in problem 1?

7. How many rods of fence will inclose the combined tracts mentioned in problem 2?

GRAPHS

511. A **graph** is a diagram illustrating some relationship. In the illustration here given it shows the relation of earnings and expenses of the X Railway.

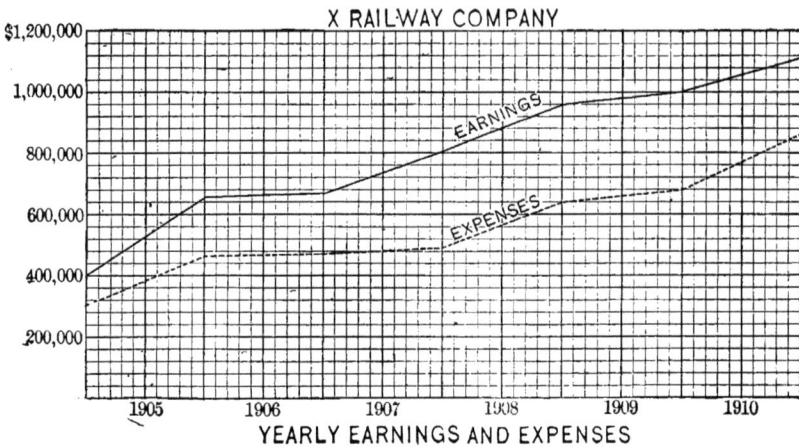

YEARLY EARNINGS AND EXPENSES

From 1905 to 1910 inclusive, the annual earnings and expenses of the **X** Railway were as follows:

The monthly earnings and expenses of the same railway for 1910 were as follows:

	EXPENSES	EARNINGS		EARNINGS	EXPENSES
1905	$460,000	$ 660,000	January .	$ 78,000	$63,000
1906	470,000	670,000	February .	73,000	61,000
1907	485,000	800,000	March . .	87,000	63,000
1908	640,000	960,000	April . .	86,000	61,000
1909	680,000	1,000,000	May .	95,000	70,000
1910	850,000	1,120,000	June . .	101,000	73,000
			July . .	115,000	82,000
			August .	119,000	87,000
			September	98,500	84,000
			October .	92,000	86,000
			November	85,500	70,000
			December	79,000	65,000

512. 1. The wheat crop of the Canadian Northwest for twenty years is indicated by the following figures (in nearest 100,000 bu.):
1889, 7,200,000 bu.; 1893, 15,600,000 bu.; 1897, 18,300,000 bu.;
1902, 67,000,000 bu.; 1906, 110,600,000 bu.; 1909, 168,400,000 bu.
Illustrate graphically.

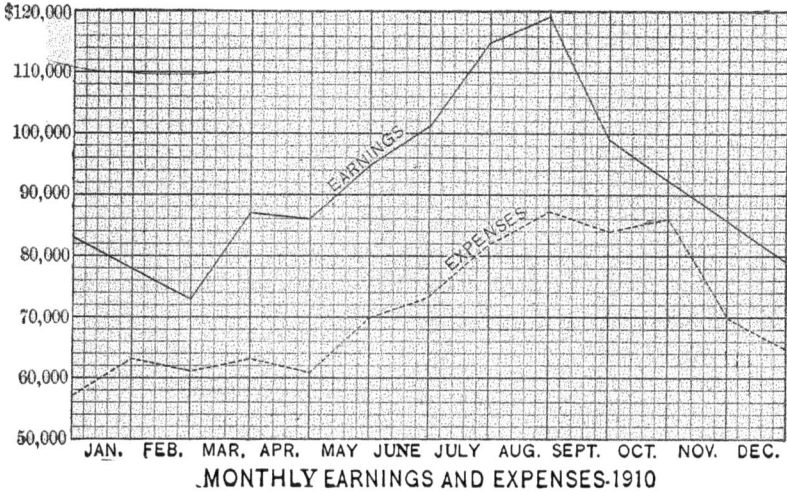

.MONTHLY EARNINGS AND EXPENSES-1910

2. Illustrate graphically: During 1910 the monthly earnings and expenses of a given railway are as follows

	EARNINGS		EXPENSES
Jan.	$19,000 *	Jan.	$20,500†
Feb.	17,000	Feb.	18,500
Mar.	22,000	Mar.	19,000
April	28,000	April	20,500
May	25,000	May	24,000
June	24,000	June	23,500
July	25,500	July	24,500
Aug.	25,000	Aug.	23,000
Sept.	24,000	Sept.	22,500
Oct.	26,000	Oct.	25,500
Nov.	20,500	Nov.	20,000
Dec.	26,000	Dec.	25,000

* Increase of $1000 over Dec., 1909.　　† Decrease of $1000 from Dec., 1909.

3. Show by diagram the relative earnings and expenses of a railway whose quarterly earnings and expenses for 1910 were as follows:

	EARNINGS		EXPENSES
Jan.-Mar.	. . . .$ 3,800*	Jan.-Mar.	.$ 3,200†
Apr.-June	 8,200	Apr.-June	8,900
July-Sept.	 15,500	July-Sept.	8,200
Oct.-Dec. .	 5,400	Oct.-Dec.	5,800

4. Make a diagram showing the production of beet sugar in the United States for five consecutive years (in nearest 1000 tons).

	LONG TONS
1903-1904	 215,000
1904-1905	 216,000
1905-1906	 279,000
1906-1907	 432,000
1907-1908	 414,000

5. Illustrate graphically the production of cotton in the United States since 1865, as shown by the following (numbers of bales given to the nearest 100,000):

	BALES (500 LB. EACH)
1865	 2,100,000
1875	 4,800,000
1885	 6,400,000
1895	 7,100,000
1905	 10,800,000
1908	 13,200,000

6. The following table shows (to the nearest 1000) the number of cattle imported into and exported from the United States for five years, beginning 1905. Show these facts graphically.

	IMPORTS	EXPORTS
1905	 28,000	568,000
1906	 29,000	584,000
1907	 32,000	423,000
1908	 92,000	349,000
1909	 139,000	208,000

7. The total corn crop of the United States for 1909 was 2,772,000,000 bu. Of this amount, seven states produced as follows:

Illinois	 370,000,000
Iowa	 290,000,000
Missouri	 214,000,000
Indiana	 197,000,000
Nebraska	 194,000,000
Kansas	 154,000,000

* Decrease from $ 5200, Oct.-Dec., 1909.　　† Increase from $ 4400, Oct.-Dec., 1909.

Find the per cent of the entire crop for each of the seven states, and for "all the other states." Show these results graphically.

EXAMINATIONS

SPEED TEST

Minimum time, twenty minutes; maximum, one hour. Deduct one credit for every minute required beyond minimum time.

513. 1. Perform the following operations:

$$\frac{\$1.25 \times 28 \times 77 \times 35}{4 \times 11 \times 7 \times 25} = ?$$

$12' \times 16'' \times 2\frac{1}{4}''$	=	board feet
$36' \times 5'6'' \times \$1.25$ a square yard	= $	
20 planks $16' \times 9'' \times 2\frac{1}{2}''$ @ \$18 per M	= $	

2. Find the total cost:

7,480 lb. bran @ \$22 per ton.
15,870 lb. hay @ 16 per ton.
12,456 lb. meal @ 24 per ton.
9,498 lb. straw @ 12 per ton.

3. Find the number of yards of carpet $\frac{3}{4}$ yd. wide, for rooms,

16' by 14'
13' 6' by 12' 4"

4. How many double rolls of paper are required for rooms,

$16' \times 12' \times 8'6''$, openings 11 ft. wide?
$18' \times 14' \times 9$, openings 15 ft. wide?

5. How many paving blocks $12'' \times 5''$ are required for streets as follows:

$1\frac{1}{4}$ mi. long, 60 ft. wide?
$\frac{7}{8}$ mi. long, 56 ft. wide?

6. Find weight of each:

Block of marble $5' \times 2' \times 1\frac{1}{2}'$, sp. gr. **2.72**.
Gold brick $8'' \times 4'' \times 2'$, sp. gr. **19.258**.

7. Find approximate contents:

Bin $30' \times 8' \times 6'$ full of wheat.
Box $6' \times 5' \times 4$ full of apples.

VAN TUYL'S BUS. ARITH.—14

8. How many gallons are there in a tank of water,

4 m. long, 2 m. wide, 1.5 m. deep ?

(1 liter = 1.0567 qt.)

9. How many bushels are there in 1280 Kg. of oats? in 3000 Kg. of wheat?   (1 Kg. = 2.2046 lb.)

10. Find the number of hectoliters of wheat that can be put into a bin 6 m. long, 3 m. wide, 2 m. deep.

WRITTEN TEST

514. 1. Find the total cost of:

    1350 ft. of pine   @ $ 32.50 per M.

    6240 ft. of hemlock @ 24.50 per M.

    3650 cedar posts   @   9.50 per C.

    4260 lb. of coal   @   7.25 per ton.

2. A railway train runs 240 rd. in $\frac{2}{3}$ of a minute. Find its velocity in miles per hour.

3. How much will it cost to construct a highway 9 mi. 64 rd. 3 yd. long at $8500 a mile ?

4. A room in a factory is 65.5 ft. long, 30 ft. wide, and $12\frac{1}{2}$ ft. high. How many cubic feet does the room contain ?

5. Fresh air enters a room through an opening 8 in. × 14 in., with a velocity of 5 ft. per second. At this rate, how many cubic feet of air would enter the room in one hour ?

6 A man had a yard 38 ft. long by 27 ft. wide; he reserved two grass plots each 8 ft. square, and had the rest paved with stone at $.45 a square yard. How much did the paving cost ?

7. How much will it cost at 32 ¢ per cubic yard to make an excavation 40 ft. long, 30 ft. wide, and 9 ft. deep ?

8. A merchant imported 18 hectoliters of almonds at a cost of 44 pesetas a hectoliter, and sold them at 12 ¢ a quart. Find his gain.

    (1 peseta = $.193. 1 liter = .908 dry quarts.)

9. A vat .95 m. deep, 1.8 m. long, and 1.2 m. wide is full of vinegar. Find its value at 20 ¢ a gallon

10. A park is 80 rd. long and 56 rd. wide. Around it is a brick walk 8 ft. wide. The exposed surface of each brick is 4″ × 8″. Find the cost of the brick at $7.50 per 1000.

PERCENTAGE

515. The term *percentage* includes those subjects in arithmetic in the computation of which $\frac{100}{100}$ is taken as the basis of comparison. All comparisons of values or quantities are expressed in hundredths. Instead of the term *hundredths*, the equivalent Latin expression *per cent.* is used.

516. Per cent means by the hundred. **The per cent sign (%) is** generally used instead of the words per cent.

Observe these equations:

$$\frac{1}{4} = \frac{25}{100} = .25 = 25\%$$

$$\frac{1}{8} = \frac{12\frac{1}{2}}{100} = .12\frac{1}{2} = 12\frac{1}{2}\%$$

$$\frac{2}{3} = \frac{66\frac{2}{3}}{100} = .66\frac{2}{3} = 66\frac{2}{3}\%$$

$$\frac{5}{13} = \frac{38\frac{6}{13}}{100} = .38\frac{6}{13} = 38\frac{6}{13}\%$$

Note that in each case a common fraction has first been reduced to hundredths, and that next the fraction is in its decimal form.

The last step shows the fraction in its " per cent " form. Note also that the per cent form of the fraction is identical with its decimal form except that the *per cent sign is used instead of a decimal point.*

Percentage problems are, therefore, only a continuation of problems in fractions under a new name.

517. Nearly all problems in percentage can be reduced to one of the three type forms explained on page 103. In the language of percentage problems, the type forms would read thus:

1. Find 75 % of $ 48.
2. $62\frac{1}{2}$ % of what number equals $ 90 ?
3. $ 75 is what per cent of $ 90?

211

When reduced to their simplest arithmetical form, the types appear thus:

1. 75 % of $ 48 = ?

2. 62½ % of ? = **$ 45.**

3. ? % of $ 90 = $ 75 ?

In each of these examples there are three elements — the multiplicand, the multiplier, and the product. The corresponding percentage terms are **base**, **rate**, and **percentage**.

. **518.** The **base** is the number or quantity represented by 100 %, and is the basis of comparison in any given problem.

519. The **rate** is the number of hundredths or per cent, and shows the *ratio* of the *percentage* to the *base*.

520. The **percentage** is the *product* of the base multiplied by the rate. It is the number or quantity whose ratio or relation to the base is expressed in the rate.

521. What number is 24 % of 300 ?

24 % of 300 = 72.

300 is the base because it is the number with which comparison is made.

24 % is the rate because it shows the ratio of the percentage to the base.

72 is the percentage because it is the product of the base and the rate. It is the number that is compared with the base.

522. If the percentage is added to the base, the **sum** is called the **amount.**

523. If the percentage is subtracted from the base, the result is called the **difference.**

524. Since any common fraction can be reduced to an equivalent decimal or per cent form, conversely, any per cent form of a fraction can be reduced to an equivalent decimal or common fraction form.

$\frac{1}{2} = \frac{50}{100} = .50 = 50\%$, and conversely, $50\% = .50 = \frac{50}{100} = \frac{1}{2}$.

$\frac{5}{8} = \frac{62\frac{1}{2}}{100} = .62\frac{1}{2} = 62\frac{1}{2}\%$, and conversely, $62\frac{1}{2}\% = .62\frac{1}{2} = \frac{62\frac{1}{2}}{100} = \frac{5}{8}$, etc.

The student must recognize the difference between $62\frac{1}{2}\%$ and $.62\frac{1}{2}\%$. The first expression equals $.62\frac{1}{2} = \frac{62\frac{1}{2}}{100} = \frac{5}{8}$, and the second expression equals $.0062\frac{1}{2} = \frac{62\frac{1}{2}}{10000} = \frac{5}{800}$.

525. When the per cent sign is used, it takes *the place* of the decimal point, or it is equivalent to moving the decimal point *two places to the right*.

To change a decimal to a rate per cent, move the point two places to the right and annex the per cent sign, and, conversely,

To change a rate per cent to a decimal, remove the per cent sign and move the point two places to the left.

$$.045 = 45\% = 4\frac{1}{2}\% \qquad 15\% = .15$$
$$1.25 = 125\% \qquad 1\frac{1}{4}\% = .01\frac{1}{4} = .0125, \text{ etc.}$$

526. Reduce each of the following to the form indicated:

To the per cent form :

1. .18	**4.** .98	**7.** $.33\frac{1}{3}$	**10.** .125
2. .25	**5.** .05	**8.** $.62\frac{1}{2}$	**11.** .025
3. .56	**6.** .01	**9.** $.56\frac{1}{4}$	**12.** .0025

To both the decimal and the per cent forms:

13. $\frac{1}{4}$	**16.** $\frac{1}{16}$	**19.** $\frac{7}{9}$	**22.** $\frac{3}{4}$
14. $\frac{1}{5}$	**17.** $\frac{5}{6}$	**20.** $\frac{5}{12}$	**23.** $\frac{2}{5}$
15. $\frac{3}{8}$	**18.** $\frac{1}{7}$	**21.** $\frac{7}{10}$	**24.** $1\frac{2}{3}$

To the decimal form :

25. 15%	**28.** 72%	**31.** $12\frac{1}{2}\%$	**34.** $\frac{1}{2}\%$
26. 29%	**29.** 125%	**32.** $66\frac{2}{3}\%$	**35.** $\frac{5}{8}\%$
27. 4%	**30.** $1\frac{1}{4}\%$	**33.** 37.5%	**36.** $\frac{2}{3}\%$

To both the decimal form and the common fraction form in its lowest terms:

37. 10%	**40.** 75%	**43.** $\frac{1}{8}\%$	**46.** 580%
38. 30%	**41.** $87\frac{1}{2}\%$	**44.** $\frac{1}{6}\%$	**47.** $137\frac{1}{2}\%$
39. 45%	**42.** $93\frac{3}{4}\%$	**45.** $\frac{9}{16}\%$	**48.** $101\frac{1}{2}\%$

527. In the chapter on Aliquot Parts many of the aliquot or fractional parts of one dollar were learned.

The aliquot or fractional parts of 100% are the same numerically as the like parts of a dollar.

528. In the following table of fractional equivalents, related rates are grouped, the more important rates being in heavy-faced type.

TABLE OF FRACTIONAL EQUIVALENTS

Half Quarters Eighths	Sixteenths	Thirds Sixths Twelfths	Sevenths Ninths	Miscellaneous Rates
$50\% = \frac{1}{2}$	$6\frac{1}{4}\% = \frac{1}{16}$	$33\frac{1}{3}\% = \frac{1}{3}$	$14\frac{2}{7}\% = \frac{1}{7}$	$1\frac{1}{4}\% = \frac{1}{80}$
	$18\frac{3}{4}\% = \frac{3}{16}$	$66\frac{2}{3}\% = \frac{2}{3}$	$28\frac{4}{7}\% = \frac{2}{7}$	$1\frac{2}{3}\% = \frac{1}{60}$
$25\% = \frac{1}{4}$	$31\frac{1}{4}\% = \frac{5}{16}$		$42\frac{6}{7}\% = \frac{3}{7}$	$2\frac{1}{2}\% = \frac{1}{40}$
$75\% = \frac{3}{4}$	$43\frac{3}{4}\% = \frac{7}{16}$	$16\frac{2}{3}\% = \frac{1}{6}$	$57\frac{1}{7}\% = \frac{4}{7}$	$3\frac{1}{3}\% = \frac{1}{30}$
	$56\frac{1}{4}\% = \frac{9}{16}$	$83\frac{1}{3}\% = \frac{5}{6}$	$71\frac{3}{7}\% = \frac{5}{7}$	$4\% = \frac{1}{25}$
$12\frac{1}{2}\% = \frac{1}{8}$	$68\frac{3}{4}\% = \frac{11}{16}$		$85\frac{5}{7}\% = \frac{6}{7}$	$5\% = \frac{1}{20}$
$37\frac{1}{2}\% = \frac{3}{8}$	$81\frac{1}{4}\% = \frac{13}{16}$	$8\frac{1}{3}\% = \frac{1}{12}$	$11\frac{1}{9}\% = \frac{1}{9}$	$6\frac{2}{3}\% = \frac{1}{15}$
$62\frac{1}{2}\% = \frac{5}{8}$	$93\frac{3}{4}\% = \frac{15}{16}$	$41\frac{2}{3}\% = \frac{5}{12}$	$22\frac{2}{9}\% = \frac{2}{9}$	$7\frac{1}{7}\% = \frac{1}{14}$
$87\frac{1}{2}\% = \frac{7}{8}$		$58\frac{1}{3}\% = \frac{7}{12}$	$44\frac{4}{9}\% = \frac{4}{9}$	$9\frac{1}{11}\% = \frac{1}{11}$
		$91\frac{2}{3}\% = \frac{11}{12}$	$55\frac{5}{9}\% = \frac{5}{9}$	$10\% = \frac{1}{10}$
			$77\frac{7}{9}\% = \frac{7}{9}$	
			$88\frac{8}{9}\% = \frac{8}{9}$	

529. The labor involved in solving percentage problems is materially diminished by using the common fractional equivalents of rate per cents wherever possible.

530. Problems in percentage are solved in accordance with the fundamental principles of multiplication and division of simple numbers, page 102. In the language of percentage they are stated thus:

PRINCIPLES: **1.** Base multiplied by the rate = percentage.

2. Percentage divided by rate = base.

3. Percentage divided by base = rate.

531. To find the percentage, the base and rate being given.

1. Find 43 % of $360.

```
$360 Base
 .43 Rate
────
1080
1440
────
$154.80 Percentage
```

43% of $360 means .43 of $360. $360 multiplied by .43 equals $154.80.

2. Find $16\frac{2}{3}\%$ of $\$540$.

$\$540$
.$16\frac{2}{3}$
———
360   Or, $16\frac{2}{3}\% = \frac{1}{6}$
3240
540                       $\frac{1}{6}$ of $\$540 = \$90$.
———
$\$90.00$

Multiplying $\$540$ by $.16\frac{2}{3}$ gives $\$90$. Or, since $16\frac{2}{3}\% = \frac{1}{6}$, take $\frac{1}{6}$ of $\$540$, which equals $\$90$.

3. Find $\frac{1}{8}\%$ of $\$1250$.

8)$\$12.50$
—————
$\$1.5625 \equiv \$1.56$

$\frac{1}{8}\%$ means $\frac{1}{8}$ of 1%. 1% of $\$1250$ is $\$12.50$; and $\frac{1}{8}$ of $\$12.50 = \$1.56$.

4. What number is $33\frac{1}{3}\%$ greater than $\$480$?

$100\% \equiv \$480$
$33\frac{1}{3}\% = \quad 160$
—————
$133\frac{1}{3}\% = \overline{\$640}$

The base is always equal to 100%. Hence $\$480$ is equal to 100%. The required number is $33\frac{1}{3}\%$, or $\frac{1}{3}$, of $\$480$ greater than $\$480$. $33\frac{1}{3}\%$ of $\$480$ is $\$160$. By adding $\$480$ and $\$160$, the desired result, $\$640$, is obtained.

NOTE. It should be observed that the base and the percentage are like numbers.

532. Performing all operations mentally where possible, find:

1. 21% of 200.
2. 35% of 500.
3. $12\frac{1}{2}\%$ of 640.
4. 25% of 700.
5. 26% of 700.
6. $66\frac{2}{3}\%$ of 950.
7. $87\frac{1}{2}\%$ of 1200.
8. 13% of 148.
9. 17% of 263.
10. 6% of 475.
11. 24% of 300.
12. $16\frac{2}{3}\%$ of 372.
13. $8\frac{1}{3}\%$ of 744.
14. $18\frac{3}{4}\%$ of 640.
15. 64% of 298.

16. 78% of 725.
17. 93% of 800.
18. 125% of 960.
19. 80% of 1200.
20. 75% of 1500.
21. $6\frac{2}{3}\%$ of 1800.
22. $14\frac{2}{7}\%$ of 2149.
23. $22\frac{2}{9}\%$ of 1881.
24. $31\frac{1}{4}\%$ of 8000.
25. $56\frac{1}{4}\%$ of 3200.
26. 250% of 440.
27. 375% of 560.
28. $2\frac{1}{2}\%$ of 1600.
29. $2\frac{1}{4}\%$ of 1260.
30. $1\frac{1}{4}\%$ of 888.

31. $\frac{3}{4}\%$ of 84.
32. $\frac{5}{8}\%$ of 500.
33. $1\frac{1}{2}\%$ of 64.
34. 100% of 128.
35. 150% of 630.
36. 300% of 25.
37. 3% of 6172.
38. $12\frac{1}{2}\%$ of 8176.
39. $58\frac{1}{3}\%$ of 7392.
40. $6\frac{1}{4}\%$ of 6500.
41. $62\frac{1}{2}\%$ of 6560.
42. $3\frac{1}{3}\%$ of 7230.
43. $33\frac{1}{3}\%$ of 7230.
44. $333\frac{1}{3}\%$ of 7230
45. $\frac{1}{10}\%$ of 8400

533. Find 28 % of $7500.

$\frac{3}{4}$ of $2800 = $2100.   By the commutative law of multiplication (page 102) the product of 28 times 75 equals the product of 75 times 28.   Therefore, 28 % of $7500 equals 75 % of $2800.   75 % $= \frac{3}{4}$, and $\frac{3}{4}$ of $2800 = $2100.

534. In like manner find, mentally, the value of each of the following:

1. 16 % of $1250.	**8.** 180 % of $750.	**15.** 39 % of 666\frac{2}{3}$.
2. 18 % of 333\frac{1}{3}$.	**9.** 32 % of $375.	**16.** 42 % of 166.66\frac{2}{3}$.
3. 26 % of $2500.	**10.** 36 % of $75.	**17.** 132 % of $2500.
4. 44 % of $2500.	**11.** 240 % of $2500.	**18.** 12 % of $25.
5. 56 % of $5000.	**12.** 320 % of $625.	**19.** 36 % of $87.50.
6. 96 % of $125.	**13.** 32 % of $3125.	**20.** 52 % of $75.
7. 160 % of $3750.	**14.** 124 % of $6250.	**21.** 48 % of $375.

PROBLEMS

535. 1. In a certain school the attendance in 1908 was 425. What was the attendance in 1909 if there was an increase of 20 % ?

2. A house and lot cost $5000. If taxes, repairs, and other expenses amount to $180 per annum, what rent per month must the owner receive in order to clear 6 % on his investment?

3. If a plant and machinery, valued at $75,000, depreciates at the rate of 5 % of its value each year, find its value at the end of the fifth year.

4. The capital stock of a manufacturing company is $150,000. If the gross receipts amount to 32 % of the capital stock, and the total expenditures amount to 75 % of the receipts, find the profit for the year.

5. If a clerk's salary is $600 the first year of service, and he gets a 10 % increase each year for 5 yr., what will be his salary the sixth year?

6. Chemical analysis shows that "Pennsylvania" Portland cement contains mainly the following ingredients in the proportion stated:

Silica	22%
Iron and Alumina	10.98%
Lime	61.50%
Magnesia	2.47%
Sulphur Anhydride	1.70%
Combustible materials	.85%

Find the amount by weight of each of the ingredients in 5 T. of cement.

7. Merchandise costing $.37½ sells so as to gain 27½%. At what price does it sell?

8. On June 1, 1906, the city of Cleveland, Ohio, covered an area of 25,620 A. If the area annexed since that time is equal to 14.727 % of its former area, how many acres does the city cover now?

9. If a young man earns $900 a year and saves 22½ % of it, how long will it take him to save enough to pay for 3 lots valued at $337.50 each?

10. A fruit dealer buys 40 doz. oranges at 24¢ a dozen. Before he can sell them, 20 % of them spoil. At what price per dozen must the remaining oranges be sold to make a profit of 16⅔ % ?

11. At $\frac{1}{10}$ % premium, what is the cost of insuring a factory for $15,000?

12. A dealer bought 860 gross (long) tons of coal at $3.86 a ton f. o. b. Freight was 6¢ a cwt., and the cost of hauling was $321.60. Find the selling price a net (short) ton to gain 22 % on entire cost.

536. To find the rate, the base and percentage being given.

What is the ratio of 6 to 24? (See page 112.)

6 is what part of 24?

6 is what per cent of 24? or

What per cent of 24 is 6?

In each of these examples, 24 is the basis of comparison, or it is the base. 6 is the number compared with 24; hence it is the percentage.

The ratio of 6 to 24 is expressed as $\frac{6}{24}$ or $\frac{1}{4}$. In each of the examples, the desired relation between the numbers is determined by dividing 6 by 24. 6 ÷ 24 = $\frac{1}{4}$ = 25 %. That is, the percentage divided by the base gives the rate (Prin. 14, page 102).

537. 1. A man's income is $1800 a year. If he saves $630
a year, what per cent of his income does he save?

? % of $1800 = $630.

$630 ÷ $1800 = .35 = 35 %.

The problem is of the third type form
as shown in the first equation. Dividing
the percentage, $630, by the base, $1800,
gives the rate, 35%.

2. $1500 is what per cent greater than $1200 ?

$1500 − $1200 = $300, amount which $1500 is greater than
$1200.

$300 = ? % of $1200 ?

$300 ÷ $1200 = $\frac{300}{1200} = \frac{1}{4} = 25$ %.

First find how much $1500 is greater than $1200. The problem then becomes,
$300 is what per cent of $1200? which, as shown in the solution, is 25%.

To the Student. Reduce each problem (either mentally or in writing) to
its proper type form. Its solution will then be entirely clear to you.

Use the fractional form of division, when possible; reduce the fraction to
its lowest terms mentally; and place the result equal to its equivalent rate
per cent.

538. Solve, mentally when possible, each of the following:

What per cent of

1. 12 is 6 ?	**5.** 25 is 30 ?	**9.** 2 bu. is 3 pk. ?
2. 18 is 12 ?	**6.** 14 is 21 ?	**10.** $24 is $2.40 ?
3. 200 is 150 ?	**7.** 450 is 225 ?	**11.** $1.75 is $ 75 ?
4. 72 is 56 ?	**8.** 4 yd. is 6 ft. ?	**12.** $15 is $.90 ?

What per cent greater than

13. 54 is 63 ?	**17.** $3000 is $4200 ?
14. 300 is 390 ?	**18.** 3 pk. is 1 bu. ?
15. 560 is 630 ?	**19.** 1 lb. Troy is 1 lb. avoir. ?
16. 270 is 300 ?	**20.** 1 oz. avoir. is 1 oz. Troy ?

What per cent less than

21. 20 is 18 ?	**25.** 1 oz. Troy is 1 silver dollar ?
22. 30 is 20 ?	**26.** A long ton is a short ton ?
23. 50 is 40 ?	**27.** $\frac{1}{2}$ is $\frac{1}{3}$?
24. 1 bu. is 3 pk. ?	**28.** $\frac{1}{3}$ is $\frac{1}{4}$?

PROBLEMS

539. **1.** If in a school there are 160 boys and 192 girls, what per cent of the enrollment are boys? girls?

2. A man asked $180 for his horse. If he accepted $160, what per cent less than his asking price did he accept?

3. A building worth $12,500 is insured for $7500. For what per cent of its value is it insured?

4. In a spelling lesson of 60 words a pupil spelled 3 words incorrectly. What per cent of the words were spelled correctly?

5. Find the per cent of increase in the population of the following cities from 1900 to 1910:

	Population, 1900	Population, 1910
Akron, Ohio	42,728	69,067
Bridgeport, Conn.	70,996	102,054
Camden, N. J.	75,935	94,538
District of Columbia	278,718	331,069
Grand Rapids, Mich.	87,565	112,571
Houston, Tex.	44,633	78,800
Reading, Pa.	78,961	96,071
St. Paul, Minn.	163,065	214,744

6. A merchant fails, having liabilities amounting to $28,500. If his total resources amount to $19,000, what per cent of his debts can he pay? He owes Thos. Higbee $1846.50. How much will Mr. Higbee receive?

7. A man bought a house for $8000. To pay for it he borrowed $4000 on bond and mortgage at 6%. The other $4000 he paid from his own money. If the house rents for $800 per annum, and he pays $200 a year for taxes, repairs, insurance, etc., besides the interest on the mortgage, what rate per cent does the investment pay him?

8. The following table shows the number of organized wage earners (males) in New York State employed one or more days during the first quarter of the year 1910. Find, to the nearest tenth per cent, the percentage for each limit of time. Prove the work.

Period	Number Employed
1–29 da.	15,497
30–59 da.	63,695
60–79 da.	189,756
80 da. or more	60,029

9. The following table shows the net registered tonnage entered at the fourteen principal ports of the world. The figures are for the latest year available, and for a date ten years earlier. Find the percentage of increase to the nearest tenth per cent for each port during the ten-year period.

TABLE

Ports	Year	Net Registered Tons
New York	1908	12,154,780
	1898	7,771,412
Antwerp	1907	11,211,803
	1897	6,181,922
London	1907	11,160,367
	1897	9,110,921
Hamburg	1907	10,888,553
	1897	6,090,510
Rotterdam	1907	10,107,155
	1897	5,409,417
Hongkong	1906	9,941,261
	1896	7,609,047
Liverpool	1907	8,167,419
	1897	5,845,384
Montevideo	1906	7,725,534
	1896	1,991,554
Marseilles	1906	6,736,603
	1896	4,032,259
Singapore	1906	6,672,789
	1896	3,992,905
Cardiff	1907	5,734,755
	1897	4,485,782
Kobé	1907	5,417,877
	1897	740,851
Genoa	1906	5,450,818
	1896	3,514,301
Buenos Ayres *	1906	5,119,291
	1897	2,866,499

10. Below are given the values (to the nearest million dollars) of the imports and exports of merchandise into and from the United States, from and to five European countries during the years ending

* Nine years.

June 30, 1906, 1907, 1908, 1909, and 1910. Find to the nearest tenth per cent the excess (or *vice versa*) of the exports over the imports of each country for each year.

From	Imports into the United States In Millions (000,000 omitted)				
	1906	1907	1908	1909	1910
France	108	128	102	108	132
Germany	135	162	143	144	169
Italy	41	50	45	49	50
Russia (in Europe)	14	17	11	11	16
United Kingdom	210	246	190	209	271
To	Exports from the United States				
France	98	114	116	109	118
Germany	235	257	277	235	250
Italy	48	62	54	59	53
Russia (in Europe)	16	19	16	16	17
United Kingdom	583	608	581	515	505

SUGGESTION. Additional drill, if desired, may be obtained by comparing each year's imports and exports with those of the preceding year.

11. The total amounts of exports of agricultural implements from the United States to France for the six years ending 1907 were as follows:

1902	$ 2,100,692	1905	$ 2,817,998
1903	2,789,256	1906	2,895,243
1904	3,063,692	1907	3,724,959

Show in per cent the annual increase or decrease of exports.

12. Out of a total of 52,021 harvesting machines imported into Germany in 1907, 43,300 were from the United States, 6358 from Canada, and 2363 from England. Show by diagram the per ᴄᴇnt from each country.

13. In May, 1910, there arrived at the port of Portland, Oreɡor ᵃ vessels having a total net registry of 82,809 net tons, as compared ⁻ ᴶ

56 vessels of 51,087 net tons in May, 1909. Find the average tonnage of each vessel, and the per cent of increase in the average for the year.

14. The first five months of 1910 there were.shipped from Chicago 61,855,300 bu. of grain. Find the average per month, and compare, in terms of per cent, with the shipments for May, when 13,244,700 bu. were shipped.

15. The imports of drugs, chemicals, and dyes into Canada for the fiscal year 1908–1909 were from the following countries. Find the per cent from each country.

FROM	AMOUNT
United States	$ 5,837,545
United Kingdom	2,016,629
Germany	259,064
France	254,485
All other countries	765,409

16. In the fiscal year 1908–1909, Canada imported watches, cases, and movements, as follows:

FROM	AMOUNT
United States	$ 538,336
United Kingdom	22,374
Switzerland . . .	. . . .	172,299

Reckon the percentage from each country and show the result graphically.

17. In 1909, Brazil imported dried fruits valued at $ 332,164, of which amount $ 7097 worth were from the United States. What per cent came from the United States ?

540. To find the base, the percentage and the rate being given.

18 is ¼ of what number ?
18 is 25 % of what number ?
18 is .25 × ?

The equation is in the form of the second type problem. The product (percentage) and one of the factors (rate) are given to find the other factor (base). $18 \div .25 = 72$, the base.

To find the base, divide the percentage by the rate.

541. 48 is $18\frac{3}{4}$% of what number?

$$18\frac{3}{4}\% \times ? = 48.$$
$$48 \div 18\frac{3}{4} = 256 \quad \textbf{Or,}$$
$$18\frac{3}{4}\% = \frac{3}{16}.$$

$\frac{3}{16}$ of the number = 48.
$\frac{1}{16}$ of the number = $\frac{1}{3}$ of 48 = **16.**
$\frac{16}{16}$ of the number = $16 \times 16 = 256.$

On reducing the problem to its simplest type form we have $18\frac{3}{4}\% \times ? = 48$. Since a product and one factor are given, the other factor (base) is found by dividing the product, 48, by the given factor, $18\frac{3}{4}\%$. The result is 256. Or,

By the use of the fractional equivalent of $18\frac{3}{4}\%$, the solution can be made mentally. $18\frac{3}{4}\% = \frac{3}{16}$.

If $\frac{3}{16}$ of the number is 48, $\frac{1}{16}$ of the number is $\frac{1}{3}$ of 48, or 16, and $\frac{16}{16}$ of the number, or the number itself, is 16 times 16, or 256.

To the Student. Wherever possible, use this fractional method of solution. It is far shorter and quicker than the solution first explained. The first method will have to be used in those problems containing rates that do not have convenient fractional equivalents.

542. Find the number of which

1. 42 is 6 %.	12. 240 is 75 %.	23. 848 is $266\frac{2}{3}$ %.
2. 18 is 2 %.	13. 280 is $62\frac{1}{2}$ %.	24. 833 is $116\frac{2}{3}$ %.
3. 25 is 20 %.	14. 224 is $87\frac{1}{2}$ %.	25. 48 is $\frac{1}{4}$ %.
4. 43 is 1 %.	15. 378 is $56\frac{1}{4}$ %.	26. 321 is $\frac{1}{2}$ %.
5. 86 is 43 %.	16. 663 is 13 %.	27. 424 is $\frac{5}{8}$ %.
6. 72 is $12\frac{1}{2}$ %.	17. 374 is 17 %.	28. 375 is $\frac{5}{8}$ %.
7. 90 is $6\frac{2}{3}$ %.	18. 836 is 19 %.	29. 15 is $\frac{1}{20}$ %.
8. 92 is 25 %.	19. 1236 is 6 %.	30. $5\frac{1}{4}$ is $\frac{1}{8}$ %.
9. 106 is $33\frac{1}{3}$ %.	20. 360 is $112\frac{1}{2}$ %.	31. $3\frac{2}{3}$ is $\frac{1}{3}$ %.
10. 125 is $16\frac{2}{3}$ %.	21. 900 is $128\frac{4}{7}$ %.	32. $7\frac{1}{2}$ is $\frac{3}{4}$ %.
11. 130 is 50 %.	22. 755 is 250 %.	33. $4\frac{1}{2}$ is $\frac{3}{8}$ %.

PROBLEMS

543. 1. A young woman, by saving $31\frac{1}{4}$% of her earnings, paid for a house and lot costing $1875 in 6 yr. How much did she earn a year?

2. A parlor suite sold for $156, which was 30 % more than it cost. How much did it cost?

3. A bookkeeper receives an increase of 10 % in his salary each year for four years. His salary for the fifth year is $2196.15. What was his salary the first year?

4. A house rents for $900 a year. The expenses are $250 a year. If the net income is $6\frac{1}{2}$ %, what was the cost of the house?

5. A house was insured for 80 % of its value at $\frac{3}{8}$ %. If the premium paid was $22.50, find the value of the house.

6. Of a consignment of eggs, 6 % were broken. For the remainder, $45.12 was received at 32 ¢ a dozen. How many dozen eggs were in the consignment?

7. A merchant's sales for October were $28,460.50, for November were $35,827.43, and for December were $46,921.07. If the sales for these three months were 35 % of the sales for the entire year, find the average monthly sales for the remainder of the year.

8. A bankrupt can pay 72 % of his debts. If his resources amount to $13,475.28, what are his liabilities? How much will a creditor receive whom he owes $2485?

9. In an orchard 40 % of the trees are winter apples; 10 % are fall apples; 5 % are harvest apples; 30 % are pear trees; and the remaining trees are peach. If there are 30 peach trees, how many trees are in the orchard?

10. A house and lot cost $6400. The house cost $66\frac{2}{3}$ % more than the lot. Find the cost of each.

11. A druggist's sales increased 15 % the second year, 20 % the third year, and 25 % the fourth year. If the fourth year's sales were $29,023.13, what were the sales for the first year?

12. A merchant's sales increased $16\frac{2}{3}$ % the second year, 20 % the third year, 25 % the fourth year, and $14\frac{2}{7}$ % the fifth year. If the sales for the five years amount to $213,244.25, what were the sales the first year?

13. A salesman's sales increased 25 % in November over October's sales, and 20 % in December over November's sales. His January sales decreased 40 % from those of December. If January's sales amounted to $3501, what were they for October?

14. In 1910 the population of Muskogee, Okla., was 25,278, an increase of 427 % in ten years. What was the population in 1900?

REVIEW

544.   RAPID DRILL EXERCISE

1. What is 25 % of 36 ?
2. What is 36 % of 25 ?
3. What is $12\frac{1}{2}$ % of 64 ?
4. What is 64 % of $12\frac{1}{2}$?
5. What is $8\frac{1}{3}$ % of 72 ?
6. What is 72 % of $8\frac{1}{3}$?
7. What is 75 % of 68 ?
8. What is 68 % of 75 ?

9. What is 10 % less than 30 ?
10. What is 30 % less than 10 ?
11. What is $18\frac{3}{4}$ % less than 48 ?
12. What is 48 % less than $18\frac{3}{4}$?
13. What is 20 % more than 44 ?
14. What is 44 % more than 20 ?
15. What is $33\frac{1}{3}$ % more than 66 ?
16. What is 66 % more than $33\frac{1}{3}$?

17. Of what number is 20, 25 % ?
18. Of what number is 25, 20 % ?
19. 40 is 10 % of what number ?
20. 10 is 40 % of what number ?
21. 30 is 15 % of what number ?
22. 15 is 30 % of what number ?
23. 60 is 75 % of what number ?
24. 75 is 60 % of what number ?
25. 50 is what per cent of 60 ?
26. 60 is what per cent of 50 ?
27. 75 is what per cent of 25 ?
28. 25 is what per cent of 75 ?
29. 20 is what per cent of 50 ?
30. 50 is what per cent of 20 ?
31. 20 is what per cent less than 30 ?
32. 30 is what per cent more than 20 ?
33. 25 is what per cent less than 30 ?
34. 30 is what per cent more than 25 ?
35. 40 is what per cent more than 25 ?
36. 25 is what per cent less than 40 ?

VAN TUYL'S BUS. ARITH. — **15**

545. PERCENTAGE DRILL

	a	b	c	d	e	f	g
1.	\$ 15	\$ 18	\$ 10	25%	10%	\$ 6	\$ 3
2.	16	22	14	20%	$12\frac{1}{2}$%	3	4
3.	21	24	15	40%	30%	4.20	6.30
4.	24	30	20	20%	$33\frac{1}{3}$%	2	8
5.	32	40	28	$16\frac{2}{3}$%	25%	6	12
6.	40	45	32	$12\frac{1}{2}$%	$11\frac{1}{9}$%	10	8
7.	48	56	36	$33\frac{1}{3}$%	25%	12	20
8.	50	60	45	25%	30%	15	18
9.	60	75	54	$16\frac{2}{3}$%	25%	20	18
10.	72	81	60	$33\frac{1}{3}$%	$16\frac{2}{3}$%	24	9
11.	80	96	70	20%	$12\frac{1}{2}$%	30	· 8
12.	96	120	84	$12\frac{1}{2}$%	$8\frac{1}{3}$%	24	16
13.	108	120	96	$16\frac{2}{3}$%	5%	9	36
14.	120	144	100	$37\frac{1}{2}$%	50%	20	12
15.	150	175	120	25%	$14\frac{2}{7}$%	50	40
16.	180	210	160	$66\frac{2}{3}$%	$12\frac{1}{2}$%	40	60
17.	240	270	210	$12\frac{1}{2}$%	25%	60	30
18.	300	360	250	20%	25%	25	100
19.	360	420	320	$33\frac{1}{3}$%	$16\frac{2}{3}$%	90	45
20.	450	540	360	20%	$16\frac{2}{3}$%	75	45

1. The numbers in column *b* are what per cent greater than those in column *a* ?

2. The numbers in column *c* are what per cent less than those in column *a*?

3. Consider column *a* as cost, and column *d* as rate of gain. Find the selling price.

4. Consider column *a* as cost, and column *e* as rate of loss. Find the selling price.

5. If column *a* is cost and column *f* is gain, find the gain per cent.

6. If column *a* is cost and column *g* is loss, find the loss per cent.

7. If column *b* is cost and column *c* is selling price, find the rate of loss.

8. If column c is cost and column b is selling price, find the rate of gain.

9. If column b is selling price and column d is rate of gain, find the cost.

10. If column b is selling price and column e is rate of loss, find the cost.

PERCENTAGE REVIEW

MENTAL

546. 1. Two railroads carry freight 200 mi. The first carries it 75 mi., and the second 125 mi. What per cent of the distance does each carry it ?

2. A man earns $1500 a year and spends $900. What per cent of his money does he spend ?

3. A boy bought a watch for $12 and a chain for $4. Later he sold them both for the cost of the watch. What per cent of the cost was the selling price ?

4. During the month of May a farmer sells 500 qt. of milk; during June he sells 600 qt. What is the per cent of increase ?

5. Of a pile of potatoes containing 120 bu., one bushel out of five is bad. What per cent are bad? How many bushels are good ?

6. From a half chest of Oolong tea containing 50 lb., 15 lb. were sold one day and 20 lb. the next day. What per cent of the tea remained unsold? What per cent was sold each day ?

7. A bankrupt's resources are $12,000, and his liabilities are $20,000. What per cent of his debts can he pay? He owes Mr. Brown $600. How much will he receive ?

8. Last year my house and lot was worth $4000. This year it is worth $4400. What per cent has it increased in value? If it increases in value at the same rate per cent next year, how much will the property then be worth ?

9. In an invoice of 4 doz. crates of oranges, 10 % are decayed. If there is an average of 8 doz. oranges to a crate, how many oranges are spoiled ?

10. If coffee loses 10 % of its weight in roasting, how many pounds of roasted coffee will there be in 500 lb. of green coffee?

11. A clerk's monthly salary was $80. Owing to hard times, his salary is reduced 5 %. How much does he now receive per month?

12. A dairyman had 1000 qt. of milk a week in July. During August he increased the amount of grain fed to his cows $3 worth per week, and obtained 6 % more milk. If he sold the milk at 8 ¢ a quart, did it pay to increase the amount of grain fed?

13. If coal has been selling at $5.50 a ton, a 10 % increase will make it sell for how much?

14. A merchant's sales increased $33\frac{1}{3}$ % the second year over the first year. If the first year's sales were $72,000, what were the sales the second year?

15. I bought 56 yd. of cloth at $2.40 a yard and sold it to gain 25 %. How much did I receive for all of it?

16. If merchandise cost $360, and increases in value $16\frac{2}{3}$ %, how much will it be worth? If later it decreases $16\frac{2}{3}$ % in value, how much will it then be worth?

17. In settling with a bankrupt, I received $2400, which was 75 % of my claim. How much did I lose?

18. A man receives $50 a month rent for a house. His expenses for a year are $150. What is the value of the house if it nets him 9 %?

19. During the second year a grocer's sales increased $5000, which was 20 % of the first year's sales. What were the sales each year?

20. If, in problem 19, the profit was 20 % of the cost of the goods, what were his profits each year? What would have been his profits if they were 20 % of the selling price?

21. How much must a young man save so that at 6 % he may have an annual income of $1200?

22. A merchant's average profit is $12\frac{1}{2}$ % of his sales. How much merchandise must he sell to make a profit of $6000?

23. If an enterprise pays 8 %, what is the investment, the income being $1600?

24. A man sold $37\frac{1}{2}\%$ of his business for $3600. What was the entire business worth?

25. I sold a house and lot at a gain of $200, which was 5% of the selling price. How much did the property cost me?

PROBLEMS

547. 1. If potatoes shrink 5% in weight during the winter, what is the shrinkage in weight of 1200 bu.?

2. A store is worth $18,000. It rents for $2000. Repairs cost 1% of the value of the store; taxes cost $1\frac{1}{2}\%$; insurance $\frac{1}{4}\%$ on a policy of $10,000; and other expenses amount to $85. The net income is what per cent of the gross rental? of the value of the store?

3. A bankrupt paid me 68% of my claim. If I received $1275.46, how much did he owe me?

4. If the duty on certain merchandise was 30%, and the present duty is 20% less than formerly, what is the present rate?

5. What per cent of the total area of the following cities is land? What per cent is water?

	LAND AREA	WATER AREA
New Haven, Conn.	11,460 acres	2,880 acres
Syracuse, N.Y.	10,639 acres	309 acres
Scranton, Pa.	12,362 acres	147 acres
Fall River, Mass.	21,862 acres	3,793 acres
Albany, N.Y.	6,914 acres	283 acres

6. In a school there are 400 girls, which is $14\frac{4}{7}\%$ more than the number of boys. How many pupils are there in the school?

7. After selling 560 sheep, a shepherd found he had $12\frac{1}{2}\%$ of his flock left. How many sheep were there in the flock?

8. What is the difference in pounds between $\frac{5}{8}$ of a ton and $\frac{5}{8}\%$ of a ton?

9. A man had $1460 and spent $365 for a year's board. What per cent of his money did his board cost him? The money he spent is what per cent of the amount he had left?

10. An agent sold three farms for $12,750. The price of the second was 50% more than the price of the first, and of the third was 30% less than the price of both the others. Find the price of each.

11. A bushel of wheat weighs 60 lb. $5\frac{1}{2}$ bu. will make a barrel of flour (196 lb.). What per cent of the wheat is flour?

12. The population of New York, N. Y., in 1900 was 3,437,202; in 1905 it was 4,000,413, and in 1910 it was 4,766,883. Find the per cent of increase for each five-year period, and for the ten-year period.

13. On an invoice of merchandise valued at $ 672, a merchant paid freight $ 12.60, and cartage $ 9.80   The charges were what per cent of the value of the goods?

14. Find the sum of $\frac{1}{2}\%$ of $ 893.68; $\frac{3}{4}\%$ of $ 723.85; $\frac{1}{20}\%$ of $ 3487.60.

15. The following table shows the taxable value of real estate in the five boroughs of the city of New York in 1910. Find the total valuation for all the boroughs, and the per cent of real estate valuation in each borough.

BOROUGH	TAXABLE VALUE OF REAL ESTATE	PER CENT
Manhattan	$ 4,743,916,785	
The Bronx	493,757,919	
Brooklyn	1,404,036,521	
Queens. . . . ·	334,563,960	
Richmond.	67,917,489	
Total		100

16. 10 % of the above total is the limit of indebtedness allowed New York, N. Y., for 1910. Find the amount of debt which the city could incur.

17. For the same year the taxable value of personal property in the entire city was $ 372,645,625. The personal property is what per cent of the real property?

18. The New York, N. Y., charter requires that a tax of $\frac{3}{10}\%$ of the total taxable values of both real and personal property shall be set aside for the use of the public schools. Using the values given in problems 15 and 17, find the school tax for 1910.

19. A stock farm pays 15 % of its value annually. What is the farm worth if the annual income is $ 22,500 ?

20. A house and lot cost $6000; the insurance is $15, taxes are $60, and repairs are $45 annually. What rent per month must be received to realize 7 % on the investment?

21. What per cent of 5 lb. avoirdupois is 7 oz.?

22. From a cask containing 45 gal., 2 gal. and 1 qt are lost by leakage. What per cent is lost by leakage?

23. An agent charged $160 as his commission for selling real estate at $2\frac{1}{2}$ %. For how much was the property sold?

24. A farmer's crop of grain this year is 15 % less than it was last year. What was this year's crop if in the two years he raised 1554 bu. of grain?

25. A drover sold 330 sheep and had 1320 sheep left. What per cent of his flock did he sell?

26. A speculator increased his real estate holdings by 44 %. He then sold 25 % of all his property, and found he had 810 A. left. How many acres had he at first?

27. A man drew 65 % of his money from the bank and with 40 % of it bought a house and lot for $5200. How much money had he remaining in the bank?

28. From his income a young man saved 20 %. Had he saved $200 more, the per cent saved would have been 30 %. What is his income?

29. A merchant having 1848 yd. of cloth, sold $33\frac{1}{3}$ % of it at $1.60 a yard, 25 % of the remainder at $1.65 a yard, and then what was left at $1.55 a yard. How much did he receive for all?

30. Multiply each of the following percentages by its relative weight, and find the general average by dividing the sum of the products by the sum of the relative weights.

PERCENTAGES	RELATIVE WEIGHTS	PRODUCTS
76.5	6	——
93.25	5	——
87	4	——
91.65	12	——
71.81	9	——
	General Average	——

TRADE DISCOUNT

548. Discount is an amount deducted from a sum owing or to be paid. The amount of discount is indicated by a rate per cent.

If goods marked $4 are sold at a discount of 25 %, for how much do they sell?

549. Discounts may be trade discounts, or time discounts.

550. Manufacturers, publishers, and wholesalers have fixed price lists for their merchandise. These price lists are printed in their catalogues. To avoid the necessity of frequently reprinting their catalogues to conform to the changes in the prices of their goods, they print the list price, usually higher than the market price, and then regulate their charges by allowing discounts. If the market price increases, the discount allowed is less; if the price decreases, the discount is increased, usually by allowing a second discount to be deducted after the first discount has been deducted. Discounts of this kind are called **trade** or **commercial discounts.**

551. An additional trade discount is frequently allowed on large orders. Two or more discounts allowed on an invoice are called a **discount series.** The "series" is usually spoken of as a "discount."

552. Dealers doing business on a wholesale basis generally allow terms of credit of 30 da., 60 da., or 90 da. A **time discount** will be allowed, in most cases, for the payment of a bill before the term of credit expires.

A payment within 10 da. is generally considered as "cash" terms.

553. The **terms** of credit and time discount, if any, are printed or written on the billheads of business houses.

554. The **gross amount** of a bill is the amount before **any dis-** counts are deducted.

555. The **net amount** of a bill is the amount to be paid after the discounts are deducted.

556. Trade discount rates are frequently aliquot parts of 100 %.

557. The order in which the discounts are deducted is immaterial, as illustrated in the following example.

Find the net amount of a bill of $640 less $33\frac{1}{3}\%$, 25%, and 10%.

3)$\$640$	Or	4)$\$640$	
$\underline{213.33} = 33\frac{1}{3}\%$		$\underline{160} = 25\%$	
4) 426.67		3) 480	
$\underline{106.67} = 25\%$		$\underline{160} = 33\frac{1}{3}\%$	
10) 320		10) 320	
$\underline{32} = 10\%$		$\underline{32} = 10\%$	
$\$288$ Net Amount.		$\$288$ Net Amount.	

The first discount is deducted from the gross amount or list price of the bill. The second discount is taken from the amount remaining after the first discount is deducted. So also the third discount is taken from what is left after subtracting the second. The net amount is $288.

By changing the order of the discounts and deducting the 25% first, the work is made a little shorter as it avoids discounts of dollars and cents, due to the fact that $640 is exactly divisible by 4, and not by 3, and that the remainder, $426.67, is not exactly divisible by 4, while the remainder, $480, in the second solution, is exactly divisible by 3. The net amount in either case is the same.

Practice alone will enable the student to determine when it is advantageous to change the order of the rates.

558. The following bill illustrates the discounts and the terms of the sale:

Sᴛ. Lᴏᴜɪs, Mo., July 30, 1910.

Mʀ. J. C. Fʟɪɴᴛ,

Kansas City, Mo.

Bought of Tʜᴇ Hᴏʏᴛ Hᴀʀᴅᴡᴀʀᴇ Cᴏᴍᴘᴀɴʏ.

Terms: 60 days; 5% 10 days.

3	Doz. Chisels	. . . . . .	$5.60	16	80		
2	" Handsaws	. . . . .	7.50	15			
4	" Hammers	. . . . .	4.80	19	20		
				51			
		Less 25%		12	75	38	25
		Less 5%				1	91
						36	34

Received payment, Aug. 9, 1910
Tʜᴇ Hᴏʏᴛ Hᴀʀᴅᴡᴀʀᴇ Cᴏᴍᴘᴀɴʏ.
Per H.

Mr. Flint, instead of waiting 60 da. to make payment, has paid the bill within 10 da., thus securing the benefit of the time discount of 5%. Had he waited 60 da. before making payment, he would have paid $38.25.

MODEL INVOICE (Paper)

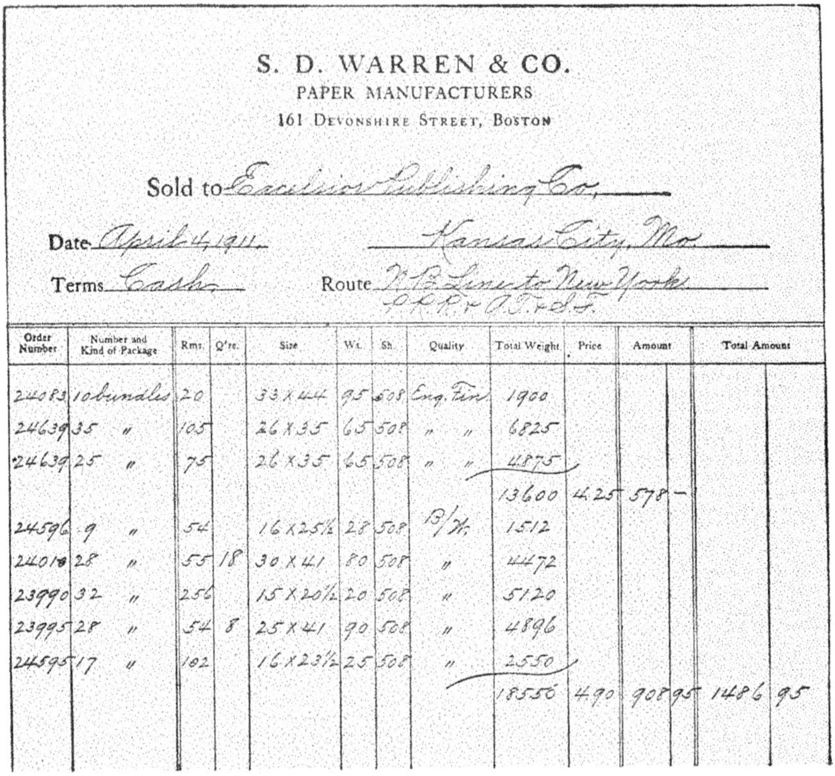

S. D. WARREN & CO.
PAPER MANUFACTURERS
161 Devonshire Street, Boston

Sold to Excelsior Publishing Co.
Date April 4, 1911. Kansas City, Mo.
Terms Cash Route N. Y. Line to New York
P. R. R. & A. T. & S. F.

Order Number	Number and Kind of Package	Rms.	Q'rs.	Size	Wt.	Sh.	Quality	Total Weight	Price	Amount	Total Amount
24083	10 bundles	20		33 x 44	95	508	Eng. Fin.	1900			
24639	35 "	105		26 x 35	65	508	" "	6825			
24639	25 "	75		26 x 35	65	508	" "	4875			
								13600	4.25	578 —	
24596	9 "	54		16 x 25½	28	508	B/W.	1512			
24018	28 "	55	18	30 x 41	80	508	"	4472			
23990	32 "	256		15 x 20½	20	508	"	5120			
23995	28 "	54	8	25 x 41	90	508	"	4896			
24595	17 "	102		16 x 23½	25	508	"	2550			
								18550	4.90	908 95	1486 95

Prove the correctness of the results in the above invoice.

559. Find the net price of each of the following:

List Price	Discount	List Price	Discount	List Price	Discount
1. \$ 8	12½ %	11. \$1.25	20 %	21. \$ 60	20 % and 16⅔ %
2. 10	20 %	12. 1.50	16⅔ %	22. 75	20 % and 33⅓ %
3. 12	25 %	13. 3.60	25 %	23. 80	25 % and 10 %
4. 6	33⅓ %	14. 4.50	10 %	24. 120	33⅓ % and 25 %
5. 9	50 %	15. 48	18¾ %	25. 150	20 % and 10 %
6. 14	14²⁄₇ %	16. 45	11⅑ %	26. 240	20 % and 12½ %
7. 16	6¼ %	17. 72	8⅓ %	27. 540	50 % and 20 %
8. 20	10 %	18. 50	4 %	28. 840	37½ % and 20 %
9. 24	16⅔ %	19. 60	3 %	29. 960	40 % and 16⅔ %
10. 30	6⅔ %	20. 75	8 %	30. 1200	50 % and 25 %

560. The following illustration shows the form of invoice used by The J. L. Mott Iron Works. Verify the results given.

GOODS ALWAYS SHIPPED AT BUYERS' RISK, NOTWITHSTANDING ANY RELEASE FOR DAMAGE THAT TRANSPORTATION COMPANIES MAY REQUIRE US TO SIGN. THESE GOODS WERE SHIPPED IN PERFECT ORDER; WE HOLD SHIPPING RECEIPT ACCORDINGLY AND IF THERE BE ANY DAMAGE ON ARRIVAL, CLAIMS MUST IMMEDIATELY BE MADE AGAINST THE TRANSPORTATION COMPANY.

Established 1828

New York, May 1, 1911.

*M*r. David J. Smith, #973 Huguenot Ave., New Rochelle, N.Y.

Bought of The J. L. Mott Iron Works,

Fifth Avenue & 17th Street,

Order No. 3/23/11 Shipped by B.& O.R.R. c/o N.Y.N.H.& H.R.R. *Terms Net Cash.*
from West Newton, Pa. Apr. 9/11.

PRICES SUBJECT TO CHANGE WITHOUT NOTICE.

1	Thermometer		$1.13
1	Altitude Gauge		1.81
15	Positive Key Air Valves	.13	1.95
		Net	$4.89
1	12 Gallon Galv. Expansion Tank complete	12.00	
	40/5/2½%		6.67
3	1½" N.P. Union Elbows	4.13	12.39
4	1¼" Ditto	3.42	13.68
	1" "	2.64	18.48
3	1½" N.P. Water Rad. Valves	7.38	22.14
4	1¼" Ditto	5.29	21.16
	1" "	3.87	27.09
			114.94
	75/10%		25.87
			37.43
	Less freight allowance 44 lb. at 28¢. per cwt. - - - -		.12
			$37.31

561. Make out a bill for each of the following purchases:

1. A. C. Bolton, Cincinnati, Ohio, bought of the Rochester Hardware Co., Rochester, N.H., Sept. 1, 1910, on account 60 da., 3 % 10 da.:

24 doz. C. S. Axes @ $10.
12 doz. Handsaws @ $14.
Less 20 %.
8 doz. Wrenches @ $12.
Less 25 % and 10 %

2. H. L. Williams, Paterson, N.J., bought of Marshall Field, Chicago, Ill., Dec. 14, 1910, on account 30 da., 1 % 10 da. :

> 8 Wilton Rugs, 8' 3" × 10' 6" @ $ 35.
> 6 Velvet Rugs, 9' × 12' @ 45.
> 210 yd. Brussels Carpet @ 1.25.

A discount of 16⅔ % and 10 % is allowed on the entire bill. Find amount due Dec. 24, 1910, and receipt the bill.

MODEL STATEMENT

SPOKANE, WASH., *May 31,* 191*1*

M*essrs Eaton & Hurlbutt,*
 San Francisco, Cal.

To **THE GREAT NORTHERN LUMBER CO.,** *Dr.*
WHOLESALE LUMBER DEALERS

May	4	To Mdse. as per bill rendered	796	45		
	15	" " " " " "	1347	63		
	19	" " " " " "	2467	38		
	26	" " " " " "	531	09	5142	55
		Cr.				
	10	By cash	500	—		
	18	" mdse. returned	125	68		
	22	" cash	1000	—	1625	68
		Balance due			3516	87

3. B. S. Shenkman & Co., Wheeling, W. Va., bought of Norfolk Grocery Co., Norfolk, Va., Dec. 1, 1910, on account 90 da., 2 % 30 da., 5 % 10 da. :

> 20 mats Java Coffee, 1500, @ 25 ¢.
> 30 bags Maracaibo Coffee, 3750, @ 20 ¢.
> 30 bags Rio Coffee, 3750, @ 15 ¢.

A discount of 15 % and 10 % is allowed on the above.

25 half-chests Japan Tea, 1875 lb., @ 25 ¢.
30 half-chests Oolong Tea, 1500 lb., @ 45 ¢.
25 half-chests Eng. Breakfast Tea, 1500 lb., @ 30 ¢.

A discount of 20 % and 10 % is allowed on the tea. Find net amount of the bill if paid on Dec. 11, 1910.

4. A. J. Cammeyer, New York, bought of New England Shoe Co., Boston, Mass., Sept. 18, 1910, on account 30 da., 5 % cash:

100 cases* Patent Calf, 3600 pr., @ $ 2.50.
80 cases Tan Calf, 2880 pr., @ 2.25.
60 cases Kangaroo, 2160 pr., @ 2.25.
120 cases French Kid, 4320 pr., @ 3.50.

If a discount of 25 % and 12½ % is allowed on the entire bill, find amount required to pay the bill at once. Receipt the bill.

5. Find the net amount of each of the following:

160 yd. Silk @ $ 3.50.
120 yd. Broadcloth @ 2.25.
240 yd. Duchess Satin @ 2.50.
80 yd. French Crépon @ 1.75.

Discount of 33⅓ %, 12½ %, and 5 %.

6. 18 grindstones at $ 2.25; 24 ice cream freezers at $ 4.50; 4 doz. jack planes at $ 7; and 12 doz. mortise locks at $ 3.50. Discount, 37½ %, 25 %, and 10 %.

7. 24 boxes lemons at $ 4.50; 30 boxes oranges at $ 5.40; 18 doz. pineapples at $ 1.25; 20 bunches bananas at $ 1.25. Discount 20%, 12½ %.

8. 1800 yd. denim at 12 ¢; 1400 yd. cashmere at 50¢; 2000 yd. kersey at $ 1.10; 3000 yd. storm serge at 62½ ¢. Discount 20 % and 12½ %.

9. 360 yd. Brussels carpet at $ 1.20; 800 yd. ingrain carpet at 75¢; 1200 sq. yd. linoleum at 62½ ¢; 900 yd. moquette carpet at $ 1.25. Discount 22½ % and 5 %.

10. One dealer offers a piano at $ 650, less 33⅓ % and 20 %. Another dealer offers the same grade of piano for $ 700 less 37½ %, 20 %, and 5 %. Which is the better offer, and how much?

* A case contains 36 pr. of shoes.

11. Which is the better offer, and how much, on a bill amounting to $1480, a discount series of $33\frac{1}{3}\%$, 25%, and 20%, or a series of 40%, 20%, and $12\frac{1}{2}\%$?

12. A bill of $1260.50 is subject to discounts as follows: $450 of the bill to 25% and $12\frac{1}{2}\%$; $375.60 to $37\frac{1}{2}\%$ and $8\frac{1}{3}\%$; and the remainder of the bill to a discount of $16\frac{2}{3}\%$. Find the net amount of the bill.

13. Find the net amount of a bill of the following amounts: $80 less 25% and 20% ; $126.50 less $16\frac{2}{3}\%$ and $12\frac{1}{2}\%$; and $172.75 less $62\frac{1}{2}\%$ and 25%. The whole bill is subject to a cash discount of 5%

562. To find a single discount equal to a discount series.

Find a single rate of discount equivalent to discounts of 10% and 10%.

Let $\qquad$ $1.00 =$ the list price.

then $\qquad$ $.10 =$ the first discount

$\qquad$ $.90 =$ remainder after discount is deducted.

$\qquad$ $.09 =$ the second discount (10% of $.90$).

$\qquad$ $.81 =$ the net amount.

$1.00 - .81 = .19 = 19\%$, the rate of discount equivalent to a discount series of 10% and 10%.

Or, $\qquad \dfrac{.10 \quad .10}{90 \times .90} = \begin{array}{l} 1.00 \text{ list price.} \\ .81 \text{ net amount.} \\ \overline{.19} \text{ single discount equal to a series of } 10\% \\ \quad \text{ and } 10\%. \end{array}$

In finding a single rate of discount equivalent to a discount series, 100% may always be used as the list price or basis of the computation. Deducting the two discounts of 10% and 10% gives the net amount of 81%. The equivalent single rate, then, must be the difference between the list price and the net price; Hence, $100\% - 81\%$ gives 19%, the single rate desired.

In the second solution, the several rates of the series are written above a horizontal line and each is deducted from 100%, the difference being written below the line. Multiplying the differences ($.90$ and $.90$) together gives $.81$ or 81% as the net amount. Deducting 81% from 100% gives 19%, the single rate equal to the series 10% and 10%.

This method may be used with any number of discounts. For its application to the solution of problems, see page 239.

563. Find a single rate of discount equal to each of the following series:

1. $33\frac{1}{3}\%$ and 25%.
2. 40% and $16\frac{2}{3}\%$.
3. $37\frac{1}{2}\%$ and 20%.
4. 25% and 20%.
5. $33\frac{1}{3}\%$ and 10%.
6. 30% and $14\frac{2}{7}\%$.
7. 30% and 20%.
8. 40% and 10%.
9. 20% and 20%.
10. 20% and 10%.
11. 20% and $12\frac{1}{2}\%$.
12. $16\frac{2}{3}\%$ and 10%.
13. 50% and 40%.
14. 50% and 30%.

15. 50% and 20%.
16. 50% and 10%.
17. 40% and 40%.
18. 40% and 30%.
19. 40% and 20%.
20. 40% and 5%.
21. $37\frac{1}{2}\%$ and 10%.
22. $37\frac{1}{2}\%$ and 5%.
23. 35% add 20%.
24. 35% and 10%.
25. 30% and 30%.
26. 30% and 10%.
27. 25% and 25%.
28. 25% and 10%.

29. $12\frac{1}{2}\%$ and $14\frac{2}{7}\%$.
30. 15% and 10%.
31. 10% and 10%.
32. 10% and 5%.
33. 20% and 5%.
34. 60% and $12\frac{1}{2}\%$.
35. 60% and 25%.
36. 60% and 10%.
37. 60% and 30%.
38. 60% and 40%.
39. 60% and 50%.
40. 55% and $33\frac{1}{3}\%$.
41. $11\frac{1}{9}\%$ and $12\frac{1}{2}\%$.
42. 75% and 20%.

43. 25%, 20%, and $16\frac{2}{3}\%$.
44. $33\frac{1}{3}\%$, 20%, and $6\frac{1}{4}\%$
45. 20%, $12\frac{1}{2}\%$, and $14\frac{2}{7}\%$.
46. 20%, 20%, and $12\frac{1}{2}\%$.

47. 25%, 20%, and 10%.
48. 20%, 10%, and 5%.
49. $37\frac{1}{2}\%$, 20%, and 10%.
50. 40%, 20%, and $16\frac{2}{3}\%$

PROBLEMS

564. 1. Find the net cost of goods listed at $\$480$, less 25%, 20%, and $16\frac{2}{3}\%$

$$\frac{\overset{80}{\cancel{\$480}} \times 3 \times 4 \times 5}{4 \times 5 \times 6} = \$240.$$

By deducting each rate in the series from 100% and reducing the remainders to their fractional equivalents, and placing the resulting fractions as shown in the solution, the problem resolves itself into one of simple cancellation. In many cases this is a very practical solution.

2. Which is the better for the purchaser, and how much per cent, a series of 30%, 20%, and 10%, or a straight discount of 50%? How much on a bill of $\$325$?

3. A bedroom suite is invoiced at $\$420$ less 20% and $14\frac{2}{7}\%$. What is the net price?

4. A gentleman bought an automobile for $2400 less $12\frac{1}{2}\%$, 10%, and 5%. What was the net cost?

5. Goods listed at $128 are sold subject to a discount of 75%, $12\frac{1}{2}\%$, and 5%. Find the net price.

6. You are offered 400 suits of ready-made clothing at $18 a suit, less 20% and 5%, by A, and at $16 a suit, less 10%, by B. If you buy of A, you pay the freight, $4.50. B offers to pay the freight. Which offer is better, and how much?

7. The list price of plows is $12. If you buy them at that price less $33\frac{1}{3}\%$ and 10%, and sell them at the same list price less 20% and 5%, how much do you make on 48 plows?

8. I can buy a bathroom outfit of one dealer for $150, less $33\frac{1}{3}\%$ and 20%. I can get the same outfit of another dealer for $125, less 25% and $12\frac{1}{2}\%$. The terms in each case are net 30 da., 2% cash. What is the least amount of ready money for which I can get the outfit?

9. A merchant bought goods at 30% and 10% less than the list price, and sold them at the same list price less 10%. Find his profit on goods listed at $1000.

10. An overcoat in a clothier's windows is marked " Was $27.50. Now $21." What is the rate of discount?

11. The merchants in a certain town gave a ticket of admission to the county fair costing fifty cents to each customer who purchased $3.50 worth of merchandise. This was equivalent to what rate of discount?

12. The fare one way between two cities is $48.75. If a round-trip ticket costs $85, what is the rate of discount?

PROFIT AND LOSS

565. If eggs are bought at 25¢ a dozen and sold at 30¢ a dozen, what is the profit? The profit is what part of the cost? what per cent of the cost?

If oranges cost 30¢ a dozen and are sold to gain $33\frac{1}{3}\%$ of the cost, how much is gained per dozen? What is the selling price per dozen?

A building lot cost $400   If it is sold so as to gain 25 %, what is the selling price?

A book cost $4   It was damaged by water and was sold at a loss of ¼ of the cost.   What was the loss? the selling price?

566. The subject of profit and loss has for its object the reckoning of gains, losses, costs, and selling prices, which arise in business transactions.

567. The **net** or **prime cost** of an article is the net amount paid for it.

568. The **gross cost** of an article is the total cost including freight, cartage, insurance, commission, etc.

569. The **gross selling price** is the total amount received for the goods sold.

570. The **net selling price** is the difference between the gross selling price and the sum of all the expenses of the sale paid by the seller.

571. The **profit** is the amount which the net selling price is greater than the gross cost.

572. The **loss** is the amount which the gross cost is greater than the net selling price.

573. In profit and loss,

> The cost is the **base**;
> The rate of gain, or loss, is the **rate**;
> The profit, or loss, is the **percentage**; and
> The selling price is the **amount** or **difference**.

574. Problems in profit and loss are solved in accordance with the principles of percentage.   In their special application to the subject the principles of percentage are stated as follows:

> Cost × rate of gain = gain.
> Cost × rate of loss = loss.
> Gain ÷ rate of gain = cost.
> Loss ÷ rate of loss = cost.
> Gain ÷ cost = rate of gain.
> Loss ÷ cost = rate of loss.
> Cost × (100 % + rate of gain) = selling price.
> Cost × (100 % − rate of loss) = selling price.
> Selling price ÷ (100 % + rate of gain) = cost.
> Selling price ÷ (100 % − rate of loss) = cost.

575. To find the gain or loss and the selling price, the cost and the gain or loss per cent being given

Find the gain or loss and the selling price in each of the following:

Cost	Rate	Cost	Rate	Cost	Rate
1. $ .15	20 % gain	**11.** $ 1.10	20 % gain	**21.** $215.00	20 % gain
2. .20	25 % gain	**12.** 1.40	20 % loss	**22.** 330.00	$66\frac{2}{3}$ % gain
3. .25	40 % loss	**13.** 4.00	15 % gain	**23.** 225.50	40 % gain
4. .24	$12\frac{1}{2}$ % loss	**14.** 5.00	30 % gain	**24.** 177.60	$12\frac{1}{2}$ % loss
5. .30	$33\frac{1}{3}$ % gain	**15.** 12.00	$16\frac{2}{3}$ % gain	**25.** 129.20	75 % gain
6. .48	$6\frac{1}{4}$ % loss	**16.** 14.00	20 % loss	**26.** 836.00	50 % gain
7. .60	20 % gain	**17.** 16.00	$2\frac{1}{2}$ % loss	**27.** 999.00	10 % loss
8. .75	40 % gain	**18.** 24.00	5 % loss	**28.** 750.00	$12\frac{1}{2}$ % gain
9. .80	$18\frac{3}{4}$ % gain	**19.** 160.00	$37\frac{1}{2}$ % gain	**29.** 660.00	$1\frac{2}{3}$ % gain
10. .96	$16\frac{2}{3}$ % gain	**20.** 180.00	$6\frac{2}{3}$ % gain	**30.** 880.00	$1\frac{1}{4}$ % loss

PROBLEMS

576. 1. A stock of goods valued at $13,500 was damaged by fire and water so as to cause a loss of 30 %. What was the loss, and for how much did the goods sell?

2. A piece of real estate costing $2400 increased in value $41\frac{2}{3}$ %, when it was sold. Find the gain and the selling price.

3. A vegetable dealer buys potatoes at 60¢ a bushel and sells them at a profit of $66\frac{2}{3}$ %. What is the gain per bushel, and the selling price per peck?

4. A fruiterer bought a carload of berries for $1380. He was obliged to sell them at a loss of $12\frac{1}{2}$ %. One retailer who purchased $375 worth of the berries on credit failed in business and paid his creditors 40¢ on the dollar. Find the wholesaler's total loss.

5. A grocer buys sugar at 5¢ a pound. He puts it up in bags of $3\frac{1}{2}$ lb. each. At what price per bag must he sell it to gain $2\frac{6}{7}$ %?

6. A grocer buys 6 bags Rio coffee, 125 lb. in a bag, at 12¢ a pound, and 9 bags Java coffee, 75 lb. in a bag, at 20¢ a pound. If he makes a mixture of the two kinds, and sells it at 90 % profit, how much does he receive per pound?

7. Find the gain on goods bought for $46.50 less 25 % and sold to gain 25 %

8. An invoice of merchandise cost $74.25 less $33\frac{1}{3}$ % and 10 %. Freight and cartage were $7.45. Find the gain and the selling price if the goods were sold to gain $37\frac{1}{2}$ %.

9. A shoe dealer buys shoes at $3 less $16\frac{2}{3}$ % and 10 % per pair. The expenses of the purchase amount to 5 ¢ a pair. At what price must he sell them to clear 40 % profit if it costs 28 ¢ per pair, on the average, to sell them?

577. To find the gain or loss per cent, the cost and the gain or loss being given.

Find the gain or loss per cent in each of the following:

	Cost			Cost			Cost	
1.	$ .15	$ .05 gain	**9.**	$ 3.50	$ .70 loss	**17.**	$603.00	$ 201.00 gain
2.	.20	.05 gain	**10.**	5.00	1.50 gain	**18.**	74.50	14.90 loss
3.	.36	.06 loss	**11.**	3.60	.90 gain	**19.**	38.50	3.85 gain
4.	45	05 gain	**12.**	4.50	90 loss	**20.**	94.50	31.50 gain
5.	.50	.05 loss	**13.**	16.00	8.00 gain	**21.**	18	22.50 s. price
6.	.80	.20 gain	**14.**	28.00	4.00 gain	**22.**	27	33.75 s. price
7.	1.00	20 loss	**15.**	450.00	75.00 gain	**23.**	40	36.00 s. price
8.	2.50	.40 gain	**16.**	425.00	170.00 gain	**24.**	64	80.00 s. price

578. The following cost and sales prices are from a "monthly statement of sales" in a hardware store. Find the rate of gain in each case.

	Cost Price	Sale Price		Cost Price	Sale Price
1.	$ 16.38	$ 20.25	**9.**	$ 8.65	$ 10.00
2.	22.50	29.45	**10.**	41.50	48.75
3.	12.50	15.00	**11.**	25.00	31.25
4.	35.25	41.50	**12.**	54.00	65.50
5.	5.65	7.50	**13.**	2.50	3.20
6.	4.50	6.00	**14.**	23.50	26.65
7.	15.25	20.00	**15.**	21.50	26.00
8.	12.50	15.00	**16.**	15.00	18.50

PROBLEMS

579. 1. Find the rate per cent of profit if goods are **bought** at $\$4.80$ less $16\frac{2}{3}\%$ and $12\frac{1}{2}\%$, and sold at $\$4.80$ less $12\frac{1}{2}\%$

$\$4.80 \times \frac{5}{6} \times \frac{7}{8} = \$3.50$, net cost.

$\$4.80 \times \frac{7}{8} = \$4.20$, net selling price.

$\$4.20 - \$3.50 = \$.70$, gain.

$\$.70 \div \$3.50 = .20 = 20\%$, rate of gain.

$\$4.80$ less the trade discounts of $16\frac{2}{3}\%$ and $12\frac{1}{2}\%$ equals $\$3.50$, the net cost. $\$4.80$ less $12\frac{1}{2}\%$ $= \$4.20$, the net selling price. The difference between $\$4.20$ and $\$3.50 = \$.70$, the gain. The gain divided by the cost equals the rate of gain.

2. Find the rate of gain when merchandise is bought for $\$6.40$ less 25% and is sold at $\$6.40$ net.

3 A sealskin coat cost $\$120$, less 30% and $16\frac{2}{3}\%$. It was marked $\$100$ by the retailer, but in order to effect a sale he gave a 10% discount. What rate of profit did he make?

4. The cost of a book is $\$8$, less a trade discount of $33\frac{1}{3}\%$ and 10% If it is sold at the same list price, less 10%, what rate of profit is made?

5 A retailer purchases hats at $\$54$ per dozen, less 20% and $16\frac{2}{3}\%$. If he sells them at $\$5$ each, what rate of profit does he make?

6. A statement of the merchandise account at the close of the year showed the following: Inventory at the beginning of the year, $\$3500$; purchases $\$22,500$; sales $\$28,000$; inventory at the close of the year, $\$6000$. Make a statement showing the net gain or loss. Find the gain or loss per cent.

Returns				
Merchandise Sales				28000
Cost				
Merchandise, inv. at beginning		3500		
Merchandise purchases		22500		
Merchandise, total cost		26000		
Less inventory at closing		6000		
Net cost of sales				20000
Merchandise gain				8000

$\$8000 \div \$20,000 = .40 = 40\%$, rate of profit.

7. At the beginning of the year a merchant had on hand merchandise amounting to \$5400. He bought during the year \$38,000 worth, and sold \$45,000 worth of merchandise. He had on hand at the close of the year an inventory of \$7400. Make a statement showing the gain or loss. Find the rate per cent of gain or loss.

8. Find the gain or loss per cent, and make a statement showing the gain or loss from the following facts: Jan. 1, 1909, inventory \$7925; Dec. 31, 1909, purchases for the year \$53,796.50; sales for the year \$54,837.50; Dec. 31, 1909, inventory \$1721.50.

9. A paper dealer buys 6 reams of "flat paper," 17" × 22", 28 lb. to the ream, at 8¢ a pound. He cuts the paper into sheets $8\frac{1}{2}$" × 11", and sells it at \$1 a ream. Find his rate per cent of gain.

10. We buy goods at \$.245 per pound and sell at \$.30625. At what price shall we sell to make the same per cent profit, if we have to pay \$.288?

580. To find the cost, the gain or loss and the per cent of gain or loss being given.

If a profit of 10¢ a dozen is made on oranges, and the profit is $\frac{1}{4}$ of the cost, what is the cost?

If a loss of 3¢ a dozen on eggs is a loss of 10% of the cost, what is the cost?

581. Find the cost and the selling price in each of the following:

	Gain	Gain		Loss	Loss		Gain	Gain
1.	\$ .06	25%	**11.**	\$ 12	25%	**21.**	\$73.00	$36\frac{1}{2}$%
2.	.07	20%	**12.**	15	$3\frac{1}{3}$%	**22.**	70.00	$17\frac{1}{2}$%
3.	.15	$33\frac{1}{3}$%	**13.**	18	$2\frac{1}{2}$%	**23.**	38.00	10%
4.	.25	$12\frac{1}{2}$%	**14.**	30	75%	**24.**	56.00	$87\frac{1}{2}$%
5.	.45	$16\frac{2}{3}$%	**15.**	160	$31\frac{1}{4}$%	**25.**	25.00	$41\frac{2}{3}$%
6.	.75	$6\frac{1}{4}$%	**16.**	220	5%	**26.**	36.00	$42\frac{6}{7}$%
7.	1.25	$14\frac{2}{7}$%	**17.**	28	$33\frac{1}{3}$%	**27.**	12.00	$22\frac{2}{9}$%
8.	2.40	$37\frac{1}{2}$%	**18.**	29	$14\frac{2}{7}$%	**28.**	1.25	$2\frac{1}{2}$%
9.	3.20	50%	**19.**	38	$66\frac{2}{3}$%	**29.**	2.50	$1\frac{1}{4}$%
10.	3.60	$56\frac{1}{4}$%	**20.**	54	9%	**30.**	37.50	$3\frac{1}{3}$%

PROBLEMS

582. 1. A house rents for $45 per month, and the expenses for taxes, insurance, repairs, etc., are $150 a year. What price can a man afford to pay for it so that he may have a $6\frac{1}{2}\%$ investment?

2. A merchant sold a sideboard at a gain of 25%, which was a gain of $18. When the merchant bought the sideboard, he was allowed discounts of 20% and 10% from the list price. Find the net cost and the list price.

3. A stock of goods was damaged 15% by fire and water. If the loss was $12,000, what was the cost of the goods? What was the selling price?

4. A speculator bought a lot and sold it so as to gain 20%. He immediately bought a second lot with the selling price of the first, which he afterward sold at a profit of $16\frac{2}{3}\%$. If his entire gain was $1400, how much did he pay for each lot?

5. In selling a quantity of oranges and pineapples a huckster gained $12 50, gaining 25% on the oranges, but losing 10% on the pineapples. If the loss on the pineapples was 20% of the net gain, find the cost of each

6. I bought a house for 10% below its estimated value and sold it for 10% more than its estimated value. My profit was $800. How much did I pay for the house? What rate of profit did I make?

7. If you buy at $.485 per pound less 25% and $12\frac{1}{2}\%$, and sell at the same list price less $16\frac{2}{3}\%$ and 10%, what discount can you allow so that you can have the same rate of profit if you have to buy at $ 485 less 20% and 10%?

· **583. To find the cost, the selling price and the rate per cent of gain or loss being given.**

If an article sells for 25¢, what is the cost if 5¢ is gained? The gain is what part of the cost? What per cent is gained?

If 25¢ is the selling price and 25% is gained, the selling price is how many per cent *more* than the cost? The selling price is what per cent of the cost? How may the cost be determined?

Let 1.00 of the cost = the cost.
.25 of the cost = the gain.
then 1.25 of the cost = the selling price.
1 25 × ? = $.25; hence, $.25 ÷ 1.25 = $.20. the cost.

584. Find the cost and the gain or loss in each of the following:

Selling Price	Rate	Selling Price	Rate	Selling Price	Rate
1. $ .36	10% loss	13. $ 6.40	25% gain	25. $ 660	20% gain
2. .36	20% gain	14. 18.00	$33\frac{1}{3}$% loss	26. 770	$16\frac{2}{3}$% gain
3. .48	20% loss	15. 22.50	25% loss	27. 1100	$8\frac{1}{3}$% loss
4. .75	25% gain	16. 37.50	25% gain	28. 1400	$12\frac{1}{2}$% loss
5. .80	$14\frac{2}{7}$% gain	17. 31.00	$3\frac{1}{3}$% gain	29. 3000	$11\frac{1}{9}$% gain
6. .90	$12\frac{1}{2}$% gain	18. 39.00	$2\frac{1}{2}$% loss	30. 3200	$11\frac{1}{9}$% loss
7. 1.05	$16\frac{2}{3}$% gain	19. 59.00	$1\frac{2}{3}$% loss	31. 4250	25% gain
8. 1.05	$12\frac{1}{2}$% loss	20. 162.00	$1\frac{1}{4}$% gain	32. 1650	20% gain
9. 1.40	30% loss	21. 182.00	4% gain	33. 1440	$14\frac{2}{7}$% gain
10. 2.25	50% gain	22. 672.00	$6\frac{2}{3}$% gain	34. 1100	10% loss
11. 3.75	$16\frac{2}{3}$% loss	23. 680.00	$6\frac{1}{4}$% gain	35. 3800	5% loss
12. 4.50	20% gain	24. 380.00	$18\frac{3}{4}$% gain	36. 4240	6% gain

PROBLEMS

585. 1. A real estate dealer sold property for $5075, gaining $16\frac{2}{3}$%. How much did the property cost?

$$\$5075 \div 1.16\tfrac{2}{3} = \$4350, \text{ cost, or } 7)\,\$5075$$

$$\begin{array}{r} 725 \\ \hline \$4350 \text{ Cost} \end{array}$$

The cost is the base, or 100%. The selling price, being at a gain, is $16\frac{2}{3}$% *more* than the cost, or $116\frac{2}{3}$% of the cost. Hence, dividing $5075 by $116\frac{2}{3}$% gives the cost, or $4350. Or,

Since $16\frac{2}{3}$% is equal to $\frac{1}{6}$, and the cost is equal to unity, or $\frac{6}{6}$, the selling price is equal to $\frac{7}{6}$ (of the cost). Dividing $5075 by 7 gives $\frac{1}{6}$ of the cost. $\frac{1}{6}$ of the cost ($725) deducted from $\frac{7}{6}$ of the cost ($5075) equals $\frac{6}{6}$ of the cost, or the cost itself.

This method, once mastered, is a great time and labor saver. Compare the amount of work required with that of the actual division in the first solution.

2. A grocer sells eggs at 32¢ a dozen and gains $6\frac{2}{3}$%. How much do the eggs cost him?

3. Find the cost of an article sold for $45, at a loss of $16\frac{2}{3}$%.

4. A hatter marked a hat $6. To sell it he gave a discount of 20%. His profit was 20%. Find the cost

5. At what price must I buy so that I can mark goods at $7.20, allow discounts of $16\frac{2}{3}$% and 10%, and still make a profit of $33\frac{1}{3}$%?

6. A merchant's selling price of an article is $65, less 20%. If he buys at the same list price ($65), what discount in addition to a discount of 25% must he be allowed in order to clear a profit of $33\frac{1}{3}$%?

7. A manufacturer's profit on a bicycle was 20%; the wholesaler's profit was $16\frac{2}{3}$%, and the retailer's profit was 25%. If the retailer's selling price is for $35, what was the first cost?

8. Two farms were sold for $4800 each. On one 20% was gained, and on the other 20% was lost. Find the net gain or loss, and the gain or loss per cent.

9. Find the cost per dozen of handkerchiefs that retail at 50¢ each if a profit of 75% is made.

10. What should be the buying list price of an article bought at discounts of 25% and $16\frac{2}{3}$%, if when sold at $64 less 25% and $16\frac{2}{3}$% a profit of $33\frac{1}{3}$% is realized?

586. PROFIT AND LOSS DRILL CHART

	COST	% GAIN	% LOSS	GAIN	LOSS	SELLING PRICE	SELLING PRICE
	1	2	3	4	5	6	7
1.	$ 16.00	$.12\frac{1}{2}$	.10	$ 4.00	$ 1.00	$ 18	$ 12
2.	18.00	$.16\frac{2}{3}$	$.11\frac{1}{9}$	6.00	2.00	24	15
3.	22.50	$.33\frac{1}{3}$	$.16\frac{2}{3}$	3.75	11.25	30	15
4.	28.00	25	$.12\frac{1}{2}$	4.00	7.00	30	20
5	36.00	$.11\frac{1}{9}$	.25	12.00	3.00	42	30
6.	48.00	$.08\frac{1}{3}$	$.16\frac{2}{3}$	4.00	2.00	64	36
7.	62.50	.20	.10	25.00	12.50	75	50
8.	96.00	$.12\frac{1}{2}$	$.37\frac{1}{2}$	6.00	24.00	128	80
9.	120.00	$.37\frac{1}{2}$	$.06\frac{2}{3}$	15.00	12.00	160	108
10.	150.00	$.33\frac{1}{3}$	.04	15.00	7.50	210	141
11.	240.00	.15	$.06\frac{1}{4}$	15.00	24.00	280	228
12.	360.00	.05	$.03\frac{1}{3}$	54.00	18.00	450	324
13.	500.00	.24	.03	25.00	10.00	575	490
14.	750.00	.18	.12	7.50	45.00	900	720
15.	900.00	.30	$.06\frac{2}{3}$	135.00	72.00	1080	837
16.	1200.00	$.08\frac{1}{3}$	$.02\frac{1}{2}$	180.00	84.00	1400	1056
17.	1500.00	$.06\frac{2}{3}$	.06	300.00	45.00	1550	1350
18.	1800.00	$.22\frac{2}{9}$	.02	450.00	90.00	2100	1500

1. Combine columns 1 and 2, and find selling price.

2. Combine columns 1 and 3, and find selling price.

3. Combine columns 1 and 4, and find gain per cent.

4 Combine columns 1 and 5, and find loss per cent.

5 Combine columns 1 and 6, and find gain per cent.

6. Combine columns 1 and 7, and find loss per cent.

7. Combine columns 2 and 4, and find cost and selling price.

8. Combine columns 3 and 5, and find cost and selling price.

9. Combine columns 4 and 6, and find gain per cent.

10. Combine columns 5 and 7, and find loss per cent.

REVIEW OF PROFIT AND LOSS
MENTAL

587. 1. If milk costs 12¢ a gallon, what per cent is gained by selling it at 6¢ a quart?

2 If eggs cost 25¢ a dozen, how many can a grocer sell for 25¢ if he gains 20%?

3. When potatoes cost wholesale 48¢ a bushel, what per cent profit does a retailer make who sells them at 15¢ a basket (8 baskets to the bushel)?

4. An article that cost $4 sells for $5. Will it pay to reduce the selling price 10%, if by so doing there are three times as many sales daily?

5. A dining table cost $20. Find the selling price to gain 20%; 25%; 15%; 10%; 40%.

6. An article cost $12. Find selling price to gain 20% of the selling price; 25% of the selling price; to lose 20% of the selling price; 25% of the selling price.

7. The cost is $48 per dozen. Find selling price of 1 to gain 25%; of 2 to gain 15%; of 3 to gain 20%; of 2 to lose 5%; of 6 to lose 10%.

8. If chairs cost $36 per dozen and 4 of them sell at $4.50 each, and the remainder sell at $3.75 each, what is the rate of profit?

9. You buy an article for $4.50 and sell it for $6. If the cost drops to $4.20, at what price will you sell to make the same rate of profit?

10 A firm has uncollected accounts of $2400. In closing their books for the year, they allow 5% for bad debts. How much do they allow for bad debts?

11. An article which cost $2.50 is sold to gain 20%. Find the gain on 50 such articles.

12. A dealer bought handsaws at $24 a dozen, less 16⅔%. What rate of profit does he make if he sells them at $2.50 each?

13. If coal costs $5 a ton wholesale, and sells at 30% above that price, what is the gain on 1000 T., if it costs 50¢ a ton to deliver it?

14. The raw material in a rocking chair costs $3.50, the labor of making costs $8.50. If the chair sells for 25% profit, for how much does it sell?

15. An American bought a horse in Canada for $130. He brought him to the United States, paying the duty,* $30, and sold him for $200. What per cent did he gain?

16. The man (in problem 15) returned to Canada and bought another horse with the $200 and paid the duty, 25%. At what price must he sell this horse to gain 20% on total cost?

17. If apples cost 90¢ a bushel, and 10% of them spoil, at what price per bushel must the remainder be sold to gain 20% on the entire cost?

18. A grocer buys 30 doz. eggs at 20¢ a dozen. If, by an accident, 20% of them are broken, what per cent does he gain or lose by selling the remainder at 25¢ a dozen?

19. I buy a carpet at $1.20 a yard, less 25%. What rate of profit do I make if I sell it at $1.20 net?

20. A room is 20 × 15', and requires 360 board feet of matched lumber to lay the floor. What is the percentage of waste?

21. 10 lb. of 15¢ coffee are mixed with 10 lb. at 20¢. At what price per pound should it sell to gain 20%?

---

* Duty on horses valued at $150 or less, $30 a head; on horses valued at more than $150, 25% of their value. (Tariff Law, 1909.)

22. Equal quantities of 30 ¢ and 40 ¢ tea are mixed. Find selling price to gain 40 %.

23. A 40 ¢ tea is mixed with a 70 ¢ grade in the ratio of 2 to 1. Find the per cent of profit if the mixture is sold at 65 ¢.

24. A clerk was told to mix Mocha at 24 ¢ with Java at 40 ¢, using 3 lb. of Mocha to 1 lb. of Java. By mistake he used 3 lb. of Java to 1 lb. of Mocha. Find the loss or gain on 100 lb. of the mixture, sold at 35 ¢.

PROBLEMS

588. 1. A merchant bought 440 yd. of cloth for $1100, less 20 % and 10 %. At what price per yard must he sell the cloth to gain 33⅓ % ?

2. Hats cost $45 a dozen, less 20 % and 16⅔ %. At what price should they retail to gain 40 % ?

3. If a wine merchant mixes 24 gal. of wine costing $2.25 a gallon with 6 gal. of water, and sells the mixture at $2.50 a gallon, what is his rate of profit?

4. I bought goods for $435, less 20 %, 12½ %, 10 %. I sold them for $560, less 25 % and 16⅔ %. What per cent of profit did I make?

5. If a desk costs $18 less 16⅔ % and 10 %, how many desks can be bought for $270?

6. A fruit dealer bought 60 doz. oranges for $12. He sold 40 doz. at 30 ¢ a dozen. Of those remaining 5 doz. spoiled. At what price must the rest be sold to gain 45 % on the cost of all?

7. A man bought a horse and sold it to gain 25 %. With the proceeds he bought a second horse and sold it at a loss of 12 %. If his gain on both was $22.50, how much did he pay for the first horse?

8. If 75 % of a stock of goods is sold for what the entire stock cost, what is the rate of gain? If the remainder of the goods is sold at cost, what is the rate of gain on the whole stock?

9. At what per cent of the list price should goods be marked that are bought at discounts of 25 % and 20 %, if a profit of 33⅓ % is to be realized ?

10. An article cost $24, less 20 %, 16⅔ %, and 6¼ %. What is the retailer's price mark if he marks it to gain 40 % ?

11. At what price shall each of the items in the following bill be sold to gain 25 % ?

MR. JAMES DOYLE,

Pittsfield, Mass.

BOUGHT OF WELLS, FOWLER & CO.

Terms : Net 60 days.

6	Roll Top Desks	$ 45.00	270					
8	Flat Top Desks	25.00	200					
12	China Closets	37.50	450					
9	Sideboards	62.50	562	50				
12	Dining Tables	25.00	300					
12	Sets Dining Chairs	20.00	240					
			2022	50				
	Less 25 % and 20 %		809		1213	50		

SUGGESTION. Find at what per cent of the list price the goods must be sold to make the required profit.

12. Find the net amount of the following items, and determine a selling price mark for each, if sold to gain $33\frac{1}{3}$ %.

> 2 doz. Smoothing Planes @ $10.50, less $14\frac{2}{7}$ %
> 4 doz. Screwdrivers @ $5.40, less 25 %
> 200 Carriage Bolts, each $\frac{5}{8}'' \times 2''$, @ $5.60
> $\frac{3}{8}'' \times 2\frac{1}{2}''$, @ $5.80
> $\frac{1}{2}'' \times 4\frac{1}{2}''$, @ $6.25,
> less $37\frac{1}{2}$ % and 20 %.
> 400 Machine Bolts, $\frac{3}{4}'' \times 6''$, @ $8.40,
> less 50 % and 10 %.

13. I bought 60 bbl. of apples at $2.25 per barrel. Upon sorting them, I found there were 35 bbl. first grade and 21 bbl. second grade, the rest being decayed. The first grade sold at $3.50, and the second grade at $2.15. If it cost 10 ¢ a barrel to sort, find the gain or loss per cent.

14. By marking goods at $4.50 each, a merchant sold on the average 4 of the articles per day. Could he afford to reduce the price to $4 if by so doing the sales increased to 8 articles a day, the cost being $3.25? What per cent would his profits be increased or decreased each day?

15. Dining tables selling for $16.50 at the rate of 6 a day are reduced to $14, whereupon the sales increase to 10 a day. If the tables cost $10.50, does the merchant gain or lose, and what per cent, by the cut in the price?

16. At a sale the price of a chiffonier was reduced from $28 to $21. The regular sales were 3 a day. During the sale there were 12 sales a day. If the cost was $16.80, were the daily profits increased or decreased, and what per cent?

17. At what price should goods costing $216 be marked to net a profit of 25 % after allowing a discount of 25 %, 20 %, and 10 %?

18. If coal can be bought at $4.75 per long ton and the cost of handling is 80¢ a short ton, what rate of profit is realized if the coal is sold at $6.50 per short ton?

19. At the beginning of a certain year a firm had on hand $16,800 worth of merchandise. They bought during the year $47,500 worth, and paid freight on it amounting to $700. They sold $62,500 worth of merchandise and paid freight on it to the amount of $850. If at the close of the year they had an inventory of $19,500, what rate of profit did they make? Prepare a statement showing all essential facts of the merchandise account.

Find the proprietor's gain or loss per cent in each of the following:

20.   PROPRIETOR

1910					1910			
Jan.	1	Liabilities	4876	50	Jan.	1	Resources	18725
Dec.	31	Profit & Loss	2562	76				

21.   PROPRIETOR

1910					1910			
Mar.	1	To withdrawal	1000		Jan.	1	By investment	8000
Oct.	1	To withdrawal	3000		June	1	By investment	4000
					July	1	By investment	2000
					Dec.	1	By profit & loss	3475

Find the gain or loss per cent in each of the following:

22. Mdse. purchases . . . . $ 13,250.00
Mdse. sales 12,687.50
Mdse. inventory at closing 4893.80

23. Mdse. inventory at beginning . . . . . $ 3890.00
Mdse. purchases 10,860.00
Mdse. sales 9640.00
Mdse. inventory at closing. 6460.00

24. Jan. 1, 1910. Inventory . . . . . . ·$ 3200.00
Purchases during the year . . . 9600.00
Sales during the year 10,500.00
Returned to creditors 450.00
Returned by debtors 600.00
Dec. 31, 1910. Inventory. 4500.00

MARKING GOODS

589. It is necessary in many lines of business to have the selling price of goods marked on ‘them. When merchants desire to keep their buying and selling prices from the public, the price marks are written in letters or characters, the interpretation of which depends upon a knowledge of the *key* to the letters or characters. If both cost and selling price are marked, a separate key is used for each.

590. The key consists of any word or phrase having ten different letters or characters.

591. If the price mark contains two or more figures alike in succession, as $ 4.55, $ 7.00, or $ 5.55, a different letter or character, called a *repeater*, is often used. The repeater makes it more difficult for strangers to interpret the marks.

Cost Key	Selling Price Key
E S R O H K C A L B	N O T S E L R A H C
1 2 3 4 5 6 7 8 9 0	1 2 3 4 5 6 7 8 9 0
Repeaters W and X	Repeaters Y and Z

The words used are black horse and Charleston, spelled backward. The repeaters are any letters not contained in the key with which they are used.

592. If the cost of an article is $4.25 and the selling price is $5.50 the markings would be as follows: $\dfrac{\text{O.SH}}{\text{E.ZC}}$, the cost mark being above and the selling price mark below a horizontal line.

593. Using the keys just given, write the cost and selling price of each of the following:

	Cost	Rate of Gain		Cost	Rate of Gain		Cost	Rate of Gain
1.	$ 12.00	25 %	7.	$ 22.50	$33\frac{1}{3}$ %	13.	$ 1.20	$16\frac{2}{3}$ %
2.	6.40	$18\frac{3}{4}$ %	8.	15.00	20 %	14.	1.50	20 %
3.	4.50	20 %	9.	4.40	25 %	15.	2.20	25 %
4.	3.30	$33\frac{1}{3}$ %	10.	6.60	$16\frac{2}{3}$ %	16.	17.00	25 %
5.	2.50	20 %	11.	8.80	$12\frac{1}{2}$ %	17.	14.00	20 %
6.	1.60	50 %	12.	1.10	10 %	18.	19.00	$12\frac{1}{2}$ %

594. Many kinds of goods are bought and sold wholesale by the dozen. The retailer sells them by the piece. Consequently, in marking such goods the cost mark shows the cost per dozen, and the selling price mark the price of one article.

In marking goods all calculations as far as possible should be mental. In dividing the price of a dozen by 12 to find the cost of one article, twelfths of a dollar from $\frac{1}{12}$ to $\frac{11}{12}$ will constantly require reducing to cents. Hence all the twelfths should be memorized. See page 16.

595. Using the words *dozen black*, repeaters *p* and *q*, as a cost key, and *what prices*, repeaters *m* and *n* as a selling price key, write the cost per dozen and the selling price per article of each of the following:

	Cost per Dozen	Rate of Gain		Cost per Dozen	Rate of Gain		Cost per Dozen	Rate of Gain
1.	$ 12.00	25 %	6.	$ 16.00	25 %	11.	$ 32.00	$12\frac{1}{2}$ %
2.	9.60	$33\frac{1}{3}$ %	7.	18.00	$33\frac{1}{3}$ %	12.	33.60	$33\frac{1}{3}$ %
3.	8.40	25 %	8.	20.00	25 %	13.	34.00	25 %
4.	14.00	20 %	9.	25.00	40 %	14.	45.00	20 %
5.	15.00	40 %	10.	27.00	$33\frac{1}{3}$ %	15.	56.00	$14\frac{2}{7}$ %

Using the same keys, write in figures the price here given:

16. $\dfrac{\text{Do.kq}}{\text{W.hp}}$ 17. $\dfrac{\text{E.zb}}{\text{te}}$ 18. $\dfrac{\text{Oq.nk}}{\text{H.ps}}$ 19. $\dfrac{\text{Zl.kq}}{\text{A.ps}}$ 20. $\dfrac{\text{Z.bk}}{\text{ac}}$

21. $\dfrac{\text{Oz.ok}}{\text{H.at}}$ 22. $\dfrac{\text{Oe.qk}}{\text{H.ps}}$ 23. $\dfrac{\text{Be.kq}}{\text{P.sn}}$ 24. $\dfrac{\text{Ok.pq}}{\text{H.sn}}$ 25. $\dfrac{\text{Dc.ok}}{\text{H.wi}}$

596. To find the marked price, having given the cost with discount, the rate of gain or loss, and the selling discount.

A jobber buys bicycles at $24, less $12\frac{1}{2}\%$. He sells to gain $33\frac{1}{3}\%$. What is his marking price if he allows 20% discount?

$\frac{1}{8}$ of $24   = $3, discount.

$24 − $3   = $21, net cost to the jobber.

$\frac{1}{3}$ of $21   = $7, gain.

$21 + $7   = $28, jobber's net selling price.

$28 ÷ 80 % ($\frac{4}{5}$) = $35, jobber's marked price.

The discount is $\frac{1}{8}$ of $24, or $3, which deducted leaves $21 as the net cost of the bicycle. $33\frac{1}{3}\%$ gain is equal to $\frac{1}{3}$ of $21, or $7, making the net selling price $28. Since the jobber allows a 20% discount, he sells for 80% of his marked price. Hence $28 ÷ .80 gives $35, the list price (Prin. 14).

PROBLEMS

597. 1. A sewing machine catalogued at $40 is sold to a jobber subject to a discount of 25% and 20%. Find the jobber's marked price if he sells to gain 25% after deducting 25% and 20%.

2. A manufacturer sells bathtubs at $35, less 40%. At what price should the retailer mark them to sell at 25% discount and still make 50%?

3. A merchant pays $10.80, less 25% and 20%, a yard for lace. Find his marking price if he sells to gain 25% after allowing a discount of 10%.

4. Find the retail selling price of each of the following items if they are sold to gain $33\frac{1}{3}\%$:

$4\frac{1}{2}$ dozen Ties @ $4.50. $2\frac{1}{2}$ dozen Ties @ $15.00.   3 dozen Ties @ $10.50. $1\frac{1}{2}$ dozen Ties @ $18.00. $1\frac{1}{2}$ dozen Ties @ $15.75. Less 20% and $16\frac{2}{3}\%$.

5. Complete the following invoice, and determine the marking price of each chiffonier to clear 20%. (Fosdick pays cash.)

CLEVELAND, OHIO, Aug. 10, 1910.

PETER R. FOSDICK,

     New Haven, Conn.

    BOUGHT OF OHIO FURNITURE COMPANY

Terms: n/60 3/10.

18	#5287 Oak Chiffoniers @ 22.50 . . .						
	Less 10% and 5%						
	Freight prepaid					13	61

6. Complete the following, and find the retail selling price to gain 30%.

KALAMAZOO, MICH., Sept. 14, 1910.

G. A. HUBBARD,

     Chicago, Ill.

    BOUGHT OF EDWARDS & CHAMBERLAIN

Terms: Net Cash

2	doz. Shovels	$12.50				
3	" Wheelbarrows	15.00				
60	ft. 2" Lead Pipe	.25				
	Less 20% and 10%					

7. If hats cost $48, less 20% and $16\frac{2}{3}$%, per dozen, what should be the list price to gain 35%, and allow $16\frac{2}{3}$% and 10%?

8. When hosiery costs $9, less 20% and $12\frac{1}{2}$% per dozen pairs, how shall they be marked per pair to gain 20%, if $12\frac{1}{2}$% discount is offered?

9. A rug $12' \times 15'$ is imported, the invoice price being £5. The duty is 90¢ a square yard, and 40% ad valorem (of the value). If a pound sterling can be bought at $4.8665, at what price must the rug sell to gain 25%?

VAN TUYL'S BUS. ARITH. — 17

COMMISSION AND BROKERAGE

598. A commission merchant is a person who buys or sells merchandise, or transacts other business, for another person.

599. The person for whom business is transacted by the commission merchant is called the **principal**.

600. Commission merchants usually charge a rate per cent on the volume of business transacted. Such a charge is called **commission** or **brokerage**. In some cases the commission is reckoned at a certain price per unit of quantity of merchandise, as the bushel, barrel, etc.

601. An additional charge called **guaranty** is sometimes made by commission merchants for undertaking to be responsible for sales on credit or for the quality and quantity of goods bought, etc.

602. Merchandise shipped by a principal (also called **consignor**) to be sold by a commission merchant is called (by the consignor) a **shipment**. The commission merchant (also called the **consignee**) calls the merchandise a **consignment**.

603. The gross proceeds of a sale or collection is the entire amount received by the commission merchant. The **net proceeds** is the amount remaining after all expenses have been paid.

604. The **prime cost** of a purchase is the first cost, or the amount actually paid for the merchandise. The **gross cost** is the prime cost plus all the expenses of the purchase.

605. An **account sales** is an itemized statement of the sales of merchandise, the charges thereon, and the net proceeds of the sale. The statement is made by the commission merchant and sent to his principal. (For illustration, see problem 22, page 260.)

606. An **account purchase** is an itemized statement of the quantity, quality, and price of merchandise purchased, the expenses incurred, and the gross cost of the purchase. It is made by the purchasing agent and sent to his principal. (For illustration, see problem 26, page 261.)

607. The principles of percentage apply in commission.

The gross sales or prime cost is the *base*.

The rate of commission is the *rate*.

The commission is the *percentage*.

The net proceeds is the *difference*.

The gross cost is the *amount*.

608. To find the commission, the cost or selling price, and the rate of commission being given.

1. Find the commission and the net proceeds of a sale of merchandise amounting to $4500; commission 5 %.

$$5 \% \text{ of } \$4500 = \$225, \text{ commission.}$$
$$\$4500 - \$225 = \$4275, \text{ net proceeds.}$$

The commission is 5% of the gross sales, $4500, which equals $225. The net proceeds is the difference between the gross sales and the expenses.

2. Find the commission and the gross cost of a purchase of merchandise amounting to $3400; commission 4 %, guaranty 2 %, cartage $20.

$$4 \% \text{ of } \$3400 = \$136, \text{ commission.}$$
$$2 \% \text{ of } \$3400 = \$68, \text{ guaranty.}$$
$$\$3400 + \$136 + \$68 + \$20 = \$3624, \text{ gross cost.}$$

The commission and the guaranty are found by taking 4 % and 2 %, respectively, of $3400. The gross cost is the sum of the prime cost, $3400, and all the charges.

609. Find the commission and the net proceeds, or the gross cost, in each of the following:

	GROSS SALES	COM.	GUAR.	OTHER EXPENSES		PRIME COST	COM.	GUAR.	OTHER EXPENSES
1.	$2400	5 %		$15	**11.**	$1030	4 %	3 %	$ 8.50
2.	3000	3 %	2 %	25	**12.**	1150	3 %	2 %	9.60
3.	1500	$6\frac{2}{3}$ %	1 %	2 %	**13.**	1250	4 %		12.50
4.	1800	$6\frac{1}{4}$ %		$20	**14.**	1440	$6\frac{1}{4}$ %		2 %
5.	2250	$3\frac{1}{3}$ %	3 %		**15.**	1880	$12\frac{1}{2}$ %		50.00
6.	3600	$8\frac{1}{3}$ %		30	**16.**	3360	$16\frac{2}{3}$ %	5 %	
7.	4500	10 %		28	**17.**	4150	5 %	$2\frac{1}{2}$ %	
8.	1600	$2\frac{1}{2}$ %	2 %	12	**18.**	9600	$2\frac{1}{2}$ %	$2\frac{1}{2}$ %	100.00
9.	2700	5 %		50	**19.**	12400	10 %	$2\frac{1}{2}$ %	350.00
10.	3200	4 %	2 %	60	**20.**	16800	$12\frac{1}{2}$ %	5 %	500.00

21. An agent sold 950 bbl. pork at $17.50 a barrel, on a commission of 2 %. The other charges were for storage $25, freight $60, and insurance $5. Find the commission and the net proceeds.

22. Find the net proceeds of the following:

ACCOUNT SALES HAY

ALBANY, N.Y., Oct. 7, 1910.

Sold for the Account of

AUGUST STADLICH, Wellsville, N.Y.

BY F. A. MEAD & CO.

1910						
Sept.	10	25 tons Hay	$ 15.00			
	14	35 tons Hay	15.50			
	20	40 tons Hay	14.75			
		CHARGES				
Sept.	6	Freight, $ 100 ; Cartage, $ 75 ;				
		Insurance, $ 5		180		
	20	Storage		10		
		Commission, 50 ¢ a ton				
		Net proceeds				

23. Prepare an account sales from the following data, using your own name as agent, and any name you wish as principal:

On August 15, you receive a consignment of 3000 bu. of corn. The same day you pay freight, 2¢ a bushel; cartage, 3¢ a bushel; insurance, $7.50. August 20, you sell 1200 bu. at 60¢; August 24, 900 bu. at 62¢; and August 28, 900 bu. at 61¢. Your commission is 2¢ a bushel, and the storage charge is $15.

24. A consignment of 460 bbl. of flour was received Nov. 1, 1910. Freight charges were $56; cartage, $14; cooperage, $3.70; advertising, $11.50. November 5, 65 bbl. at $6.40 were sold; November 7, 185 bbl. at $6.45; November 9, 50 bbl. at $6.35; and November 16, the remainder at $6.50. The storage charge was $15.80; commission 2%; guaranty, 1%. Prepare an account sales from the foregoing data.

25. If it requires 5½ bu. of wheat to make 1 bbl. of flour, and the cost of grinding is 15¢ a barrel, and the barrels cost 35¢ each, find the gain or loss per cent to the consignor in problem 24, wheat costing 80¢ a bushel.

26. Find the gross cost of the following:

ACCOUNT PURCHASE

PHILADELPHIA, PA., Aug. 4, 1910.

Purchase for the account of

J. L. COOK, Watkins, N.Y.

BY H. M. DONOVAN & CO.

	25 M White Pine Lumber	$ 36				
	30 M Georgia Pine Ceiling	45				
	18 M Hemlock Boards	16				
	22 M White Oak	48				
	CHARGES					
	Cartage		125			
	Commission 2½%					
	Amount charged to your account					

27. Prepare an account purchase from the following facts:

A. B. Carver & Co., Baltimore, Md., bought for the account of Howard Cole of Scranton, Pa., 100 gal. oysters at 75¢; 200 doz. clams at 8¢; 600 lb. bluefish at 5¢; and 800 lb. white fish at 8¢. Charges are 3% commission and drayage $4.50.

COMMISSION REVIEW

MENTAL

610. 1. An agent sold $5000 worth of clothing on a commission of 5%. What was his commission?

2. I bought for my principal sugar worth $1250 on a commission of 4%. What was the total cost of the sugar?

3. A lawyer collected 75% of a claim of $1600 and charged 10% for collecting. What was his commission?

4. A book agent's commission was 40%. He sold 120 books at $2.50 each. How much did he earn?

5. An auctioneer received 5% of his sales, which amounted to $2400. Find his commission.

6. My agent sold for me 100 tubs of butter averaging 50 lb. each at 25¢ a pound, and charged 2%. How much did I receive for the butter?

Find the proceeds in each of the following:

7. 20 T. hay at $15; commission 8 % ; other expenses $25.

8. 12 T. cabbage at $12.50; commission 6 % ; other expenses $11.

9. 300 bbl. flour at $6; commission 3 % ; storage 3¢ a barrel; freight 20¢ a barrel.

10. 20,000 ft. lumber at $18 per M; commission 2½ % ; freight $30.

11. I bought through an agent 300 yd. of carpet at $1.50, paying him 2 %. How much did the carpet cost me?

12. An agent buys 1000 bu. of wheat at $1.10, and charges ½¢ a bushel. What is his commission, and the total cost?

13. Find the gross cost of 4000 brick at $9.50 per M; commission 50¢ a thousand; other charges $2.50.

Find the gross cost of the following purchases:

14. 1200 bu. corn at 75¢; commission 3 % ; freight 1½¢ a bushel.

15. 700 T. coal at $5; commission 2 % ; freight 50¢ a ton.

16. 800 yd. silk at $2.50; commission ¾ % ; freight $2.50.

17. 10 half chests Young Hyson, 700 lb., at 25¢; commission 2 % ; freight $1.50.

18. 10 bbl. sugar, 3200 lb., at 5¢; commission 4 % ; freight and drayage $1.25 per barrel.

19. A broker bought for me 800 lb. of leather and charged $8, commission being at 4 %. What was the price of the leather, and what was the gross cost including freight at 25¢ a hundred pounds?

20. My agent charges 2 % commission, and 2 % guaranty, for buying coffee at 20¢ a pound. If the commission and guaranty amount to $40, how many pounds were bought, and what was the gross cost?

21. Which is better for the agent, a commission of 1½¢ a bushel or 2 %, when a bushel sells for 80¢?

22. I bought a job lot of muslin consisting of 1500 yd. bleached and 800 yd. unbleached for $120; commission 5 % ; and other expenses $9. If I sell the bleached muslin at 7¢ a yard and the unbleached at 5¢ a yard, what is my profit?

23. I have 75 bbl. of flour which cost me $360. I send them to my agent, who sells them at $6.20, and charges 2 % commission, and $5.70 for other expenses. What per cent profit do I make?

24. Find the per cent profit realized by buying potatoes at 40¢ a bushel, and shipping them to New York, if I pay 5¢ a bushel for freight, 3¢ a bushel commission, 2¢ a bushel for other expenses, and the potatoes sell at 60¢.

REVIEW PROBLEMS

611. 1. An agent remitted to me $247.38 after retaining a commission of 5 % for collection. What sum did he collect? What was the amount of his commission?

2. The gross cost of a purchase of sweet potatoes was $864.55. The expenses were for drayage $12.50, cooperage $3.35, and commission $2\frac{1}{2}$ %. If the potatoes cost $.72 a bushel, how many bushels were bought?

3. An agent charged his principal $106.25 (commission being $2\frac{1}{2}$ %), for buying 5000 bu. of wheat; the freight charges, etc., amounted to $43.75. How much a bushel did the wheat cost the principal?

4. An agent sold 3000 bu. of wheat, and after deducting his commission of $2\frac{1}{2}$ %, sent his principal the proceeds, $2808. For how much a bushel was the wheat sold?

5. A commission merchant sold a consignment of cottonseed oil for $12,500 on a commission of 4 %. Other expenses of the sale amounted to $375. What amount was due the principal?

6. My agent purchased for me 180 bbl. of sugar, averaging 275 lb. each, at $5\frac{1}{4}$¢ a lb. He charged 2 % commission and $71.77 for other expenses. Find the gross cost of the sugar. If it is sold at 6¢ a pound, what rate of gain is realized?

7. An agent sold for his principal 6500 bu. of potatoes at 60¢ a bushel, charging 3¢ a bushel for selling. The freight charges were 2¢ a bushel, and other expenses amounted to $85. He purchased for his principal an invoice of silks for $2475, charging 5 % for buying, and remitted the proceeds. What was the amount of the remittance?

8. At what price shall I mark an article that cost me $18, so that I can instruct my agent to allow 25% discount on the marked price, pay him 15% commission for selling, allow 5% of sales for bad debts, and still make a profit of $33\frac{1}{3}$% ?

9. To manufacture a certain style of automobile costs $800. What must be the catalogue price of this machine, so that the manufacturer can give the retailer 25% discount on the list price, pay an agent $16\frac{2}{3}$% commission for selling, and still make a profit of 25% ?

10. A broker sold for me 400 bales of cotton (500 lb. in a bale) at 12.52 ¢ a pound on a commission of $5 per hundred bales. He invested the proceeds of the sale in flour at $5 a barrel, charging 2% for buying. How many barrels were purchased, and what sum was left unexpended in the broker's hands ?

Suggestion : — Find gross cost of 1 barrel of flour before dividing.

11. An agent received $7500 to invest in apples at $2.40 a barrel after paying all expenses of the purchase. Charges were as follows: commission 3% ; guaranty 2% ; drayage 5¢ a barrel; and freight 12¢ a barrel. Find the agent's commission, the number of barrels purchased, and the unexpended balance.

12. An agent sells $8000 worth of hardware on a commission of $2\frac{1}{2}$%. He pays freight charges, $162.50; drayage $68.50; and charges $2\frac{1}{2}$% for guaranteeing payment. What are the net proceeds of the sale ?

13. You are a commission merchant and have received, Sept. **1,** 1910, 840 bbl. of flour to be sold for the Twin City Milling Co. You pay the freight, $131.50; drayage $210; advertising $27.50; and insurance $\frac{1}{4}$% of gross sales. September 8, you sell 225 bbl. at $6.75; September 14, 300 bbl. at $6.85; and September 21, the balance at $6.80. Storage charge is $16.74, and your charge for commission is $2\frac{1}{4}$%. Prepare an account sales, showing all facts here given, and the net proceeds.

14. You buy through a purchasing agent 42 Ideal refrigerators at $17.50; 24 dining tables at $26.40; and 18 dressers at $32.75. The agent's commission charge is 5%, and he pays $15 for drayage. Prepare an account purchase using the current date and any name of agent and place you wish.

EXAMINATIONS

SPEED TEST

(Minimum time, 20 minutes; maximum, 1 hour.) Deduct one credit for each minute over minimum time.

612. **1.** Find the selling price of each of the following:

	Cost		Selling Price
(a)	$ 0.16	$37\frac{1}{2}$% gained	——
(b)	6.00	20 % lost	——
(c)	56.00	25 % gained	——
(d)	4.50	$16\frac{2}{3}$% lost	——
(e)	828.00	$\frac{1}{4}$% gained	——

2. Find in each of the following the number of hours and minutes from the time commenced to time finished:

	Time Commenced	Time Finished	Hours	Minutes
(a)	9 : 50 A.M.	4 : 13 P.M.	——	——
(b)	9 : 16 A.M.	3 : 55 P.M.	——	——
(c)	9 : 45 A.M.	12 : 16 P.M.	——	——
(d)	10 : 15 A.M.	1 : 00 P.M.	——	——
(e)	9 : 14 A.M.	10 : 02 P.M.	——	——

3. Divide the numbers in column 1 by the numbers in column 2, and express your answer as per cent correct to two figures only:

	Column 1	Column 2	Per Cent
(a)	245,326	981,304	——
(b)	5,253	30,900	——
(c)	12,596	59,923	——
(d)	257,515	476,878	——
(e)	18,487	55,460	——

4. Multiply each of the following percentages by its relative weight, and find the general average by dividing the sum of the products by the sum of the relative weights:

	Percentages	Relative Weights	Products
(a)	97.51	8	——
(b)	81.33$\frac{1}{3}$	12	——
(c)	65.	5	——
(d)	69.16	7	——
		General average	——

5. Perform the following indicated operations:

$\frac{1}{4}$ of 6876.88 =
$79,487.98 × 6% =
9569.64 ÷ $\frac{1}{4}$ =
$6734.17 × .5% =

6. Find the gain per cent in each of the following:

	Cost	Selling Price	Gain %
(a)	$4.50	$4.95	——
(b)	7,500.00	8,750.00	——
(c)	5,280.00	7,040.00	——
(d)	9,600.00	10,200.00	——

7. Find the amount to be remitted to the employer in each case:

	Amount Collected	Rate of Commission	Amount Remitted
(a)	$3,400	5 %	——
(b)	91,500	3¼%	——
(c)	81,600	2¼%	——
(d)	30,501	¾%	——

8. Find net cost of each item and total net cost of all:

	List Price	Discount	Net Cost
(a)	$270.00	25%	——
(b)	480.00	25 % and 20%	——
(c)	6.40	12½% and 10%	——
(d)	960.00	83⅓% and 25%	——
		Total	

9. Canadian exports for two years; find the total for each year:

Article	1908	1909
Mineral Products	$36,840,044	$38,669,008
Fish	14,435,023	14,863,343
Forest Products	38,504,738	46,716,480
Animal Produce	53,019,843	53,040,391
Agricultural Products	75,883,951	84,921,684
Manufactures	28,892,297	30,807,548
Miscellaneous	54,913	118,756

10. Find the excess of each kind, and the total excess of exports for 1909 over those for 1908 in problem 9.

WRITTEN TEST

613. 1. A retired merchant has an income of $25 per day   If his property is invested at 6 %, how much is he worth?

2. Oranges are bought at $1.80 a gross, and sold at the rate of two for 5 ¢.   Find the gain per cent.

3. A certain lot of goods was sold for $189 after discounts of 12½ % and 10 % were allowed.   What was the list price?

4. A man commenced business with $3000 capital.   The first year he gained 22½%, which he added to his capital; the second year he gained 30 % of the whole sum, and put the gain into his business; the third year he lost $16\frac{2}{3}\%$ of his entire capital. How much did he make in 3 yr.?

5. An agent receives $1092.42 with which to buy oats at 42¢ a bushel after deducting his commission of 2 % on the sum expended. How many bushels can he buy?

6. A man sold 28 yd. of cloth at $1.25 per yard, thereby gaining $3.50. Find the rate per cent of gain.

7. An insolvent merchant's assets are $3027.90 and his liabilities are $4974. How much ought a creditor to receive whose bill against the merchant is $627?

8. A peddler bought 491 yd. of cloth at 81¢ a yard; he spoiled 29 yd., and sold the remainder at 95¢ a yard. Did he gain or lose, and what per cent?

9. A. M Orton bought of The Excelsior Hardware Co., Dec. 15, 1910, terms 2/10 net 30, 600 pairs barn-door hangers at $2.75, less 25 % and 10 %; 175 doz. strap hinges at $3.25, less 20 % and $12\frac{1}{2}\%$; and 15 doz. wrought iron wrenches at $14.50. Find the net cost if paid for Dec. 24, 1910. At what price should these items be sold retail to gain 25 %?

10. A company paid $3000 for a hall. Within a given period they rented it 148 evenings at $15 an evening. Expenses were: coal, 15 T., at $6.25; lighting $31.40; repairs $78.60; insurance $15.20; and taxes $47.80. What per cent profit did they make?

WRITTEN TEST

614. 1. Russian cotton mills used in 1908 $125,000,000 worth of raw cotton. $61,365,340 worth was imported, $49,900,925 worth of it being American cotton. What per cent of the total is grown in Russia? What per cent of the imports is from the United States?

2. Exports of leather and leather goods from the United States to Porto Rico for 1907 were $854,017; 1908, $643,295; 1909, $960,672. Compare the exports of 1908 and 1909 with those of 1907, showing the per cent of increase or decrease.

3. In 1909 Germany imported 3,879.3 metric tons of leather, of which amount 249.7 T. were from the United States. What per cent of it came from the United States?

4. Imports of diamonds into the United States were valued at $37,060,237 in 1906; $27,691,877 in 1907; $11,538,130 in 1908; and $36,159,054 in 1909. Show these values graphically.

5. Copy the following table and fill the vacant spaces with the proper figures:

DATE	MILES OF RAILROAD IN UNITED STATES	DECENNIAL INCREASE	
		In miles	In per cent
1830	40		
1840	2,755		
1850	8,571		
1860	28,920		
1870	49,168		
1880	87,724		
1890	163,562		
1900	193,344		
1909*	236,868		

6. From the following table find for each seaport its per cent of the total receipts of grain:

RECEIPTS OF GRAIN AT SIX DOMESTIC SEAPORTS, MAY, 1910

SEAPORT	BUSHELS
Boston	2,416,011
New York	6,162,786
Philadelphia	2,632,337
Baltimore	2,114,599
New Orleans	1,154,625
San Francisco	1,058,922

7. The largest mill in Russia for the production of cottonseed products uses 180,560 lb. of cottonseed a day. From cottonseed there is produced 15 % oil, 38 % oil cake, and $1\frac{1}{2}$ % linter, and the remainder is shell, which is used as fuel.

If the oil sells for $2.25 for 36 lb., the oil cake for $.20 for 36 lb., and the linter for $4.50 for 36 lb., find the amount received from their annual product (reckon 300 da. for a year).

* Mileage not available for 1910.

8. The following table shows the aggregate of resources of all the banks and trust companies doing business in the state of Washington on the twenty-seventh day of November, 1908, and the sixteenth day of November, 1909.

Find the increase or decrease and the per cent of increase or decrease in each of the amounts for 1909 as compared with those for 1908.

RESOURCES		
	1908	1909
Loans on real estate	$ 9,866,248.36	$ 10,923,556.28
Other loans and discounts	39,854,730.02	52,670,093.45
Stocks, bonds, and warrants	7,160,352.16	10,302,660.93
Due from banks	17,279,753.54	19,774,462.22
Banking house, furniture and fixtures .	4,874,160.59	4,944,242.83
Other real estate	1,216,982.44	1,410,577.46
Expenses	987,027.45	616,693.72
Overdrafts	798,028.66	693,965.47
Profit and loss	508.94	1,022.93
Other resources	192,285.33	2,452,647.73
Cash on hand	9,970,568.73	11,078,421.91
Trust investments	791,765.99	
Totals		

9–10. For the first quarter of 1910 the following facts regarding organized male wage workers were reported in the state of New York. Arrange them in a table with columns showing: (1) the per cent of workers in each industry employed; (2) the average number of days worked by each man employed; (3) the average wage per day; and (4) the average earnings for the quarter for each person employed.

In the building industry there were 116,755 men reported, of whom 100,304 were employed an aggregate of 5,376,294 da., earning $20,161,104. In transportation, 60,367 were reported, of whom 53,549 were employed an aggregate of 4,396,373 da., earning $12,155,623. In clothing and textiles, 46,481 were reported, of whom 45,353 were employed an aggregate of 2,820,957 da., earning $7,392,539 In metals and machinery, 30,428 were reported, of whom 29,850 were employed an aggregate of 2,092,485 da., earning $6,537,150. In printing, binding, etc., 23,995 were reported, of whom 23,216 were employed an aggregate of 1,694,768 da., earning $5,943,946.

INTEREST

615. If a man lends $1000 for 1 yr. and receives at the end of the year $1060, why is he paid $60 more than he lent?

616. The compensation paid for the use of money is called interest.

617. That sum of money for the use of which interest is paid is called the **principal**. In the above illustration, $1000 is the *principal* and $60 is the *interest*.

618. The amount of interest to be paid is reckoned as a certain rate per cent per annum. The most common legal rate is 6 %. That means that for the use of $1 for a year a man pays 6 % of $1, or $.06. For the use of $1 for 2 yr. a man pays 2 times 6 % of $1, or $.12.

619. The **rate of interest** is the fractional part of $1 (expressed as hundredths, or as a per cent) that is paid for the use of $1 for 1 yr.

620. The **unit of time** used as a basis in interest calculations is 1 yr. Hence, the expressions 4 %, 5 %, and 6 % interest mean, respectively, 4 %, 5 %, and 6 % interest for 1 *year's* use of a given principal.

621. The sum of the principal and the interest is called the amount.

622. The principles of percentage apply to all solutions of interest problems.

623. Since the rate of interest is a rate for 1 yr., a new element — time — has to be taken into consideration in interest computations. For the purpose of solving problems the rate is equal to the rate per cent of interest multiplied by the time expressed as years. That is, from a percentage standpoint, if the rate of interest is 6 %, and the time is 2 yr. and 6 mo., the rate is $2\frac{1}{2}$ times 6 %, or 15 %. Hence,

The **principal** is equivalent to the *base.*

The **rate of interest** multiplied by the time in years is equivalent to the *rate.*

The **interest** is equivalent to the *percentage.*

The **amount** is equivalent to the *amount.*

624. The legal rate of interest in the several states is determined by their legislatures. Valid contracts to pay more than the legal rate of interest cannot be made, except in certain cases provided for by statute.

625. Charging more than the legal rate is called **usury.** Most, if not all, the states impose a penalty for charging usury.

626. There are several kinds of interest, viz., **ordinary, accurate,*** and **compound.***

627. Ordinary interest is simple interest reckoned on a 360-day-year basis.

METHODS

628. There are a number of good methods of reckoning ordinary interest. One method is explained in this work, viz., the **banker's 6o-day method** — that being the best adapted to all kinds of problems.

For a preliminary discussion of this method, see pages 26–29.

629. To find the interest for any number of days.

Find the interest at 6 % on $ 1660 for 45 da.

$$\begin{array}{rl} \$\,16|60 =& \text{int. for 60 da. at 6 \%.} \\ 4|15 =& \text{int. for 15 da. at 6 \%.} \\ \hline \$\,12|45 =& \text{int. for 45 da. at 6 \%.} \end{array}$$

Interest for 60 da. is found by pointing off 2 places in the principal, which gives $16.60. 45 da. is 15 da. less than 60 da.; hence, the interest for 45 da. is $\frac{1}{4}$ less than the interest for 60 da. $\frac{1}{4}$ of $ 16.60 is $ 4.15, which, taken from $ 16.60, leaves $ 12.45.

DRILL EXERCISE

630. The student should practice until he can find the total interest in each of the following groups in four minutes or less.

* For definitions, see pages 281 and 286, respectively.

631. Find the total amount of interest at 6 % on:

1.	2.	3.
$ 1400 for 90 da.	$ 2900 for 75 da.	$ 7340 for 6 da.
$ 1600 for 75 da.	$ 3200 for 45 da.	$ 8970 for 12 da.
$ 1560 for 80 da.	$ 2800 for 15 da.	$ 5980 for 3 da.
$ 1300 for 20 da.	$ 2400 for 21 da.	$ 3890 for 9 da.
$ 1200 for 24 da.	$ 3300 for 25 da.	$ 9660 for 10 da.
$ 875 for 50 da.	$ 4400 for 33 da.	$ 4520 for 7 da.
$ 930 for 65 da.	$ 1800 for 85 da.	$ 3360 for 5 da.
$ 870 for 40 da.	$ 1200 for 55 da.	$ 2180 for 9 da.
$ 1100 for 150 da.	$ 1000 for 72 da.	$ 1890 for 10 da.
$ 1370 for 180 da.	$ 2000 for 84 da.	$ 3280 for 18 da.
$ 920 for 36 da.	$ 3000 for 80 da.	$ 4170 for 2 da.
$ 1670 for 42 da.	$ 4000 for 48 da.	$ 7320 for 1 da.

4.	5.	6.
$ 1172 for 40 da.	$ 720 for 14 da.	$ 80 for 15 da.
$ 1260 for 30 da.	$ 840 for 16 da.	$ 90 for 60 da.
$ 1440 for 8 da.	$ 960 for 18 da.	$ 60 for 90 da.
$ 1240 for 3 da.	$ 1200 for 21 da.	$ 72 for 45 da.
$ 890 for 45 da.	$ 1500 for 27 da.	$ 45 for 72 da.
$ 680 for 48 da.	$ 1800 for 28 da.	$ 40 for 45 da.
$ 840 for 65 da.	$ 3600 for 36 da.	$ 300 for 20 da.
$ 960 for 55 da.	$ 7200 for 54 da.	$ 200 for 30 da.
$ 1020 for 70 da.	$ 5400 for 45 da.	$ 180 for 21 da.
$ 1210 for 90 da.	$ 6300 for 75 da.	$ 210 for 18 da.
$ 1620 for 80 da.	$ 4500 for 90 da.	$ 360 for 40 da.
$ 740 for 3 da.	$ 7500 for 12 da.	$ 500 for 12 da.

632. To find the interest for years, months, and days.

1. Find the interest at 6 % on $ 840.50 for 5 mo. 10 da.

$ 8|40.50 = int. for 2 mo. at 6 %.
16|81 00 = int. for 4 mo. at 6 %
4|20 25 = int. for 1 mo at 6 %.
1|40 08 = int. for 10 da. at 6 %.
$ 22|41 33 = int. for 5 mo. 10 da. at 6 %.

Interest for 2 mo. is found by pointing off *two places* in the principal, 4 mo. is twice 2 mo., and the interest for 4 mo. is twice $8.405, or $16.81. Similarly, the interest for 1 mo. is one half of $8.405, or $4.2025, and the interest for 10 da. is ⅓ of $4.2025, or $1.4008. On adding, $22.41 is found to be the interest for the entire time.

2. Find the interest at 6 % on $1360.75 for 1 yr. 7 mo. 25 da.

$13|60.75 = int. for 2 mo. at 6 %

$\overline{136|075}$ = int for 20 mo. at 6 %.

$\underline{1|1339}$ = int. for 5 da. at 6 %

$134|9411 = int. for 1 yr. 7 mo 25 da. at 6 %

1 yr. 7 mo. 25 da. = 19 mo. 25 da. = 20 mo. less 5 da. From the interest for 2 mo., it is easily found for 20 mo. to be $136.075. 20 mo. is 5 da. too much; hence, find the interest for 5 da., $1.1339, and deduct it from $136.075, leaving $134.94, the desired result.

3. Find the interest at 6 % on $1500 for 2 yr. 3 mo., 23 da.

$15|00 = int. for 2 mo at 6 %.

$\overline{195|00}$ = int. for 26 mo. at 6%.

7|50 = int. for 1 mo at 6 %.

5|00 = int. for 20 da at 6 %.

$\underline{|75}$ = int. for 3 da. at 6 %.

$208|25 = int. for 2 yr. 3 mo. 23 da. at 6 %.

The time is 27 mo. 23 da. 13 times the interest for 2 mo. gives the interest for 26 mo. The interest for the remaining 1 mo. is $\frac{1}{2}$ of $15, or $7.50. 20 da. is $\frac{1}{3}$ of 60 da.; hence, the interest is $\frac{1}{3}$ of $15, or $5. For 3 da. the interest is $\frac{1}{10}$ of the interest for 1 mo., or $.75. The total is $208.25.

NOTE. Keep partial results to the fourth decimal place to insure accurate results

633. Find the interest at 6 % on the following:

	PRINCIPAL	TIME		PRINCIPAL	TIME
1.	$ 38.40	2 mo. 15 da.	14.	$560	45 da.
2.	$126.75	4 mo. 20 da.	15.	$640	29 da.
3.	$248.50	6 mo. 12 da.	16.	$832	24 da.
4.	$360	8 mo 15 da.	17.	$975	18 da.
5.	$375.60	9 mo. 20 da.	18.	$760	19 da.
6.	$890.50	11 mo 10 da.	19.	$480	1 mo. 18 da.
7.	$429.60	1 yr. 4 mo. 15 da.	20.	$760	3 mo. 20 da.
8.	$387.50	2 yr. 6 mo. 20 da.	21.	$1840	9 da.
9.	$840	90 da.	22.	$1920	28 da.
10.	$660	100 da.	23.	$1500	42 da.
11.	$930	4 mo.	24.	$7290	48 da.
12.	$740	6 mo.	25.	$4000	54 da.
13.	$425	30 da.	26.	$3000	50 da.

PRINCIPAL	TIME	PRINCIPAL	TIME
27. $125	5 mo. 15 da.	40. $582	1 yr. 3 mo. 5 da.
28. $115.75	9 mo. 13 da	41. $2135	2 yr. 6 mo.
29. $825.50	7 mo. 14 da.	42. $4240	10 mo. 11 da.
30. $1260	3 mo. 23 da.	43. $1325	9 mo. 8 da.
31. $1340	5 mo. 29 da.	44. $2150	5 mo. 13 da.
32. $1696	1 mo. 20 da.	45. $6322	8 mo. 15 da.
33. $1440	70 da.	46. $4178	132 da.
34. $5000	65 da.	47. $7500	96 da.
35. $2840	36 da.	48. $8800	84 da.
36. $1000	75 da.	49. $8100	78 da.
37. $2375	90 da.	50. $5400	154 da.
38. $1024	95 da.	51. $4300	150 da.
39. $2717	7 mo. 15 da.	52. $3300	165 da.

634. To find the interest at any rate per cent.

What is the interest on $4800 for 60 da. at 6%? at 5%? at 4%? at 7%?

(a) $48|00 int. at 6%	(b) $48|00 int. at 6%	(c) $48|00 int. at 6%
8|00 int. at 1%	16|00 int. at 2%	8|00 int. at 1%
$40|00 int. at 5%	$32|00 int. at 4%	$56|00 int. at 7%

In solution (a), the interest at 6% is found by pointing off two places. Since 5% is 1% less than 6%, the interest at 5% is found by subtracting the interest at 1% from the interest at 6%. To find the interest at 1%, divide the interest at 6% by 6. On subtracting, the interest at 5%, $40, is found.

In solution (b), the interest at 2% is subtracted from the interest at 6% to find the interest at 4%. The interest at 2% is found by dividing $48 by 3. Why?

In solution (c), the interest at 1% is added to the interest at 6% to find the interest at 7%; and so on.

NOTE. The student should observe that in each case the interest is first found for the given time at 6%. Then the interest at 6% is increased or diminished in proportion as the required rate is greater or less than 6%.

635. The interest at rates other than 6% may be found as follows:

To find the interest at:

$6\frac{1}{2}$% . . . add $\frac{1}{12}$ to the interest at 6%.
$6\frac{3}{4}$% . . add $\frac{1}{8}$ to the interest at 6%.
7 % . . . add $\frac{1}{6}$ to the interest at 6%.
$7\frac{1}{5}$% . . . add $\frac{1}{5}$ to the interest at 6%.
$7\frac{1}{2}$% . . . add $\frac{1}{4}$ to the interest at 6%.
8 % . . . add $\frac{1}{3}$ to the interest at 6%.
9 % . . . add $\frac{1}{2}$ to the interest at 6%.
10 % . . . divide the interest at 6% by 6, and move the decimal point one place to the right.
12 % . . . multiply the interest at 6% by 2.
$5\frac{1}{2}$% . . . deduct $\frac{1}{12}$ from the interest at 6%.
$5\frac{1}{4}$% . . . deduct $\frac{1}{8}$ from the interest at 6%.
5 % . . deduct $\frac{1}{6}$ from the interest at 6%.
$4\frac{4}{5}$% . . . deduct $\frac{1}{5}$ from the interest at 6%.
$4\frac{1}{2}$% . . . deduct $\frac{1}{4}$ from the interest at 6%.
4 % . . . deduct $\frac{1}{3}$ from the interest at 6%.
3 % . . . divide the interest at 6% by 2.
2 % . . . divide the interest at 6% by 3.
1 % . . . divide the interest at 6% by 6.

To find the interest at any other rate, divide the interest at 6% by 6, to find the interest at 1%, and then multiply the interest at 1% by the given rate.

636. Find the interest at $6\frac{1}{2}$% and at 7% in Art. 633, problems 1–12 inclusive; at $7\frac{1}{2}$% and at $4\frac{1}{2}$% in problems 13–24 inclusive; at $5\frac{1}{2}$% and at 5% in problems 25–38 inclusive; and at 10% and 4% in problems 39–52 inclusive.

637. Pointing off two places in any principal gives the interest at rates other than 6% as follows:

At 1 % for 1 yr.
At 2 % for 6 mo. = 180 da.
At 3 % for 4 mo. = 120 da.
At 4 % for 3 mo. = 90 da.
At $4\frac{1}{2}$% for . . . . . 80 da.
At 5 % for . . . . . 72 da.

At $7\frac{1}{2}$% for 48 da.
At 8 % for 45 da.
At 9 % for 40 da.
At 10 % for 36 da.
At 12 % for 30 da.
At 15 % for 24 da.

Note. The figures here given may be very easily recalled if it is observed and remembered that the number of days in each case is the quotient of 360 da. divided by the rate (expressed integrally).

638. Find the interest on:

1. $875 at 2 % for 6 mo	**11.** $8900 at 12 % for **90 da.**
2. $980 at 8 % for 4 mo.	**12.** $7400 at 10 % for **72 da.**
3. $1260 at 5 % for 72 da.	**13.** $6900 at 9 % for **160 da.**
4. $1485 at 7½ % for 48 da.	**14.** $7870 at 7½ % for **96 da.**
5. $3475 at 8 % for 45 da.	**15.** $8840 at 6 % for **180 da.**
6. $6890 at 9 % for 80 da.	**16.** $9870 at 5 % for **216 da.**
7. $8960 at 10 % for 108 da.	**17.** $8865 at 4 % for **90 da.**
8. $5475 at 4 % for 180 da.	**18.** $7182 at 8 % for **240 da.**
9. $6400 at 5 % for 144 da.	**19.** $6275 at 4½ % for **80 da.**
10. $5600 at 4½ % for 80 da.	**20.** $8785 at 15 % for **24 da.**

SHORT METHODS

639. What is the interest at 6 % on $485 for 60 da.?

What is the interest at 6 % on $485 for 600 da.?

What is the interest at 6 % on $485 for 6000 da.?

How does the interest for 600 da. compare with the interest for 60 da.?

How many places are pointed off in the principal to find the interest for 60 da.? to find the interest for 600 da.?

How does the interest for 6000 da. compare with the interest for 600 da.? How many places are pointed off to find the interest for 6000 da.?

640. Find the interest at 6 % on $580 for 600 da.; for 6000 da.

$5.80 = int. for 600 da.

$58. = int. for 6000 da.

600 da. is 10 times 60 da.; hence, the interest for 600 da. is 10 times the interest for 60 da. Since the interest for 60 da. is found by pointing off two places in the principal, the interest for 600 da. is found by pointing off one place (Prin. 1 page 101).

Because 6000 da. is 10 times 600 da., the interest for 6000 da. is found by pointing off no places in the principal. That is, the interest at 6 % equals the principal in 6000 da.

641. Summary of Principles:

1. *Pointing off 3 places in the principal gives the interest at 6 % for 6 da.*

2. *Pointing off 2 places in the principal gives the interest at 6 % for 60 da.*

3. *Pointing off* 1 *place in the principal gives the interest at* 6 % *for* 600 *da.*

4. *Pointing off no places in the principal gives the interest at* 6 % *for* 6000 *da.*

642. Find the interest at 6 % on :

1. $ 492 for 600 da.
2. $ 1486 for 1200 da.
3. $ 1590 for 1800 da.
4. $ 1840 for 6000 da.
5. $ 2460 for 3000 da.
6. $ 1230 for 1500 da.
7. $ 6280 for 4500 da.
8. $ 9684 for 2000 da.
9. $ 8346 for 900 da.
10. $ 9471 for 200 da.

11. $ 1760 for 3600 da.
12. $ 1840 for 4200 da.
13. $ 1932 for 4800 da.
14. $ 750.75 for 6000 da.
15. $ 916.73 for 900 da.
16. $ 813.25 for 12,000 da.
17. $ 532.37 for 6000 da.
18. $ 928.50 for 800 da.
19. $ 4328 for 300 da.
20. $ 7388 for 1500 da.

INTERCHANGING PRINCIPAL AND TIME

643. Time is rarely, if ever, expressed in days, except for parts of a year. Hence, the problems in the preceding exercise are not of direct practical importance. Indirectly the solutions are valuable. By the commutative law of multiplication, the interest on $ 6000 for 89 da. is the same as the interest on $ 89 for 6000 da., as shown in the following illustration.

Find the interest at 6 % on $ 6000 for 89 da.

(a) $ 60|00 = int. for 60 da.   (b) $ 89 = int. for 6000 da.
      20|00 = int. for 20 da.
       6|00 = int. for 6 da.
       3|00 = int. for 3 da.
    ────────────
    $ 89|00 = int. for 89 da.

In solution (a) the direct solution is given by finding the interest first for 60 da., then for 20 da., 6 da., and 3 da., the sum of the several amounts being $ 89, for 89 da.

In solution (b) the principal, $ 6000, and the time, 89 da., are interchanged so that the problem would read " Find the interest on $ 89 for 6000 da." But the interest at 6 % on any principal for 6000 da. is equal to the principal. Therefore the interest is $ 89.

638. Find the interest on:

1. $875 at 2 % for 6 mo
2. $980 at 3 % for 4 mo.
3. $1260 at 5 % for 72 da.
4 $1485 at 7½ % for 48 da.
5. $3475 at 8 % for 45 da.
6. $6890 at 9 % for 80 da.
7. $3960 at 10 % for 108 da.
8. $5475 at 4 % for 180 da.
9. $6400 at 5 % for 144 da.
10. $5600 at 4½ % for 80 da.

11. $3900 at 12 % for 90 da.
12. $7400 at 10 % for 72 da.
13. $6900 at 9 % for 160 da.
14. $7870 at 7½ % for 96 da.
15. $8340 at 6 % for 180 da.
16. $9870 at 5 % for 216 da.
17. $8365 at 4 % for 90 da.
18. $7182 at 3 % for 240 da.
19. $6275 at 4½ % for 80 da.
20. $3785 at 15 % for 24 da.

SHORT METHODS

639. What is the interest at 6 % on $485 for 60 da.?

What is the interest at 6 % on $485 for 600 da.?

What is the interest at 6 % on $485 for 6000 da.?

How does the interest for 600 da. compare with the interest for 60 da.?

How many places are pointed off in the principal to find the interest for 60 da.? to find the interest for 600 da.?

How does the interest for 6000 da. compare with the interest for 600 da.? How many places are pointed off to find the interest for 6000 da.?

640. Find the interest at 6 % on $580 for 600 da.; for 6000 da.

$58.0 = int. for 600 da.
$580. = int. for 6000 da.

600 da. is 10 times 60 da.; hence, the interest for 600 da. is 10 times the interest for 60 da. Since the interest for 60 da. is found by pointing off two places in the principal, the interest for 600 da. is found by pointing off one place (Prin. 1, page 101).

Because 6000 da. is 10 times 600 da., the interest for 6000 da. is found by pointing off no places in the principal. That is, the interest at 6% equals the principal in 6000 da.

641. Summary of Principles:

1. *Pointing off 3 places in the principal gives the interest at 6 % for 6 da.*

2. *Pointing off 2 places in the principal gives the interest at 6 % for 60 da.*

3. *Pointing off* 1 *place in the principal gives the interest at* 6 % *for* 600 *da.*

4. *Pointing off no places in the principal gives the interest at* 6 % *for* 6000 *da.*

642. Find the interest at 6 % on :

1.	$ 492 for 600 da.	**11.**	$1760 for 3600 da.
2.	$1486 for 1200 da.	**12.**	$1840 for 4200 da.
3.	$ 1590 for 1800 da.	**13.**	$1932 for 4800 da.
4.	$ 1840 for 6000 da.	**14.**	$ 750.75 for 6000 da.
5.	$ 2460 for 3000 da.	**15.**	$ 916.73 for 900 da.
6.	$ 1230 for 1500 da.	**16.**	$ 813.25 for 12,000 da.
7.	$ 6280 for 4500 da.	**17.**	$ 532.37 for 6000 da.
8.	$ 9684 for 2000 da.	**18.**	$ 928.50 for 800 da.
9.	$ 8346 for 900 da.	**19.**	$ 4328 for 300 da.
10.	$ 9471 for 200 da.	**20.**	$ 7388 for 1500 da.

INTERCHANGING PRINCIPAL AND TIME

643. Time is rarely, if ever, expressed in days, except for parts of a year. Hence, the problems in the preceding exercise are not of direct practical importance. Indirectly the solutions are valuable. By the commutative law of multiplication, the interest on $ 6000 for 89 da. is the same as the interest on $ 89 for 6000 da., as shown in the following illustration.

Find the interest at 6 % on $ 6000 for 89 da.

(*a*) $ 60|00 = int. for 60 da.   (*b*) $ 89 = int. for 6000 da.
      20|00 = int. for 20 da.
       6|00 = int. for 6 da.
       3|00 = int. for 3 da.
   ─────────
   $ 89|00 = int. for 89 da.

In solution (*a*) the direct solution is given by finding the interest first for 60 da., then for 20 da., 6 da., and 3 da., the sum of the several amounts being $ 89, for 89 da.

In solution (*b*) the principal, $ 6000, and the time, 89 da., are interchanged so that the problem would read " Find the interest on $ 89 for 6000 da." But the interest at 6 % on any principal for 6000 da. is equal to the principal. Therefore the interest is $ 89.

Proof of the Principle

Find the interest on $1800 for 47 da. at 6 %.

$1800 at 6 % int. for 47 da. = $47 at 6 % int. for 1800 da.

$$\frac{\$1800 \times .06 \times 47}{360} = \frac{\$47 \times .06 \times 1800}{360} = \$14.10.$$

By the cancellation method of finding interest the principal multiplied by the rate per annum, multiplied by the time expressed as years or a fraction of a year, gives the interest. In the illustration here given it is to be observed that both the principal and the number of days are above the line, — that is, they are each a part of a dividend to be divided by the number below the line. Since both principal and time (in days) are factors of the same product, their relative order in the list of factors is immaterial (Prin. 13, page 102). Therefore, the principal and the time in days can be interchanged without altering the amount of the interest.

644. By interchanging principal and time, find the interest at 6 % on:

1.	$6000 for 29 da.	19.	$60 for 142 da.
2.	$3000 for 47 da.	20.	$90 for 101 da.
3.	$600 for 83 da.	21.	$150 for 140 da.
4.	$1200 for 19 da.	22.	$1500 for 64 da.
5.	$1500 for 37 da.	23.	$2100 for 44 da.
6.	$2000 for 81 da.	24.	$2700 for 14 da.
7.	$12,000 for 43 da.	25.	$6600 for 31 da.
8.	$9000 for 53 da.	26.	$7200 for 25 da.
9.	$8000 for 79 da.	27.	$7800 for 17 da.
10.	$1000 for 33 da.	28.	$8400 for 13 da.
11.	$180 for 55 da.	29.	$540 for 38 da.
12.	$120 for 74 da.	30.	$630 for 41 da.
13.	$300 for 71 da.	31.	$72 for 35 da.
14.	$480 for 92 da.	32.	$30 for 80 da.
15.	$360 for 103 da.	33.	$240 for 70 da.
16.	$200 for 110 da.	34.	$15 for 100 da.
17.	$900 for 111 da.	35.	$15,000 for 160 da.
18.	$600 for 132 da.	36.	$12,600 for 131 da.

645. 1. Find the interest on $750 at 6% from July 8, 1910, to Sept. 4, 1910. (Exact time.)

July, 23 da.	$5\|80 = 600$ days' int.
Aug., 31 da.	$1\|45 = 150$ days' int.
Sept., 4 da.	$7\|25 = 750$ days' int.
58 da.	

The exact time is found by counting all the days after July 8 to and including Sept. 4. It is 58 da. Interchange principal and time ; the interest is found to be $7.25.

2. Find the interest at 6% on $897.50 from May 15, 1909, to Aug. 30, 1910. (Compound time.)

yr.	mo.	da.	
1910	8	30	$8\|97.50 =$ int. for 2 mo.
1909	5	15	$62\|82\ 5 =$ int. for 14 mo.
1	3	15	$4\|48\ 75 =$ int. for 1 mo.
			$2\|24\ 37 =$ int. for 15 da.
			$69\|55\ 62$

$69.56 =$ int. for 1 yr. 3 mo. 15 da.

The compound time is 1 yr. 3 mo. 15 da. ; the interest is found as explained in Art. 632.

3. Find the interest and the amount at 6% on $920 from April 14, 1910, to Sept. 2, 1910. (Bankers' time.)

From April 14 to Aug. 14 is 4 mo.
From Aug. 14 to Sept. 2 is 19 da.
Therefore the time is 4 mo. 19 da.

$9\|20 =$ int. for 2 mo.
$18\|40 =$ int. for 4 mo.
$2\|30 =$ int. for 15 da.
$\|46 =$ int. for 3 da.
$\|1533 =$ int. for 1 da.
$21\|3133$

$21.31 =$ int. for 4 mo. 19 da.
$920 + $21.31 = $941.31, amount.

The time is found by counting the whole months for the months and the exact days for the fraction of a month. The time is 4 mo. 19 da. Having the time, the interest is found as previously explained. The amount is equal to the principal plus the interest.

646. Find the interest and the amount at 6%, (1) for compound time, (2) for exact time, and (3) for bankers' time on each of the following :

1. $420 from Apr. 4, 1908, to Jan. 10, 1909.
2. $640 from Aug. 7, 1908, to Feb. 4, 1909. .
3. $560 from May 8, 1909, to Dec. 1, 1909.

4. $727.50 from Mar. 7, 1909, to Nov. 8, 1909.

5. $643.75 from Jan. 8, 1909, to Oct. 5, 1909.

6. $798.40 from Feb. 28, 1909, to Sept. 13, 1909.

7. $1260 from July 1, 1908, to Mar. 14, 1909.

8. $1375 from May 10, 1909, to Feb. 10, 1910.

9. $1760 from June 1, 1909, to Dec. 30, 1909.

10. $1840 from Oct. 17, 1909, to July 10, 1910.

11. $475.80 from Aug. 31, 1909, to May 15, 1910.

12. $768.50 from Dec. 8, 1909, to June 4, 1910.

13. $978.30 from Nov. 17, 1909, to Mar. 24, 1910.

14. $7485 from Oct. 8, 1908, to Aug. 6, 1909.

15. $7382 from Aug. 12, 1909, to Jan. 2, 1910.

16. $6600 from May 15, 1909, to Feb. 4, 1910.

17. $7500 from June 9, 1909, to Oct. 1, 1909.

18. $8000 from Mar. 1, 1909, to Dec. 10, 1909.

19. $12,000 from Feb. 14, 1909, to Oct. 1, 1910.

20. $1000 from Aug. 13, 1909, to June 20, 1910.

21. $900 from Dec. 18, 1909, to July 30, 1910.

22. $770 from Oct. 15, 1908, to Jan. 2, 1910.

23. $990 from May 16, 1909, to Nov. 24, 1910.

24. $650 from Jan. 22, 1909, to Dec. 13, 1909.

25. $1860 from Feb. 1, 1908, to Jan. 8, 1909.

26. $3600 from Mar. 4. 1908, to Oct. 17, 1909.

27. $4200 from June 30, 1909, to Jan. 1, 1910.

28. $426.74 from May 8, 1909, to Dec. 4, 1909.

29. $587.50 from Mar. 4, 1908, to Dec. 25, 1908.

30. $736.40 from Aug. 17, 1909, to July 1, 1910.

31–40. Find the interest and the amount at 5 % and at 7 % on problems 1–10 inclusive.

41–50. Find the interest and the amount at $5\frac{1}{2}$ % and at $6\frac{1}{2}$ % on problems 11–20 inclusive.

51–60. Find the interest and the amount at $7\frac{1}{2}$ % and at 4 % on problems 21–30 inclusive.

ACCURATE INTEREST

647. Accurate interest is computed for exact time in days, and on a basis of 365 da. to the year. It is used chiefly by the United States government, and by some bankers. Business men and bankers generally find the computation of accurate interest too inconvenient, though some merchants and bankers use it with the help of interest tables.*

648. Since by the "ordinary" interest methods, 360 da. are counted as a year, 360 days' interest by the ordinary method is greater than 360 days' interest by the exact interest method, by as much as $\frac{360}{360}$ is greater than $\frac{360}{365}$, or $\frac{1}{72}$. That is, for any given number of days, the ordinary interest is $\frac{1}{72}$ greater than the accurate interest at the same rate per cent. Hence, to change accurate interest to ordinary interest, add $\frac{1}{72}$ of the accurate interest to itself; and to change ordinary interest to accurate interest, deduct $\frac{1}{73}$ of the ordinary interest from itself.

649. In practice, however, accurate interest is computed as shown in the following illustrative example:

1. Find the accurate interest on $790 at 5% for 140 da.

$$\frac{\$790 \times .05 \times \overset{28}{\cancel{140}}}{\underset{73}{\cancel{365}}} = \frac{1106}{73} = \$15.15, \text{ the accurate interest.}$$

Multiplying $790 by 5% gives the interest for one year. Dividing that product by 365 gives the interest for 1 da., which multiplied by 140 gives the interest for 140 da.

650. Find the accurate interest on:

1. $750 for 148 da. at 6%.
2. $675 for 213 da. at 4%.
3. $876 for 219 da. at 5%.
4. $1314 for 58 da. at 7%.
5. $1325 for 64 da. at 6½%.
6. $1600 for 13 da. at 4½%.
7. $1500 for 18 da. at 3%.
8. $1700 for 38 da. at 5%.
9. $4380 for 51 da. at 6%.
10. $3825 for 47 da. at 5%.
11. $4564 from Aug. 11, 1908, to May 15, 1909, at 5%.
12. $3870 from June 1, 1909, to Dec. 30, 1909, at 6%.

*Bankers and others use interest tables for both ordinary and accurate interest. The tables are too long to permit of illustration in this book.

13. $4330 from Jan. 3, 1909, to Oct. 14, 1909, at 9 %.

14. $5786 from Feb. 23, 1908, to Jan. 4, 1909, at 6 %.

15. $3342 from May 16, 1909, to Aug. 1, 1909, at 10 %.

16. $5540 from Oct. 21, 1909, to Mar. 1, 1910, at 12 %.

17. $48.75 from Jan. 5, 1906, to Aug. 4, 1910, at 6 %.

18. $125.50 from Dec. 3, 1908, to Nov. 15, 1910, at 5 %.

19. $456.73 from Oct. 18, 1908, to Jan. 31, 1909, at 4 %.

20. $592.15 from June 11, 1909, to Dec. 7, 1909, at 6 %.

21. $678.25 from Sept. 29, 1908, to Sept. 13, 1909, at 5 %.

22. $431.25 from Apr. 19, 1908, to Aug. 10, 1909, at 6 %.

23. $598.80 from Mar. 31, 1907, to May 1, 1910, at 6 %.

24. $163.25 from July 3, 1908, to Dec. 24, 1909, at 5 %.

25. Find the interest of $25,000 U.S. bonds at 4 % for 87 da.

26. The United States government settled a claim of $12,500 with a creditor 78 da. after it became due. At 5 % interest what amount was due?

27. The interest-bearing debt of the United States, July 31, 1910, was as follows:

Title of Loan	Authorizing Act	Rate of Interest	When Redeemable	Amount
Consols of 1930	Mar. 14, 1900	2 %	After Apr. 1, 1930	$646,250,150
Loan of 1908–1918	June 13, 1898	3 %	After Aug. 1, 1908	63,945,460
Loan of 1925	Jan. 14, 1875	4 %	After Feb. 1, 1925	118,489,900
Panama Canal Loan:				
Series 1906 {	June 28, 1902 & Dec. 21, 1905	2 %	After Aug. 1, 1916	54,631,980
Series 1908 {	June 28, 1902 & Dec. 21, 1905	2 %	After Nov. 1, 1918	30,000,000

Find the total interest for 1 da.; for the month of August.

28. Find the average rate of interest paid on the United States debt.

PROBLEMS

651. 1. A man buys 150 shares of stock for $18,000. He receives a semiannual dividend of $2.50 a share. How much better is this than 4% interest, if the interest is paid semiannually?

2. A savings bank pays 4% interest ou its deposits. It loans on bond and mortgage at $5\frac{1}{2}$%. If its deposits amount to $16,475,000, and its loans are $14,850,000, what is its gross gain in interest per annum, interest in each case being paid semiannually?

3. A banker pays 3% interest on a deposit of $45,000. He loans it 3 mo. at 6%; 4 mo. at $5\frac{1}{4}$%; and for 48 da. at 7%. If it lies idle the remainder of the year, does he gain or lose, and how much?

4. A merchant bought a bill of goods amounting to $13,875; terms 3 mo., 5% cash. He accepted the cash terms. How much did he save if money was worth 6%?

5. A young man had $1600 saved. He purchased a house and lot for $4000, borrowing $2400 at a savings bank, giving as security a mortgage on the house and lot. The house rented for ¡$37.50 a month. He paid $5\frac{1}{2}$% interest on the mortgage; $\frac{1}{4}$% premium on an insurance policy of $3000; 1.8% tax on a valuation of $3500; $14 water rent; and $50 for repairs. What rate of interest did he receive on his investment?

6. Aug. 1, 1909, a wholesaler sold a bill of goods amounting to $1875.28, terms net 30 da. If the bill was not paid until Jan. 8, 1910, what amount was due, interest being 6%? (Exact time.)

7. What amount can a man afford to pay for a house and lot that rents for $65 a month so that his investment shall net him 6% interest, if he pays for taxes, insurance, repairs, and all other expenses, $360 a year?

8. On July 8, 1909, a man borrowed money at 6% interest and bought 3000 bu. of wheat at $1.08 a bushel. He sold the wheat Aug. 17, 1909, at $1.21 a bushel and returned with interest the money he borrowed. What was his gain?

9. A piece of property valued at $8000 is sold on the following terms: 20% of the selling price is to be paid in cash; for the remainder five notes falling due in one, two, three, four, and five years, respectively, with interest at 6 %, are given. If each note is paid at maturity, what is the total amount paid for the property?

10. What is the difference between the ordinary interest and the accurate interest at 6 % on $10,000 for 144 da.

11. A merchant's cash price of an article is $35; his price at 4 months' credit is $37.50. If money is worth 6 % per annum, how much extra profit does he make when selling on credit?

12. A real estate broker offers me a house and lot for $7500, guaranteeing an 8 % investment. The property is assessed for $5000 on which a tax of $1\frac{3}{4}$ % has to be paid. Insurance and other expenses will amount to $212.50 a year. At what price must the property rent per month to make good the broker's guarantee?

13. An invoice of merchandise was bought on the following terms: 4 mo. or 5 % cash. The cash discount is equivalent to what rate of interest per annum on the gross amount? on the net amount?

14. In what time will any sum of money double itself at 5 %? at 6 %? at 8 %? at 9 %?

15 At what rate per cent of interest will any sum of money double itself in 25 yr.? in $14\frac{2}{7}$ yr.? in 10 yr.? in 8 yr. 4 mo.?

16. A clerk is instructed to accept the best terms on all bills, allowing that money is worth 15 % to his employer. Under these conditions, which terms should the clerk accept in the following invoices?

$428.75, terms, 30 da., 2 % cash.
$769.50, terms, 90 da., 4/30, 5 % cash.
$1287.50, terms, 4 mo., 5/30, 6 % cash.
$1760.80, terms, 30 da., 1 % cash.

17. On an invoice of $4378.60, a merchant is offered 4 mo. credit or a discount of 5 % for cash. Not having the ready money, he accepts the credit terms. What rate per cent of interest does he pay on the net amount of the bill? How much would he have saved if he had borrowed the money at 6 % and paid cash?

18. A man bought $35,000 U.S. Panama Canal Bonds bearing 2 % interest Forty-six days later he sold them. How much interest did they earn while he owned them?

652. COMPOUND INTEREST

TABLE SHOWING AMOUNT OF $1 AT COMPOUND INTEREST IN ANY NUMBER OF YEARS, NOT EXCEEDING TWENTY-FIVE

Yr.	2 per cent	2½ per cent	3 per cent	3½ per cent	4 per cent	4½ per cent
1	1.0200 0000	1.0250 0000	1.0300 0000	1.0350 0000	1.0400 0000	1.0450 0000
2	1.0404 0000	1.0506 2500	1.0609 0000	1.0712 2500	1.0816 0000	1.0920 2500
3	1.0612 0800	1.0768 9062	1.0927 2700	1.1087 1787	1.1248 6400	1.1411 6612
4	1.0824 3216	1.1038 1239	1.1255 0881	1.1475 2300	1.1698 5856	1.1925 1860
5	1.1040 8080	1.1314 0821	1.1592 7407	1.1876 8631	1.2166 5290	1.2461 8194
6	1.1261 6242	1.1596 9342	1.1940 5230	1.2292 5533	1.2653 1902	1.3022 6012
7	1.1486 8567	1.1886 8575	1.2298 7387	1.2722 7926	1.3159 3178	1.3608 6183
8	1.1716 5938	1.2184 0290	1.2667 7008	1.3168 0904	1.3685 6905	1.4221 0061
9	1.1950 9257	1.2488 6297	1.3047 7318	1.3628 9735	1.4233 1181	1.4860 9514
10	1.2189 9442	1.2800 8454	1.3439 1638	1.4105 9876	1.4802 4428	1.5529 6942
11	1.2433 7431	1.3120 8666	1.3842 3387	1.4599 6972	1.5394 5406	1.6228 5305
12	1.2682 4179	1.3448 8882	1.4257 6089	1.5110 6866	1.6010 3222	1.6958 8143
13	1.2936 0663	1.3785 1104	1.4685 3371	1.5639 5606	1.6650 7351	1.7721 9610
14	1.3194 7876	1.4129 7081	1.5125 8972	1.6186 9452	1.7316 7645	1.8519 4492
15	1.3458 6834	1.4482 9817	1.5579 6742	1.6753 4883	1.8009 4351	1.9352 8244
16	1.3727 8570	1.4845 0562	1.6047 0644	1.7339 8604	1.8729 8125	2.0223 7015
17	1.4002 4142	1.5216 1826	1.6528 4763	1.7946 7555	1.9479 0050	2.1133 7681
18	1.4282 4625	1.5596 5872	1.7024 3306	1.8574 8920	2.0258 1652	2.2084 7877
19	1.4568 1117	1.5986 5019	1.7535 0605	1.9225 0132	2.1068 4918	2.3078 6031
20	1.4859 4740	1.6386 1644	1.8061 1123	1.9897 8886	2.1911 2314	2.4117 1401
21	1.5156 6634	1.6795 8185	1.8602 9457	2.0594 3147	2.2787 6807	2.5202 4116
22	1.5459 7967	1.7215 7140	1.9161 0341	2.1315 1158	2.3699 1879	2.6336 5201
23	1.5768 9926	1.7646 1068	1.9735 8651	2.2061 1448	2.4647 1555	2.7521 6635
24	1.6084 3725	1.8087 2595	2.0327 9411	2.2833 2849	2.5633 0417	2.8760 1383
25	1.6406 0599	1.8539 4410	2.0937 7793	2.3632 4498	2.6658 3633	3.0054 3446

Subtract $1 from the amount in this table to find the interest.

AMOUNT OF $1 AT COMPOUND INTEREST IN ANY NUMBER OF YEARS, NOT EXCEEDING TWENTY-FIVE

Yr.	5 per cent	6 per cent	7 per cent	8 per cent	9 per cent	10 per cent
1	1.0500 000	1.0600 000	1.0700 000	1.0800 000	1.0900 000	1.1000 000
2	1.1025 000	1.1236 000	1.1449 000	1.1664 000	1.1881 000	1.2100 000
3	1.1576 250	1.1910 160	1.2250 430	1.2597 120	1.2950 290	1.3310 000
4	1.2155 063	1.2624 770	1.3107 960	1.3604 890	1.4115 816	1.4641 000
5	1.2762 816	1.3382 256	1.4025 517	1.4693 281	1.5386 240	1.6105 100
6	1.3400 956	1.4185 191	1.5007 304	1.5868 743	1.6771 001	1.7715 610
7	1.4071 004	1.5036 303	1.6057 815	1.7138 243	1.8280 391	1.9487 171
8	1.4774 554	1.5938 481	1.7181 862	1.8509 302	1.9925 626	2.1435 888
9	1.5513 282	1.6894 790	1.8384 592	1.9990 046	2.1718 933	2.3579 477
10	1.6288 946	1.7908 477	1.9671 514	2.1589 250	2.3673 637	2.5937 425
11	1.7103 394	1.8982 986	2.1048 520	2.3316 390	2.5804 264	2.8531 167
12	1.7958 563	2.0121 965	2.2521 916	2.5181 701	2.8126 648	3.1384 284
13	1.8856 491	2.1329 283	2.4098 450	2.7196 237	3.0658 046	3.4522 712
14	1.9799 316	2.2609 040	2.5785 342	2.9371 936	3.3417 270	3.7974 983
15	2.0789 282	2.3965 582	2.7590 315	3.1721 691	3.6424 825	4.1772 482
16	2.1828 746	2.5403 517	2.9521 638	3.4259 426	3.9703 059	4.5949 730
17	2.2920 183	2.6927 728	3.1588 152	3.7000 181	4.3276 334	5.0544 703
18	2.4066 192	2.8543 392	3.3799 323	3.9960 195	4.7171 204	5.5599 173
19	2.5269 502	3.0255 995	3.6165 275	4.3157 011	5.1416 613	6.1159 090
20	2.6532 977	3.2071 355	3.8696 845	4.6609 571	5.6044 108	6.7275 000
21	2.7859 626	3.3995 636	4.1405 624	5.0338 337	6.1088 077	7.4002 499
22	2.9252 607	3.6035 374	4.4304 017	5.4365 404	6.6556 004	8.1402 749
23	3.0715 238	3.8197 497	4.7405 299	5.8714 637	7.2578 745	8.9543 024
24	3.2250 999	4.0489 346	5.0723 670	6.3411 807	7.9110 832	9.8497 327
25	3.3863 549	4.2918 707	5.4274 326	6.8484 752	8.6230 807	10.8347 059

Subtract $1 from the amount in this table to find the interest.

653.

Table showing Amount of \$1 Deposited Annually at Compound Interest for any Number of Years not Exceeding Twenty-Five

Periods	2 per cent	3 por cent	4 per cent	4½ per cent	5 per cent	6 per cent
1.	1.02	1.03	1.04	1.045	1.05	1.06
2.	2.0604	2.0909	2.1216	2.137025	2.1525	2.1836
3.	3.121608	3.183627	3.246464	3.278191	3.310125	3.374616
4.	4.204040	4.309136	4.416323	4.470710	4.525631	4.637093
5.	5.308121	5.468410	•5.632975	5.716892	5.801913	5.975319
6.	6.434283	6.662462	6.898294	7.019152	7.142008	7.393838
7.	7.582969	7.892336	8.214226	8.380014	8.549109	8.897468
8.	8.754628	9.159106	9.582795	9.802114	10.026564	10.491316
9.	9.949721	10.463879	11.006107	11.288209	11.577893	12.180795
10.	11.168715	11.807796	12.486351	12.841179	13.206787	13.971643
11.	12.412090	13.192030	14.025805	14.464032	14.917127	15.869941
12.	13.680332	14.617790	15.626838	16.159913	16.712983	17.882138
13.	14.973938	16.086324	17.291911	17.932109	18.598632	20.015066
14.	16.293417	17.598914	19.023588	19.784054	20.578564	22.275970
15.	17.639285	19.156881	20.824531	21.719337	22.657492	24.672528
16.	19.012071	20.761588	22.697512	23.741707	24.840366	27.212880
17.	20.412312	22.414435	24.645413	25.855084	27.132385	29.905653
18.	21.840559	24.116868	26.671229	28.063562	29.539004	32.759992
19.	23.297370	25.870374	28.778079	30.371423	32.065954	35.785591
20.	24.783317	27.676486	30.969202	32.783137	34.719252	38.992727
21.	26.298984	29.536780	33.247970	35.303378	37.505214	42.392290
22.	27.844963	31.452884	35.617889	37.937030	40.430475	45.995828
23.	29.421862	33.426470	38.082604	40.689196	43.501999	49.815577
24.	31.030300	35.459264	40.645908	43.565210	46.727099	53.864512
25.	32.670906	37.553042	43.311745	46.570645	50.113454	58.156383

. **654.** Compound interest is interest on the principal and its unpaid interest, the principal and interest being combined at regular intervals. The intervals may be a year, six months, or three months, etc.

655. While compound interest is not generally collectible by law, its acceptance by a creditor is not usurious.

656. Savings banks generally pay compound interest on deposits. Premiums in life insurance are determined by the application of the principles of compound interest. Returns from an investment in bonds are reckoned by compound interest.

657. To find compound interest.

Find the compound interest on $1000 for 3 years at 6 %.

6 % of $1000 = $60,	first year's interest.
$1000 + $60 = $1060,	new principal, second year.
6 % of $1060 = $63.60,	second year's interest.
$1060 + $63.60 = $1123.60,	new principal, third year.
6 % of $1123.60 = $67.42,	third year's interest.
$1123.60 + $67.42 = $1191.02,	amount at end of third year.
$1191.02 − $1000 = $191.02,	compound interest.

Or, by using the table:

1000 × $1.191016 = $1191.016 = $1191.02, amount.

$1191.02 − $1000 = $191.02, compound interest.

Find the interest on $1000 for the first year, and add it to the $1000, making a new principal of $1060 for the second year. Find the interest on $1060 for the second year and add it to $1060, which gives $1123.60, the new principal for the third year, and so on. From the total amount due at the expiration of the three years deduct the original principal, $1000; the remainder is the compound interest.

By using the compound interest table, the labor of computation is much lessened. First find the amount of $1 for 3 yr. at 6 %, by looking in the 6 % column and at the number opposite the third year, as indicated in the margin. The number is $1.191016. Since $1 amounts to $1.191016, $1000 will amount to 1000 times $1.191016, or $1191.02. Deducting the principal gives $191.02, the compound interest.

PROBLEMS

658. Find, without the use of the table, the compound interest on:

1. $800 for 5 yr. at 5 %.
2. $900 for 6 yr. at 6 %.
3. $1200 for 4 yr. at 8 %.
4. $2000 for 3 yr. at 6 %, compounded semiannually.*
5. $4000 for 2 yr. at 8 %, compounded quarterly.

* Find the interest for ½ yr., and add to the principal.

If $100 is deposited annually at **5%** compound interest, how much will it amount to in **20 yr.?**

$1 amounts to $34.7193.

$100 \times $34.7193 = $3471.93, amount.

The table, page 286, shows that $1 deposited each year for 20 yr. at 5% amounts to $34.7193. $100 will amount to 100 times $34.7193, or $3471.93.

Use the table in the following:

6. A boy 16 yr. of age has $1500 deposited in a savings bank for him. The savings bank pays 4% interest, compounded semi-annually. What amount will he have to his credit when he is 21 yr. of age, no withdrawals nor any more deposits having been made?

7. A young man 25 yr. of age has his life insured for $2000 by taking out a 20-yr. endowment policy for which he pays annually $96.30. If, at the expiration of the 20 yr., he receives the face value of the policy, find the gain to the insurance company if money is worth 4% compound interest to them. (See table, page 286.)

8. If the young man in problem **7** had died at the age of 35, would the insurance company have gained or lost, and how much?

9. A boy on his twelfth birthday started a savings bank account by depositing $25. If he deposits $25 every 6 mo. thereafter until he is 21 yr. of age, what amount will he have to his credit, the bank paying 4% interest compounded every 6 mo.?

10. What sum deposited in a savings bank paying 4% compound interest will amount to $5000 in 25 yr.?

SINKING FUNDS

659. In financing large undertakings, as the building of railways or the erection of large buildings, etc., immediate funds are provided by borrowing the money and issuing bonds therefor. The bonds are payable in 10, 15, 20, or more years, as the case may be. To meet these bonds at maturity, a sufficient sum of money is invested each year to amount, with compound interest, to the face value of the bonds. The sum thus set aside is called a sinking fund.

NOTE. All problems in sinking funds are based on compound interest.

660. TABLE SHOWING AMOUNT OF ANNUITY* OF $1 AT END OF EACH PERIOD

Periods	3 per cent	4 per cent	4½ per cent	5 per cent	6 per cent
1.	1.	1.	1.	1.	1.
2.	2.03	2.04	2.045	2.05	2.06
3.	3.0909	3.1216	3.137025	3.1525	3.1836
4.	4.183627	4.246464	4.278191	4.310125	4.374616
5.	5.309136	5.416323	5.470710	5.525631	5.637093
6.	6.468410	6.632975	6.716892	6.801913	6.975319
7.	7.662462	7.898294	8.019152	8.142008	8.393838
8.	8.892336	9 214226	9.380014	9.549109	9.897468
9.	10.159106	10.582795	10.802114	11.026564	11.491316
10.	11.463879	12.006107	12.288209	12.577893	13.180795
11.	12.807796	13.486351	13.841179	14.206787	14.971643
12.	14.192030	15.025805	15.464032	15.917127	16.869941
13.	15.617790	16.626838	17.159913	17.712983	18.882138
14.	17.086324	18.291911	18.932109	19.598632	21.015066
15.	18.598914	20.023588	20.784054	21.578564	23.275970
16.	20.156881	21.824531	22.719337	23.657492	25.672528
17.	21.761588	23.697512	24.741707	25.840366	28.212880
18.	23.414435	25.645413	26.855084	28.132385	30.905653
19.	25.116868	27.671229	29.063562	30.539004	33.759992
20.	26.870374	29.778079	31.371423	33.065954	36.785591

661. To find an annuity which will amount to a given debt.

1. The city of **X** put in a system of city waterworks at an expense of $150,000, and sold municipal bonds maturing in 10 yr. to pay for the works. What sum must the city set aside each year at 4% to redeem the bonds at maturity?

By referring to the above table, an annuity of $1 for 10 yr. at 4% is found to amount to $12.006107. If $1 amounts to $12.006107 in ten years, it will require as many dollars to amount to $150,000 as $12.006107 is contained times in $150,000, or $12,493.64.

---

* A series of equal payments made at regular periods is called an annuity. If an annuity is invested, so that, with its interest accumulation it will be equal to a given amount or debt, it is called a sinking fund.

VAN TUYL'S BUS. ARITH. —19

2. Jan. 1, 1911 a man borrows $1000 for 5 yr. at 4% and agrees to pay principal and interest (compounded annually) in five equal payments. What annual payment is required?

1000 × $1.2166529 = $1216.6529, amount of $1000 in 5 yr.

$1216.65 ÷ $5.416323 = $224.627, or $224.63.

Reference to the compound interest table, page 285, shows the amount of $1 at 4% for 5 yr. to be $1.2166529. $1000 amounts to 1000 times $1.2166529, or $1216.65.

By the table on page 289, an annuity of $1 for 5 yr. at 4% amounts to $5.416323. Dividing the total amount due by $5.416323 gives the annual payment as $224.627, or $224.63.

Proof of this solution is shown in the following:

Schedule of Amortization

Dates	Annual Payment	Interest on Balance	Amortization	Principal Unpaid
Jan. 1, 1911				$1000.00
Dec. 31, 1911	$224.63	$40.00	$184.63	815.37
Dec. 31, 1912	224.63	32.62	192.01	623.36
Dec. 31, 1913	224.63	24.94	199.69	423.67
Dec. 31, 1914	224.63	16.95	207.68	215.99
Dec. 31, 1915	224.63	8.64	215.99	000.00
	$1123.15 — $123.15 = $1000.00			

Each annual payment is composed of two items: first, the interest on the unpaid principal; and second, a payment on account of the principal. The column headed "Amortization" shows the amount of the several payments on account of the principal. The total amount paid equals the sum of the interest and the principal.

Note The apparent discrepancy between the total amount paid ($1123.15), and the amount of $1000 at 4% for 5 yr. ($1216.65), is due to the fact that $1216.65 is the amount that would have been due and payable if no payments had been made till Dec. 31, 1915.

PROBLEMS

662. 1. For the erection of new school buildings, the city of M issued $75,000 municipal bonds bearing 4½% interest, and payable in 15 years. What amount of taxes should be levied each year to provide a sinking fund which, at 4 % compound interest, will be sufficient to redeem the bonds at maturity?

2. City bonds to the amount of $10,000,000, maturing in 20 yr., were issued Oct. 1, 1909. To pay the bonds at maturity a sinking fund earning $4\frac{1}{2}$ % is provided. What should be the annual investment in the sinking fund?

3. The government of Ontario, Canada, loans money to farmers in even hundreds of dollars to be used in draining their land. The loan and 4 % compound interest is to be repaid in 20 equal annual installments. Find the annual payment per $100 required to cancel the loan. Prepare a schedule of amortization showing the amount paid on account of the interest and of the principal each year.

4. The Southern Power Company has $3,000,000 of 5 % first mortgage bonds outstanding. If these bonds are dated Mar. 1, 1910, and are payable on Mar. 1, 1930, what amount must be set aside each year at $4\frac{1}{2}$ % to meet them at maturity?

NEGOTIABLE PAPER

663. Negotiable paper is business paper which may be transferred by indorsement and delivery, or by delivery only, the transferee taking good title thereto. The principal forms of negotiable paper are checks, notes, drafts, and certificates of deposit.

664. A check is a written order by a depositor on a bank or banker payable on demand.

665.   FORM OF A CHECK

666. In this check S B Kooper is the *drawer;* Walter G. Lindsay is the *payee;* and the bank is the *drawee.*

667.   FORM OF A BANK DRAFT

668. A **bank draft** is a check drawn by one bank on another bank. In the above form the Second National Bank is the drawer; James E. Winter is the payee; and the Mercantile National Bank is the drawee.

669. A **draft** is a written order by one party on another directing the payment of a specified sum of money at a certain time to a person named therein, to order or to bearer.

670. In the following draft, F. G. Huxley is the **drawer**. He is the person who wrote or " drew " the draft. He has addressed the draft to C. D. Evarts, who is called the **drawee**. The drawee is the person drawn on, — the one who is directed to pay. L. M. Norton is the **payee**. Huxley wrote the draft and gave it to Norton, who will present it to Evarts for payment when it is due.

671.   FORM OF A TIME DRAFT (AFTER DATE)

672. FORM OF A TIME DRAFT (AFTER SIGHT)

It should be observed that in the first of the preceding draft forms, the time is "Sixty days *after date*," and in the second it is "Thirty days *after sight*." In the first form the time is reckoned from the date of the draft, Aug. 30, 1911, while in the second form it is reckoned from the date of acceptance, June. 5, 1911.

673. Acceptance is the written promise (an oral acceptance is legal and binding) of the drawee to pay the draft when it is due. Since the time (in drafts of the second form) is reckoned from the time of acceptance, these drafts must be accepted, and the acceptance must be dated. Drafts of the first form do not require acceptance.

674. Until the drawee accepts a draft, he is in no way liable for its payment. When he accepts it he becomes liable for its payment the same as the maker of a note.

675. FORM OF A SIGHT DRAFT

676. A **sight draft** is payable on demand, the expression "at sight" meaning when the draft is presented by the payee to the drawee, for payment.

677. A **promissory note** is a written promise to pay to a person named in the note, to order or to bearer, a specified sum of money.

678.　　　　FORM OF A PROMISSORY NOTE

679. In the above note S. L. Newhouse is the *maker*, and F. J. Gildersleeve is the *payee*.

680. A **certificate of deposit** is a written statement issued by a bank to a person who deposits money for safe keeping. It contains the promise of the bank to return the specified sum on the return of the certificate. The certificate is of the nature of a promissory note.

681.　　　　FORM OF A CERTIFICATE OF DEPOSIT

682. Indorsement is anything written on the back of commercial paper, which has reference to the paper itself. Indorsement is for one of three purposes. It may be for the purpose: (1) of transferring the paper; (2) of adding to its security; or (3) of recording a partial payment of the paper.

For the transfer of paper, indorsements are of several kinds, the most important of which are:

(1) **Blank Indorsement,** written,   Robert C. Ogden.

It consists of only the name of the indorser.

(2) **Special Indorsement,** written,   Pay to the order of
                                                    J. J. Astor,
                                                             A. T. Stewart.

It directs payment to some specified person, to order, or to bearer.

(3) **Qualified Indorsement,** written,

Without Recourse.   or   Pay to the order of
W. C. Mooney   W. C. Mooney
                               without recourse.
                               F. R. Crook.

The words "without recourse" have the effect of excusing the indorser from further liability of payment.

683. When the indorsement is for the purpose of adding to the security of the paper, it is generally written as a blank indorsement.

684. Each indorser, except those who use the qualified indorsement, is liable for the payment of the paper if the party primarily liable fails to pay.

685. An indorsement of a partial payment is a record of the amount and date of the payment. Thus:

Received May 4, 1910, $500
Received Dec. 1, 1910, $600
Received Oct. 1, 1911, $1000

686. If negotiable paper is not accepted when presented for acceptance, or is not paid at maturity, it is said to be **dishonored.**

687. Dishonored negotiable paper should be protested. **Protest** is the official act of a notary public in certifying to the dishonor of a negotiable instrument

688. A notice of protest is sent immediately by the notary public to all parties conditionally liable for payment of the paper.

689. In case of failure to give notice of protest, all parties conditionally liable for payment are excused from their liability.

690. A non-negotiable instrument is one made payable to the payee only.

691.   FORM OF NON-NEGOTIABLE NOTE

$ 525.00.   NEW YORK, N.Y., Sept. 25, 1909.

  Three months after date I promise to pay to------------------------

---------------------------HENRY WEBSTER------------------------

Five hundred twenty-five.~~~~~~~~~~~~~~~~~~~~~~~~~~$\frac{00}{000}$ Dollars.

Value Received.

               SAMUEL PARKER.

---

692. The absence of the words " order of " or " or order," renders the instrument **non-negotiable**.

BANK DISCOUNT

693. Bank discount is a deduction made from the amount due at maturity on notes and drafts in consideration of their being cashed or bought before maturity

694. If a merchant has received from one of his debtors a draft payable in 60 da., and does not wish to wait for payment till it matures, he can take it to a bank and sell it, or discount it, as it is called. The *discount* is determined by reckoning the simple interest on the amount of the draft from the day it is discounted to the day of maturity. The merchant will receive the **proceeds**, which is the difference between the amount due on the draft at maturity and the discount

695. As a rule banks charge interest, **or** discount, for the *exact number of days* from the day of discount to the day of maturity. This time is called the **term of discount.**

NOTES. (1) In a few states, viz. Ark., Ky., La., Md., Mo. (in large cities), N.C., Pa., Utah, and Va., the day of discount, as well as the day of maturity, is included in the term of discount, thus adding one day to the time.

(2) In Miss. and Tex., notes, acceptances, etc., falling due on Sunday or a legal holiday, are payable the preceding business day. In all other states they are payable the next succeeding business day.

(3) In Ariz., Calif., Ga., Ky., Maine, Miss., N. C., Okla., S. Dak., Tex., and Wis., notes, acceptances, etc., falling due on Saturday, are payable the same day. In all other states they are payable the next succeeding business day.

(4) In discounting paper falling due on Saturday, Sunday, or a legal holiday, bankers charge discount for the time till the "next succeeding business day" (except where payable on the same or the preceding day).

(5) Answers to problems in bank discount in this book are prepared in accordance with the most common method used by bankers. It is suggested that the student follow the custom of his own state.

696. **Days of grace** are three days allowed by law, for the payment of notes and drafts, beyond the expiration of the time specified in the paper. The practice of allowing three days of grace, which was formerly very common, has now been abolished in most states.

697. The solution of problems in bank discount is based on the principles of simple interest, or of percentage.

The *value* of the note at maturity corresponds to the *principal* in *interest*, or to the *base* in *percentage.*

The *rate of discount* corresponds to the *rate per cent* in *interest* and in *percentage.*

The *bank discount* corresponds to the *interest* or the *percentage;* and the *proceeds* to the *difference.*

698. To find the date of maturity and the term of discount.

1. Find the date of maturity and the term of discount of a 3 months' note dated Apr. 14, and discounted May 3.

Apr. 14 + 3 mo. = July 14, date of maturity.

From May 3 to July 14 = 72 da., term of discount.

The date of maturity is found by counting forward 3 mo. from Apr. 14, which gives July 14. The term of discount is found by counting all the days from the date of discount, May 3, to the date of maturity, July 14. There are 28 da. remaining in May, 30 da. in June, and 14 da. in July, the sum of which is 72 da. (See Art. 449, page 182.)

2. Find the date of maturity and the term of discount of a 60 days' sight draft, dated Aug. 18, accepted Aug. 21, and discounted Sept. 4.

Aug. 21 + 60 da. = Oct. 20, date of maturity.

From Sept. 4 to Oct. 20 = 46 da., term of discount.

A draft drawn at 60 days' sight* does not mature until 60 da. after it is accepted, hence, the day of maturity in this draft is 60 da. after Aug. 21, which is Oct. 20. Exact days must be counted when the time is expressed in days. (See Art. 455, page 185.) The term of discount is the exact number of days from Sept. 4 to Oct. 20, which is 46 da.

699. Copy the following and fill in the dates of maturity and the terms of discount:

	DATE OF PAPER	TIME	DATE OF ACCEPTANCE	DUE DATE	DATE OF DISCOUNT	TERM OF DISCOUNT
1.	Jan. 15	4 months	†		Mar. 10	
2.	Feb. 23	90 days			Mar. 1	
3.	June 10	3 months			July 18	
4.	May 8	90 days after sight	May 14		May 26	
5.	Oct. 7	2 months			Oct. 21	
6.	Aug. 11	30 days after sight	Sept. 1		Sept. 10	
7.	July 8	60 days			Aug. 14	
8.	May 1	45 days			May 14	
9.	Apr. 3	2 months after date	May 1		May 2	
10.	Feb. 18	90 days			Mar. 31	
11.	June 17	3 months			June 24	
12.	July 27	60 days' sight	Aug. 4		Aug. 7	
13.	Sept. 19	4 months			Nov. 10	
14.	Mar. 7	1 month			Mar. 9	
15.	Apr. 13	60 days after date	May 1		May 4	
16.	Jan. 28	3 months after sight	Feb. 1		Mar. 8·	
17.	June 15	5 months			Sept. 30	
18.	Aug. 11	90 days			Sept. 30	
19.	May 21	40 days' sight	June 1		June 10	
20.	June 16	2 months			July 1	
21.	Dec. 13	3 months			Feb. 7	
22.	Nov. 30	3 months			Jan. 2	
23.	Nov. 30	90 days			Dec. 1	
24.	Oct. 9	90 days after date	Oct. 12		Nov. 2	
25.	Mar. 14	30 days' sight	Mar. 20		Mar. 24	

* See Art. 672, page 293.   † Blank spaces in this column are *not* to be filled in.

700. To find the bank discount and the proceeds of a note or draft.

Find the date of maturity, the term of discount, the bank discount, and the proceeds of a 90-da. note of $1500, dated May 1 and discounted May 16, at 6 %.

May 1 + 90 days = July 30, date of maturity.

From May 16 to July 30 = 75 days, term of discount.

$$\$15|00 = \text{discount for 60 days.}$$
$$\underline{3|75} = \text{discount for 15 days.}$$
$$\$18|75 = \text{discount for 75 days.}$$

$1500 — $18.75 = $1481.25, proceeds.

The date of maturity and the term of discount are found as in the preceding exercise. The bank discount is the simple interest on the value of the note for the term of discount, 75 days, which is $18.75. The proceeds is the difference between the bank discount and the value of the note ; $1500 — $18.75 = $1481.25.

701. Find the date of maturity, the term of discount, the bank discount, and the proceeds of the following notes and drafts:

	Face	Time	Date of Paper	Date of Acceptance	Date of Discount	Rate of Discount
1.	$1800.00	4 months	Aug. 1		Aug. 11	6%
2.	2000.00	3 months	July 11		Aug. 1	6%
3.	1640.00	90 days	May 13		June 4	6%
4.	1150.00	60 days after sight	Mar. 18	Mar. 24	Apr. 7	5%
5.	1990.00	30 days' sight	Apr. 27	May 3	May 6	5%
6.	2870.00	5 months	June 19		Aug. 1	6%
7.	3675.00	2 months	July 17		Aug. 1	7%
8.	9400.00	60 days' sight	Oct. 11	Oct. 18	Nov. 1	6%
9.	8478.00	30 days' sight	May 28	June 4	June 11	4½%
10.	7200.00	3 months	Aug. 13		Aug. 19	4%
11.	5900.00	1 month	Jan. 31		Feb. 4	5½%
12.	8212.00	2 months	July 1		Aug. 14	6%
13.	4750.00	4 months	Oct. 6		Nov. 6	7½%
14.	5125.00	6 months	June 30		Oct. 18	6%
15.	6000.00	90 days after date	Sept. 5	Sept. 25	Oct. 11	6%
16.	5400.00	60 days after sight	Aug. 4	Aug. 11	Aug. 23	8%
17.	4500.00	30 days' sight	July 5	July 11	July 11	7½%
18.	3100.00	5 mo. after date	Jan. 12	Mar. 11	Mar. 13	6%
19.	412.50	3 months	Apr. 12		May 1	9%
20.	368.75	90 days	May 10		June 11	10%

702. Bankers often use a table like the following to find the number of days between two dates:

BANKERS' TIME TABLE

FROM ANY DAY OF	TO THE SAME DAY OF THE NEXT											
	Jan.	Feb.	Mar.	Apr.	May	June	July	Aug.	Sept.	Oct.	Nov.	Dec.
Jan.	365	31	59	90	120	151	181	212	243	273	304	334
Feb.	334	365	28	59	89	120	150	181	212	242	273	303
Mar.	306	337	365	31	61	92	122	153	184	214	245	275
Apr.	275	306	334	365	30	61	91	122	153	183	214	244
May	245	276	304	335	365	31	61	92	123	153	184	214
June	214	245	273	304	334	365	30	61	92	122	153	183
July	184	215	243	274	304	335	365	31	62	92	123	153
Aug.	153	184	212	243	273	304	334	365	31	61	92	122
Sept.	122	153	181	212	242	273	303	334	365	30	61	91
Oct.	92	123	151	182	212	243	273	304	335	365	31	61
Nov.	61	92	120	151	181	212	242	273	304	334	365	30
Dec.	31	62	90	121	151	182	212	243	274	304	335	365

703. How to use the table.

The table gives at a glance the exact number of days from any day of any month to the corresponding day of any other month for periods of time not greater than one year. Thus, the number of days from the 18th of May to the 18th of the next January is found by taking the number to the right of the month of May in the "January column," which is 245 da. From the 9th of April to the 9th of December is 244 da., found on the "April line" and the December column.

From Mar. 15 to July 28 is found thus: Take from the table the number of days from Mar. 15 to July 15, which is 122 da., and add to it the number of days from July 15 to July 28. From July 15 to July 28 is 13 da., which, added to 122 da., equals 135 da.

For the time from Oct. 25 to Apr. 10, find the time from Oct. 25 to Apr. 25, which is 182 da., and deduct the number of days from Apr. 10 to Apr. 25, which is 15 da. 182 da. less 15 da. leaves 167 da., the time from Oct. 25 to Apr. 10.

704. Find, by using the table, the number of days from:

1. Aug. 4 to Mar. 4.
2. Jan. 8 to Dec. 8.
3. Apr. 11 to Aug. 11.
4. Mar. 3 to Jan. 3.
5. May 30 to Nov. 30.
6. July 1 to Jan. 1.

7. Jan. 1 to July 1.
8. Oct. 7 to May 12.
9. Aug. 13 to Dec. 25.
10. June 24 to Nov. 12.
11. Mar. 17 to Dec. 13.
12. Apr. 18 to Jan. 4.
13. May 29 to Apr. 1.

14. Jan. 14 to Dec. 20.
15. June 17 to Oct. 20.
16. May 11 to Dec. 18.
17. July 3 to Nov. 24.
18. Feb. 15 to Dec. 10.
19. Sept. 1 to Oct. 19.
20. Sept. 19 to June 10.

705. When a bank discounts a note or draft that is payable in some other town or city, an additional charge called collection is often made. In large cities having a clearing house the collection charge is uniform for all banks that are members of the clearing house. Banks that are not members of the clearing house, and banks in small cities, have their own regulations concerning collection charges. The charge may be either a per cent of the value of the paper collected, or it may be an arbitrary amount for each note or draft.

706. The following are the charges made for collecting out-of-town paper by banks that are members of the New York Clearing House.

PAR POINTS	$\frac{1}{10}$ OF 1 PER CENT		$\frac{1}{4}$ OF 1 PER CENT		
Greater New York . . .	* Conn.	* N. Y.	Ala.	Kans.	Okla.
Albany, N. Y.	Del.	N. H.	Ariz.	La.	Oregon
Baltimore, Md.	D. C.	* N. J.	Ark.	Minn.	S. C.
Bayonne, N. J.	Ill.	Ohio	Calif.	Miss.	S. Dak.
Boston, Mass.	Ind.	* Pa.	Colo.	Mont.	Tenn.
Hoboken, N. J.	Ky.	* R. I.	Fla.	Nebr.	Tex.
Newark, N. J.	Maine	Vt.	Ga.	Nev.	Utah
Philadelphia, Pa. . . .	* Md.	Va.	Idaho	N. Mex.	Wash.
Providence, R. I. . . .	* Mass.	W. Va.		N. C.	Wyo.
Troy, N. Y.	Mich.	Wis.	Iowa	N. Dak.	Can.
	Mo.				

* Except cities and banks listed as par. Minimum charge, ten cents.

In addition to the above " par points " there is a specified list of banks and trust companies in about 175 cities and villages in Connecticut, Massachusetts, New Jersey, New York, and Rhode Island on which items will be received at par in New York City.

707. To find the proceeds of a note or draft when collection is charged.

Find the proceeds of a 90-da. note of $3800 dated June 3, 1910, if discounted July 1, 1910, at 6 %; collection $\frac{1}{10}$ %.

June 3 + 90 da. = Sept. 1, due date.

From July 1 to Sept 1 = 62 da., term of discount.

$$\$38|00 = \text{discount for 60 da.}$$
$$\underline{1|27} = \text{discount for 2 da.}$$
$$\$39|27 = \text{discount for 62 da.}$$

$\frac{1}{10}$ % of $3800 = $3.80, collection.

$39.27 + $3.80 = $43.07, the bank's total charge.

$3800 − $43.07 = $3756.93, the proceeds.

The date of maturity, the term of discount, and the bank discount are found as previously explained (pages 297–299). The collection charge is reckoned on the value of the note; $\frac{1}{10}$% of $3800 is $3.80. The bank's total charge is the sum of the bank discount, $39.27, and the collection charge, $3.80, or $43.07. The proceeds is the difference between $3800 and $43.07, or $3756.93.

708. Find the date of maturity, the term of discount, the bank discount, and the proceeds of the following:

	FACE	TIME	DATE	DATE OF DISCOUNT	RATE OF DISCOUNT	RATE OF COLLECTION
1.	$5000	4 mo.	Jan. 10	Mar. 8	6 %	$\frac{1}{20}$%
2.	4500	3 mo.	Mar. 15	Apr. 13	6 %	$\frac{1}{10}$%
3.	2500	90 da.	June 8	July 1	6 %	$\frac{1}{8}$%
4.	1600	60 da.	Oct. 17	Nov. 10	6 %	$\frac{3}{4}$%
5.	1748	60 da.	Aug. 12	Sept. 1	6 %	
6.	1372	6 mo.	Oct. 11	Nov. 4	5 %	$\frac{1}{4}$%
7.	4382.75	90 da.	Nov. 17	Nov. 29	6 %	$\frac{1}{10}$%
8.	526.80	2 mo.	Dec. 8	Dec. 13	5 %	$\frac{1}{4}$%
9.	2600	3 mo.	Mar. 8	Mar. 13	6 %	
10.	557.68	2 mo.	Dec. 4	Dec. 13	5 %	$\frac{3}{4}$%
11.	943.75	4 mo.	Oct. 1	Oct. 2	6 %	
12.	321.60	30 da.	Jan. 31	Feb. 4	$7\frac{1}{2}$%	$\frac{1}{10}$%
13.	416.50	60 da.	Mar. 31	Apr. 13	$4\frac{1}{2}$%	$\frac{1}{4}$%
14.	531.60	3 mo.	Nov. 30	Jan. 3	5 %	$\frac{1}{4}$%

15.

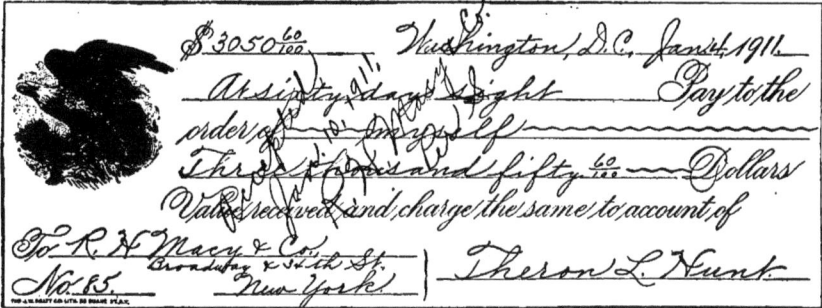

Discounted Jan. 13 at 6%, collection ⅛%.

709. To find the proceeds of an interest-bearing note.

Find the proceeds of a 4 months' 6% interest-bearing note of $3200, dated May 15, 1910, and discounted June 10, 1910 at 6%; collection ⅛%.

May 15 + 4 mo. = Sept. 15, due date.

From June 10 to Sept. 15 = 97 da., term of discount.

$32|00 = int. for 2 mo. $3200 + $64 = $3264, value of the
  64|00 = int. for 4 mo. note at maturity.

  32|64 = dis. for 60 da. ⅛% of $3264 = $4.08, collection.
  16|32 = dis. for 30 da. $52.77 + $4.08 = $56.85, bank's total
   3|264 = dis. for 6 da. charge.
    |544 = dis. for 1 da. $3264 − $56.85 = $3207.15, proceeds.
$52|768 = dis. for 97 da.

The due date and the term of discount are found first as already explained on page 297, Art. 698. Next find the interest on the face of the note, $3200, for the *entire time* the note has to run, 4 mo. Add the interest, $64, to the face of the note, $3200, obtaining $3264, the value of the note at maturity. The discount is then reckoned for 97 da. on the full value of the note, and amounts to $52.77. The collection charge is reckoned on the value of the note at maturity, $3264. ⅛% of $3264 equals $4.08. The bank's total charge is equal to the sum of the bank discount and the collection, which is $56.85. $3264 − $56.85 = $3207.15, proceeds of note.

710. Find the proceeds of the following notes:

	Face	Time	Date	Interest	Date of Discount	Rate of Discount	Collection
1.	$1400	2 mo.	Aug. 1	6 %	Aug. 10	6 %	$\frac{1}{8}$ %
2.	1550	3 mo.	May 15	5 %	June 1	5 %	
3.	2160	90 da.	Feb. 15	8 %	Mar. 17	7 %	$\frac{1}{8}$ %
4.	560	60 da.	Dec. 10	6 %	Dec. 20	8 %	$\frac{1}{4}$ %
5.	940	5 mo.	July 1	6 %	Sept. 10	5 %	$\frac{1}{4}$ %
6.	1120	4 mo.	Oct. 20	5 %	Dec. 1	6 %	
7.	2875	90 da.	June 12	6 %	June 27	4 %	$\frac{1}{10}$ %
8.	3125	75 da.	Apr. 20	$7\frac{1}{2}$ %	May 5	$4\frac{1}{2}$ %	$\frac{1}{8}$ %
9.	4700	40 da.	May 10	8 %	May 14	6 %	
10.	5100	1 mo.	Feb. 28	6 %	Mar. 1	6 %	$\frac{1}{20}$ %

11.

$6380.00   BINGHAMTON, N.Y., July 10, 1910.

Ninety days after date I promise to pay to the order of -------------.

-------------------------------- JAMES HASKINS ----------------------------

Sixty-three hundred eighty~~~~~~~~~~~~~~~~~~~~~~~~~~~~~~~~~$\frac{00}{100}$ Dollars.

Value received, with interest at 6 %.

LEO BOUTELL.

Discounted Aug. 1 at 6 % ; collection $\frac{1}{8}$ %.

12.

$4728.40   ROCHESTER, N.Y., June 3, 1910.

Two months after date we promise to pay to the order of---------------

-----------------------C. J. HERMANS & Co.-----------------------

Forty-seven hundred twenty-eight~~~~~~~~~~~~~~~~~~~~~~$\frac{40}{100}$ Dollars,

with interest at 6 %   Value received.

F. F. JONES & Co.

Discounted June 8 at 5 % ; collection $\frac{1}{8}$ %.

PRESENT WORTH AND TRUE DISCOUNT

711. How much will $1 amount to in 1 yr. at 6%?

At the same rate, how many dollars will, in 1 yr. amount to $3.18? $5.30? $10.60? $21.20? $212? $848?

How much will $1 amount to in 2 mo. at 6%?

At the same rate, how many dollars will, in 2 mo., amount to $2.02? $5.05? $10.10? $808? $2020? $5050?

712. A man owes a debt of $2550 due in 4 mo. How many dollars should he pay his creditor to-day, so that if put at interest at 6%, the payment will amount to the face of the debt at maturity?

$1.00 + $.02 = $1.02, amount of $1 at 6% for 4 mo.

$2550 ÷ $1.02 = $2500, amount of payment, or present worth of the debt.

If $1 is placed at interest for 4 mo. at 6%, the interest will be $.02, or the dollar will amount to $1.02. Since $1 in 4 mo. at 6% amounts to $1.02, it will take as many dollars to amount to $2550 as $1.02 is contained times in $2550, or 2500 times. That is, a payment of $2500 placed at interest for 4 mo. at 6% will amount to $2550.

713. The present worth of a debt is a sum which, if put at interest, will amount to the value of the debt at maturity.

In the above example $2500 is the present worth of the debt because $2500 placed at interest for 4 mo. will amount to $2550, the value of the debt when it is due.

714. The true discount is the difference between the present worth of a debt and the amount due at maturity.

In the solution just given, the true discount is $50.

The true discount is equal to the interest on the present worth of the debt for the time for which the debt is discounted.

PROBLEMS

715. Find the present worth and true discount of each of the following debts:

	DEBT	TIME	RATE OF DIS.		DEBT	TIME	RATE OF DIS.
1.	$520	8 mo.	6%	**3.**	$824	4 mo.	9%
2.	$618	6 mo.	6%	**4.**	$1656	7 mo.	6%

5.	$1212	3 mo.	4 %	**8.**	$4100	10 mo.	**3 %**
6.	$1020	4 mo.	6 %	**9.**	$972	1 yr. 4 mo.	6 %
7.	$1260	1 yr.	5 %	**10.**	$1100	2 yr. 6 mo.	4 %

11. What payment to-day will cancel a debt of $840 due in 10 mo. if money is worth 6 % ?

12. Find the true discount on a debt of $5100 if paid 4 mo. before maturity, interest 6 %.

13. A merchant sells an article for $52.50 on 4 mo. credit. At what cash price should he sell the same article if, to him, money is worth 15 % ?

14. Find the difference between the simple interest and the true discount on $1000 for 6 mo. at 6 %.

15. A merchant's cash price of a given article is $75, and his price at 3 mo. credit is $80. At 6 %, how much better is the cash price for the buyer ?

16 If a merchant's 4 months' credit price of a piano is $420, and his cash price is $400, at what rate per annum does he reckon money is worth to him ?

17. I am offered a bill of goods invoiced at $1500 on 4 months' credit. In payment I give my note with interest at 6 % for a sum which, at maturity, will cancel the debt. Find the face of the note.

18. Compare the true discount with the simple interest on $4080 for 4 mo at 6 %. How does the difference between the true discount and the interest compare with the interest on the true discount for the given time and rate ? Why ?

19. Find the present worth of a note of $800 drawn at 4 mo. from Aug. 1, if discounted Sept. 15, at 7 %.

20 Apr. 10, I bought a bill of merchandise valued at $5600 on 5 months' credit June 20, I paid $3000 on account. Aug. 1, I paid the present worth of the balance of the debt. Reckoning interest at 6%, what was the amount of my final payment? (Exact time.)

21. A father wishes to provide for a fund of $10,000 for his son when he shall become 21 yr. of age. What sum should he invest at 5% simple interest on the son's fifteenth birthday, that it may amount to the $10,000?

PARTIAL PAYMENTS

716. When part of note, draft, or other obligation is paid, such payment is called a **partial payment.**

717. Any partial payment of a note or draft should be indorsed on the note or draft, thus:

Rec'd on this note

Apr. 1, 1910, $500.00

July 10, 1910, $675.50

Sept. 15, 1910, $853.75

UNITED STATES RULE.

718. The United States Rule (so called because it has been approved by the United States Supreme Court) is based on the following principles:

1. Any payment must first be applied to the paying of accrued interest If the payment exceeds the amount of interest due, the principal sum is reduced by the amount of the excess.

2. Interest must not be charged upon interest.

719. This rule is generally used when partial payments are made on interest-bearing notes having more than one year to run. (Use compound time.)

720. To find the balance due by the United States Rule.

$3000.00   NEW YORK, N.Y., Sept. 8, 1909.

Two years after date, I promise to pay to the order of A. C. LOSEY, Three thousand $\frac{00}{100}$ Dollars, with interest at 6%. Value received.

JOHN CARPENTER.

The following payments were indorsed on the above note: Jan. 20, 1910, $1200; Nov. 10, 1910, $75; and Apr. 4, 1911, $500 What was the balance due at maturity?

Principal sum due	$3000.00
Int. from Sept. 8, 1909, to Jan. 20, 1910 (4 mo. 12 da.) .	66.00
Amount due Jan. 20, 1910	3066.00
Payment, Jan. 20, 1910	1200.00
Balance due, to draw int. from Jan. 20, 1910 . . .	1866.00
Int. from Jan. 20, 1910, to Nov. 10, 1910 (9 mo. 20 da.)	90.19

Since the payment is less than the accrued interest, the balance due remains unchanged.

Int. on $1866 from Nov. 10, 1910, to Apr. 4, 1911
(4 mo. 24 da.) 44.78
Amount due Apr. 4, 1911 . . . . $ 2000.97
Payments, Nov. 10, 1910, and Apr. 4, 1911 ($ 75 + $ 500) 575.00
Balance due, to draw int. from Apr. 4, 1911 . . . . 1425.97
Int. from Apr. 4, 1911, to Sept. 8, 1911 (5 mo. 4 da.) . 36.60
Balance due Sept. 8, 1911, date of maturity . . . . . $ 1462.57

The principal sum, $3000, draws interest from the date of the note, Sept. 8, 1909, to the date of the first payment, Jan. 20, 1910, a period of 4 mo. 12 da. The interest is $66, making a total amount due of $3066. Deducting the payment of $1200, leaves a balance of $1866 to draw interest from Jan. 20, 1910. From Jan. 20, 1910, to the date of the next payment, Nov. 10, 1910, is 9 mo. 20 da., and the interest accrued in that time is $90.19. But the payment made on Nov. 10, 1910, is only $75, not large enough to cancel the interest; hence the balance due, $1866, remains unchanged, and continues to draw interest to the date of the next payment, Apr. 4, 1911, or 4 mo. 24 da. The interest on $1866 for 4 mo. 24 da. is $44.78, making a total amount due of $2000.97. Deducting the sum of the two payments, $75 and $500, leaves $1425.97, the balance to draw interest till the date of maturity, Sept. 8, 1911. From Apr. 4, 1911, to Sept. 8, 1911, is 5 mo. 4 da., and the interest on $1425.97 for 5 mo. 4 da. is $36.60, which makes the amount due at maturity equal to $1462.57.

PROBLEMS

721. Find the balance due at maturity on each of the following notes, payments having been indorsed as indicated:

	DATE	TIME TO RUN	FACE	INT.	INDORSEMENTS
1.	May 1, 1908	3 yr.	$2500	6 %	Dec. 1, 1908, $ 500 June 21, 1909, .300 Jan. 15, 1910, 1000
2.	Aug. 7, 1907	4 yr.	$5000	5 %	Sept. 1, 1908, $1200 Aug. 15, 1909, 1500 Dec. 20, 1910, 1000
3.	Jan. 10, 1908	2 yr. 6 mo.	$4000	7 %	Apr. 20, 1908, $ 500 July 30, 1908, 50 Dec. 10, 1909, 1000

MERCHANTS' RULE

722. The Merchants' Rule is frequently used to find the balance due when partial payments are made on interest-bearing notes in which the time is less than one year.   It is based on the following principles:

1. *The face of the note draws interest from its date to the time of settlement.*

2. *Each payment draws interest from its date to the time of settlement.*

723. To find the balance due by the Merchants' Rule.

$ 1800.00   ·   NEW YORK, N.Y., Oct. 8, 1909.

Ten months after date I promise to pay to the order of J. K. RAMSEY, Eighteen hundred $\frac{00}{100}$ Dollars, with interest at 6 %.   Value received.

THEODORE A. BIDWELL.

On the above note the following payments were indorsed: Dec. 20, 1909, $300; Feb. 1, 1910, $600; June 1, 1910, $400   Find the balance due at maturity.

COMPOUND TIME

DATE 1909	FACE	TIME	INT.	DATES 1909	PAYMENTS	TIME	INT.
Oct. 8.	$1800	10 mo.	$90	Dec. 20	$300	7 mo. 18 da.	$11.40
				1910			
				Feb. 1	600	6 mo. 7 da.	18.70
				June 1	400	2 mo. 7 da.	4.47
	$1800		$90		$1300		$34.57

$1800 + $90 = $1890; $1300 + $34.57 = $1334.57.
$1890 − $1334.57 = $555.43, balance due at maturity.

EXACT TIME

DATE 1909	FACE	TIME	INT.	DATES 1909	PAYMENTS	TIME	INT.
Oct. 8	$1800	304 da.	$91.20	Dec. 20	$300	231 da.	$11.55
				1910			
				Feb. 1	600	188 da.	18.80
				June 1	400	68 da.	4.53
	$1800		$91.20		$1300		$34.88

$1800 + $91.20 = $1891.20; $1300 + $34.88 = $1334.88.
$1891.20 − $1334.88 = $556.32.

Compound Time. The solution of problems in Partial **Payments by the Mer-chants'** Rule is a direct application of the principles involved in Cash Balance. (See page 391.) The face of the note constitutes the debit side of the account and the several payments the credit side. The due date is Aug. 8, 1910. The time of the note is 10 mo., and the interest amounts to $90, making a total amount due of $1890. From the time of the first payment, Dec. 20, 1909, to the due date, Aug. 8, 1910, is 7 mo. 18 da., and the interest on the payment, $300, for that time is $11.40. In the same way the time and interest of the second payment is found to be 6 mo. 7 da. and $18.70, and for the third payment, 2 mo. 7 da. and $4.47, respectively. The total credit is the sum of the payments, $1300, plus the sum of the interest items, $34.57, or $1334.57. The balance due, Aug. 8, 1910, is the difference between $1890 and $1334.57, or $555.43.

Exact Time. The solution by exact time is the same as by compound time ex-cept that the several interest periods are found by exact time. This makes a slight difference in the interest items, as shown in the solution.

NOTE. There is no law making the Merchants' Rule a legal method. Hence, when it is to be used in practice, it should be agreed upon by the parties interested.

Some business men use exact time, while others use compound time. The answers in this text are given for both methods.

· **PROBLEMS**

724. Find the balance due at maturity on each of the following notes, payments having been indorsed as indicated:

	DATE	TIME TO RUN	FACE	INT.	INDORSEMENTS
1.	July 1, 1910	11 months	$1000	6%	Sept. 20, 1910, $200
					Dec. 18, 1910, 300
					Apr. 1, 1911, 300
2.	Oct. 8, 1910	9 months	1500	.5%	Dec. 15, 1910, 200
					Jan. 20, 1911, 400
					June 14, 1911, 500
3.	Mar. 1, 1910	10 months	1860	6%	May 12, 1910, 75
					June 15, 1910, 90
					Aug. 20, 1910, 120
					Oct. 11, 1910, 250
					Nov. 4, 1910, 300
4.	Apr. 30, 1910	8 months	980	7%	June 10, 1910, 80
					July 13, 1910, 125
					Sept. 15, 1910, 250
					Nov. 20, 1910, 200

725. The Improved Real Estate Company of New York, N. Y., offers for sale houses and lots on the following terms:

Cash, 20 % of the selling price.

First mortgage at 5½ % interest for 50 % of the selling price.

Second mortgage at 6 % interest for 30 % of the selling price.

The second mortgage is payable in 10 equal semiannual payments (plus the interest).

On the above terms find the total outlay for each year for five years on a house bought for $11,000, assuming that taxes, water rents, and insurance are $200 per annum.

First Year

Interest on first mortgage	$ 302.50
Interest on second mortgage	188.10
Taxes, water rents, insurance	200.00
Paid on account of principal of second mortgage	660.00
Total, first year	$ 1350.60

Second Year

Interest on first mortgage	$ 302.50
Interest on second mortgage	148.50
Taxes, etc.	200.00
Paid on account of principal of second mortgage	660.00
Total, second year	$ 1311.00

etc.

The interest on the first mortgage is 5½% of $5500, or $302.50. The interest on the second mortgage is 3% of $3300, or $99 plus 3% of $2970 ($3300–$330), or $89.10, which is $188.10. These amounts taken with the $200 for taxes, water rents, and insurance, and the two semiannual payments on the second mortgage, of $330 each, make a total of $1350.60, the total outlay for the first year.

The second year the charges are the same except the interest on the second mortgage. The balance due on the second mortgage is $3300 − $660, or $2640. The semiannual interest on $2640 is 3% of $2640, or $79.20. $2640 − $330 = $2310, balance due after the third payment is made. The next semiannual payment of interest is 3% of $2310, or $69.30. $79.20 + $69.30 = $148.50, total interest paid on second mortgage the second year. This amount, taken with the other charges, makes the total outlay the second year, $1311. And so on.

Assuming that the purchaser had been paying rent at $1200 per annum, how long after the second mortgage is paid will it take to pay the first mortgage and interest thereon by applying to such payment the difference between the former annual rental and the carrying cost of the property as estimated? (Reckon to the nearest whole year, and make payments annually.)

First Year

Face of first mortgage		$5500.00
Former rental	$1200	
Carrying cost first year		
Interest	$302.50	
Taxes, water rent, insurance	200.00	502.50
Amount to apply on account mortgage		697.50
Balance due on mortgage		$4802.50

Second Year

Former rental	$1200	
Carrying cost second year		
Interest	$264.14	
Taxes, etc.	200.00	464.14
Amount to apply on account mortgage		735.86
Balance due on mortgage		$4066.64

Third Year

Out of the $1200 which was formerly paid for rent, there has to be paid the interest on the mortgage (5½% of $5500), which is $302.50; also the taxes, water rents, and insurance, estimated at $200, making a total carrying cost of $502.50. Deducting $502.50 from $1200 leaves $697.50 to apply on the mortgage, which leaves $4802.50, the balance due at the beginning of the second year.

Proceed in the same way for each succeeding year until the mortgage is paid.

Note that the carrying cost grows less each year, and consequently the payment on the mortgage becomes correspondingly greater.

PROBLEMS

726. If properties are bought on the terms given above, find: (1) the total outlay for the first five years; (2) the number of years after the second mortgage is paid, required to pay the first mortgage by applying thereto the difference between the former rental paid by the purchaser and the carrying cost of the property; and (3) the total outlay by the purchaser from the date of purchase till both mortgages are paid.

	PURCHASE PRICE OF PROPERTY	FORMER ANNUAL RENTAL PAID BY THE PURCHASER	TAXES, WATER RENTS, REPAIRS, INSURANCE, ETC.
1.	$ 4,000	$ 420	$ 75
2.	13,000	1,320	225
3.	3,500	400	70
4.	16,000	1,800	300
5.	2,500	250	60
6.	4,000	420	100

TAXES

727. Who pays the salary of the governor, the legislators, the mayor, the policemen, and the teachers, in your state and city, or town?

How is the money raised for such purposes?

Ask the assessors of your town to explain to you the method of raising money for the above-named purposes.

728. Money charged against persons and their property for the payment of public expenses is called a **tax.**

729. Taxes are of two kinds — direct and indirect.

730. A direct **tax** is a tax levied on a person, his property, or his business.

731. A tax on a person is called a **poll tax.**

732. A tax on property, real or personal, is called **a property tax.**

733. A tax on a person's business is called a **license fee.**

734. An indirect **tax** is a tax (called a duty) on imported goods (see U. S. Customs, page 376) or a tax (called excise or internal revenue) on the manufacture of liquors and tobacco products. The tax need not be paid on tobacco and liquors that are exported.

HOW TAXES ARE APPORTIONED

735. Suppose that in the state of X there are 4 counties, whose property values are as follows:

County A	$ 15,000,000
County B	25,000,000
County C	27,000,000
County D	33,000,000
Total	$ 100,000,000

Suppose, also, that County A is divided into 3 towns, whose values are:

Town E	$ 5,000,000
Town F	4,000,000
Town G	6.000,000
Total	$ 15,000,000

The amount to be raised in the state for state expenses, exclusive of the revenue received from licenses, permits, etc., is, say, $100,000. This amount is apportioned among the several counties in the ratio of their valuation. The valuation of County A is 15 % of that of the entire state; hence, County A must raise 15 % of $100,000, or $15,000, for its share of the state tax.

The county tax is $45,000, to which the state tax of $15,000 is added, making a total of $60,000 to be raised in County A for state and county expenses. This $60,000 is apportioned among the towns in the ratio of their valuation. The value of Town E is $\frac{1}{3}$ that of County A; hence, Town E is charged with $\frac{1}{3}$ of $60,000, or $20,000, its share of state and county taxes. To this amount is added the town tax of $70,000, which makes $90,000 to be raised in the town of E. To find the *rate of tax*, $90,000 is divided by the valuation of the town, $5,000,000, which gives .018 $= 1\frac{4}{5}$ %. Each person owning taxable property in the town is then charged with a tax equal to $1\frac{4}{5}$ % of the valuation placed on his property.

736. The persons appointed to make an estimate of the value of each person's property, and to apportion the taxes in proportion to the value of each person's property, are called **assessors.**

737. The assessors make what is called an **assessment roll.** It is a list showing the names of property owners, a brief description of the property owned, its assessed valuation, and the tax thereon.

738. The person who collects the tax from the property owners is called the **collector.**

739. Solutions of problems in taxes depend upon the principles of percentage.

The **assessed valuation** is the *base.*

The **tax rate** is the *rate.*

The **tax** is the *percentage.*

740. The tax rate is frequently expressed as **$1.80** per **$100,** instead of **1.8 %.**

EXERCISES

741. 1. A tax of $.003 on a dollar is how much on $100? on $1000?

2. What is the tax on $4500 at the rate of $.65 per $100?

3. At $8.50 per $1000, what is the tax on $7800?

4. Find the tax on $15,400 at 1.365 %.

Find the tax on the following:

5. $13,500 at .0065.
6. $34,500 at .01467.
7. $44,800 at .00895.
8. $18,500 at .0102.
9. $125,000 at .00882.
10. $135,500 at .004.
11. $500,000 at .0115.

12. $240,000 at $1.64 per $100.
13. $15,900 at $1.845 per $100
14. $32,600 at $13.26 per $1000.
15. $16,400 at $11.15 per $1000.
16. $69,500 at $.9875 per $100.
17. $51,600 at $1.065 per $100.
18. $12,600 at $12.853 per $1000.

Oftentimes the collector is allowed a commission on all taxes collected, instead of a salary.

742. 1. Find the total tax paid by a man whose property is assessed for $13,800, at 1.25 %, the collector's commission being 1 %.

1.25 % of $13,800 = $172.50, tax.

1 % of $172.50 = ___1.73, collector's commission.

Total tax, $174.23.

The tax is 1¼ % of $13,800, or $172.50. The collector's commission of 1 % is reckoned on the amount of tax, and then added to the tax, and charged against the property owner.

2. The town of Wellsville has an assessed valuation of $2,118,000. The total amount of tax to be collected is $18,917. There are 910 polls at $1 each, and $2200 highway tax not to be charged on the valuation. Find the rate of taxation and the amount of tax paid by a farmer whose property is assessed for $6400, who pays for 1 poll, and $3 highway tax.

$910 + $2200   = $3110, to be deducted from total tax.

$18,917 − $3110   = $15,807, tax levied on property.

$15,807 ÷ $2,118,000 = $.007463 = 7.463 mills on $1.

$6400 × .007463   = $47.76, property tax.

$47.76 + $1 + $3   = $51.76, total tax.

To find the amount of the tax to be levied on the property of the town, deduct the sum of the poll tax and the highway tax, which leaves $15,807. The tax on the property divided by the value of the property, gives the rate, which is 7.463 mills to the dollar. The property tax on a farm valued at $6400 is found by multiplying $6400 by .007463, which gives $47.76, to which is added $1, poll tax, and $3, highway tax, making a total tax of $51.76.

PROBLEMS

743. 1. The assessed valuation of the real and personal property in a certain town is $1,040,000. The total amount of tax to be collected is $7136. There are 206 polls at $1 each, and $1600, highway tax. Find the rate of taxation, and the amount of tax paid by a man whose property is assessed at $10,500, and who pays for 1 poll and a highway tax of $8.

2. The city of Albany, N.Y., appropriated $320,684 for the support of the public schools. If the taxable property of the city is assessed at $74,718,000, what was the rate of the school tax?

3. The tax rate in Albany for the same year was $1.94 per $100. What was the amount raised by tax? The school tax was what per cent of the entire tax?

In Albany, a discount of 1 % is allowed on all taxes paid before the 10th of February; if paid on or after the tenth day of February and before the first day of March, $\frac{1}{2}$% discount is allowed; if paid on or after Mar. 1 and before Apr. 1, no discount is allowed; if paid on or after Apr. 1, $\frac{1}{2}$% is added on the first day of each month for the remainder of the year. Under these conditions find the amount of tax paid on the following:

4. House assessed at $2900, tax paid Feb. 8.

5. House assessed at $6300, tax paid Feb. 10.

6. Store assessed at $15,000, tax paid Mar. 31.

7. Factory assessed at $45,500, tax paid July 30.

8. Store assessed at $38,700, tax paid Dec. 20.

9. In 1910, the values of real and personal property, **and the tax** levy, in the several boroughs of New York City, were as follows:

BOROUGH	ASSESSED VALUATION (Real Estate)	ASSESSED VALUATION (Personal)	TAX LEVY
Manhattan	$4,743,916,785	$298,030,483	$88,632,391
Bronx	493,757,919	7,716,550	8,815,420
Brooklyn	1,404,036,521	59,332,625	26,560,000
Queens	334,563,960	5,358,480	6,155,282
Richmond	67,917,489	2,207,487	1,314,850
Totals			

Find the tax rate in each borough.

10. Find the per cent of the total valuation in each borough.

11. What per cent of the tax is levied in each borough ?

In the city of New York one half of the tax on real estate and all the tax on personal property are due and payable on and after May 1. The other half of the real estate tax is due and payable on and after Nov. 1. A discount of 4 % per annum is allowed on the second half of the real estate tax if paid before Nov. 1, provided the first half has been paid. Interest at 7 % per annum from May 1 is added to all payments of the first half of the real estate tax and all personal taxes paid on and after June 1. Interest at 7 % per annum from Nov. 1 is added to all payments of the second half of real estate tax on and after Dec. 1.

Find the amount of tax due and payable in the city of New York on the following properties (discounts and penalties as stated above):

Property	Assessed Valuation	Borough	Rate	Date of Payment	
				First Half	Second Half
12. Metropolitan Life Bld'g	$ 12,415,000	Manhattan	1.78	June 18, '14	Oct. 10, '14
13. N. Y. C. & H. R. R.	17,605,400	Bronx	1.77	July 1, '14	Dec. 30, '14
14. Long Island Railroad Co.	16,756,500	Queens	1.80	May 31, '14	Nov. 18, '14
15. Apartment house	75,000	Brooklyn	1.84	Aug 15, '14	Mar. 21, '15
16. House and lot	12,500	Richmond	1.90	Sept. 10, '14	Sept. 10, '14

In St. Louis, Mo., the tax rate for 1909 was, for state purposes, 17¢; for schools, 60¢; and for city purposes, $1.45 per $100. Taxes are due Sept. 1. A rebate of 8 % per annum from day of payment to Dec. 31 is allowed on city taxes only, if paid before Oct. 1. Taxes become delinquent after Dec. 31. The penalty on all delinquent taxes is 1 % a month from date of delinquency till paid, provided they are paid before the Saturday preceding the first Monday in March. After said date all costs of collection are added.

Under the above conditions find the state, school, and city tax for

the year 1909 on each of the following properties, if taxes were paid
on the dates given:

	Description of Property	Valuation	Date Tax was Paid
17	4-story apartment house . . .	$ 25,000	Sept. 25, 1909
18.	Lot 75' × 100'	5,500	Feb. 18, 1910
19.	12-story office building . . .	1,200,000	Sept. 29, 1909
20.	2-story frame dwelling . . .	5,000	Dec. 30, 1909

In Denver, Colo., taxes are paid in two payments of 50% each.
Taxes are due Jan. 1 of each year. The first half is delinquent and
draws interest at 1 % a month on and after Mar. 1. The second half is
delinquent and draws interest at $1\frac{1}{4}$ % a month on and after Aug. 1.

If the state tax is 21 ¢, and the city tax is $1.38 per $100, find
the total tax paid on the following:

21. Store valued at $ 45,000, both installments paid July 20.

22. Dwelling valued at $12,500, first installment paid Feb. 27;
second installment paid Dec. 1.

23. Hotel valued at $500,000, first installment paid July 20;
second installment paid Oct. 10.

On the following page is a form of tax bill used in San Francisco,
Cal. Real estate valuation is higher for state purposes than for city
and county, owing to a raise of 10% by the State Board of Equaliza-
tion. Study this form carefully and verify the results given.

Using the information given in the tax bill on the next page,
find the tax paid in San Francisco on the following properties,
the valuations being those given by the county assessor.

	Valuation	First Installment Paid	Second Installment Paid
24.	$1,960	Nov. 20, 1909	Apr. 10, 1910
25.	22,500	Apr. 20, 1910	Apr. 20, 1910
26.	13,600	June 1, 1910	June 1, 1910
27.	5,500	May 15, 1910	June 15, 1910
28.	3,200	Jan. 30, 1910	Feb 8, 1910

DAVID BUSH, Tax Collector
City and County of San Francisco
STATE OF CALIFORNIA

TAX COLLECTOR NOT RESPONSIBLE FOR ERRORS

REAL ESTATE TAXES

Received the sum of $ _143_ _03_
For 1st Installment only.

FISCAL YEAR 1909

RATE—City, County, State ____ STATE 364
C & CO. 1.60.

1st Installment due October 11, 1909
, 15 pe. cent added November 29, 1909, at 6 P. M.
5 per cent added April 25, 1910, at 6 P. M.
2nd Installment due January 3, 1910.
5 per cent added April 25 1910, at 6 P. M.
And 50c additional for costs,

Received the sum of $ _143_ _03_
For 2nd Installment.

EXAMINE THIS BILL CAREFULLY
And see that all your Property is Included
and Correctly Described

POSITIVELY NO CHECKS RECEIVED AFTER THE 224 DAY OF NOVEMBER, 1909, AND THE 18th DAY OF APRIL, 1910   NO CHECKS ACCEPTED UNLESS PAYABLE IN SAN FRANCISCO. POSTAGE STAMPS WILL
NO RECEIPT VALID UNLESS PAID AT THE OFFICE, OFFICIALLY STAMPED, DATED, RECEIPTED AND CREDIT TAG OR TAGS TORN OFF.
N. B.—Tax payers have the option of paying both Installments when the 1st Installment is payable.

Real Estate Assessed to _Allison B. Ware_

VOL.	PAGE	BLOCK	SUB.	NO.	REAL ESTATE VALUATION	FIRST INSTALLMENT OR FIRST ½ OF TAX	SECOND INSTALLMENT OR LAST ½ OF TAX
3	43	1114	2	519			
Lot W line of Grant Av. St,					15730	28 63	28 63
68 9/12 ft S. from Clay					14300	114 40	114 40
St., th S. 22 9/12 ft by W. 110 ft.						143 03	143 03
					%		5%
							Costs
Real Estate $9300 Imp., $5000 Total, $14300							
Less Mortgage							

BRING THIS BILL WHEN PAYING 2d INSTALLMENT

DO NOT DETACH THESE TAGS, THEY BELONG TO THE TAX COLLECTOR. NOT CREDITED ON ROLL WITHOUT THESE TAGS.

Assessed to _Allison B. Ware_		15730	**1909**	28 63	
PAGE	BLOCK	SUB.	NO.	14300	114 40
3	43	114	2	519	5% 143 03
Second Installment Tag by					Costs

Personal Property		15720	28 63	**1909**
Real Est		14300	114 40	
First Installment Tag by			143 03	
Assessed to _Allison B. Ware_		%		H. C. Hughes Co., 73 Folsom St.
PAGE	BLOCK	SUB.	NO.	
3	43	114	2	519

The rates in the table on the following page apply to the $100 assessed valuation, which is one third of the full value.

Taxes for 1910 are collectible in 1911.

On general taxes a penalty of 1% is charged after May 1, 2% after June 1, 3% after July 1, and so on, until paid   In addition to this, advertising and copying costs of 19 ⨍ for each lot or part of a lot, and 29 ⨍ for each tract of land, are added to unpaid general taxes and special assessments, early in May.

The following table shows how the tax rate is apportioned in Chicago, Ill.:

TAX RATES FOR 1910

Town	State	County	City	School	Sani-tary	Park	Town	Lake Shore Pro.	Blvd. and Park	Total
West Town	.30	.53	1.41	1.55	.34	.77			.05	4.95
South Town	.30	.53	1.41	1.55	.34	.51				4.64
North Town	.30	.53	1.41	1.55	.34	.54	.13	.08		4.88
Hyde Park	.30	.53	1.41	1.55	.34	.51				4.64
Lake	.30	.53	1.41	1.55	.34	.51				4.64
Lake View	.30	.53	1.41	1.55	.34	.76	.12			5.01
Jefferson	.30	.53	1.41	1.55	.34					4.13

Using the rates and information given above, find the tax for each purpose and the total tax on the following properties:

	Description of Property	Assessed Valuation	Town	Date of Payment
29.	Office building	$150,000	North Town	April 20, 1911
30.	3-story brown stone dwelling	90,000	Hyde Park	May 20, 1911
31.	Vacant lot 100' x 100' . .	15,000	Jefferson	July 30, 1911
32.	A tract of land containing 17 acres	17,000	Lake	June 25, 1911

INSURANCE

744. Insurance is an agreement by one party, for a consideration, to indemnify another party for losses arising from certain stipulated causes.

The **insurance company**, or its authorized agent, is the party agreeing to pay in case of loss.

745. The agreement or contract between the parties is called the **policy**.

746. The consideration paid by the insured is called the **premium**. It may be a certain per cent of the face of the policy, or it may be a specified amount per $100.

747. The **term of insurance** may be any length of time, but is generally one or more years.

748. Of the many kinds of insurance, only fire insurance and life insurance will be discussed in this text.

FIRE INSURANCE

749. **Fire insurance** treats of agreements to indemnify the insured for loss by fire.

750. "Loss by fire" is held to cover not only loss of property actually burned, but also loss or damage resulting from the use of chemicals or water used in extinguishing the fire, and from smoke. Fire caused by lightning comes within the meaning of the term "loss by fire."

751. The fire insurance policies of all companies in Connecticut, New Jersey, New York, and Pennsylvania, are uniform, and contain the "New York Standard (80 %) Average Clause," which reads in part as follows: "This Company shall not be liable for a greater proportion of any loss or damage to the property described herein than the sum hereby insured bears to eighty per centum (80 %) of the actual cash value of said property at the time such loss shall happen." These policies also contain a

"Waiver Clause," which reads : " In case of loss if the value of the property described herein does not exceed $ 2500, the 80 % average or co-insurance clause shall be waived."

In other states there is the ordinary policy, which stipulates the amount of loss for which the insurance company is liable. Under the ordinary policy the company pays any loss not exceeding the face of the policy.

752. In some states the policy contains a co-insurance clause which states that only such a part of the loss will be paid as the face of the policy bears to the value of the property insured.

In all the states the fire insurance policies are quite similar.

753. If more than one company insures the same property, each company pays only its pro rata share of any loss.

754. Two general kinds of fire insurance policies are the valued policy and the open policy.

755. In the **valued policy,** the value is stipulated and the company agrees to pay on that specified amount.

756. The **open policy** is used to cover goods in storage and elsewhere, the amount of the policy varying as the quantity of goods is increased or diminished by additions or withdrawals. The addition and withdrawal of goods are recorded in an "Open Policy Book" retained by the company. Upon each receipt of goods, their nature and value are recorded in the policy book, and the premium charged is based on the annual rate.

In case of withdrawal in less than 1 yr. the company returns the unearned premium.

757. A policy may be canceled at any time, either by the insurer or by the insured, the company giving the insured 5 days' notice. If the insurance company cancels the policy, it will return to the insured such a part of the premium as the unexpired time of the policy is of the entire term of the policy. If, however, the insured cancels the policy, the company will return to him only the amount by which the premium paid is in excess of the premium reckoned at the short rate. The short rate is a correspondingly higher rate for a short period of time.

758. Insurance companies frequently issue a policy for 3 yr. for a premium equal to two and one half times the annual premium, and for 5 yr. for four annual premiums.

759. The principles of percentage apply in insurance.

The **value of the policy** is the *base*.

The **rate of premium** is the *rate*.

The **premium** is the *percentage*.

760. To find the premium.

If property is insured for $18,000 at 20¢ per $100 per annum, what is the annual premium?

$$\$18,000 = 180 \text{ hundreds of dollars.}$$
$$180 \times \$.20 = \$36, \text{ the annual premium.}$$

Since the premium rate is on $100, first find the number of hundreds of dollars by pointing off 2 places in the $18,000, which gives 180 hundreds dollars. If the premium on $100 is $.20, on 180 hundreds it is 180 times $.20, or $36.

PROBLEMS

761. Find the premium on each of the following policies:

	FACE OF POLICY	RATE OF INSURANCE		FACE OF POLICY	RATE OF INSURANCE
1.	$13,200	22¢ per $100	8.	24,600	$1\frac{1}{8}\%$, less 10 %
2.	9,600	37½¢ per $100	9.	55,000	$\frac{7}{8}\%$, less 10 % and 5 %
3.	16,400	45¢ per $100	10.	48,000	55¢ per $100, less 10 %
4.	27,500	65¢ per $100	11.	20,000	$\frac{3}{8}\%$
5.	8,500	$\frac{1}{4}\%$	12.	24,500	27½¢ per $100
6.	11,500	$\frac{5}{8}\%$, less 10 %	13.	38,000	$\frac{1}{4}\%$, less 20 %
7.	17,800	$\frac{3}{4}\%$, less 15 %	14.	75,000	44¢ per $100, less 10 %

762. To find the amount paid by the insurer.

1. Property valued at $25,000 is insured for $15,000 at $\frac{3}{8}\%$ per annum. Fire and water cause a loss of $12,000. What amount would be paid by the insurance company (*a*) under an ordinary policy? (*b*) under a co-insurance clause policy? (*c*) under the New York Standard (80 %) Average Clause Policy?

(*a*) $12,000, amount paid under an ordinary policy.

(*b*) $\frac{15000}{25000} = \frac{3}{5}$.

$\frac{3}{5}$ of $12,000 = $7200, amount paid under a co-insurance policy.

(c) 80% of $\$25,000 = \$20,000$.

$\frac{15000}{20000} = \frac{3}{4}$.

$\frac{3}{4}$ of $\$12,000 = \$9000$, amount paid under New York policy.

(a) Under an ordinary policy the company pays the full amount of loss up to the amount of the policy; hence, $\$12,000$ would be paid in this case.

(b) Under a co-insurance, or average clause policy, the part of the loss paid is the ratio of the face of the policy, $\$15,000$, to the value of the property $\$25,000$, or $\frac{15000}{25000}$, which is equal to $\frac{3}{5}$. Hence in this case the company would pay $\frac{3}{5}$ of $\$12,000$, or $\$7200$.

(c) Under the 80% average clause policy, the company pays such a part of the loss as the ratio of the face of the policy, $\$15,000$, is to 80% of the value of the property, 80% of $\$25,000 = \$20,000$. The ratio of $\$15,000$ to $\$20,000$ is $\frac{15000}{20000} = \frac{3}{4}$. Hence the company would pay $\frac{3}{4}$ of $\$12,000$, or $\$9000$.

2. A stock of merchandise is insured in Company A for $\$5000$, in Company B for $\$7000$, and in Company C for $\$8000$. In case of damage amounting to $\$10,000$, how much will each company pay?

$\$5000 + \$7000 + \$8000 = \$20,000$, total amount of insurance.

$\frac{5000}{20000} = \frac{1}{4}$; $\frac{1}{4}$ of $\$10,000 = \$2500$, paid by Company A.

$\frac{7000}{20000} = \frac{7}{20}$; $\frac{7}{20}$ of $\$10,000 = \$3500$, paid by Company B.

$\frac{8000}{20000} = \frac{2}{5}$; $\frac{2}{5}$ of $\$10,000 = \$4000$, paid by Company C.

The total insurance is $\$20,000$. Each company pays such part of the loss as its risk is of the total risk. Hence, Company A pays $\frac{1}{4}$ of the loss, or $\$2500$; Company B pays $\frac{7}{20}$ of the loss, or $\$3500$; and Company C pays $\frac{2}{5}$ of the loss, or $\$4000$.

PROBLEMS

763. 1. A dwelling valued at $\$9600$ was insured for $\frac{2}{3}$ of its value at $\frac{3}{4}\%$ and contents valued at $\$3200$, for $\frac{3}{4}$ of its value at $\frac{4}{5}\%$. Fire causes a total loss of the building and a loss of $\$1400$ on the contents. Under an ordinary policy, how much will the insurance company pay?

2. Property insured under an ordinary policy for $\$15,000$ at 60 ¢ per $\$100$ per term of 3 yr. is damaged by fire to the amount of $\$8000$. What is the insurance company's net loss if they have held the insurance 15 yr., interest 6%?

3. A store and its contents were insured in Company X for $\$22,500$ at 60 ¢ per $\$100$; in Company Y for $\$30,000$ at $\frac{5}{8}\%$; and in Company Z for $\$40,000$ at 65 ¢ per $\$100$. The property was damaged by fire and water to the amount of $40,000. What was each company's net loss, they having held the insurance 7 yr., if money is worth 5 %?

4. A man has his furniture insured for $\frac{2}{3}$ of its value, in a policy containing an average clause. In case of a loss of $2400, how much should the company pay?

5. A house valued at $15,000, and contents valued at $7500, are insured for $\frac{2}{3}$ of their value. Fire causes a loss of $4000 on the house and $1500 on the contents. Find the amount payable under an 80 % average policy.

764. The following tables show the short rates adopted by the "Western Union":

Standard Short Rate Scale for computing Premiums for Terms less than One Year, as adopted by the "Western Union"

RULE. Take the percentage indicated in scale opposite the number of days risk is to run, on the premium for one year at given rate, and the result will be the premium earned in case of cancellation, or to be charged in case of short risks.

1 Day	2 per cent annual prem.	55 Days	29 per ct. an'l prem.
2 Days	4 per cent annual prem.	60 Days	30 per ct. an'l prem.
3 Days	5 per cent annual prem.	65 Days	33 per ct. an'l prem.
4 Days	6 per cent annual prem.	70 Days	36 per ct. an'l prem.
5 Days	7 per cent annual prem.	75 Days	37 per ct. an'l prem.
6 Days	8 per cent annual prem.	80 Days	38 per ct. an'l prem.
7 Days	9 per cent annual prem.	85 Days	39 per ct. an'l prem.
8 Days	9 per cent annual prem.	90 Days or 3 mo.	40 per ct. an'l prem.
9 Days	10 per cent annual prem.	105 Days	45 per ct. an'l prem.
10 Days	10 per cent annual prem.	120 Days or 4 mo.	50 per ct. an'l prem.
11 Days	11 per cent annual prem.	135 Days	55 per ct. an'l prem.
12 Days	12 per cent annual prem.	150 Days or 5 mo.	60 per ct. an'l prem.
13 Days	13 per cent annual prem.	165 Days	65 per ct. an'l prem.
14 Days	13 per cent annual prem.	180 Days or 6 mo.	70 per ct. an'l prem.
15 Days	14 per cent annual prem.	195 Days	73 per ct. an'l prem.
16 Days	14 per cent annual prem.	210 Days or 7 mo.	75 per ct. an'l prem.
17 Days	15 per cent annual prem.	225 Days	78 per ct. an'l prem.
18 Days	16 per cent annual prem.	240 Days or 8 mo.	80 per ct. an'l prem.
19 Days	16 per cent annual prem.	255 Days	83 per ct. an'l prem.
20 Days	17 per cent annual prem.	270 Days or 9 mo.	85 per ct. an'l prem.
25 Days	19 per cent annual prem.	285 Days	88 per ct. an'l prem.
30 Days	20 per cent annual prem.	300 Days or 10 mo.	90 per ct. an'l prem.
35 Days	23 per cent annual prem.	315 Days	93 per ct. an'l prem.
40 Days	26 per cent annual prem.	330 Days or 11 mo.	95 per ct. an'l prem.
45 Days	27 per cent annual prem.	360 Days or 12 mo.	100 per ct. an'l prem.
50 Days	28 per cent annual prem.		

765. To find the cost of insurance at short rates, or the amount of premium to be returned if policy is canceled.

1. Find the cost of insuring a stock of merchandise for $15,000 for 3 mo. if the annual rate is $\frac{7}{8}\%$.

$\frac{7}{8}\%$ of $15,000 = $131.25, annual premium.

40% of $131.25 = $52.50, premium for 3 mo.

By the short rate scale, page 325, the premium for 3 mo. is 40% of the annual premium. The annual premium is $131.25; hence, the premium for 3 mo. is 40% of $131.25, or $52.50.

2. Property valued at $24,000 is insured for $\frac{3}{4}$ of its value for **1** yr. at 60¢ per $100. How much of the premium should be returned if the policy is canceled at the expiration of 8 mo. (*a*) by the insured? (*b*) by the insurance company?

(*a*) $\frac{3}{4}$ of $24,000 = $18,000.

180 × $.60 = $108, annual premium.

20% of $108 = $21.60, amount to be returned if insured cancels the policy.

(*b*) 4 mo. : 12 mo. :: $? : $108.

$\frac{4 \times \$108}{12} = \$36$, amount to be returned if insurer cancels the policy.

(*a*) By the short rate scale the rate for 8 mo. is 80% of the annual rate. That is, the company will return 20% of the premium paid. The annual premium is $108; hence, the amount returned is 20% of $108, or $21.60.

(*b*) When the insurer cancels the policy, the amount returned bears the same ratio to the entire premium paid as the unexpired time bears to the entire time; hence, the proportion 4 mo. : 12 mo. :: $? : $108, the solution of which gives $36, the amount to be returned.

PROBLEMS

766. 1. A policy of insurance for $10,000 at $\frac{3}{4}\%$ per annum was dated Apr. 7, 1910. 5 mo. later it was canceled by the insured. How much of the premium was returned?

2. May 10, 1910, I took out a policy of insurance on my furniture for $2400 at 40¢ per $100 per annum. Feb. 19, 1911, I canceled the policy. How much of the premium should be returned to me?

3. A stock of merchandise was insured for 1 yr. for $ 25,000 at $\frac{7}{8}$ %. At the end of 7 mo. the insurance company canceled the policy. How much of the premium was returned? (Solve by proportion.)

4. A merchant insured his store and contents for 1 yr. for $16,500 at $\frac{3}{5}$ %. At the end of 5 mo. and 15 da. the insurer canceled the policy. What was the return premium? (Solve by proportion.)

5. A policy of $ 32,000 at $\frac{5}{8}$ % for 1 yr. was canceled by the insured at the expiration of 8 mo. Find the amount of the return premium. If the company had canceled the policy, what amount of premium would have been returned?

6. The Ætna Insurance Company insured an office building for $ 45,000 at $\frac{1}{2}$ % for 1 yr. Find the return premium if the policy was canceled at the end of 6 mo. (a) by the company; (b) by the insured.

7. Find the cost of insuring a stock of goods for $ 12,000 for 6 mo. at 75 ¢ per $ 100 per annum.

REVIEW PROBLEMS

767. 1. Find the cost of insuring a house for $ 8000 for 3 yr. if the annual premium is $\frac{1}{4}$ %.

2. At 60 ¢ per $ 100 per annum, what is the cost of insuring a shipment of merchandise for $ 6500 for 30 da. ?

3. Property worth $ 75,000 is insured as follows: In the Hartford Insurance Company for $ 20,000 at $1\frac{1}{4}$ %, and in the Albany Insurance Company for $30,000 at $1\frac{1}{8}$ %. There is a $ 30,000 loss by fire during the first term of insurance. Find each company's net loss under (a) an ordinary policy; (b) an average clause policy.

Find the amount to be paid in each of the following cases, supposing each policy contains the Standard 80 % Average Clause:

	VALUE OF PROPERTY	FACE OF POLICY	DAMAGE BY FIRE	PAID BY INSURANCE CO.
4.	$ 20,000	$ 12,000	$ 8,000	?
5.	28,000	15,000	15,000	?
6.	12,000	5,000	12,000	?
7.	16,500	10,000	3,000	?
8.	45,000	20,000	25,000	?

9. An open policy of insurance was issued on merchandise stored, or to be stored, in a warehouse, the premium to be 80¢ per $100 per annum. Goods withdrawn inside of 1 yr. were to be charged the short rate. Find the total premium paid, and the total return premium on the following receipts and withdrawals:

RECEIPTS		WITHDRAWALS	
Date	Amount	Date	Amount
Apr. 1, 1910	Silk, $ 8,400	Oct. 1, 1910	Woolens, $5000
Apr. 15, 1910	Furs, 12,800	Oct. 10, 1910	Furs, 9000
May 10, 1910	Woolens, 11,500	Oct. 20, 1910	Silk, 8400
June 1, 1910	Gloves, 6,500	Nov. 10, 1910	Gloves, 4500
June 20, 1910	Hosiery, 5,600	Dec. 1, 1910	Woolens, 6500

LIFE INSURANCE

768. Life insurance is a contract in which the insurer, for a consideration, agrees to pay a specified sum of money at the death of the insured, or at some fixed time.

769. The consideration is called the **premium** and may be paid in one sum at the beginning of the contract, or in annual payments in advance for a series of years.

770. Life insurance policies may generally be classified as follows: (1) whole life; (2) term; (3) endowment

771. Whole life policies are those in which the sum insured is payable at death only. The premiums may be paid annually during the life of the insured, or may be limited to 10, 15, 20, or 25 annual payments. In the latter case the policy becomes *paid up* for life, after the specified number of payments have been made.

772. In term policies, the sum insured is payable only in the event of death during a fixed term. At the expiration of the term the insurance ceases.

773. Endowment policies provide for the payment of the sum insured at a fixed date if the insured is then living; or, in case of death before that date, the amount of the policy is paid to the beneficiary.

774. The beneficiary is the person named in the policy to receive payment upon the death of the insured.

775. A portion of every premium is required by law to be set aside and properly invested for the purpose of providing a reserve fund from which death losses and maturing policies are to be paid.

776. If the insured discontinues his policy, he is entitled to the reserve set aside to meet his policy when it matures. He may receive the cash value of the reserve, or he may take a paid-up policy for the sum which the reserve will purchase, or he may use the reserve to *extend the time* of the policy for such a time as the reserve will purchase.

777. The following table of values shows the options and their value, in one insurance company, to which the insured is entitled in case of discontinuance of the payment of premiums.

TABLE OF VALUES *

AT END OF YEAR	LOAN OR CASH SURRENDER VALUE	IN CASE OF LAPSE OF POLICY		
		Paid-up Insurance on Surrender	Or Extended Insurance without any Notice from the Insured	
3d	$ 25.96	$ 150	5 years	40 days
4th	39.05	200	7	14
5th	52.71	250	9	30
6th	67.00	300	11	112
7th	81.92	350	13	250
8th	97.52	400	16	51
9th	113.83	450	18	184
10th	130.88	500	20	261
11th	148.71	550	22	279
12th	167.37	600	24	199
13th	186.89	650	26	66
14th	207.33	700	27	248
15th	228.75	750	29	27
16th	251.18	800	30	144
17th	274.70	850	31	308
18th	299.37	900	33	135
19th	325.25	950	35	31
20th	352.42	1000 Policy	Fully Paid	
25th	402.60			
30th	458.53			

* Age of insured 22 yr. Annual premium, $ 25.68 per $ 1000, payable for 20 yr. Amount of the policy payable at the death of the insured.

778. The following tables ε the annual premium on $1000 of insurance on different kinds ʝolicies and for different ages of the insured in a leading insuraɔ ompany.

	RATES FOR $1000 INSURANCE					RATES FOR $1000 INSURANCE			
	Whole Life Policies, with Premiɩ payable annually during Life, or iɩ or 20 Years, or by a Single Payment					Endowment Insurance Policies, payable at the End of the Terms Stated, or on Prior Death; Annual Premiums.			
Age	Annual Premiums	Ten Annual Premiums	Twenty Annual Premium-	ʂ ɪː ʝ ʌɔ	Age	10 Years	15 Years	20 Years	25 Years
20	$18.00	$46.75	$27.76	ʂ ːː	20	$106.30	$67.79	$48.92	$37.92
21	18.40	47.43	28.17	ːː	21	106.34	67.83	48.07	37.98
22	18.80	48.13	28.60	.ʝ	22	106.37	67.88	49.03	38.05
23	19.23	48.86	29.04	ːʝ	23	106.41	67.92	49.08	38.12
24	19.67	49.60	29.50	ːʝ	24	106.45	67.97	49.14	38.20
25	20.14	50.38	29.98	ːʝ	25	106.49	68.02	49.21	38.28
26	20.63	51.18	30.47	4 ɿ	26	106.53	68.08	49.28	38.38
27	21.15	52.00	30.98	4	27	106.58	68.14	49.36	38.48
28	21.69	52.86	31.51	4 ɩ	28	106.63	68.21	49.45	38.59
29	22.26	53.74	32.06	4	29	106.69	68.28	49.54	38.71
30	22.85	54.65	32.62	4 4	30	106.75	68.36	49.64	38.85
31	23.48	55.59	33.21	4 0	31	106.82	68.45	49.76	39.00
32	24.14	56.56	33.83	4	32	106.90	68.55	49.89	39.18
33	24.84	57.56	34.47	4 ɿ1	33	106.98	68.65	50.03	39.37
34	25.58	58.60	35.13	ɿ ɿ7	34	107.06	68.77	50.18	39.58
35	26.35	59.67	35.82	ɿ ɩ5	35	107.16	68.90	50.36	39.82
36	27.17	60.78	36.54	ɿ ɿ4	36	107.27	69 04	50.56	40.09
37	28.04	61.92	37.30		37	107.39	69.20	50.78	40.39
38	28.95	63.11	38.08	39	38	107.52	69.39	51.03	40.72
39	29.92	64.33	38.91	65	39	107.67	69.59	51.30	41.10

ʾAMPLES .

779. 1. What is the annu ɔost of a whole life policy of $5000, age 30 yr., if payments are to made during the life of the insured?

In the table showing rates on ɭe life policies, in the first money column opposite age 30 is the amount, $: 5, the rate on $1000. On $5000, the cost is 5 × $22.85, or $114.25.

2. Find the annual cost ɼ a 20-yr. endowment policy of $300ʾ age 25.

In the table showing rates ɼ endowment policies, in the column "20 years" opposite age 25, is f ɩd $49.21, the annual cost of $100ɼ costs 3 × $49.21, or $147.63.

The annual premium is $15 \times \$6\ldots$, or $\$1033.50$.

778. The following tables show the annual premium on $1000 of insurance on different kinds of policies and for different ages of the insured in a leading insurance company.

	RATES FOR $1000 INSURANCE Whole Life Policies, with Premiums payable annually during Life, or in 10 or 20 Years, or by a Single Payment.					RATES FOR $1000 INSURANCE Endowment Insurance Policies, payable at the End of the Terms Stated, or on Prior Death; Annual Premiums.			
AGE	ANNUAL PREMIUMS	TEN ANNUAL PREMIUMS	TWENTY ANNUAL PREMIUMS	SINGLE PAYMENT	AGE	10 YEARS	15 YEARS	20 YEARS	25 YEARS
20	$18.00	$46.75	$27.76	$372.54	20	$106.30	$67.79	$48.92	$37.92
21	18.40	47.43	28.17	377.36	21	106.34	67.83	48.07	37.98
22	18.80	48.13	28.60	382.33	22	106.37	67.88	49.03	38.05
23	19.23	48.86	29.04	387.46	23	106.41	67.92	49.08	38.12
24	19.67	49.60	29.50	392.74	24	106.45	67.97	49.14	38.20
25	20.14	50.38	29.98	398.20	25	106.49	68.02	49.21	38.28
26	20.63	51.18	30.47	403.83	26	106.53	68.08	49.28	38.38
27	21.15	52.00	30.98	409.63	27	106.58	68.14	49.36	38.48
28	21.69	52.86	31.51	415.61	28	106.63	68.21	49.45	38.59
29	22.26	53.74	32.06	421.78	29	106.69	68.28	49.54	38.71
30	22.85	54.65	32.62	428.14	30	106.75	68.36	49.64	38.85
31	23.48	55.59	33.21	434.70	31	106.82	68.45	49.76	39.00
32	24.14	56.56	33.83	441.45	32	106.90	68.55	49.89	39.18
33	24.84	57.56	34.47	448.41	33	106.98	68.65	50.03	39.37
34	25.58	58.60	35.13	455.57	34	107.06	68.77	50.18	39.58
35	26.35	59.67	35.82	462.95	35	107.16	68.90	50.36	39.82
36	27.17	60.78	36.54	470.54	36	107.27	69 04	50.56	40.09
37	28.04	61.92	37.30	478.36	37	107.39	69.20	50.78	40.39
38	28.95	63.11	38.08	486.39	38	107.52	69.39	51.03	40.72
39	29.92	64.33	38.91	494.65	39	107.67	69.59	51.30	41.10

EXAMPLES .

779. 1. What is the annual cost of a whole life policy of $5000, age 30 yr., if payments are to be made during the life of the insured?

In the table showing rates on whole life policies, in the first money column opposite age 30 is the amount, $22.85, the rate on $1000. On $5000, the cost is 5 × $22.85, or $114.25.

2. Find the annual cost of a 20-yr. endowment policy of $3000, age 25.

In the table showing rates on endowment policies, in the column headed "20 years" opposite age 25, is found $49.21, the annual cost of $1000. $3000 costs 3 × $49.21, or $147.63.

3. A 15-yr. endowment policy of $15,000 was taken by a man 35 yr. of age. If he lives till the policy matures, how much less will he receive from the insurance company than he would have had if he had placed his annual premiums in a savings bank and received 4%, interest compounded annually?

The annual premium is 15 × $68.90, or $1033.50.

A payment of $1 each year, for 15 yr., will amount to $20.824531 (see page 286), if compounded annually at 4%. $1033.50 will amount to 1033.50 × $20.824531, or $21,522.15, amount if deposited in savings bank. $21,522.15 − $15,000 = $6522.15 less from the insurance company.

If the insured in example 3 had died at the age of 40 yr., how much more would the beneficiary have received from the company than the amount of the annual premiums at 4% compound interest?

Each annual premium is $1033.50, and has been paid for 5 yr. At 4% compound interest the amount would be 1033.50 × $5.632975 (see page 286), or $5821.68.

$15,000 − $5821.68 = $9178.32, excess received from the company.

PROBLEMS

780. 1. Find the annual cost of a $5000 policy on the life of a person 39 yr. of age if premiums are to be paid during life.

2. At age 39, a person's expectation of life is 25 yr. If the person insured in problem 1, lives and pays the premium during his expectation of life, what sum will he actually have paid to the insurance company? At 4% compound interest the amount of his payments exceed by how much the face value of the policy?

3. A man 22 yr. of age takes out a whole life policy of $5000 with premiums payable annually for 20 yr. At the age of 35 he ceases making payments. What is the surrender value of his policy? What amount of paid-up insurance has he to his credit? For what length of time will insurance for the face value of the policy be extended if he so chooses? (See page 329.)

4. A man takes a 15-yr. endowment policy of $20,000 on his life at the age of 30 yr. If he survives the 15-yr. period, how much more will he have paid the company than the face of his policy? How much would his payments have amounted to at 3% compound interest?

SAVINGS BANKS

781. A **savings bank** is a bank under the control of state laws, for the purpose of receiving small deposits of money and paying interest thereon.

782. Interest is usually paid on any number of dollars from $1 to $3000, on condition that such sum has been on deposit for an entire interest term.

783. The **interest term** is the period of time between the dates on which interest payments are due.

784. If interest payments are due Jan. 1 and July 1, the interest term is 6 mo. If due Jan. 1, Apr. 1, July 1, and Oct. 1, the interest term is 3 mo., etc. In some states interest begins to accrue from the first of each calendar month on all sums on deposit at that time. A withdrawal of funds cancels the interest on the sum withdrawn for the entire interest term.

785. When interest becomes due it may be drawn out of the bank or it will be placed to the credit of the depositor, in which case it draws interest the same as the other deposits. Savings banks, therefore, pay compound interest.

786. Each depositor receives a small pass book in which are entered all deposits and withdrawals. Checks cannot be drawn against a savings bank account.

787. To find the balance due on a savings bank account.

Find the balance due July 1, 1910, on the following: The account was opened Sept. 30, 1908, by depositing $50. Dec. 31, 1908, deposited $20; Apr. 8, 1909, withdrew $25; June 20, 1909, deposited $75; July 8, 1909, withdrew $30; Dec. 20, 1909, deposited $100; June 5, 1910, withdrew $40. Interest at 4% per annum is allowed from the first of each quarter, and is credited Jan. 1 and July 1.

DRAFTS	DEPOSITS	DATE	BALANCE
		1908	
	$ 50.00	Sept. 30	$ 50.00
	20.00	Dec. 31	70.00
		1909	
	Int. .50	Jan. 1	70.50
$ 25.00		Apr. 8	45.50
	75.00	June 20 ·	120.50
	Int. .90	July 1	121.40
30.00		July 8	91.40
	100.00	Dec. 20	191.40
		1910	
	Int. 1.82	Jan. 1	193.22
40.00		June 5	153.22
	Int. 3.06	July 1	156.28

The simplest method of solution is to use a form like the above, and to enter the deposits, withdrawals, and interest credits in their proper chronological order.

Since interest is allowed from the first of each quarter, the original deposit of $50 draws interest from Oct. 1, 1908 to Jan. 1, 1909,—3 mo. Interest is not allowed on the deposit of $20 made Dec. 31, 1908, because it has not been on deposit during the quarter. The interest on $50 for 3 mo. is 50 ¢ which makes a total amount on deposit of $70.50. On Apr. 8, 1909, a withdrawal of $25 reduces the balance to $45.50. The deposit made June 20 increases the balance to $120.50, but draws no interest until after July 1, 1909. The amount on which to reckon interest for the interest term from Jan. 1, 1909, to July 1, 1909, is $45 (interest is not paid on parts of a dollar), the smallest balance during the term. The interest is 90 ¢ which makes the balance $121.40. The draft of $30, July 8, 1909, leaves a balance of $91.40. The deposit Dec. 20, 1909, of $100 increases the balance to $191.40. The interest on $91, the smallest balance of the term, is $1.82, making the balance Jan. 1, 1910, $193.22. The withdrawal of $40 June 5, 1910, leaves $153.22 on deposit. The interest on $153 for 6 mo. is $3.06, making a balance due of $156.28.

PROBLEMS

788. Assuming that interest at 4 % per annum is allowed from the beginning of each quarter, and is credited Jan. 1 and July 1, prepare a statement similar to the illustration on this page for each of the following savings bank accounts:

1. Deposited Apr. 1, 1908, $125; May 15, 1908, $75; withdrew July 3, 1908, $25; deposited, Oct. 1, 1908, $50; deposited Dec. 28, 1908, $100; withdrew June 1, 1909, $60; deposited Aug. 30, 1909, $125; deposited Mar. 30, 1910, $200. How much is due July 1, 1910?

2. Balance Jan. 1, 1907, $387.50; deposited Oct. 1, 1907, $225; withdrew May 1, 1908, $100; deposited Apr. 1, 1909, $300; withdrew Oct. 20, 1909, $150. How much was due Jan. 1, 1910?

3 Balance Oct 1, 1909, $575; withdrew Jan. 4, 1910, $150; deposited July 1, 1910, $400; deposited Sept. 15, 1910, $200; withdrew Nov. 10, 1910, $50; deposited Feb. 18, 1911, $500. Find balance due July 1, 1911.

4–6. Find the balance due on problems 1–3 if interest is allowed from the first of each month, credit being given Jan. 1 and July 1

POSTAL SAVINGS BANKS

789. The postal savings system was established in the United States January 3, 1911, at 48 second-class post offices, one in each state and territory.

790. Deposits are evidenced by postal-savings certificates issued in fixed denominations of $1, $2, $5, $10, $20, $50, $100, $200, and $500, each bearing the name of the depositor, the number of his account, the date of issue, the name of the depository office, and the date on which interest begins.

791. Interest is allowed on all deposits at the rate of 2 per cent per annum, computed on each savings certificate separately, and payable annually. No interest is paid on money which remains on deposit for a fraction of a year only

792. Deposits bear interest from the first day of the month next following that in which deposited.

793. Compound interest is not allowed on an outstanding certificate, but a depositor may withdraw interest payable and include it in a new deposit, which bears interest at the regular rate.

794. A depositor is permitted to exchange his deposits for registered or coupon United States postal savings bonds, issued in denominations of $20, $100, and $500, bearing interest at the rate of $2\frac{1}{2}$ per cent .per annum, payable semiannually. The exchange may be made under date of Jan. 1 and July 1 of each year, provided such bonds are then available.

PROBLEMS

795. 1. On Feb. 28, 1911 (after 2 months' operation), the number of banks was 48, the total number of depositors was 3664, and the amount on deposit was $133,869. Dec. 30, 1911, there were 5132 depositories, 162,697 depositors, and $10,614,676 on deposit. Show the per cent of increase in the number of depositories and of depositors, and in the amount of deposits as compared with Feb. 28, 1911.

2. March 31, 1914, the total deposits amounted to $42,300,000. Of this amount there was held at the New York office $4,073,504; at Chicago, $2,178,528; at Brooklyn, $1,404,198; and at Boston, $1,070,269. What per cent of the total deposits was held by all the offices named? Show graphically the relative amounts held by these four offices as compared with each other.

3. Beginning May 20, 1911, a man made a monthly deposit of $25 in a postal savings bank. On Jan. 1, 1913, he exchanged his total deposits for a postal savings bond. Assuming that he made no deposits after Jan. 1, 1913, and that he drew his interest as it became due, how much interest did he receive on his deposits and bond to and including July 1, 1915?

4. On the last business day of each month, a man deposits $10 in a postal savings bank, for five years. If no interest has been drawn, what amount will he then have to his credit?

EXAMINATIONS

SPEED TEST

796. Minimum time thirty minutes; maximum, one hour. Deduct one credit for each minute beyond minimum time.

1. Find the interest in the following cases:

$500 for 2 yr. 2 mo. 6 da. at 6%.
$300 for 5 yr. 6 mo. 18 da. at 6%.
$930 for 6 yr. 8 mo. at 5%.
$750 for 4 yr. 6 mo. at $3\frac{1}{4}$%.

2. Find the proceeds of the following discounted notes:

$350	90 da.	6%.
$750	75 da.	6%.
$960	45 da.	4%.
$150	72 da.	5%.

3. Find balance due July **1,** 1910, on the following (United States Rule):

Face of note, $3000; date, July **1,** 1909; Interest, 6%.
Payments, Nov. 1, 1909, $1000.
Apr. 1, 1910, $1500.

5. Find the present worth of each of the following notes:

$ 515   due in 6 mo. at 6%.
$2020   due in 72 da. at 5%.
$ 858.50 due in 60 da. at 6%.
$1540.25 due in 90 da. at 4%.

7. Find the total tax on the following:

Farm valued at $2400 at .0175.
House and lot valued at $18,000 at .01875.
City lot valued at $5500 at $1.60 per $100.
Store valued at $30,000 at $1.7246 per $100.

4. Find the accurate interest in the following cases:

$3650 for 125 da. at 5%.
$7300 for 116 da. at 6%.
$1095 for 17 da. at 4%.
$ 550 for 146 da. at $3\frac{1}{2}$%.

6. Find the total insurance on the following properties:

Store worth $25,000 at $\frac{1}{3}$% premium.
Mdse. worth $40,000 at $1\frac{1}{4}$% premium.
House worth $15,000 at 60¢ per $100 premium.
Furniture worth $5000 at 80¢ per $100 premium.

8. At what price each shall these articles be sold to gain the rates given?

Cost per Dozen	Rate of Gain
$18	20%
$24	40%
$15	$33\frac{1}{3}$%
$14.40	25%

9-10. Add vertically and horizontally and find grand total:

16832	21265	17234	18875
15293	17294	16872	5570
13846	17286	9237	18765
16450	18642	27436	18778
8287	18643	21299	17694

WRITTEN TEST

797. 1. In a certain county containing 25,482 taxable inhabitants, a tax of $103,294.60 is assessed for town, county, and state purposes: a part of this sum is raised by a tax of 30¢ on each poll; the entire valuation of property on the assessment roll is $38,260,000. What is the per cent of tax, and what is a person's tax who pays for 1 poll and whose property is valued at $9470?

2. Find the interest on $ 738.72 for **1** yr. **3** mo. 22 da., at 5 %.

3. I bought a horse for $360 and sold it for a .note at 60 da. for $418. I discounted the note at 6% the day it was made. What was my gain ?

4. A store was insured for $12,000 at $\frac{3}{4}$% and the goods for $15,000 at $1\frac{1}{4}$%. What was the entire premium ?

5. A grocer sold 50 bbl. of wine, each containing 31 gal. 2 qt. at $2.40 a gallon, receiving in payment a note at 90 da., with interest at 6 %. What would be the proceeds of this note when discounted at a bank at $7\frac{1}{2}$% per annum ?

6. For the installation of a system of city waterworks a city issues bonds to the amount of $1,500,000 payable in 20 yr. What amount must be set aside each year as a sinking fund at 4 % compound interest to pay the bonds when they are due ?

7. A man deposited $15,000 in a bank June 1, 1910, receiving therefor a certificate of deposit which stated that 3 % interest would be paid if the money was left on deposit 6 mo. or more. The bank loaned the money at 5%. Dec. 20, 1910, the man asked for his money and interest. If the bank pays exact interest and charges ordinary interest, find the gain to the bank, the money having been loaned all the time.

8. A merchant bought a bill of goods invoiced at $5475.76, less 10 % and 5 %, terms net 60 da. or 5 % 10 da. Not having the ready money, he borrowed it at 6 % interest and accepted the 10-da. terms. Find his gain by so doing.

9. On a note of $4000 dated June 1, 1909, and drawing interest at 6%, the following payments were made: Sept. 16, 1909, $1000; Jan. 19, 1910, $1500; and Apr. 1, 1910, $1200. How much was required to settle the balance Aug. 1, 1910 ?

10. A factory was insured under a policy containing the standard eighty-per-cent-average clause, as follows: Building for $22,000 at $1\frac{1}{4}$%; machinery for $25,000 at $1\frac{1}{8}$%; and the stock at $15,000 at $1\frac{3}{8}$%. The building is worth $30,000, the machinery $35,000, and the stock $20,000. Fire broke out causing a damage as follows: to the building, $8000; machinery, $6000; and to the stock, $12,000. Find the amounts paid by the company.

STOCKS

798. Stock, or **capital stock,** is the name given to the capital of a corporation.

799. A corporation is a person composed of several real persons. It is formed only by the consent of the state, and hence, has only those rights and powers that are granted to it by the state as set forth in its charter. The list of powers, rights, and duties are stated in writing and issued to the corporation. This written instrument is called a charter.

800. The capital stock of a corporation is divided into shares, usually $ 100 each, though shares may be of various values. Reading Railroad shares are $ 50 par value. Mining stocks are at all prices from 10¢ to $ 100.

801. A stockholder is any person who owns one or more shares of stock. As evidence of ownership of stock, each stockholder receives a written statement called a **stock certificate,** showing the number of shares he owns and their par value.

802. Stock is either common or preferred. **Common stock** is the kind most usually issued by a corporation, and carries no guaranty of a specified dividend. **Preferred stock** carries with it a guaranty of a definite dividend even to the exclusion of dividends on the common stock. Preferred stock is often issued in times of emergency to provide working capital or to pay debts.

803. The **par** value of stock is the value stated on the stock certificate. **The market value is** the price for which the stock can be sold.

804. The market value of stock depends chiefly upon the rate of dividends paid or earned by the company. If the dividends are small, the market value is below par (less than face value, or at a discount), and if the dividends are large, the market value is above par (or at a premium).

805. ILLUSTRATION OF A STOCK CERTIFICATE

806. Shares of stock are transferred by assignment. The assignee, then, has all the rights and privileges of the original stockholder.

807. FORM OF ASSIGNMENT

808. The following prices of certain stocks were printed in the *New York Journal of Commerce*, Jan. 7, 1911.

NEW YORK STOCK EXCHANGE

STOCK EXCHANGE SALES AND QUOTATIONS JAN. 6, 1911

Sales	Open	High	Low	Close	Net Chg.
15,600 Amal Copper	$63\frac{5}{8}$	$64\frac{7}{8}$	$63\frac{3}{4}$	$64\frac{7}{8}$	$+1\frac{3}{8}$
600 Am Cotton Oil	59	59	$58\frac{3}{4}$	$58\frac{5}{8}$	$-\frac{5}{8}$
300 Am Locomotive	$39\frac{1}{2}$	$39\frac{5}{8}$	$39\frac{1}{2}$	$39\frac{5}{8}$	$+\frac{1}{2}$
500 Am Sugar	114	114	113	113	-2
900 Am Tel & Tel	$141\frac{1}{4}$	$141\frac{1}{4}$	$141\frac{3}{8}$	$141\frac{3}{8}$	..
2,900 Atchison	$102\frac{1}{4}$	$102\frac{3}{4}$	$102\frac{1}{2}$	$102\frac{1}{4}$	$+\frac{1}{4}$
2,100 Ches & Ohio	$82\frac{1}{4}$	$82\frac{1}{4}$	$81\frac{7}{8}$	82	$-\frac{1}{8}$
3,600 C M & St P	$124\frac{3}{4}$	$125\frac{1}{2}$	$124\frac{3}{4}$	125	$+\frac{1}{4}$
500 Chi & N W	$142\frac{3}{4}$	$142\frac{3}{4}$	$142\frac{1}{2}$	$142\frac{1}{2}$	..
250 Col & South 1st pfd	76	76	$75\frac{3}{4}$	$75\frac{3}{4}$	$+\frac{1}{4}$
6,600 Consol Gas	$140\frac{1}{2}$	$140\frac{1}{2}$	$139\frac{3}{4}$	$140\frac{1}{2}$	$+\frac{1}{4}$
200 Del & Hudson	166	$166\frac{1}{2}$	166	$166\frac{1}{2}$	$-\frac{1}{8}$
500 Den & R G	$29\frac{1}{4}$	$29\frac{3}{4}$	$29\frac{1}{4}$	$29\frac{5}{8}$	$+\frac{3}{8}$
600 Erie	28	28	$27\frac{5}{8}$	$27\frac{5}{8}$	$+\frac{1}{8}$
100 do 1st pfd	$46\frac{1}{8}$	$46\frac{1}{8}$	$46\frac{1}{8}$	$46\frac{1}{8}$	$-\frac{3}{8}$
200 Illinois Central	$132\frac{7}{8}$	$132\frac{7}{8}$	$132\frac{1}{2}$	$132\frac{1}{2}$	..
300 Louis & Nash	145	$145\frac{1}{4}$	145	145	..
100 Mer Marine pfd	16	16	16	16	$+\frac{3}{8}$
400 Missouri Pacific	48	48	$47\frac{5}{8}$	48	..
200 National Biscuit	$118\frac{1}{8}$	$118\frac{1}{8}$	118	118	-1
100 New York Central	112	112	$111\frac{1}{4}$	$111\frac{1}{4}$	$-\frac{1}{4}$
100 Northern Pacific	$118\frac{1}{4}$	$118\frac{1}{4}$	$117\frac{3}{4}$	118	..
800 Pacific Tel & Tel	45	$45\frac{1}{4}$	$44\frac{3}{4}$	$45\frac{1}{4}$	$+\frac{1}{4}$
3,700 Pennsylvania R R	$129\frac{1}{4}$	$129\frac{5}{8}$	129	$129\frac{1}{8}$	$-\frac{1}{8}$
86,100 Reading	$153\frac{3}{4}$	$154\frac{1}{4}$	$153\frac{1}{4}$	154	$+\frac{5}{8}$
4,900 Rock Island	$30\frac{1}{4}$	$30\frac{7}{8}$	$30\frac{1}{8}$	$30\frac{1}{2}$	$+\frac{1}{4}$
3,300 Southern Pacific	$116\frac{3}{4}$	$116\frac{5}{8}$	$116\frac{1}{8}$	$116\frac{1}{4}$	$-\frac{3}{8}$
900 Tenn Copper	$34\frac{7}{8}$	$35\frac{1}{4}$	$34\frac{7}{8}$	$35\frac{1}{4}$	$+1\frac{1}{4}$
38,000 Union Pacific	$173\frac{5}{8}$	$174\frac{1}{4}$	$173\frac{1}{4}$	$173\frac{3}{4}$	$+\frac{3}{8}$
48,100 U S Steel	74	$74\frac{3}{4}$	$73\frac{7}{8}$	$74\frac{1}{8}$	$+\frac{1}{4}$
1,100 do pfd	$117\frac{1}{2}$	118	$117\frac{1}{2}$	118	$+\frac{1}{2}$
1,200 Wisconsin Central	59	$60\frac{3}{8}$	59	60	$+1\frac{1}{4}$

809. The profits of a corporation, after deductions are made for surplus, reserve for bad debts, sinking fund, etc., are divided among the stockholders as follows: To the preferred stockholders (if any) is paid the dividend guaranteed on their stock; to the common stockholders, the balance in proportion to the number of shares they own.

NOTES. (1) In some cases, however, the preferred stockholders share equally with the common stockholders if the profits are sufficient to pay on all stock more than the guaranteed rate on the preferred stock.

(2) **Dividends** are always reckoned as a certain rate per cent of the par value of the issued stock of the company.

810. An **assessment** is a certain per cent of the par value of the stock levied against the stockholders to pay losses, etc.

811. Stocks are bought and sold by **stockbrokers**, who make a business of buying and selling stocks for others. For their services they charge a commission, or brokerage, usually $\frac{1}{8}$ % of the par value of the stocks handled.

812. **Watered stock** is stock based on no real value. Certificates of stock are issued which do not represent capital paid in. The chief purpose of issuing watered stock is to conceal from the public the true rate of profit earned by the company.

813. Stocks are bought and sold under various conditions, some of which are:

1. "Cash," meaning that the stock is to be delivered the day of the sale.

2. "Regular way," which means that the stock is to be delivered the day after the sale.

3. "Buyer 3, 10, 20, 30, 60," in which case the buyer has the right to call for the delivery of the stock any time within the number of days specified. If the buyer's option is for more than 3 da., he is charged interest on the purchase price of the stock unless "flat" (without interest) is specified.

4. "Seller 3, 10, 20, 30, 60," sold deliverable at the seller's option within the number of days specified. Interest is charged the buyer, if the option is for more than 3 da., unless sold "flat."

814.   **EXPLANATION OF STOCK EXCHANGE TERMS**

1. **Bear.** A broker who operates for declining prices. He has sold "short" (see 6), and wishes to cover his short sale at a lower price.

2. **Bull.** A broker who operates for rising prices. He has bought stocks (is "long"—see 5) and wishes to sell them at higher prices.

3. **Collaterals.** Stocks or bonds deposited as security for a loan.

4. **Ex-dividend.** A sale ex-dividend is a sale in which the dividend just declared or about to be declared does not go to the purchaser. Sales are usually made ex-dividend from the day the transfer books of a corporation are closed to the day they are opened.

5. **Long.** A trader who buys stocks for a rise in price is long.

6. **Short.** When a trader has sold more stock than he owns, he is said to be short.

7. **Margin.** A deposit made with a broker by a person who wishes to buy more stock than he can pay for, with the expectation of a rise. The broker furnishes or borrows the balance of the cost of the stocks. The margin protects the broker in case of a decline.

815. To find the cost of a purchase, or the proceeds of a sale, of stock.

1. Find the cost, including brokerage, of 250 shares of Amalgamated Copper at the lowest price quoted on page 340.

$$\$63\tfrac{3}{8} + \$\tfrac{1}{8} = \$63\tfrac{1}{2}, \text{ gross cost of 1 share.}$$
$$250 \times \$63\tfrac{1}{2} = \$15{,}875, \text{ cost.}$$

The quotation is $63\tfrac{3}{8}$ per share. The brokerage is $\tfrac{1}{8}\%$ of the par value of the stock, but in all cases in which the par value of the stock is $100, the brokerage is equal to $\$\tfrac{1}{8}$ a share, and hence increases the cost of a share by $\$\tfrac{1}{8}$. 250 shares cost 250 times $63\tfrac{1}{2}$, or $15,875.

2. Find the proceeds of 400 shares of Canadian Pacific at 199.

$$\$199 - \$\tfrac{1}{8} = \$198\tfrac{7}{8}, \text{ proceeds of 1 share.}$$
$$400 \times \$198\tfrac{7}{8} = \$79{,}550, \text{ proceeds.}$$

The brokerage is deducted from the quotation, giving $198\tfrac{7}{8}$ as the proceeds of 1 share. The proceeds of 400 shares is 400 times $198\tfrac{7}{8}$, or $79,550.

816. Using the quotations on page 340, find the cost, including brokerage, at the lowest price, and the proceeds, including brokerage, at the highest price, on the following:

Find the cost of:

1. 350 Am Locomotive.
2. 125 Am Tel & Tel.
3. 50 Chi & N W.
4. 175 Consol Gas.
5. 225 Del & Hudson.

6. 325 Louis & Nash.
7. 500 Mer Marine pfd.
8. 750 New York Central.
9. 600 Northern Pacific.
10. 800 U S Steel.

Find the proceeds of:

11. 200 Illinois Central.
12. 275 Missouri Pacific.
13. 175 National Biscuit.

14. 450 Rock Island.
15. 375 Southern Pacific.
16. 650 Tenn Copper.

17. 500 Atchison.   19. 1250 Union Pacific.
18. 400 Pacific Tel and Tel.   20. 200 Wisconsin Central.

Find the gain or loss on the following stocks if bought at the lowest and sold at the highest price quoted, allowing for brokerage each way:

21. 150 Am Cotton Oil.   26. 350 Den & R G.
22. 200 Am Sugar.   27. 500 Erie 1st pfd.
23. 300 Ches & Ohio.   28. 450 Pennsylvania R R.
24. 400 C M & St. P.   29. 800 Reading.*
25. 250 Col and South. 1st pfd.   30. 1000 U S Steel pfd.

817. To find the number of shares.

How many shares of Manhattan Elevated at $140\frac{1}{2}$ can be bought for $33,750, brokerage $\frac{1}{8}$ % ?

$$\$140\frac{1}{2} + \$\frac{1}{8} = \$140\frac{5}{8}, \text{ gross cost 1 share.}$$
$$\$33,750 \div \$140\frac{5}{8} = 240, \text{ no. shares.}$$

The quotation, $140½, plus the brokerage, $⅛, gives $140⅝, the gross cost of 1 share. $33,750 will buy as many shares as $140⅝ is contained times in $33,750, or 240 times.

PROBLEMS

818. 1. At 35, how many shares of Texas and Pacific can be bought for $9835, including brokerage?

2. Western Union Telegraph is quoted at $76\frac{1}{4}$. How many shares can be bought for $12,260, usual brokerage?

3. I sold stock through a broker at $112\frac{1}{2}$, and received therefrom $13,485. How many shares were sold?

4. To meet a debt of $18,550, a man sold through his broker Pacific Telephone and Telegraph stock at $46\frac{1}{4}$. How many shares did he have to sell?

5. A broker bought on his own account 1000 shares of Pressed Steel Car stock at $49\frac{3}{4}$. When the price advanced to $51\frac{1}{4}$, he sold enough of the stock to gain $1200. How many shares did he sell?

* Par value $50 a share.

819. To find the dividend on stocks.

1. A man invested $27,720 in American Locomotive pfd. at 115⅜. If the stock paid 6 % dividends, what was his annual income ?

$115⅜ + $⅛ = $115½, gross cost of 1 share.

$27,720 ÷ $115½ = 240 no. shares.

Value of 240 shares = $24,000, par value of the stock.

6 % of $24,000 = $1440, annual income.

Since the dividends are always based on the par value of the stock, it is necessary, first, to find the number of shares of stock (see art. 817), which is 240. The par value of the stock is, then, $24,000 (240 × $100). 6 % of the par value gives the annual income, $1440.

2. The capital stock of a corporation is $5,000,000. The gross earnings are $1,250,000; the operating expenses are $900,000. $50,000 is carried to the surplus fund. What per cent dividend can be declared? If A owns 250 shares, how much will he receive?

$900,000 + $50,000 = $950,000, to deduct from gross earnings.

$1,250,000 — $950,000 = $300,000, total dividend.

$300,000 ÷ $5,000,000 = .06 = 6 %, rate of dividend.

Value of 250 shares = $25,000, par value of A's stock.

6 % of $25,000 = $1500, A's dividend.

From the gross earnings there are deducted the operating expenses, and the amount set aside for surplus fund, which leaves $300,000 as the total dividend. Dividing by the capital stock, $5,000,000, gives 6%, the rate of dividend. The par value of A's stock is $25,000. 6% of $25,000 gives $1500, A's dividend.

PROBLEMS

820. 1. An investment of $39,183.75 was made, through a broker, in Louisville and Nashville at 145. The stock pays regular dividends of 8 %. Find the dividend.

2. A gentleman drew $44,550 from a bank paying 3½% interest annually, and invested it in U. S. Steel at 74⅛, brokerage ⅛ %, paying 4 % dividends. Find the increase or decrease in his income.

3. A man sells 25 shares of Union Pacific stock, paying 8 %, at 199½, brokerage ⅛ %, and deposits the proceeds in savings banks paying 4 % interest. Does he gain or lose in his annual income ?

4. The Delaware & Hudson Railroad Company has a capital stock of $42,400,000. The gross income for 1 yr. was $44,829,555.20; the total expenses were $38,363,381.79. Find the largest whole per cent dividend that could be declared, and the balance left over for the reserve fund. What amount of dividend would be received by a man who owned 1500 shares of the stock?

821. To find the investment required to produce a given income.

How much must be invested in Amalgamated Copper at the last price (page 340) to yield an annual income of $1500 if the stock pays 5 %?

5% of $100 = \$5$, income from one share.

$\$1500 \div \$5 = 300$, no. shares.

$300 \times \$65 = \$19,500$, investment.

5% stock pays $5 a share in dividends (when the par value is $100). To produce $1500, requires 300 shares. One share costs $64⅞ + $⅛, or $65. 300 shares cost 300 times $65, or $19,500.

822. 1. What investment in American Smelting & Refining pfd. at 111⅝ will yield $1680 annual income, if the stock pays 6 %?

2. If Chicago, Milwaukee & St. Paul pfd. pays 8 % dividends, what investment at 173½ will yield $1800 annually?

3. When Illinois Central stock paying 6 % sells at 147, what investment will yield $2250 per annum?

4. If Rock Island Company pfd. costs 79⅞, when it pays 5 % what investment will produce a semiannual return of $750?

823. To find what per cent an investment pays.

If Denver & Rio Grande pfd. pays 5 %, and sells at 84, what rate per cent does the stock pay on the investment?

5% of $100 = \$5$, income from 1 share.

$\$84 + \$⅛ = \$84⅛$, cost of 1 share.

$\$5 \div \$84⅛ = .0594 = 5.94\%$, rate on the investment.

Since the income from one share is $5, and the cost of a share is $84⅛, the rate of the investment is found by dividing the income by the cost, which gives 5.94%.

NOTE. The above example stated in its simplest type form would be, "What % of $84⅛ = $5?" Hence, $5 ÷ $84⅛ equals the rate per cent required (Prin. 14, page 102)

824. 1. If 6 % stock is bought at 120, no brokerage, what per cent on the investment does the stock pay?

2. When a stock costs 210, and pays quarterly dividends of 3 %, what rate does it pay on the investment (no brokerage)?

3. Which is better, and how much per cent, to buy stock paying semiannual dividends of 4 % at 159⅞, or to buy a 4½ % stock at 87⅞, allowing for brokerage in each case?

4. An investor is offered 5 % stock at 89⅞, and 3½ % semiannual dividend stock at 124⅞. Which is the better investment, and how much per cent, usual brokerage being allowed?

5. A broker owns 600 shares of Manhattan Elevated stock, which pays annual dividends of 7 %. He sells it at 140½, and buys with the proceeds Atchison, Topeka & Santa Fe pfd., paying 6 %, at 105⅜. Is his income increased or diminished, and what per cent?

825. To find the investment to yield a given rate per cent.

At what price will a 4 % stock pay 5 % on the investment?

$$4 \% \text{ of } \$100 = \$4, \text{ income from 1 share.}$$
$$\$4 \div .05 = \$80, \text{ cost of 1 share.}$$

Since this is to be a 5 % investment, the dividend, $4, must be equal to 5 % of the cost of the share. Hence, $4 ÷ .05 gives $80, the cost of 1 share.

Note. In its simplest type form, this problem is: 5 % of what sum equals $4? (Prin. 14, page 102.)

826. 1. A man desires a 5% investment. What price can he afford to pay for 6 % stock?

2. A 7 % stock pays only 4 % on the investment. What is the market value (usual brokerage)?

3. What is the market value of stock paying dividends of 4½ %, if an investment in them, including brokerage, pays 5 %?

4. Find the cost, including brokerage, of 350 shares of stock paying 6 % dividends, if the rate on the investment is 4⅘ %.

5. A broker has an option on 500 shares of 5 % stock at a price to yield 8 % on the investment, or a sufficient quantity of stock paying 2 % semiannually at 50, to require the same investment. He accepts the latter stock. Is his income more or less, and how much, than it would have been on the other stock, if money is worth 6 % per annum?

BONDS

827. A **bond** is a secured obligation, promising to pay, for value received, a sum of money at a certain time, with interest, payable at stated intervals and at a fixed rate.

828. Bonds are known by the name of the government, municipality, or corporation issuing them, by the rate of interest they bear, the date of maturity, the purpose for which they were issued, and by the type of security back of them. For example, U. S. 3's, 1908–1918, means United States bonds paying 3 % interest, due in 1918, payment of the bond being optional on any interest date after 1908. City of New York $4\frac{1}{2}$'s, 1957, means bonds of the City of New York bearing $4\frac{1}{2}$ % interest, due in 1957. Chicago, Milwaukee & St. Paul General Mortgage 4's, 1989, means bonds bearing 4 % interest, issued by the Chicago, Milwaukee & St. Paul Railway, secured by a general mortgage on the property of the company, and due in 1989.

Bonds have many names, some of which are First Mortgage 5's, General Mortgage 4's, Prior Lien $4\frac{1}{2}$'s, Refunding and Extensions 5's, etc. If issued by a private corporation, they are generally secured by mortgage on the property of the company.

English Consols (consolidated stock) are government bonds of England bearing $2\frac{1}{2}$ % interest payable quarterly, Jan. 5, April 5, July 5, and Oct. 5. These bonds have no definite date of ma-

829.   ILLUSTRATION OF A BOND

turity, but cannot be paid before April 5, 1923.   They are one of the so-called perpetual loans.

French Rentes are government bonds of France, and bear interest at 3 % payable quarterly.   These loans are also perpetual loans.

These bonds are issued in small denominations in order that persons with small means may purchase them. ·

830. There are two general classes of bonds — coupon and registered.

831. Coupon bonds have interest coupons attached.   The coupons are made payable to bearer, and are dated at regular intervals (usually semiannually) and are, in effect, negotiable notes promising to pay the interest on the bond as it becomes due.

832. Registered bonds take their name from the fact that the name of the owner is registered with an agent of the maker of the bond. On "fully registered" ("registered as to principal and interest") bonds, interest is paid by check or draft mailed to the registered holders.   Bonds "registered as to principal only" carry interest coupons like an ordinary coupon bond.

833.   ILLUSTRATION OF A COUPON

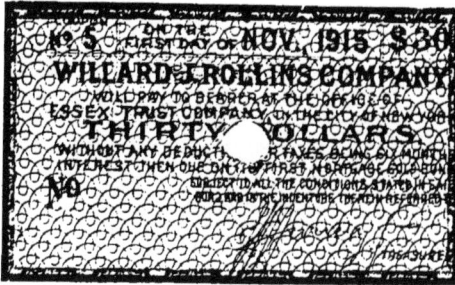

834. The terms, *par value, market value, above par, premium, below par,* and *discount,* have the same signification as they have when applied to stocks.

835.   STOCKS AND BONDS COMPARED

STOCKS	BONDS
1. Owner is part owner of corporation.	1. Owner is a creditor of the corporation.
2. Owner generally has a vote in the election of directors.	2. Owner generally has no vote

Stocks	Bonds
3. Dividends not due until declared by the directors.	3. Dividends (interest) are payable regularly.
4. Dividends depend on earning power of the corporation.	4. Interest is fixed.
5. Purchased, many times, for speculation.	5. Purchased, generally, for investment.

836. Bond quotations are not subject to sudden fluctuations, as are stock quotations. The price of bonds depend chiefly upon four things: 1st, the security back of the bonds; 2d, the time the bond has to run; 3d, interest rates of money; and 4th, the earning power of the issuing corporation.

Notes. Bonds are redeemable at maturity at face value unless otherwise specified.

If interest payments on bonds are not made when due the company is said to default in the interest. To default in the bonds is to fail to redeem them at maturity.

837. Following is a partial list of bond sales on the New York Stock Exchange, Friday, Jan. 6, 1911. Note the variations in price of bonds paying the same rate of interest.

Sales of Bonds

Adams Ex 4s	C M & St P 4s	N. Ry Mex $4\frac{1}{2}$s	So Pac ref 4s
500...... $90\frac{1}{4}$	5,000...... $99\frac{1}{2}$	50,000...... $84\frac{5}{8}$	6,000...... $94\frac{3}{4}$
Allis Chal 5s	8,000...... $\frac{3}{8}$	City of N Y 4s	17,000...... $\frac{5}{8}$
1,000...... $76\frac{1}{2}$	1,000...... $\frac{1}{2}$	1958 15,000. $99\frac{3}{4}$	5,000...... $\frac{1}{2}$
Am Tob 4s	Del & Hud fd	N Y H H & H cv	U S Rubber 6s
14,000...... $80\frac{3}{8}$	12,000......$100\frac{3}{8}$	$3\frac{1}{2}$s 12,500...$100\frac{1}{8}$	5,000......$103\frac{3}{8}$
13,000...... $\frac{1}{2}$	4,000......$101\frac{1}{2}$	Nor Pac 3s	1,000...... $\frac{1}{4}$
2,000...... $\frac{3}{4}$	Ga & Ala ref 5s	3,000...... $70\frac{3}{4}$	9,000......103
Am Tob 6s	2,000......$104\frac{1}{2}$	1,000...... $\frac{5}{8}$	U S Stl s f 5s
9,000......$105\frac{3}{4}$	5,000......114	Nor Pac pr 4s	6,000 .....$103\frac{7}{8}$
2,000...... $\frac{5}{8}$	Mo Pacific 4s	500 .....$100\frac{1}{4}$	3,000...... $\frac{3}{4}$
17,000......106	21,000...... $77\frac{1}{2}$	Rio G & W 4s	3,000...... $\frac{7}{8}$
Cen Pac 1st 4s	Mo Pac con 6s	1,000...... 91	West Un $4\frac{1}{2}$s
5,000...... $96\frac{3}{4}$	1,000..... $109\frac{1}{2}$		10,000......$95\frac{3}{4}$

838. To find the cost of bonds.

1. What is the cost of \$25,000 City of New York $4\frac{1}{2}$'s, 1957, at $109\frac{1}{2}$ (no brokerage)? $1.09\frac{1}{2} \times \$25,000 = \$27,375$.

The quantity of bonds is expressed in dollars par value. The quotation is to be understood as "per cent of the par value." Hence, the cost is $109\frac{1}{2}\%$ of \$25,000, or \$27,375.

2. Find the cost of $10,000 Baltimore & Ohio 4's, on Feb. 14, 1910, at 102 and interest.

102% of $10,000 = \$10,200$, cost, flat.

Jan. 1 to Feb. 14 = 1 mo. 13 da.

Int. on $10,000 for 1 mo. 13 da. at $4\% = \$47.78$.

$\$10,200 + \$47.78 = \$10,247.78$, cost.

At 102 flat, the bonds would cost $10,200. Interest days are Jan. 1 and July 1; hence, interest at 4% has been accumulating on these bonds from Jan. 1 to Feb. 14, a period of 1 mo. 13 da., and at the rate of interest the bonds pay, amounts to $47.78, which, added to $10,200, gives the cost as $10,247.78.

839. Using the prices quoted on page 350, find the cost, including interest, of the following bonds:

AMOUNT AND NAME	INT. DATES	DATE OF PURCHASE
1. $6,000 Adams Express 4's	Mar.–Sept.*	Oct. 18
2. 8,000 American Tobacco 4's	Feb.–Aug.	Sept. 20
3. 4,000 Western Union 4½'s	May–Nov.	Nov. 15
4. 12,000 Missouri Pacific con. 6's	May–Nov.	Jan. 4
5. 25,000 Missouri Pacific 4's	Jan.–July	Apr. 3
6. 15,000 U. S. Rubber 6's	June–Dec.	Aug. 25
7. 20,000 Chicago, Milwaukee & St. Paul General 4's	Jan.–July	June 10
8. 30,000 N. Y., N. H. & H. con. deb. 3½'s	Jan.–July	June 30
9. 40,000 American Tobacco 6's	Apr.–Oct.	July 18
10. 35,000 So. Pac. Railroad first refunding 4's	Jan.–July	Aug. 11

NOTE. Bankers use Interest Tables for finding accrued interest on bonds. The tables are too complicated for insertion here.

840. To find the cost of a bond to yield a given rate.

At what price will a 5 % bond maturing in 3 yr. yield 4 % ?

$\$1025.00 \div 1.02 = \$1004.90$, cost if maturity is in 6 mo.

$\$1004.90 + \$25 = \$1029.90$, cost plus 6 months' int. on the cost.

$\$1029.90 \div 1.02 = \$1009.70$, cost if maturity is in 1 yr.

$\$1009.70 + \$25 = \$1034.70$, cost plus 6 months' int. on the cost.

$\$1034.70 \div 1.02 = \$1014.41$, cost if maturity is in 1½ yr.

* Interest is due on the first day of the months named unless otherwise specified.

$1014.41 + $25 = $1039.41, cost plus 6 months' int. on the cost

$1039.41 ÷ 1.02 = $1019.03, cost if maturity is in 2 yr. .

$1019.03 + $25 = $1044.03, cost plus 6 months' int. on the cost.

$1044.03 ÷ 1.02 = $1023.55, cost if maturity is in 2½ yr.

$1023.55 + $25 = $1048.55, cost plus 6 months' int. on the cost.

$1048.55 ÷ 1.02 = $1027.99, cost if maturity is in 3 yr.

Hence, the quotation is 102.80.

If the bond matured in 6 mo., the cost would be a sum which in 6 mo. would amount to $1025 (the face of the bond plus 6 months' interest) at 5% interest; that is, the cost is the present worth of $1025 due in 6 mo. at 4%, which is $1004.90. If the bond were to mature in 1 yr. instead of 6 mo., another semiannual payment of interest would be received. Adding the semiannual interest, $25, to $1004.90, gives $1029.90, which is the sum of the cost of the bond maturing in 1 yr. plus the interest for 6 mo. at 4% on that cost. The cost, then, is the present worth of $1029.90, due in 6 mo. at 4%, or $1009.70.

Continuing the process as shown in the solution, the price to yield 4% in 3 yr. is shown to be $102.80.

NOTE. The above solution is based on the theory that each semiannual interest payment of $25 is made up of two separate items, viz., interest on the investment at 4% per annum, and a portion of the premium paid for the bond. For illustration, take the cost of the bond when maturing in 3 yr. The interest for 6 mo. at 4% on $1027.99 equals $20.56. The interest on the bond is $25. The difference between $25 and $20.56 equals $4.44. The $4.44 is considered as a repayment on account of the premium paid for the bond. Deducting $4.44 from $1027.99 leaves $1023.55, which is the cost of the bond if maturing in 2½ yr.

. BOND TABLES

341. Bankers use bond tables to determine the rate of interest an investment in bonds will pay, or the price to pay to secure a given rate on the investment. The following is a reproduction of one page from the "Bond Tables" published by N. W. Harris & Co., Bankers, New York and Boston, and is printed with their permission:

The tables cover periods of time from ½ yr. to 100 yr., and show the net annual rates of interest which bonds paying 7%, 6%, 5%, 4½%, 4%, 3½%, and 3%, respectively, will pay for 64 different quotations on bonds bearing the above rates of interest.

How to use Bond Tables

842. **1.** If a 4% bond maturing in 20 yr. costs 80.64, how much will it net the purchaser?

Find the 4% column in the table and follow down the column until 80.64 is reached. Then follow the horizontal line in which 80.64 is found to the extreme left-hand column, and there find $5\frac{5}{8}$ in the column marked "net per annum." The bond will net $5\frac{5}{8}$% on the cost.

2. If a man wishes a $4\frac{1}{2}$% investment, what price can he afford to pay for a 6% bond maturing in 20 yr.?

In the left-hand column find $4\frac{1}{2}$, and follow the horizontal line in which $4\frac{1}{2}$ is, to the right until the column having 6% at its top is reached. The number found in both the horizontal line and in the vertical column, viz., 119.65 is the price he can pay.

20 YEARS—Interest Payable Semi-annually.

Net per Annum.	BONDS BEARING INTEREST AT THE RATE OF						
	7%	6%	5%	$4\frac{1}{2}$%	4%	$3\frac{1}{2}$%	3%
4	141.03	127.36	113.68	106.84	100.00	93.16	86.32
4.10	139.32	125.76	112.20	105.42	98.64	91.86	85.09
$4\frac{1}{8}$	138.90	125.37	111.84	105.07	98.31	91.54	84.78
4.20	137.63	124.19	110.75	104.03	97.31	90.59	83.87
$4\frac{1}{4}$	136.80	123.42	110.04	103.35	96.65	89.96	83.27
4.30	135.98	122.65	109.33	102.66	96.00	89.34	82.68
$4\frac{3}{8}$	134.75	121.51	108.27	101.65	95.04	88.42	81.80
4.40	134.35	121.14	107.93	101.32	94.72	88.11	81.51
$4\frac{1}{2}$	132.74	119.65	106.55	100.00	93.45	86.90	80.35
4.60	131.16	118.18	105.19	98.70	92.21	85.72	79.22
$4\frac{5}{8}$	130.77	117.82	104.86	98.38	91.90	85.42	78.94
4.70	129.61	116.74	103.86	97.43	90.99	84.55	78.11
$4\frac{3}{4}$	128.84	116.02	103.20	96.80	90.39	83.98	77.57
4.80	128.08	115.32	102.55	96.17	89.79	83.40	77.02
$4\frac{7}{8}$	126.95	114.27	101.59	95.24	88.90	82.56	76.22
4.90	126.58	113.92	101.27	94.94	88.61	82.28	75.95
5	125.10	112.55	100.00	93.72	87.45	81.17	74.90
5.10	123.65	111.20	98.76	92.53	86.31	80.09	73.86
$5\frac{1}{8}$	123.29	110.87	98.45	92.24	86.03	79.82	73.61
5.20	122.22	109.87	97.53	91.36	85.19	79.02	72.85
$5\frac{1}{4}$	121.51	109.22	96.93	90.78	84.64	78.49	72.34
5.30	120.81	108.57	96.33	90.21	84.09	77.97	71.85
$5\frac{3}{8}$	119.77	107.60	95.44	89.36	83.27	77.19	71.11
5.40	119.42	107.28	95.14	89.07	83.01	76.94	70.87
$5\frac{1}{2}$	118.06	106.02	93.98	87.96	81.94	75.92	69.90
$5\frac{5}{8}$	116.38	104.47	92.55	86.59	80.64	74.68	68.72
$5\frac{3}{4}$	114.74	102.95	91.15	85.26	79.36	73.46	67.57
$5\frac{7}{8}$	113.13	101.46	89.78	83.95	78.11	72.27	66.43
6	111.56	100.00	88.44	82.66	76.89	71.11	65.33
$6\frac{1}{4}$	108.50	97.17	85.84	80.18	74.51	68.85	63.19
$6\frac{1}{2}$	105.55	94.45	83.34	77.79	72.24	66.69	61.14
$6\frac{3}{4}$	102.72	91.83	80.95	75.50	70.06	64.62	59.17
7	100.00	89.32	78.04	73.31	67.97	62.63	57.29

PROBLEMS

843. Using the table on this page, find the price at which bonds maturing in 20 yr. can be bought to produce as follows:

1. 5% bond to yield 4%.

2. $4\frac{1}{2}$% bond to yield 5%.

3. 6% bond to yield 4.6%.

4. $3\frac{1}{2}$% bond to yield $4\frac{1}{2}$%.

5. 5% bond to yield $5\frac{1}{2}$%.

6. 7% bond to yield $5\frac{3}{8}$%.

Find the rate on the investment on bonds maturing in 20 yr. and bought as follows:

7. $4\frac{1}{2}$% bonds bought at 96.80.

8. 4% bonds bought at 88.90.

9. 6% bonds bought at 117.82.

10. 5% bonds bought at 101.27.

11. $3\frac{1}{2}$% bonds bought at 71.11.

12. 7% bonds bought at 126.95.

EXCHANGE

844. The system by which merchants in different places discharge their debts to each other without the transmission of money is called exchange.

845. Exchange is of two kinds, viz.: Inland or Domestic, and Foreign.

DOMESTIC EXCHANGE

846. Domestic exchange treats of the payment of debts between merchants in the same country, without the sending of money.

For instance, if a merchant of New York owes a manufacturer of St. Louis for a bill of merchandise, he can pay his debt without sending money, in any one of the following ways:

1. By his personal check.
2. By a bank draft.
3. By a commercial draft on one of his debtors.
4. By a postal money order.
5. By an express money order.
6. By a telegraphic money order.

NOTE. For the forms of check, bank draft, commercial draft, and note, see the chapter on Negotiable Paper, page 291.

If he sends his check, the manufacturer in St. Louis will deposit it in his bank. In due time the check will be returned to the bank on which it is drawn, whereupon the amount will be charged against the merchant's account.

If the merchant sends a bank draft, he must buy and pay cash in advance for it. The draft is then sent in the same manner as a check.

NOTE. Bankers in the City of New York seldom draw drafts on other cities. They sell cashier's or manager's checks, drawn on their own bank.

If the merchant draws and sends a draft on one of his debtors, the manufacturer must present the draft to the drawee for payment, or he must have his bank collect it for him, and place the proceeds to his credit. Or, if it is a time draft, the bank will buy (discount) it, after acceptance by the drawee, and credit him at once with the proceeds. The bank will then collect the amount of the draft from the drawee at maturity, to reimburse itself for the credit given the indorser.

847. New York, Chicago, and San Francisco are the principal exchange centers in the United States.

848. An exchange center is a recognized money center. In it are located the correspondent banks of the banks in the surrounding cities and towns.

849. A postal money order is a written order given by the postmaster in one place, to the postmaster in another place, to pay a specified sum of money to a certain person or to his order.

850. To obtain a postal money order, one has to make a written application stating

1. The amount of the order wanted.
2. The name and address of the payee.
3. The name and address of the purchaser of the money order.

851. The largest amount for which a single postal money order can be obtained is $100. The rates charged are as follows:

$2.50 or less . . . 3¢	$30.01 to $40.00 . . . 15¢
2.51 to $5.00 . . . 5¢	40.01 to 50.00 . . 18¢
5.01 to 10.00 . . . 8¢	50.01 to 60.00 . . . 20¢
10.01 to 20.00 . . . 10¢	60.01 to 75.00 . . . 25¢
20.01 to 30.00 . . . 12¢	75.01 to 100.00 . . . 30¢

852. More than one indorsement on a postal money order is prohibited by law, except indorsements by banks.

853. Postal money orders may be presented for payment at the post office on which they are drawn, or at any bank. If not presented for payment within one year from their date, or if lost, the Post Office Department at Washington will upon request and presentation of evidence of loss, issue a duplicate.

854. An express money order is a written order of an agent of an express company directing another agent of the company to pay a certain sum of money to a specified person or to his order.

855. Express money orders are similar to postal money orders. The rates charged for them are the same. Express money orders may be indorsed and transferred any number of times, in the same manner as checks and drafts.

FORM OF EXPRESS MONEY ORDER

856. A telegraph money order is an order by telegraph sent by an agent of a telegraph company at one place to a telegraph agent at another place, directing him to pay a specified sum of money to the person named in the telegram, with or without identification, as directed by the sender.

857. The cost of sending a telegraphic money order, in addition to the regular charge for a fifteen-word message between the two places, is as follows:

For $ 25 or less . . . . 25 ¢
Over $ 25 and not over $ 50 . . . . 35 ¢
Over $ 50 and not over $ 75 . . . . 60 ¢
Over $ 75 and not over $ 100 . . . . 85 ¢

After the first $ 100, up to and including $ 3000, add 20 cents for each $ 100 or part thereof. (In addition there is a war tax of 5 cents regardless of amount. This is to be disregarded in these problems.)

NOTE. A telegraphic order between the depositing and paying agents is a message in cipher, the interpretation of which is known only to the company and its agents.

858. Domestic rates of exchange fluctuate slightly, depending on the conditions of supply and demand. If in New York there is a large demand for drafts on Chicago without an equivalent demand in Chicago for drafts on New York, the supply (that is, the funds of

New York banks on deposit in banks in Chicago) will become exhausted. To check or remedy this condition of affairs, the price of exchange on Chicago rises. There will be a corresponding drop in Chicago in the price of exchange on New York.

859. The following rates were published in the New York *Wall Street Journal*, Jan. 7, 1911. All quotations are in cents per $1000 premium or discount.

Boston — New York Exchange at par.
Chicago — New York Exchange 10 cents premium.
Montreal — New York Exchange 31¼ cents premium.
San Francisco — New York Exchange 80 cents premium.
St. Louis — New York Exchange 15 cents premium.
St. Paul — New York Exchange 70 cents premium.

860. To find the cost of a draft or money order.

1. At ½% premium, find the cost of a draft on Chicago for $4000.

½% of $4000 = $20, premium. Since exchange is selling at a premium of ½%, the premium amounts
$4000 + $20 = $4020, cost. to ½% of $4000, or $20. The cost is the sum of $4000 and $20, or $4020.

2. Find the cost of a draft of $3500, when exchange is selling at 50¢ discount.

50¢ discount means 50¢ discount on $1000.

3.500 × $.50 = $1.75, discount on $3500.

$3500 − $1.75 = $3498.25, cost.

Since the discount on $1000 is 50¢, on $3500 it is 3½ times $.50, or $1.75. $3500 less $1.75 gives $3498.25, the cost.

PROBLEMS

861. Using the rates quoted above, find the cost of drafts on New York as follows:

1. At Chicago for $1500.

2. At Montreal for $3875.

3. At San Francisco for $4580.

4. At St. Paul for $547.50.

5. At Montreal for $12,875.

6. At St. Louis for $1350.

7. At Chicago for $15,728.70.

8. Find the cost of postal money orders for $17.20; $21.60; $32.85; and $63.10.

9. A man desires to make a payment of $78.70 by means of express money orders. Find the cost.

10. A traveling salesman telegraphs for expense money. His principal sends a telegraphic money order for $150. The regular charge on the telegram being $.40, find the cost of the order.

11. Compare the costs of sending $175 by postal money order, express money order, and by telegraphic money order, if a fifteen-word telegram costs $1.15.

12. When exchange is quoted at $1.005, which is the cheaper way to make a payment of $500, by draft, or by a telegraphic money order, if the regular rate of a fifteen-word telegram is 35 ¢?

13. A merchant pays the following bills: $39.75 by express money order; $78.50 by postal money order; $382.50 by telegraphic money order (telegram 50 ¢); and $1837.50 by draft purchased at $.9975. Find the total cost of making settlement.

14. A man wishes to remit $1890 in settlement of a debt. Which is better, and how much, to send his check or to send a bank draft purchased at $.9985?

862. To find the proceeds of a draft.

1. Find the proceeds of a sight draft of $2840, collection and exchange being $\frac{1}{8}$ %.

$\frac{1}{8}$ % of $2840 = $3.55, collection and exchange.

$2840 − $3.55 = $2836.45, proceeds.

Since the bank charges $\frac{1}{8}$% for collection, the charge amounts to $\frac{1}{8}$% of $2840, or $3.55. The proceeds is the difference between $2840 and $3.55, or $2836.45.

2. Find the proceeds of a 60-da. commercial draft of $1500, if sold on its date at $\frac{1}{2}$ % discount, money being worth 6 %.

$15.00 = the interest for 60 da.

$\frac{1}{2}$ % of $1500 = $7.50, discount.

$15 + $7.50 = $22.50, total discount.

$1500 − $22.50 = $1477.50, proceeds.

This solution is identical with that of bank discount. The interest (bank discount) is found for the unexpired time of the draft (term of discount). The $\frac{1}{2}$% discount is equivalent to a collection charge, and is reckoned on the face value of the draft. The total discount is the sum of the two discounts, $15 and $7.50, or $22.50. The proceeds is the face of the draft less the total discount.

863. Find the proceeds of the following sight drafts:

	FACE	COLLECTION AND EXCHANGE		FACE	COLLECTION AND EXCHANGE
1.	$ 784.00	$\frac{1}{2}\%$	11.	$3981.43	$\frac{1}{4}\%$
2.	1260.00	$\frac{1}{4}\%$	12.	4289.64	$\frac{1}{8}\%$
3.	2500.00	$\frac{1}{8}\%$	13.	1562.50	$\frac{1}{4}\%$
4.	3675.00	$\frac{1}{10}\%$	14.	5283.25	$\frac{1}{10}\%$
5.	3875.00	$\frac{1}{2}\%$	15.	6374.90	$\frac{1}{20}\%$
6.	4272.50	$\frac{1}{4}\%$	16.	9743.85	$\frac{1}{20}\%$
7.	1761.40	$\frac{1}{8}\%$	17.	8498.75	$\frac{1}{10}\%$
8.	2364.82	$\frac{1}{4}\%$	18.	4964.80	$\frac{1}{8}\%$
9.	4152.87	$\frac{1}{20}\%$	19.	3829.34	$\frac{1}{4}\%$
10.	7387.95	$\frac{1}{8}\%$	20.	4382.86	$\frac{1}{2}\%$

21. Find the proceeds of a 90-da. draft of $968.40 sold at $\frac{1}{4}\%$ discount, money being worth 6 %.

22. A jobber purchased a bill of goods invoiced at $6483.28, less 5 % for cash. To make immediate payment, he discounted a draft of $1864.50 due in 48 da. at $\frac{1}{2}\%$ discount, and 6 % interest; a note of $2487.25 due in 33 da., at $\frac{1}{4}\%$ discount and 6 % interest; and sent a bank draft bought at $1.0025 for the balance. How much was saved by discounting the invoice and paying cash?

FOREIGN EXCHANGE

864. Foreign exchange treats, broadly speaking, of the settlement of international indebtedness by way of bills of exchange (checks, etc.), telegraphic transfers, or the actual shipment of gold (or silver). It also treats of instruments of credit in the form of letters of credit, travelers' checks, etc.

865. A bill of exchange is a draft for a stipulated amount drawn in one country on another country, payable, according to its tenor either on presentation (as checks, demand and sight bills) or after a stated period of days or months. It may be a documentary bill of exchange; i.e. have documents attached (bills of lading, insurance, inspection, weighing, certificates, invoice, etc.), in evidence of the underlying transaction, or it may be a clean bill of exchange; i.e. have no documents attached. In the case of a documentary draft, the documents, according to a memorandum attached, may be delivered to the drawee either on actual payment of the draft or on his mere promise to pay the draft at maturity (called acceptance), which takes the form of a written notice across the face of the draft, "Accepted, payable on (date)," with the signature of the acceptor. Before the consignee of the underlying goods can get possession of his bills of lading and other documents, he must, therefore, either pay or accept—as the case may be—the bill of exchange drawn on him.

866. Bills of exchange are generally drawn in sets of two, called either "First and Second of Exchange," or in the case of checks and demand drafts, "Original and Duplicate." These are sent by different mails, but the payment of one of them cancels the other.

867.   A SET OF EXCHANGE

NOTE. Checks on Great Britain having "and Co." written crosswise on their face, are payable only through the mediation of a bank, necessitating identification of the payee, and preventing misuse.

868. Foreign bills of exchange are generally negotiable instruments covering a multitude of international transactions, the most important of which are:

1. The import or export of merchandise from one country into another.

2. The purchase or sale of securities — public or private — of one country, from or to the other.

3. The investment or loaning of funds in countries (markets) giving a higher interest return, and the reflux of such funds upon a change in monetary conditions.

4. The providing of funds or settling of expenses of domestic travelers in foreign countries, and the remittances of aliens to their mother countries.

869. Gold is the universal standard of value in international financial transactions, because in each country a fixed value has been given to a given weight of gold of a given fineness.

870. The **par of exchange** is the value of the pure metal contained in the monetary unit of one country expressed in terms of the monetary unit of another country. To illustrate: One pound sterling (£1) contains 113.0016 gr. of fine gold. $1 contains 23.22 gr. of fine gold. $113.0016 \div 23.22 = 4.8665$. Hence, £1 = $4.8665, called the *par of exchange* between the United States and England.

871. The rates of exchange fluctuate by certain fixed amounts or fractions. Thus, sterling rates vary by $.0005 **and** multiples of $.0005.

872. The mark rates vary by sixteenths, eighths, and quarters of one cent. In large transactions the mark rates are modified by adding or subtracting fractions of 1 %. Thus, $94\frac{3}{4} + \frac{1}{32}$, or $94\frac{3}{4} - \frac{1}{32}$ equals $.94\frac{3}{4} + (\frac{1}{32}\% \text{ of } .94\frac{3}{4})$, or $.94\frac{3}{4} - (\frac{1}{32}\% \text{ of } .94\frac{3}{4}) = .94\frac{3}{4} + .0002960937 = .9477960937$, or $.94\frac{3}{4} - .0002960937 = .9472039063$.

873. The franc rates generally vary by $\frac{5}{8}$ of a centime, because $\frac{5}{8}$ of a centime is approximately equal to $\frac{1}{8}$ of one cent. Thus, the next higher and lower rates than $5.18\frac{1}{8}$ would be $5.17\frac{1}{2}$ and $5.18\frac{3}{4}$, respectively. These rates, on large transactions, are also modified by adding or subtracting $\frac{1}{16}$, $\frac{1}{32}$, $\frac{1}{64}$, or $\frac{3}{64}$, etc., of one per cent.

874. Since $5.17\frac{1}{2}$ is a higher rate than $5.18\frac{1}{8}$, a minus fraction must be added to the rate, and *vice versa*. For example, $5.17\frac{1}{2} - \frac{1}{32}$

$= 5.17\frac{1}{2} + (\frac{1}{32}\% \text{ of } 5.17\frac{1}{2}) = 5.17\frac{1}{2} + .001617 = 5.176617;$ and $5.17\frac{1}{2}$
$+ \frac{1}{32} = 5.17\frac{1}{2} - (\frac{1}{32}\% \text{ of } 5.17\frac{1}{2}) = 5.17\frac{1}{2} - .001617 = 5.173383.$

875. The most important exchange rates are quoted in the follow-ing manner:

Sterling	£1	= $4.8665
French	$1	= F. 5.18⅛
German	M. 4	= $.95
Holland	1 guilder	= $.40

876. A **bill of exchange**, or revenue **stamp** (frequently called a "bill stamp"), is required on foreign drafts as follows:

1. Great Britain. All drafts above 3 days' sight are subject to $\frac{1}{20}\%$ bill stamp.

2. France. All drafts, not checks, are subject to $\frac{1}{20}\%$ bill stamp. In order to exempt checks from bill stamp, it is necessary to spell the date with letters; for instance, "December eighth, 1910."

3. Germany. All drafts, not checks, that are not drawn on a German bank or banker, are subject to $\frac{1}{20}\%$ bill stamp; if drawn above 90 days' sight, to $\frac{1}{10}\%$ bill stamp. In order to make checks exempt from bill stamp, the following clause is necessary: "Drawn against balance due me."

877. The principal exchange centers of Europe are **London**, **Paris**, and **Berlin**.

878. A **cable transfer** is a cabled order to pay a specified sum of money to a certain firm or person in a foreign country. In cable transfers between private parties, the cost is the cable rate of ex-change plus $\frac{1}{4}\%$ commission, plus the telegraph and cable charges. In cable transfers in trades between bankers, if above a minimum amount of about $50,000, there is generally no commission or cable charge.

879. A **letter of credit** is a circular letter issued by a bank or banker, introducing the beneficiary and authorizing him to draw money on demand from a large number of institutions all over the world up to the inscribed face value of the letter of credit.

880. A foreign letter of credit is usually issued in pounds sterling, this being the currency most readily negotiated the world over.

881. When wishing to draw funds, the traveler presents his letter of credit to the most convenient bank or banker, states the amount he desires to draw, and is required to sign a demand draft in pounds sterling on the London banker addressed in his credit. After careful comparison of his signature with the one appearing in his letter of credit, the foreign banker pays over to him the exact amount wanted or the equivalent thereof and hands the credit back to the traveler. All amounts thus drawn are written off by the paying banker on the reverse side of the letter of credit.

882. LETTERS OF CREDIT

FRONT.   REVERSE.

883. A commission of **1%** is usually charged by the issuing banker in addition to the face value of the letter of credit issued.

884. TRAVELERS' CHECK

885. A travelers' check is a circular check which is made payable for a stipulated amount in the currency of the foreign countries enumerated on the face of the check. Such checks are issued in denominations of $10, $20, $50, $100, and $200, and the equivalents in foreign currency are plainly printed on them.

886. A commission of $\frac{1}{2}\%$ is usually charged by the issuing banker or company, with a minimum charge of 30 ⊄.

887.   VALUES OF FOREIGN COINS

COUNTRY	LEGAL STANDARD	MONETARY UNIT	VALUE IN TERMS OF U. S. MONEY
Argentine Republic . .	Gold	Peso	$ 0.9650
Austria-Hungary . . .	Gold	Crown	.2030
Belgium	Gold and silver	Franc	.1930
Brazil	Gold	Milreis	.5460
Canada.	Gold	Dollar	1.0000
Central Amer. States :			
Costa Rica	Gold	Colon	.4650
British Honduras . .	Gold	Dollar	1.0000
Guatemala . . . . ⎫			
Honduras . . . . ⎟	Silver	Peso	.4030
Nicaragua . . . . ⎬			
Salvador . . . . ⎭			
Chile	Gold	Peso	.3650
China	Silver	Tael	.004 to .673
Denmark	Gold	Crown	.2680
Egypt	Gold	Pound (100 piasters)	4.9430
Finland	Gold	Mark	.1930
France	Gold and silver	Franc	.1930
German Empire . . .	Gold	Mark	.2380
Great Britain. . . .	Gold	Pound sterling	4.8665
Greece	Gold and silver	Drachma	.1930
Italy	Gold and silver	Lira	.1930
Japan	Gold	Yen	.4980
Mexico	Gold	Peso	.4980
Netherlands	Gold	Florin	.4020
Norway.	Gold	Crown	.2680
Panama	Gold	Balboa	1.0000
Philippine Islands . . .	Gold	Peso	.5000
Portugal	Gold	Milreis	1.0800
Russia	Gold	Ruble	.5150
Spain	Gold and silver	Peseta	.1930
Sweden	Gold	Crown	.2680
Switzerland	Gold	Franc	.1930
Venezuela.	Gold	Bolivar	.1930

888. To find the cost of a bill of exchange.

1. Find the cost of a draft of £ 675 9*s.* 8*d.* at $4.86\frac{1}{2}$.

$$£\,675\ 9s. = £\,675.45.$$
$$675.45 \times \$4.86\tfrac{1}{2} = \$3286.06.$$
$$\$3286.06 + \$.16\ (8d. = \$.16) = \$3286.22, \text{ cost of draft.}$$

9 shillings is $\frac{9}{20}$ or .45 of a pound; hence £ 675 9*s.* = £ 675.45. Since £ 1 costs \$ $4.86\frac{1}{2}$, £ 675.45 will cost $675.45 \times \$4.86\frac{1}{2}$, or \$ 3286.06. Each penny is worth two cents; hence, 8*d.* are worth \$.16, which, added to \$ 3286.06, equals \$3286.22, the cost of the draft.

2. How much will a draft of F. 10,000 cost at $5.18\frac{3}{4} + \frac{1}{32}\,\%$?

$$\text{F. } 10{,}000 \div \text{F. } 5.18\tfrac{3}{4} = 1927.71;\ \$1927.71 = \text{cost at } 5.18\tfrac{3}{4}.$$
$$\tfrac{1}{32}\,\% \text{ of } \$1927.71 = \$.60.$$
$$\$1927.71 + \$.60 = \$1928.31, \text{ cost of draft.}$$

First find the cost of F. 10,000 at $5.18\frac{3}{4}$. Since \$ 1 will buy F. $5.18\frac{3}{4}$ it will cost as many dollars to buy F. 10,000 as F. $5.18\frac{3}{4}$ is contained times in 10,000, or \$1927.71. Next add $\frac{1}{32}\%$ of \$ 1927.71, which gives \$ 1928.31 as the cost of the draft.

3. Find the cost of cabling £ 2500 to London if the cable rate is 4.8780, and the charges are commission $\frac{1}{4}\%$, and cablegram of 12 words at 25¢ a word.

$$2500 \times \$4.8780 = \$12{,}195, \text{ prime cost.}$$
$$\tfrac{1}{4}\,\% \text{ of } \$12{,}195 = \$30.49, \text{ commission.}$$
$$12 \times \$.25 = \$3.00, \text{ cost of a cablegram.}$$
$$\$12{,}195 + \$30.49 + \$3.00 = \$12{,}228.49, \text{ total cost.}$$

The cable rate of exchange being 4.8780, the first cost is $2500 \times \$4.8780$, or \$ 12,195. Adding the commission and cost of cablegram gives the entire cost as \$ 12,228.49.

889. The following foreign exchange quotations were printed in the New York *Wall Street Journal,* Jan. 7, 1911:

	CABLES	DEMAND	60 DAYS	90 DAYS
Sterling . . .	4.8590 @ 4.8595	4.8550 @ 4.8555	$4.82\frac{1}{4}$	$4.81\frac{1}{8}$ — $4.81\frac{1}{4}$
Francs . . .	$5.20 - \frac{1}{32}$	$5.20\frac{5}{8} + \frac{1}{32}$	$5.24\frac{3}{8} + \frac{1}{16}$	
Marks . . .	$.95 - \frac{1}{32}$	$.91\frac{7}{8} + \frac{1}{32}$	$93\frac{15}{16}$	

Paris exchange on London F. 25 29 c.
Berlin exchange on London M. 20 47 pf.

In the following problems based on the above rates a figure (1) will indicate that the first of the two rates quoted is to be used, and a figure (2), the second of the two rates.

890. Using the above quotations, find the cost of:

1. A demand bill for £750. (1.)

2. A cable for £1500. (2.) $\frac{1}{4}\%$ com. 14 words at 25 ¢.

3. A 60-da. bill for £985 10s.

4. A demand bill for £1362 8s. 4d. (2.)

5. A demand bill for F. 12,500.

6. A cable for F. 24,500. $\frac{1}{4}\%$ com. 15 words at 25 ¢

7. A 60-da. bill for F. 18,728.64.

8. A 60-da. bill for F. 14,725.

9. A cable for M. 16,000. $\frac{1}{4}\%$ com. 13 words at 25 ¢.

10. A demand bill for M. 8250.

11. A 60-da. bill for M. 23,250.

12. A cable for M. 19,600. $\frac{1}{4}\%$ com. 14 words at 25 ¢.

13. A demand bill for £1440 15s. 9d. (2.)

14. A cable for £2463 18s. 7d. (1.) $\frac{1}{4}\%$ com. 16 words at 25 ¢.

15. A 60-da. bill for F. 1825.

16. A demand bill for M. 3475.

17. A New York firm owed for imported merchandise as follows: From London £789 5s.; from Paris F. 11,843.50; from Hamburg M. 7284. They bought exchange in payment therefor, and drew their check for the entire cost. If the rates were 4.87$\frac{1}{4}$, 5.16$\frac{1}{4}$, and 95$\frac{3}{4}$, respectively, what was the amount of the check?

891. Conversion tables are used by foreign exchange bankers, which reduce to a minimum the labor of finding the cost or the proceeds of a bill of exchange. The complete table as used by the banker covers two pages for each rate. The upper part of the table is for pounds and the lower part for shillings and pence.

892.

TABLE FOR THE CONVERSION OF STERLING INTO DOLLARS AND CENTS, AND VICE VERSA. RATE 4.865.*

	0	100	200	300	400	500
	$ ¢	$ ¢	$ ¢	$ ¢	$ ¢	$ ¢
0		486.50.0	973.00.0	1,459.50.0	1,946.00.0	2,432.50.0
1	4.86.5	491.36.5	977.86.5	1,464.36.5	1,950.86.5	2,437.36.5
2	9.73.0	496.23.0	982.73.0	1,469.23.0	1,955.73.0	2,442.23.0
3	14.59.5	501.09.5	987.59.5	1,474.09.5	1,960.59.5	2,447.09.5
4	19.46.0	505.96.0	992.46.0	1,478.96.0	1,965.46.0	2,451.96.0
5	24.32.5	510.82.5	997.32.5	1,483.82.5	1,970.32.5	2,456.82.5
6	29.19.0	515.69.0	1,002.19.0	1,488.69.0	1,975.19.0	2,461.69.0
7	34.05.5	520.55.5	1,007.05.5	1,493.55.5	1,980.05.5	2,466.55.5
8	38.92.0	525.42.0	1,011.92.0	1,498.42.0	1,984.92.0	2,471.42.0
9	43.78.5	530.28.5	1,016.78.5	1,503.28.5	1,989.78.5	2,476.28.5
10	48.65.0	535.15.0	1,021.65.0	1,508.15.0	1,994.65.0	2,481.15.0

s	0	1	2	3	4	5
d	$ ¢	$ ¢	$ ¢	$ ¢	$ ¢	$ ¢
0		.24.3	.48.6	.72.9	.97.3	1.21.6
1	.02.0	.26.3	.50.6	.75.0	.99.3	1.23.6
2	.04.0	.28.3	.52.7	.77.0	1.01.3	1.25.6
3	.06.0	.30.4	.54.7	.79.0	1.03.3	1.27.7
4	.08.1	.32.4	.56.7	.81.0	1.05.4	1.29.7
5	.10.1	.34.4	.58.7	.83.1	1.07.4	1.31.7
6	.12.1	.36.4	.60.8	.85.1	1.09.4	1.33.7

Suppose it is desired to find the cost of a bill of exchange for £ 410 5s. 4d.

In the first part of the table, find the column marked 400, then in the column at the extreme left find the number 10. Follow the line in which 10 is, to the right, until the column marked 400 is reached. The number there, $1994.65, is the cost of £ 410.

Next, in the second part of the table, follow down the column marked 5, and to the right on the line marked 4, until the line and the column cross, which gives $1.297, the cost of 5s. 4d. Adding $1994.65 and $1.29.7, gives $1995.947 = $1995.95, the cost of £ 410 5s. 4d. at 4.865.

How much sterling exchange at 4.865 can be bought for $2467.10 ?

Find in the first part of the table the number nearest in value to, but less than, $2467.10, which is $2466.555 in column 500 and on line 7. This is the

* By permission of the International Textbook Company, Scranton, Pa.

cost of £ 507. Deduct $ 2466.555 from $ 2467.10, which leaves $.545. In the second part of the table find the number nearest in value to $.545, which is $.547 in column 2 and on line 3, and hence is the value of 2s. 3d. Therefore the exchange that can be bought for $ 2467.10 is £ 507 2s. 3d.

893. Using the conversion table, find the cost of bills of exchange for the following amounts :

1. £ 108 3s. 4d.   3. £ 400 3s.   5. £ 309 1s. 2d.

2. £ 206 2s. 1d.   4. £ 505 4s. 4d.   6. £ 407 3s. 2d.

How much sterling exchange can be bought for the following amounts ?

7. $ 511.19.   9. $ 1951.35.   11. $ 993.21.

8. $ 1479 75.   10. $ 2477.52.   12. $ 1018.

894. Tables are also used for computing the cost of 30-, 60-, and 90-da. sight exchange on the basis of the cost of demand exchange.

Interest on Sterling Exchange

Computed on basis of $ 4.85 per pound Stg. (363 days) with 3 days' grace and bill stamp $\frac{1}{20}\% = 0.24 \phi$.

Rate	33 Days & Stp.*	63 Days & Stp.	93 Days & Stp.	Rate	33 Days & Stp.	63 Days & Stp.	93 Days & Stp.
½	.46	.66	.86	2⅞	1.50	2.65	3.80
⅝	.52	.77	1.01	3	1.56	2.75	3.95
¾	.57	.87	1.17	3⅛	1.61	2.86	4.10
⅞	.63	.98	1.32	3¼	1.67	2.96	4.26
1	.68	1.08	1.48	3⅜	1.72	3.07	4.41
1⅛	.74	1.18	1.63	3½	1.78	3.17	4.57
1¼	.79	1.29	1.79	3⅝	1.83	3.28	4.72
1⅜	.85	1.39	1.94	3¾	1.89	3.38	4.88
1½	.90	1.50	2.10	3⅞	1.94	3.49	5.03
1⅝	.96	1.60	2.25	4	2.00	3.59	5.19
1¾	1.01	1.71	2.40	4⅛	2.05	3.70	5.34
1⅞	1.06	1.81	2.56	4¼	2.11	3.80	5.49
2	1.12	1.92	2.71	4⅜	2.16	3.90	5.65
2⅛	1.17	2.02	2.87	4½	2.22	4.01	5.80
2¼	1.23	2.13	3.02	4⅝	2.27	4.11	5.96
2⅜	1.28	2.23	3.18	4¾	2.33	4.22	6.11
2½	1.34	2.34	3.33	4⅞	2.38	4.32	6.27
2⅝	1.39	2.44	3.49	5	2.43	4.43	6.42
2¾	1.45	2.54	3.64				

* Revenue, or " bill stamp."

INTEREST ON FRENCH EXCHANGE

Computed on basis of F. 5.20 per Dollar (365 days) and bill stamp $\frac{1}{20}$%–0.26 centime.

RATE	30 DAYS & STP.	60 DAYS & STP.	90 DAYS & STP.	RATE	30 DAYS & STP.	60 DAYS & STP.	90 DAYS & STP.
$\frac{1}{2}$	.48	.69	.91	$3\frac{1}{8}$	1.61	2.97	4.32
$\frac{5}{8}$	.53	.80	1.07	$3\frac{1}{4}$	1.67	3.08	4.49
$\frac{3}{4}$	.59	.91	1.24	$3\frac{3}{8}$	1.72	3.19	4.65
$\frac{7}{8}$	.64	1.02	1.40	$3\frac{1}{2}$	1.78	3.29	4.81
1	.69	1.13	1.56	$3\frac{5}{8}$	1.83	3.40	4.97
$1\frac{1}{8}$	.75	1.24	1.72	$3\frac{3}{4}$	1.89	3.51	5.14
$1\frac{1}{4}$	.80	1.34	1.89	$3\frac{7}{8}$	1.94	3.62	5.30
$1\frac{3}{8}$	.86	1.45	2.05	4	1.99	3.73	5.46
$1\frac{1}{2}$	.91	1.56	2.21	$4\frac{1}{8}$	2.05	3.84	5.62
$1\frac{5}{8}$	.96	1.67	2.37	$4\frac{1}{4}$	2.10	3.94	5.79
$1\frac{3}{4}$	1.02	1.78	2.54	$4\frac{3}{8}$	2.16	4.05	5.95
$1\frac{7}{8}$	1.07	1.89	2.70	$4\frac{1}{2}$	2.21	4.16	6.11
2	1.13	1.99	2.86	$4\frac{5}{8}$	2.26	4.27	6.27
$2\frac{1}{8}$	1.18	2.10	3.02	$4\frac{3}{4}$	2.32	4.38	6.44
$2\frac{1}{4}$	1.24	2.21	3.19	$4\frac{7}{8}$	2.37	4.49	6.60
$2\frac{3}{8}$	1.29	2.32	3.35	5	2.43	4.59	6.76
$2\frac{1}{2}$	1.34	2.43	3.51	$5\frac{1}{8}$	2.48	4.70	6.92
$2\frac{5}{8}$	1.40	2.54	3.67	$5\frac{1}{4}$	2.54	4.81	7.09
$2\frac{3}{4}$	1.45	2.64	3.84	$5\frac{3}{8}$	2.59	4.92	7.25
$2\frac{7}{8}$	1.51	2.75	4.00	$5\frac{1}{2}$	2.64	5.03	7.41
3	1.56	2.86	4.16				

If sterling exchange on demand is quoted at 4.8685, find the corresponding 90-da. quotation, interest $3\frac{1}{4}$ %.

SOLUTION

Demand quotation 4.8685
Interest 93 days, $3\frac{1}{4}$ %, and bill stamp (from table on ster-
    ling exchange) 0426
Cost of 90-da. exchange 4.8259

Quoted as 4.8260

895. Using the table, find the time quotation corresponding to the demand quotation as follows:

DEMAND QUOTATION	TIME	INTEREST
1. 4.865	30 days	$2\frac{3}{8}$ %
2. 4.8675	60 days	$2\frac{7}{8}$ %

VAN TUYL'S BUS. ARITH. — 24

3. 4.8680	90 days	3%
4. 4.8665	60 days	$2\frac{1}{2}$%
5. $5.16\frac{7}{8}$	90 days	$4\frac{1}{4}$%
6. $5.16\frac{1}{4}$	60 days	$3\frac{3}{4}$%
7. $5.17\frac{1}{2}$	30 days	$3\frac{7}{8}$%
8. $5.18\frac{1}{8}$	60 days	$4\frac{1}{2}$%

896. To find the face of a bill of exchange.

1. How large a bill of exchange on London can be bought for $5000 at 4.85?

$5000 ÷ $4.85 = 1030.928, no. of pounds in $5000.

928 × 20s.   = 18.56s.

.56 × 12d.   = 6.72d = 7d.

Hence the face of the bill is £1030 18s. 7d.

Since £1 cost $4.85, $5000 will buy as many pounds as $4.85 is contained in $5000, or £1030.928. Reducing the decimal of a pound to shillings and pence gives the face of the bill as £1030 18s. 7d.

2. How many francs can be bought for $3000 at $5.17\frac{1}{2}$?

$3000 × F $5.17\frac{1}{2}$ = F. 15,525.

Since $1 buys F. $5.17\frac{1}{2}$, $3000 will buy 3000 times F. $5.17\frac{1}{2}$, or F. 15,525.

3. At $95\frac{1}{4}$, how many marks can be bought for $1000?

$1000 ÷ .95\frac{1}{4}$ = 1049.87

1049.87 × M. 4 = M. 4199.48

M. 4 cost $.95\frac{1}{4}$. Dividing $1000 by $.95\frac{1}{4}$ gives the number of times M. 4 that can be bought, or 1049.87;

hence, the number of marks is 1049.87 times 4 marks, or M. 4199.48.

897. Find the amount of exchange that can be bought for the following amounts at the prices quoted:

1. $3500 at $4.84\frac{1}{2}$.

2. $4500 at $4.87\frac{1}{4}$.

3. $5400 at 4.8785.

4. $7500 at 4.8725.

5. $2350 at $5.16\frac{7}{8}$.

6. $3475 at $5.20\frac{5}{8}$.

7. $6400 at $5.19\frac{3}{8}$.

8. $10,000 at $5.16\frac{1}{4}$.

9. $3000 at $94\frac{1}{4}$.

10. $4000 at $95\frac{1}{16}$.

11 $1250 at $95\frac{1}{2}$.

12. $1400 at $95\frac{1}{4}$.

13. $2500 at $5.20\frac{5}{8} - \frac{1}{32}$.

2500 × F. $5.20\frac{5}{8}$ = F. 13,015.625.

$\frac{1}{32}$% of F. 13,015.625 = F. 4.067.

F. 13,015.625 + F. 4.067 = F. 13,019.69.

At 5.20⅝, $2500 will buy F. 13,015.625. Since the quotation is *less* than 5.20⅝ by $\frac{1}{32}$%, it follows that more exchange can be bought than at 5.20⅝. Hence F. 13,015.625 must be increased by $\frac{1}{32}$% of F. 13,015.625, which gives F. 13,019.69,

Note.— When a rate is quoted in foreign money, as the franc rate, a change from 5.19 to 5.20 is a *fall*, and from 5.19 to 5.18 is a *rise*, in the rate. 5.20⅝ — $\frac{1}{32}$ = 5.20⅝ + ($\frac{1}{32}$% of 5.20⅝) = 5.207877. 2500 × F. 5.207877 = F. 13,019.69.

How much exchange can be bought for:

14. $1800 at 5.16⅞ $-\frac{3}{64}$?   **18.** $4375 at $95\frac{7}{16}+\frac{1}{32}$?

15. $2300 at 5.16⅞ $+\frac{1}{32}$?   **19.** $7500 at $95\frac{1}{4}+\frac{1}{64}$?

16. $2150 at 5.17½ $-\frac{1}{16}$?   **20.** $1275 at $95\frac{7}{8}-\frac{1}{32}$?

17. $3750 at $95\frac{5}{16}-\frac{1}{64}$?   **21.** $1448 at 5.18⅛ $+\frac{1}{32}$?

898. To find the proceeds of a bill of exchange.

1. Find the proceeds of a bill of exchange for F. 4200 at 5.17½ $-\frac{1}{32}$.   F. 4200 ÷ F. 5.17½ = 811.59; $811.59 = proceeds at 5.17½.
$\frac{1}{32}$% of $811.59 = $.25.
$811.59 — $.25 = $811.34, proceeds.

The principles involved here are the same as explained on page 365.

2. What are the proceeds of a documentary bill of exchange on London for £ 528 8s. 6d., at 60 days after sight, if sold to a banker, the demand rate being 4.87½, interest 4%, revenue stamp $\frac{1}{20}$%, and commission ¼%?

Quotation at 4.87½, ... $4.875
Interest 63 da., $4.85 × .04 × $\frac{63}{365}$ = $.03348
Revenue stamp (bill stamp), $\frac{1}{20}$%   .00244 $.03592
(Compare with table, page 368.)
Commission ¼%, ...01219 $ .04811
Proceeds of £ 1, ... $4.82689

£ 528 8 s. = £ 528.4.
528.4 × $4.82689 = $2550.53.
$2550.53 + $.12 (6 d.) = $2550.65, proceeds.

First find the proceeds of £1. Interest is on a uniform value of $4.85 per pound, 365 days to the year, and for 3 days of grace. Revenue stamp is $\frac{1}{20}$% of the quotation. (Interest and revenue stamp are here calculated separately to show the entire solution. In practice, tables like those on pages 368 and 369 are used.) To the interest and stamp charge is added the commission at ¼%, making a total of $.04811 to be deducted from $4.875, which leaves $4.82689, the proceeds of £1. Multiplying the proceeds of £1 by the face of the draft gives the proceeds of the draft.

3. Find the proceeds of a 90-days' sight exchange on Paris for F. 5548.75, demand drafts being at $5.18\frac{1}{8} + \frac{3}{32}$, interest * 3 %, revenue stamp $\frac{1}{20}$ %, and commission $\frac{1}{8}$ %.

Quotation $(5.18\frac{1}{8} + \frac{3}{32})$ . . . . F. 5.17639
Interest and bill stamp (table) 04160
Commission $\frac{1}{8}$ % 00647
Proceeds equivalent to $1 . . . . F. 5.22446
     (Quoted at $5.22\frac{1}{2} + \frac{1}{64}$)
F. 5548.75 ÷ F. 5.22446 = 1062.07 ; $1062.07 = proceeds.

The fraction $+\frac{3}{32}$ % *increases* the rate from $5.18\frac{1}{8}$ to 5.17639. The interest and bill stamp charge is .0416 (see table page 369), and the commission is .00647, making the quotation 5.22446, or in the banker's term, $5.22\frac{1}{2} + \frac{1}{64}$. Dividing F. 5548.75 by F. 5.22446 gives 1062.07 ; or $1062.07 is the proceeds of the draft.

899. Find the proceeds of the following bills : (A charge for revenue stamps, $\frac{1}{20}$ %, is to be understood in each case. See page 362.)

	FACE OF BILL	TIME	INT.	COM.	DEMAND QUOTATION
1.	£ 1,750	30 days' sight	3 %	$\frac{1}{8}$ %	$4.85\frac{7}{8}$
2.	F. 11,000	60 days' sight	4 %	$\frac{1}{4}$ %	$5.18\frac{1}{8}$
3.	M. 9,000	90 days' sight	$3\frac{1}{2}$ %	$\frac{1}{8}$ %	† $95\frac{15}{16} + \frac{1}{32}$
4.	£ 943 7s.	60 days' sight	4 %	$\frac{1}{8}$ %	4.8625
5.	F. 3,328.40	90 days' sight	3 %	$\frac{1}{8}$ %	$5.16\frac{7}{8} + \frac{3}{64}$
6.	M. 4,726.85	60 days' sight	4 %	$\frac{1}{4}$ %	$95\frac{1}{4} + \frac{1}{32}$

900. To find the cost of exporting gold.

Find the market quotation equivalent to the cost of shipping gold bars from New York to London, if London pays 77s. 10$\frac{3}{4}$d. for 1 oz. of gold $\frac{11}{12}$ fine.

£ 1 = 240d.
77s. 10$\frac{3}{4}$d. = 934$\frac{3}{4}$d. = 1 oz. standard gold ($\frac{11}{12}$ fine, English Standard).
934$\frac{3}{4}$d. ÷ $\frac{11}{12}$ = 1019.72727d. = 1 oz. fine gold.
1 oz. fine gold = $20.67183 (U. S. Treasury buying and selling price).
    Therefore, the parity of exchange is,
($20.67183 ÷ 1019.72727) × 240 = $4.8653 = £ 1.
Parity of exchange 4.8653
Plus Freight, about $\frac{1}{8}$ % 0061
   Insurance, about $\frac{1}{20}$ %. 0024
   Boxing, carting, loss of int., etc. about $\frac{1}{20}$ % 0024
Bar charge of U. S. Treasury when selling gold, 40 ¢ per $1000 . . .0019
    Against this shipment of gold, demand drafts could be sold at 4.8781

---

* Interest on a uniform value of F. 5.20 to the dollar, 365 days, and no grace.
† Interest on a basis of 95 ¢ per 4 marks. 360 days to the year, and no grace.

The minimum figure at which the Bank of England buys gold bars $\frac{11}{12}$ fine is 77s. 9d. per ounce. In this problem the price is 77s. 10$\frac{3}{4}$d., equal to 934$\frac{1}{4}$d. If gold $\frac{11}{12}$ fine is worth 934$\frac{3}{4}$d., fine gold is worth 1019.72727d. (per ounce). But an ounce of fine gold is worth $20.67183. Therefore, when London pays 77s. 10$\frac{3}{4}$d. per ounce $\frac{11}{12}$ fine, the parity of exchange is 4.8653. Adding the charges (freight, insurance, boxing, etc., and the bar charge) makes a total cost of exporting £1 equal to $4.8781, quoted as $4.8780.

Therefore as soon as demand drafts on London sell for more than 4.8781, it would be cheaper to export gold at 77s. 10$\frac{3}{4}$d. per ounce.

The cost of exporting gold *coin* is reckoned in the same way, except that there is no "bar charge," but a small allowance (about $\frac{1}{10}$ %) is necessary on account of abrasion.

The Bank of England's minimum buying price of **U. S. gold coin,** .900 fine, is 76s. 4d. per ounce.

PROBLEMS

901. 1. A banker wishes to remit £150,000 to London. Exchange is at 4.8855. The Bank of England is paying 77s. 10$\frac{1}{2}$d. per ounce for gold bars $\frac{11}{12}$ fine. How much will he save by exporting gold (usual charges)?

2. N. W. Harris & Co. of New York export U. S. gold coin to London to settle a balance of £125,000. If London pays 76s. 4$\frac{1}{2}$d. per ounce, how much does the shipment cost them if charges are as given above?

3. When London is paying 77s. 9d. per ounce for gold bars $\frac{11}{12}$ fine, and 76s. 4d. per ounce for U. S. gold coin, which is it better to export, and how much better, in settling an account of £200,000?

902. Arbitrage is the name given to the process of making an indirect transfer of money.

For example, if a banker wished to remit exchange on Paris, he might, instead of buying direct on Paris at 516$\frac{1}{4}$ — $\frac{1}{16}$, buy London exchange at 4.8785, and cable this to his London agent and instruct him to buy Paris exchange at 25.22$\frac{1}{2}$ francs to the pound sterling. The result would be as follows:

Cost of demand exchange on London.	London buying rate on Paris.	Equivalent New York rate on Paris.
4.8785 ÷	25.22$\frac{1}{2}$ =	.1934 or 5.1706
Exchange direct on Paris costs 5.16$\frac{1}{4}$ — $\frac{1}{16}$ = ...		5.1657
Resulting in a gain to the banker of		.0049, called 49 points.

903. To find the cost of indirect exchange

Which is the better when remitting to London, to buy £ 5000 direct at 4.88¼ or to buy on Paris at 5.17½, and on London at.Paris at 25.13 ?

5000 × $4.88¼ .   . = $ 24,412.50, cost direct.

5000 × F. 25.13 . . . . = F. 125,650, cost of £ 5000 in Paris.

F. 125,650 ÷ F. 5.17½ . . = 24,280.19, no. dollars cost at Paris.

$ 24,412.50 — $ 24280.19 . = $ 132.31, gain by indirect exchange.

PROBLEMS

904. 1. A of New York owes B of London £ 8000. Exchange on London is 4.88½. How much, if anything, can A save by buying Paris exchange at 5.18⅛, and on London at Paris at 25.12 ?

2. Exchange on Hamburg is quoted at 95¾, and on Paris at 5.15⅝. How much will be gained or lost on a remittance to Hamburg of 15,000 marks via Paris at F. 1.22 to the mark ?

3. At New York exchange on Bremen is at 94¾; on Paris, at Bremen F. 1.24 to the mark, and on London at Paris F. 25.10 to the pound. If sterling is quoted in New York at 4.88¼, find the gain or loss by remitting £ 12,500 to London via Bremen and Paris

REVIEW PROBLEMS

905. 1. An exporter sold to a broker the following bills of exchange: £ 3472 8s. at 4.85¾; 7425.50 francs at 5.19⅜; and 17,480 marks at 95½. Find the total net proceeds, if the broker charged ⅛ % brokerage.

2. A manufacturer in Liverpool drew a sight draft on an importer in Philadelphia for £ 859 13s. 8d. How much does it cost the importer to pay the draft at 4.8650 ?

3. A grain exporter in New York drew a bill of exchange on a Paris firm for $ 27,580. If exchange in Paris on New York is quoted at 5.17½, how many francs are required to meet the draft ?

4. A wholesaler has drawn on him against an importation of woolens a 30-days' sight draft for £ 583 7s. If exchange is at 4.8555, what does it cost him to pay the draft ?

5. A New York merchant drew a 60-days' sight draft on one his debtors in Paris for F. 18,700. What should a banker in New York pay for the draft, allowing for interest at $3\frac{5}{8}\%$, revenue stamp $\frac{1}{20}\%$, and a commission of $\frac{1}{4}\%$, if demand exchange is at $5.18\frac{3}{4}$?

6. Find the total cost of travelers' checks as follows: 20 checks of $10 each; 15 at $20 each; 4 at $50 each; and 3 at $100 each, if the issuing bank charges $\frac{1}{2}\%$ commission.

7. Using the printed amounts on the specimen travelers' check on page 363 as a guide, find how much English money would be received for a $50 check, a $20 check, and a $10 check. How much German money?

8. You buy of Redmond & Co., New York, a letter of credit for £600. If exchange is at 4.865, and they charge you a commission of 1%, how much does the draft cost you?

9. In Paris you draw on the above-mentioned letter of credit £75. If exchange on London at Paris is 25.08, how much French money do you receive?

10. In Berlin you need £150 more, and obtain the equivalent amount in marks at 20.375. How many marks do you receive?

11. Using the table on page 367, find the cost of a bill of exchange on London for £508 4s. 6d.

12. From the table, page 367, find how large a draft can be bought for $2433.75.

13. Will it pay a banker to remit to Paris via London under the following conditions: exchange on London at New York is 4.8625; on Paris at London is 25.14 francs to the pound; and on Paris at New York is $5.16\frac{7}{8} + \frac{1}{16}$?

14. J. P. Morgan & Co. remit to London via Paris when rates are: on Paris $5.19\frac{3}{4}$, and on London at Paris 25.04. Find their gain on a remittance of £50,000, if direct exchange on London is at 4.8675.

15. If the Bank of England is paying 77s. $9\frac{1}{2}d.$ per ounce for gold bars $\frac{11}{12}$ fine, find the highest market quotation at which it will be profitable to buy exchange on London instead of shipping gold, the charges being as given on page 372, Art. 900.

16. Find the cost of exporting gold amounting to £200,000, if U. S. gold coin is shipped to London at 76s. $5\frac{1}{2}d.$ per ounce. (For charges, see page 372, Art. 900.)

375

⌐es, are taxes levied by the government on ⌐se.

⌐urpose of the tax is twofold.

⌐ protection of American industries.

For revenue. The expenses of the United States Government are about one million nine hundred thousand dollars a day. The income derived from customs is applied toward the payment of the nation's expense.

908. Customs or duties are of two kinds — specific and ad valorem.

909. "Duties" is the more commonly used term, and will be used in this work.

910. A specific duty is a specified amount levied on each article — pound, ton, bushel, gallon, or upon each square foot or square yard, etc., as the case may be.

911. An ad valorem duty is a certain per cent levied on the appraised market value of merchandise in the country from which it is imported.

912. Specific duties are levied on some imported articles, ad valorem duties on others, and both are assessed on other kinds of merchandise. Specific duties are not reckoned on fractions of a unit when the duty on the unit is fifty cents or less. Fractions of such units equal to or greater than one half are considered as a whole unit. Fractions less than one half are rejected. At some ports if the duty on the unit is more than fifty cents, the duty is charged on the exact fraction.

Ad valorem duties are not computed on fractions of a dollar. Fifty cents or more is reckoned as a dollar ; less than fifty cents is rejected.

913. Tare is an allowance made for the weight of the box, bag, or other covering material. Leakage is an allowance made for the loss of liquids imported in barrels. Breakage is an allowance for the loss of liquids imported in bottles. (Wines and liquors excepted.)

914. The **long ton** of 2240 pounds is the legal ton in reckoning import duties.

915. A **customhouse** is a place designated by the government where importers of merchandise are to make entry of it and pay the duties chargeable thereon, and where vessels are entered and cleared.

916. A vessel is officially *entered* when the manifest is filed with the collector of customs.

917. A vessel is said to be *cleared* when the master has received his clearance, or permit to leave the port.

The United States is divided into collection districts. In each district there is a customhouse. The port at which the customhouse is located is called a **port of entry.** Vessels and cargoes are entered only at those ports having a customhouse.

A **port of delivery** is any port at which imported merchandise may be delivered after it has been entered at a port of entry. All ports of entry are ports of delivery.

918. At each port of entry is an officer, called the **collector of customs** (popularly called **collector of the port**), whose duty it is to receive entrance papers, issue permits for the clearance of vessels, collect the duties, etc.

919. The **public store** is a place provided for the examination of imported merchandise.

920. All invoices are made out in the weights, measures, and currency of the country from which the merchandise is imported. Invoices of more than $100 must be certified by a United States consul.

921. There are two principal forms of entry of merchandise, consumption entry and warehouse entry.

922. A **consumption entry** is made for all merchandise on which the duty is paid at the time of entry.

923. A **warehouse entry** is made for all merchandise to be stored in a bonded warehouse.

924. A **bonded warehouse** is a building for the storage of merchandise on which the duty has not been paid. It derives its name from the fact that the owner is required to give a bond to the government that he will not deliver from the warehouse any merchandise until the duty has been paid. (With certain exceptions. See Art. 927.)

925. The importer of the merchandise also gives a bond obligating himself to pay the duty and to remove the goods within three years of the date of entry. Merchandise not removed within the three years may be sold for duty and storage charges.

926.

IMPORT   INVOICE DATED Paris, December 25, 1910   C. C. No. 25,000

ENTRY FOR CONSUMPTION OF MERCHANDISE IMPORTED BY John Doe IN THE French s.s. Niagara

WHEREOF............IS MASTER, ARRIVED January 13, 1911   FROM Havre. NEW YORK, January 13, 1911

Marks	Nos.	PACKAGES AND CONTENTS Specifications as per Accompanying Invoices		35%	45%			Total
J.D.	1/10	Ten Cases						
		Furniture, Mfr. Wood.	francs	650.00				
		Hardware, Mfr. Metal.			130.00		francs	780.00
				$126.--	$25.--			$151.--
					(Signed)	John	Doe,	
						1 Broadway,	N.Y.	

The above illustration shows the form of entry used when merchandise is imported for immediate consumption. There are ten cases of furniture valued at F. 650, or $126, the duty thereon being 35 %; and 10 cases of hardware valued at F. 130, or $25, the duty on which is 45%.

927. Merchandise may be withdrawn from a bonded warehouse without the payment of the duty for such purposes as:

1. Exportation.
2. Transportation and exportation.
3. Transportation in bond to another warehouse in another district.
4. Transportation to a manufacturing bonded warehouse. The manufactured product must be exported.

If merchandise in bond, on which the duty amounting to $50 or more has been paid, is exported, the duty less one per cent will be returned. The amount so returned is called a *drawback*.

928. A tariff is a list of dutiable articles with the legal rate of duty on each.

929. A free list is a schedule of articles on which there is no duty.

930. The value in United States money of foreign currency proclaimed by the secretary of the treasury must be used in determining the duties. If the value is not so fixed, a consular certificate showing the value must accompany the invoice.

931. For value of foreign currency see page 364.

932. The following rates of duty are from the Tariff Act of 1913.

IMPORT DUTIES

Articles	Specific Duty	Ad Valorem Duty
Brussels carpet.		25%
Butter	2½ ¢ a pound	
Cheese		20%
China, decorated		55%
Cotton cloth, unbleached.		20%
Cotton clothing, ready made		30%
Cotton stockings		20%
Envelopes		15%
Hay.	$2 a ton	
Horses		10%
Lead pencils	36 ¢ a gross, but not less than 25%	
Leather gloves, ladies', not over 14 inches long	$2 a dozen	
Over 14 inches long	$2 a dozen plus 25 ¢ a dozen additional for each inch in excess of 14 inches	
Leather gloves, men's	$2.50 a dozen	
Lumber, mahogany.		10%
Oilcloth		20%
Pens, steel	8 ¢ a gross	
Plate glass, not exceeding 384 sq. in.	6 ¢ a square foot	
Above 384 sq. in., not exceeding 720 sq. in.	8 ¢ a square foot	
All above 720 sq. in..	12 ¢ a square foot	
Silk handkerchiefs		50%
Silk ribbons		45%
Woolen clothing, ready made		35%
Writing paper.		25%

933. To find a specific duty.

What is the duty on 250 dozen men's leather gloves?

$250 \times \$2.50 = \$625$, specific duty.

According to the table, page 379, the duty on men's leather gloves is $2.50 a dozen. 250 dozen cost 250 times $2.50, or $625. (Short method, $\frac{1}{4}$ of $2500 = $625; or $25^2 = 625$, hence $625.)

EXERCISES

934. Using the rates given in the table on page 379, find the duty on:

1. 28,400 lb. of hay at $12 per ton.

2. 2476 lb. of butter valued at 24 ¢ a pound.

3. 400 doz. ladies' leather gloves, 17 inches long.

4. 250 doz. ladies' leather gloves, 19 inches long.

5. 180 gross lead pencils, valued at $1.25 per gross.

6. 44 boxes plate glass, each containing 18 pieces 20 in. × 18 in.

7. 6 boxes plate glass, each containing 12 pieces 48 in. × 60 in.

8. 28 gross steel pens.

9. 50 doz. men's leather gloves.

10. 800 doz. ladies' leather gloves, 14 inches long.

935. To find an ad valorem duty.

Find the ad valorem duty on merchandise imported from England valued at £560 8s. 10d., if the duty is 28 %.

$$£560\ 8s. = £560.4$$
$$560.4 \times \$4.8665 = \$2727.19$$
$\$2727.19 + \$.20\ (10d. = 20\ ¢) = \$2727.39$, value in U. S. money.

28 % of $2727 = 763.56, ad valorem duty.

The value of the invoice must first be reckoned in United States money. It amounts to $2727.39. The duty is reckoned on the nearest dollar in the value; hence the duty is 28 % of $2727, which equals $763.56.

EXERCISES

936. Using the rates given in the table, page 379, find the duty on.

1. Envelopes invoiced at M. 370.

2. Silk ribbons invoiced at F. 875.

3. Cotton stockings invoiced at £328 14s. 8d.

4. Cotton clothing invoiced at £649 7s. 6d.

5. Decorated china invoiced at F. 2372.

6. 1600 sq. yd. oilcloth invoiced at £ 30 8s. 10d.

7 800 gross lead pencils invoiced at M. 5000.

8. 1800 yd. Brussels carpet valued at F. 3 per yard.

9. 1400 lb. of cheese valued at F. 720.

10. Decorated china valued at M. 3275.

11. Ten horses valued at $175 each.

12. 250 gross lead pencils invoiced at M. 12 per gross.

13. 1500 lb. of writing paper invoiced at F. 2.50 per pound.

14. Woolen clothing valued at £ 387 12s.

15. 2500 envelopes invoiced at F. 125.

PROBLEMS

937. 1. A merchant imported merchandise from Paris amounting to F. 13,878. The duty was 15%. What was the total cost, if in remitting he bought exchange at $5.17\frac{1}{2}$?

2. A stationer imported 100 reams of writing paper, weighing $4\frac{1}{2}$ lb. to the ream, at F. 3 per ream, and 350 boxes of envelopes (250 in a box) at F. 2 per box. What was the cost, including duty, exchange being at par? At what price per ream for paper and per box for envelopes must he sell to gain 50%?

3. An importer purchased 300 gross lead pencils in London at 16s. 6d. per gross. At what price per dozen should he sell them to gain 40%, if he paid for the pencils with a bill of exchange bought at 4.8560?

4. An invoice of Brussels carpet contains 3480 yd. at M. 3 per yard. Transportation and other charges except duty amount to $125. At what price per yard must the carpet be sold to net a profit of $33\frac{1}{3}$%, exchange being $95\frac{1}{4}$?

5. A clothier imported ready-made woolen clothing from England invoiced at £ 728 13s. 9d. Find the entire cost of the clothing if he remitted at 4.8585.

6. If an importation of cotton stockings valued at £ 382 7s. contains 2250 doz. pairs, at what price per pair must they be sold to gain $37\frac{1}{2}$%, exchange being at $4.86\frac{1}{2}$?

7. A merchant ordered from Paris 900 doz. pairs ladies' leather gloves, 14 in. long, at F. 35 per dozen; 800 doz. pairs, 17 in. long, at F. 42 per dozen; and 1200 doz. pairs, 19 in. long, at F. 48 per dozen. Transportation and other charges except duty amount to $960 to be charged equally against the three grades of gloves. At what price per pair must he sell each grade to realize a profit of 25 % on gross cost, if he buys exchange in settlement at 5.20 ?

8. An invoice of men's leather gloves valued at £ 360 contains 150 doz. pairs. At what price per pair must they sell to gain 30 %, exchange being 4.8650?

9. If 4800 yd. of unbleached cotton cloth 30 in. wide costs 2*d.* per yard, at what price per yard will it sell if a gain of $33\frac{1}{3}$ % is realized, exchange being at par ?

10. A Boston firm imported 75 doz. silk handkerchiefs from Paris at F. 22 per dozen, and 800 yd. of silk ribbon at F. 1.50 per yard. Freight and other expenses were: on the handkerchiefs F. 81.61, and on the ribbon F. 64.25. If they sell the handkerchiefs at 75¢ each, and the ribbon at 60 ¢ a yard, do they gain or lose, and what per cent on each, if exchange is at par ?

11. R. H. Macy & Co., of New York, imported decorated china valued at M. 1275.48. Commission charges were 5 %, and other expenses amounted to M. 250. Find the entire cost, exchange being at .95$\frac{3}{4}$.

12. An importation of woolen clothing invoiced at M. 14,728 weighed 1275 lb. The expenses of the purchase were: commission $3\frac{1}{2}$ %, and freight to port of export M. 1.25 per 100 lb. Find total cost, reckoning exchange at .95$\frac{1}{4}$.

STORAGE

938. Storage is the keeping or placing of articles in a warehouse or other place for safekeeping or awaiting favorable markets. The term storage is also applied to the price paid for storing goods.

939. If storage rates are not fixed by statute, or by ordinances, they may be determined by agreement of the parties interested.

940. The rate is generally a specified sum per unit of merchandise for a given period of time, usually a week, 10 days, or 30 days.

941. The time for which a certain rate is charged is called the **term of storage.** A fraction of a term usually counts as a whole term.

942. 1. Find the storage charge at 8 ¢ a month per barrel on the following :

<table>
<tr><td colspan="2" align="center">RECEIVED</td><td colspan="2" align="center">DELIVERED</td></tr>
<tr><td align="right">Aug. 1,</td><td>375 bbl.</td><td align="right">Sept. 15,</td><td>300 bbl.</td></tr>
<tr><td align="right">10,</td><td>225 bbl.</td><td align="right">Oct. 1,</td><td>300 bbl.</td></tr>
<tr><td align="right">20,</td><td>200 bbl.</td><td align="right">Oct. 20,</td><td>200 bbl.</td></tr>
</table>

Received Aug. 1, 375 bbl.
                10, 225 bbl.
                20, 200 bbl.
Total, . . .   800 bbl.
Delivered Sept. 15, 300 bbl. stored 45 da. (2 terms) at 8¢ . . . **= $48**
                     500 bbl. balance.

Oct. 1, 300 bbl. stored $\left(\begin{array}{l} 75 \text{ bbl. 2 mo.} \\ 225 \text{ bbl. 52 da.} \end{array} \right)$ (2 terms) at 8¢= 48

         200 bbl. balance.

Oct. 20, 200 bbl. stored 2 mo. (2 terms) at 8¢ . . . . = 32

                          Total storage **$128**

PROBLEMS

943. 1. At 5¢ per barrel for the first 10 da. and 3¢ for each succeeding 10 da. or fraction thereof, find the cost of the storage on the following:

<table>
<tr><td colspan="2" align="center">RECEIVED</td><td colspan="2" align="center">DELIVERED</td></tr>
<tr><td align="right">June 1,</td><td>80 bbl.</td><td align="right">June 15,</td><td>150 bbl.</td></tr>
<tr><td align="right">12,</td><td>100 bbl.</td><td align="right">30,</td><td>270 bbl.</td></tr>
<tr><td align="right">24,</td><td>240 bbl.</td><td></td><td></td></tr>
</table>

2. A merchant imported 150 cases of shoes and stored them in a bonded warehouse on the 15th of January, 1911. He took them from storage Mar. 10, 1911. If the storage was 18 ¢ per case for a period of 30 da. or fraction thereof, the labor charge 18 ¢ per case, and the cartage 20 ¢ per case, what was the total charge ?

3. A jobber imported, and placed in storage Apr. 15, 1911, 750 packages of firecrackers. The charges were, storage 4 ¢ a package for a month of 30 da. or fraction thereof, labor 4 ¢ a package, and cartage 5 ¢ a package. Find the total charge if the firecrackers were removed June 20, 1911.

EQUATION OF ACCOUNTS

944. If you buy a bill of merchandise valued at $1200 on a credit of 60 days from Aug. 1, when is the bill due? If you pay the face value of the bill Aug. 31, how many days' discount could you claim? If you do not pay the bill till Oct. 30, how many days' interest could the seller claim?

If you pay $600 on account Aug. 31, how long after the maturity ought you to be allowed to retain the other $600 of the bill, to balance your loss by paying $600 Aug. 31?

945. A merchant bought goods Apr. 10 worth $400, terms cash; and $400, terms 20 days. On what date may both bills be paid in one amount without loss of interest to either party?

946. Equation of accounts is the process of finding the date when (1) several items due at different dates may equitably be paid in one amount, or (2) the balance of an account may be equitably settled.

947. Accounts are generally divided into two classes, viz.: those having items on but one side, the equating of which is called **simple equation**, or **equation of payments**, and those having both debit and credit items, the equating of which is called **compound equation**, or **equation of accounts**.

EQUATION OF PAYMENTS

948. The theory on which equation of payments is based is that the interest on overdue items in an account is canceled by the discount on other items in the account paid before they are due.

To illustrate the theory, suppose a man owes $200 due May 10, and $600 due May 30. By equation, the entire amount may be paid May 25, because the interest on $200 from May 10 to May 25 is equal to the discount on $600 from May 25 to May 30.

384

949. To find the equated date of a one-sided account.

The following example will illustrate several terms and their definitions:

Find the equated date of the following bills:

July 8, $600; July 20, $1200; Aug. 17, $1800.

July 8,	$ 600	40 da.	$4.00, interest
July 20,	$1200	28 da.	$5.60, interest.
* Aug. 17,	$1800	0 da.	$0.00, interest.
	$3600		$9.60, interest.

Interest on $3600 for 1 da. = $3.600 ÷ 6 = $.60.

$9.60 ÷ $.60 = 16 da., average term of credit.

Aug. 17 − 16 da. = Aug. 1, equated date.

Arrange the several items and their respective due dates as shown in the solution. For the purpose of the solution, the latest date, Aug. 17, is assumed as the date of settlement. The assumed date of settlement is called the *focal date.** If the $600 due on July 8 is not paid till Aug. 17, it is overdue 40 da., and the accrued interest amounts to $4. So, also, if the $1200 due July 20 is not paid till Aug. 17, it is overdue 28 da., and the interest is $5.60. If the $1800 is paid on its due date, Aug. 17, there is no interest charged or discount allowed. Now, if the total amount of the bill, $3600, is paid on Aug. 17, the seller loses $9.60 interest on the overdue items. To avoid this loss the account should have been settled earlier.

The interest on $3600 for 1 da. is $.60; hence, if the payment had been made one day earlier, the loss to the seller would have been $.60 less. Therefore, the account should have been settled as many days before Aug. 17, as $.60, the interest on the debt for 1 da., is contained times in $9.60, the amount of interest due, which is 16 times, or 16 da. Counting back 16 da. from Aug. 17, gives Aug. 1, the *equated date.†*

To prove the correctness of the work and to show there is neither loss nor gain to either party, find the interest on all items due before Aug. 1 from the date they are due to Aug. 1. Then find the discount (simple interest) on the item due after Aug. 1 from Aug. 1, to its due date. Thus:

* Any date may be used as a focal date. In practice either the earliest or latest date is generally used, owing to the fact that they are the most convenient dates. In this book, the latest date is used.

† The equated date, the average date of payment, the average due date, and similar expressions are synonymous, and are the names given the date of settlement as determined by equation.

The interest on $600 from July 8 to Aug. 1, 24 da., is $2.40
and the interest on $1200 from July 20 to Aug. 1, 12 da., is $2.40
                 Total interest due $4.80

The discount on $1800, from Aug. 1 to Aug. 17, 16 da., is $4.80; that is, the interest on the overdue items is equal to the discount on the item that was pre-paid. Hence, the work is correct, and Aug. 1 is the average due date.

PROBLEMS

950. Find the average due date of the following bills:

1910	**1910**	**1910**
1. Mar. 1, $300	2. Sept. 15, $800	3. Dec. 1, $1200
Mar. 10, $600	Sept. 30, $900	Dec. 8, $1500
Mar. 20, $900	Oct. 15, $300	Dec. 24, $3000

1910	**1910**	**1910**
4. May 1, $1500	5. June 18, $560	6. Aug. 10, $360
May 20, $600	June 30, $720	Aug. 20, $480
June 8, $300	July 15, $900	Aug. 28, $660
June 20, $900	July 25, $1000	Sept. 15, $840

7. On what date may the following items be paid in one amount without loss of interest to either party?

Jan. 3, 1910, Mdse., 30 da., $600
Jan. 14, 1910, Mdse., 1 mo., $900
Jan. 20, 1910, Mdse., 60 da., $800
Feb. 10, 1910, Mdse., 1 mo., $300

1910	1910			
Jan. 3 + 30 da. =	Feb. 2,	$600	47 da.	$4.70
Jan. 14 + 1 mo. =	Feb. 14,	$900	35 da.	$5.25
Jan. 20 + 60 da. = *Mar. 21,		$800	0 da.	$.00
Feb. 10 + 1 mo. =	Mar 10,	$300	11 da.	$.55
		$2600		$10.50

The interest on $2600 for 1 da. = $2.600 ÷ 6 = $.4333.
$10.50 ÷ $.433 = 24.2; 24 da.,* average term of credit.
Mar. 21 − 24 da. = Feb. 25, equated date.

When various terms of credit are given on the items to be equated, it is necessary first to find the due date of the separate items. Having found the several due dates, we perform the solution as in the preceding exercises.

* A fraction of a day less than ½ is rejected; a fraction greater than ½ is called 1 da.

Find when the following bills may be equitably paid in one amount. Prove the work.

1911	8			1910	9		
Apr. 1,	Mdse., 15 da.,	$500		July 1,	Mdse., 1 mo.,	$856.60	
Apr. 10,	Mdse., 1 mo.,	$800		July 18,	Mdse., 2 mo.,	$1260.00	
May 1,	Mdse., 30 da.,	$1000		July 30,	Mdse., 30 da.,	$1000.00	
May 10,	Mdse., Net,	$300		Aug. 10,	Mdse., 10 da.,	$1268.40	

10. Daniel Reardon bought of Lord & Taylor merchandise on 60 days' credit as follows: Aug. 12, 1910, a bill for $1388.75; Sept. 10, a bill for $1564.25; and Sept. 20, a bill for $975.50. Reardon wishes to give a 2-months' interest-bearing note* in settlement of the entire amount. When should the note be dated? Write the note.

EQUATION OF ACCOUNTS

951. To find the equated date of a two-sided account.

1. What is the equated date of payment of the balance of the following account?

1910				1910			
May 1,	Mdse.,	$1840		May 10,	Cash,	$600	
June 1,	Mdse.,	$1200		May 20,	Cash,	$900	

Due	Amt.	Days	Int.	Due	Amt.	Days	Int.
1910				1910			
May 1,	$1840	31	$9.506	May 10	$600	22	$2.20
*June 1,	$1200	0	0.000	May 20	$900	12	$1.80
	$3040		$9.506		$1500		$4.00

$3040 − $1500 = $1540, balance due on the account.

$9.506 − $4.00 = $5.506, balance of interest due.

Interest on $1540 for 1 da. = $1.540 ÷ 6 = $.2566.

$5.506 ÷ $.2566 = 21.4, no. da.; 21 da.

21 da. back from June 1 = May 11, the equated date.

All the debit items and all the credit items are arranged as shown in the solution. The latest date on either side is taken as a focal date. If the bill of $1840 were not paid until June 1, it would have been 31 da. overdue, and there would

* An account becomes interest bearing at the equated date.

have been $9.506 interest due on it. There would be no interest due on the bill of $1200 if paid on June 1. The total interest due, then, by settling the account on June 1, would be $9.506. But two payments have been made. The $600 paid on May 10 was paid 22 da. before the assumed date of settlement, hence the total amount of interest due as shown on the debit side is reduced by the interest on $600 for 22 da., or $2.20. In like manner the $900 paid on May 20, or 12 da. before June 1, reduces the debit interest by $1.80 more. The total credit interest is $4, leaving a balance of $5.506 interest due. The balance of the account is $1540. The interest on $1540 for 1 da. is $.2566, hence if the assumed date of settlement had been 1 da. earlier, the balance of interest would have been $.2566 less. To find the number of days earlier than June 1, which the account should have been settled, divide $5.506 by $.2566, which gives 21 da. Therefore the equated date is 21 da. before June 1, or May 11.

PROOF

Due	Amt.	Days	Int.	Dis.	Due	Amt.	Days	Int.	Dis.
1910					1910				
May 1,	$1840	10	$3.066		May 10,	$600	1	$.10	
June 1,	$1200	21 (Dis.)		$4.20	May 20,	$900	9 (Dis.)		$1.35
					May 11,	$1540	0	.00	
	$3040		$3.066	$4.20		$3040		$.10	$1.35

$4.20 — $3.066 = $1.134, debit discount.

$1.35 — $ .10 = $1.25, credit discount.

$1.25 — $1.134 = $.116, less than half a day's interest on the balance, $1540.

Write the balance, $1540, with equated date on the credit side of the account as shown. Equating the account with May 11 as the focal date, and balancing the items of interest and discount, makes the credit side of the account $.116 more than the debit side. Since the interest on $1540 for 1 da. is $.2566, the difference between the two sides is less than half the interest for 1 da. Inasmuch as the law does not recognize a fraction of a day, the account as it now stands is considered as being in balance, and the solution proves.

2. Equate the following account:

CHARLES R. GOODRICH

1910					1910			
June	10	Mdse. 60 da.	800		June	30	Cash	1000
	25	Mdse. 3 mo.	1260		July	10	Note, 2 mo.	1000
July	1	Mdse. 1 mo.	900					

Due	Amt.	Days	Int.	Due	Amt.	Days	Int.
1910				1910			
Aug. 9,	$800	47	$6.267	June 30,	$1000	87	$14.50
*Sept. 25,	$1260	0	0.00	Sept. 10,	$1000	15	$2.50
Aug. 1,	$900	55	$8.25				
	$2960		$14.517		$2000		$17.00

$2960 − $2000 = $960, balance of the account.

$17.00 − $14.517 = $2.483, credit balance of interest.

Interest on $960 for 1 da. = $.960 ÷ 6 = $.16.

$2.483 ÷ $.16 = 15.5; 16 da.

Sept. 25 + 16 da. = Oct. 11, equated date.

First find the due dates of the several items having terms of credit. 60 da. after June 10 is Aug. 9; 3 mo. after June 25 is Sept. 25; and 1 mo. after July 1 is Aug. 1. The several amounts due are written opposite their respective due dates. On the credit side of the account, the cash was paid June 30, but the note is not due until Sept. 10.

The latest date on either side is taken as a focal date, which is Sept. 25. If the account were not settled until Sept. 25, the $800 due Aug. 9 would be overdue 47 da., the interest on which would be $6.267. So, also, the $900 due Aug. 1 would be 55 da. past due, with accrued interest of $8.25. Against these charges of interest there stand on the credit side $1000 paid June 30, or 87 da. before the assumed date of settlement, the interest on which is $14.50, and the note of $1000 due and paid Sept. 10, 15 da. before Sept. 25, the interest on which is $2.50.

By balancing the account it is found that while there is $960 due the holder of the account, he is owing a balance of $2.483 interest on payments made before the assumed date of settlement. Hence, the debtor has a right to keep the balance of the account, $960, as many days after Sept. 25 as it will take $960 to earn the balance of interest, $2.483. $960 earns $.16 a day and to earn $2.438 will require 16 da., which added to Sept. 25 gives Oct. 11 as the equated date.

The proof is the same as in the preceding example.

952. Counting back or forward from the focal date.

1. When to count back.

If the balance of the account and the balance of interest are on the same side of the account, that is, both on the debit side, or both on the credit side, the person who owes the balance of the account owes the balance of interest also; hence he should have settled the account *earlier*.

2. When to count forward.

If the balance of the account and the balance of interest are on opposite sides of the account, that is; one balance on the credit side, and the other on the debit side, the person who owes the balance of the account has the balance of interest due him; hence he has the right to keep the balance of the account *after* the focal date long enough to earn the balance of interest

953. Equate the following accounts:

1.   C. H. STURGIS

1909				1909			
Sept.	3	Mdse.	720	Sept.	6	Cash	600
	18	"	960		25	"	1000
Oct.	20	"	600				

2.   D. H. OGLETHORPE & SON

1910				1910			
Jan.	2	Mdse.	1150	Jan.	10	Cash	500
	24	"	900		15	"	600
Feb.	8	"	780	Feb.	1	"	1000
	16	··	1500				

3.   L. D . THORNBURG & Co.

1909				1909			
Nov.	1	Mdse., 60 da.	840	Dec.	7	Cash	1500
	15	" 30 da.	560	1910			
Dec.	1	" 3 mo.	1860	Jan.	4	"	1000
	15	" 1 mo.	1000		10	Note, 10 da.	500

4.   D. L. SUTHERLAND

1910				1910			
Feb	1	Mdse., net	125	Feb.	8	Cash	100
	11	" 30 da.	450		15	Note, 30 da.	500
	20	" 60 da.	750	Mar.	1	" 60 da.	
	28	" 30 da.	1000			with int.	1200

CASH BALANCE

954. Cash balance, or accounts current, treats of finding the balance due on an account on a given date.

955. Interest accrues, and is collectible, on overdue accounts. The debtor is entitled to discount on all items paid before maturity.

Retailers seldom charge interest on overdue accounts except on balances brought down quarterly, semiannually, or annually, and then by agreement. Wholesalers generally do charge interest on overdue accounts.

956. If an account has been equated, the balance due on a given date is the balance of the account plus the interest or less the discount for the time between the equated date and the date of settlement, according as the date of settlement is after or before the focal date. Generally, however, the cash balance of an account is determined as shown in the following example:

957. If money is worth 6 %, find the cash balance of the following account Jan. 1, 1911.

JOHN R. CALDWELL

1910					1910			
July	8	Mdse., 2 mo.	1260		Sept.	15	Cash,	1000
Sept.	1	" 3 mo.	1848	75	Oct.	10	Cash,	1200
Nov.	10	" 2 mo.	600		Dec.	1	Note, 10 da.	500

	AMT.	DAYS	INT.	DIS.		AMT.	DAYS	INT.
1910					1910			
Sept. 8	$1260	115	$24.15		Sept. 15	$1000	108	$18.00
Dec. 1	$1848.75	31	$9.55		Oct. 10	$1200	83	$16.60
1911					Dec. 11	$500	21	$1.75
Jan. 10	$600	9 dis.		$.90				
	$3708.75		$33.70	$.90		$2700		$36.35

$3708.75 + $33.70 − $.90 = $3741.55, total debit.

$2700 + $36.35 = $2736.35, total credit.

$3741.55 − $2736.35 = $1005.20, balance due.

Write all items, both debit and credit, with their respective due dates as shown in the solution. Find the number of days from the due date of each item to Jan. 1, 1911. Thus, the time from Sept. 8 to Jan. 1 is 115 da., and the interest on $1260 for 115 da. is $24.15. The $1848.75, due Dec. 1, is overdue 31 da., and $9.55 interest has accrued. The $600 due Jan. 10 is due 9 da. after Jan. 1, and the debtor is entitled to $.90 discount if it is paid Jan. 1. On the credit side, $1000 was paid 108 da. before the day of settlement, and in that time has earned $18 interest. The $1200 was paid 83 da. before settlement, and has earned $16.60 interest, and the $500 has earned $1.75 interest.

The gross amount due, including interest and discount, is $3741.55, against which there has been paid $2700 plus the total credit interest, $36.35, making $2736.35. The balance of the account is, therefore, $3741.55 − $2736.35 or $1005.20.

NOTE. Compare this solution with that of partial payments by the Merchants' Rule, page 309.

PROBLEMS

958. **1.** How much is due on the following account July 1, 1911, interest at 6 %?

L. M. BOARDMAN

1911					1911			
Jan.	1	Balance	1328	75	Feb.	1	Cash,	1000
Mar.	10	Mdse., 3 mo.	1475		Apr.	10	Note, 1 mo.	1200
Apr.	1	" 30 da.	875		June	10	Cash,	800

2 Find balance due Oct. 1, 1910, at 5 %.

S. L. PUTZMAN

1910					1910			
May	1	Balance,	2896	50	May	14	Note, 3 mo.	2500
June	1	Mdse., 2 mo.	1824	70	June	20	Cash,	1000
July	12	" 10 da.	1022	60	Sept.	1	"	1000

3. How much is due Apr. 1, 1910, interest at 7 %?

M. N. ORSON

1910					1910			
Jan.	18	Mdse., 2 mo.	4827	50	Feb.	20	Cash,	4000
Feb.	12	" 1 mo.	3600		Feb.	20	Note, 2 mo.	3000
Mar.	1	" 10 da.	1200					

PARTNERSHIP

959. A **partnership** is an association of two or more persons for the purpose of carrying on a legal business and dividing the profits.

960. The **profits**, or **losses**, of a partnership are shared according to an agreement made in advance. The agreement may be to share equally, according to investment, according to average investment, or according to some other specified proportion.

961. A **general partner** is one who takes an active part in the business and who is known to the public as a partner.

962. A **nominal partner** is one who is held forth as a partner, and as such is liable to innocent third parties for the liabilities of the firm.

963. A **dormant**, or **secret partner** is a general partner with the exception that his connection with the partnership is unknown to the public. If his membership in the firm becomes known, he is liable for the debts of the partnership.

964. A **special, or limited partner** is one whose liability for the debts of the firm is limited to the amount of his investment.

965. The **capital** of a firm consists of all money or other form of property invested in the business.

966. The **resources** of a firm consist of all money and property belonging to the firm, and all debts or obligations owing to the firm.

967. The **liabilities** of a firm consist of all its debts or obligations to others.

968. The **present worth** of a firm is the excess of its resources over all its liabilities. It is also the sum of the net investment and the net gain.

969. The **net investment** of a firm is the difference between the total investments and the total withdrawals.

970. The insolvency of a firm is the excess of its liabilities over its resources.

971. The average investment of a firm is a sum which, if placed at interest for a given unit of time (day, month, or year) will produce the same amount of interest as the various amounts invested by the partners for different lengths of time.

972. The net gain is the excess of gains over losses

973. The net loss is the excess of losses over gains.

974. Apportionment of gains and losses when investments are made for equal periods.

EXAMPLE

1. A and B each invest $5000. They gain $3000 in one year. Find gain of each

$5000 + $5000 = $10,000, total investment.

$\frac{5000}{10000} = \frac{1}{2}$, each partner's share of the gains.

$\frac{1}{2}$ of $3000 = $1500, each partner's gain.

PROBLEMS

975. 1. Three partners invest $3000, $4000, and $5000, respectively, for two years. Their gain is $15,000. Find the present worth of each at closing.

2. C and D begin business with a joint capital of $12,000. At the end of 4 years their present worth is $28,000. Find each partner's gain if they share equally.

3. Jan. 1, 1909, a firm's liabilities exceeded its resources by $1200. Two years later its present worth was $15,000 If there were three partners sharing equally, with what amount of gain should each be credited?

4. Two men began business together with $3000 borrowed money. In three years' time their resources exceeded their liabilities by $12,000. What was their average annual gain?

5. Three partners' interests in a business were in the ratio of 4, 5, and 6. During a certain year they gained $13,500. Their present worth at the end of the year was $9000. What was the condition of each partner's account at the beginning of the year?

6. Two partners, whose interests were in the ratio of 2 and 3 suffered a loss of $8000 in the year 1910. Dec. 31, 1910, the firm was insolvent $3000. What was the condition of each partner's account Jan. 1, 1910?

7. E and F share equally in gains and losses. E has invested $6000, and F $7000. At the end of one year their resources amount to $24,000, and their liabilities to $5000. Find each partner's present worth.

8. The net gain of a firm for one year is 25 % of the original investment. At the close of the year the present worth of the firm is $25,000. If there are two equal partners, find the net gain of each.

9. Messrs. Brown & Brooks form a partnership with an investment of $16,000, one half of which is contributed by each partner. Their net gains and net losses for three years are as follows: first year, net gain $6000; second year, net loss $2000; third year, net loss $10,000. Find the condition of each partner's account at the close of each year.

10. Three partners, A, B, and C, invest $8000, $9000, and $10,000, respectively. They are to share the gains and losses in proportion to their investments. Their total gains for a year are $16,000; their losses are $5200. Find each partner's present worth at the end of the year.

11. X, Y, and Z, are partners sharing gains and losses according to investment. X invests $5000; Y $8000; and Z $7000. At the end of their first year of business the firm's present worth is $32,000. Find each partner's gain.

12. D, E, and F are partners. D is to receive ⅖ of the gains, E ⅖, and F ⅕. Their total investment is $16,000. At dissolution they have $14,800 in the bank, $8000 worth of merchandise, and notes on hand $2200. How shall the net gain be divided?

13. Parker, Rogers, and Robinson share gains and losses according to investment. Parker invests $7000, Rogers $8000, and Robinson $9000. At the close of the first year, their ledger shows gains and losses as follows: merchandise, gain $13,600; expense, loss $4700; interest, gain $221.60; freight, loss $87.50; real estate, gain $1234.30. What is each partner's present worth?

976. Apportionment of gains and losses according to average investment.

A and B enter into partnership agreeing to share gains and losses according to average investment. A invests, Jan. 1, 1910, $4000; Apr. 1, $3000; and withdraws, Sept. 1, $2000. B invests, Jan. 1, 1910, $5000; June 1, $4000; and withdraws, Oct. 1, $2000. Dec. 31, 1910, they have a net gain of $8940. Find each partner's present worth.

A

1910				1910			
Sept.	1	*Present Worth*	2000	Jan.	1		4000
Dec.	31		9020	Apr.	1		3000
				Dec.	31	Net Gain	4020
			11,020				11,020
				1910			
				Jan.	1	Present Worth	9020

B

1910				1910			
Oct.	1	*Present Worth*	2000	Jan.	1		5000
Dec.	31		11920	June	1		4000
				Dec.	31	Net Gain	4920
			13,920				13,920
				1910			
				Jan.	1	Present Worth	11,920

A invests	$4000 for 3 mo. =	$12,000 for 1 mo.
	7000 for 5 mo. =	35,000 for 1 mo.
	5000 for 4 mo. =	20,000 for 1 mo.
	which =	$67,000 for 1 mo.

B invests	$5000 for 5 mo. =	$25,000 for 1 mo.
	9000 for 4 mo. =	36,000 for 1 mo.
	. 7000 for 3 mo. =	21,000 for 1 mo.
	which =	$82,000 for 1 mo.

$67,000 + $82,000 = $149,000, firm's total investment for 1 mo.

A's share of the gain $= \frac{67}{149}$ of $8940, or $4020.

B's share of the gain $= \frac{82}{149}$ of $8940, or $4920.

A's investment, Jan. 1, 1910, of $4000 remains unchanged until Apr. 1, a period of 3 mo., which is equivalent to an investment of $12,000 for 1 mo.

Apr. 1, he invests $3000 more, making his total investment $7000, which remains unchanged until Sept. 1, or 5 mo., and is equivalent to an investment of $35,000 for 1 mo. Sept. 1, he withdraws $2000, leaving a balance of $5000 invested in the business for the remainder of the year, or 4 mo., and equivalent to $20,000 for 1 mo. A's average investment, then, is equivalent to an investment of $67,000 for 1 mo.

In the same way B's average investment is found to be equivalent to an investment of $82,000 for 1 mo. Hence, the firm's average investment for the year is equivalent to $149,000 invested for 1 mo. Therefore,

A's share of the gain is $\frac{67}{149}$ of $8940, or $4020, and B's share of the gain is $\frac{82}{149}$ of $8940, or $4920.

The simplest method of determining the condition of the partners' accounts is to write them in the form of ledger accounts, as shown in the solution. Each partner is credited with his investments and charged with his withdrawals. He is then credited with his net gain (or charged with his net loss) and the account is closed the same as in a ledger.

PROBLEMS

977. 1. Three men are in partnership and make a profit of $16,350. A invests, Jan. 1, 1910, $3000, and July 1, $2000. B invests, Jan. 1, 1910, $4000, May 1, $2000, and Sept. 1, $1000. C invests, Jan. 1, 1910, $8000, Mar. 1, $3000, and withdraws, Aug. 1, $4000. Find the present worth of each partner.

2. C and D are partners; C invests $1200 for 5 mo., and then increases his investment to $2000 for 5 mo. D invests $1500 for 6 mo., and then withdraws $500 for 4 mo. At the expiration of 10 mo. they dissolve partnership, having resources amounting to $8800. To how much of the resources is each partner entitled?

3. E and F begin business as partners, Jan. 1, 1908, each investing $10,000. Jan. 1, 1909, G is admitted into the firm with an investment of $12,000. E increases his investment $5000, and F invests $8000 more on the day G is taken into the business. Jan. 1, 1910, they have resources amounting to $78,000 and liabilities, $7000. Find the condition of each partner's account.

4. A and B engage to perform a given piece of work for $580, the work to be completed in 40 da. A has furnished 3 men for 22 da., and B 4 men for 19 da. Only 10 da. of time remain, and in order to finish the work on time, A engages 5 extra men, and B 4 extra men. How shall the $580 be divided?

5. X and Y engaged in business with a capital of $9000, of which X furnished $\frac{2}{3}$ and Y $\frac{1}{3}$. At the end of 1 yr. X increased his investment to $10,000, and Y made an additional investment of $4000. At the end of the second year they had resources to the amount of $13,000, and liabilities to the amount of $36,582. What was each partner's net insolvency?

6. Two men engaged in partnership for 2 yr. A invested at first $5000; 8 mo. later he put in $2000; and at the end of 18 mo. withdrew $1500. B invested at first $8000; at the end of 1 yr. he withdrew $1000; and 4 mo. later put in $3000. At the expiration of the term of the copartnership their debts were all paid, and they had resources amounting to $21,675. Divide the net gain or loss in proportion to average investment, and find each partner's present worth.

7. Three men engaged in partnership for 5 yr. from Jan. 1, 1905, on which date X made an investment of $6000, Y $8000, and Z $5000. July 1, 1906, X invested $4000 more; Y withdrew $2000, and Z put in $3000. Jan. 1, 1908, each partner increased his investment to $12,000. X was to have a salary of $1200; Y, of $1500; and Z, of $1800 per annum. If no salaries had been drawn and no profits divided until the expiration of the five years, what was each partner's present worth at that time, their total resources being $87,500, and their liabilities $12,500?

PARTNERSHIP SETTLEMENTS

978. 1. F and G enter into a copartnership Jan. 1, 1910. At that time F invests $45,000, but withdraws $5000 July 1. G invests $50,000 on Jan. 1, and withdraws $3000 on May 1, and $4000 on Aug. 1. Interest at 6 % per annum is to be allowed on investments and charged on withdrawals. F is to have $\frac{5}{11}$ and G $\frac{6}{11}$ of the gains, the losses being shared in the same ratio. Dec. 31, 1910, their ledger shows resources and liabilities as follows: Cash, $42,000 ; merchandise, $65,000; accounts receivable, $24,000; bills receivable, $18,000; bills payable, $16,000; and accounts payable, $11,000. They make an allowance of 5 % for bad debts on bills receivable and accounts receivable. Determine the present worth of each at the time of closing.

2. C and D began business July 1, 1909. C invested $10,000, and D invested $9000. Nov. 1, 1909, C withdrew $2000, and D

invested $1000. Jan. 1, 1910, E was admitted to the partnership with an investment of $5000. Apr. 1, 1910, E invested $3000 more, and D withdrew $1000. Dec. 31, 1910, D and E purchased C's interest in the business. On that date their books showed the following resources and liabilities: Cash, $18,750, merchandise, $17,000; bills receivable, $9600; accounts receivable, $8425; interest due on bills receivable, $237.50; real estate, $5000; accounts payable, $13,500; bills payable, $4800; interest on the same, $131.25; and sundry unpaid bills, $362.50. Good will was estimated at $3000, and $500 was reserved as a fund to meet bad debts. Each partner was entitled to interest at 6 % on his net investment. Gains and losses were to be shared in the proportion of C $\frac{3}{8}$, D $\frac{3}{8}$, and E $\frac{2}{8}$. D and E were to pay C such amounts that their investments would be equal. How much should each pay C?

3. Following is given the balance of each account in John W. Dolson's ledger, and the inventories, Dec. 31, 1910. Make a trading account, a profit and loss account, and a balance sheet.

John W. Dolson (investment account) Cr.	$30,910.67
John W. Dolson (private account) Dr.	2,040.00
Merchandise (inventory, Jan. 1, 1910)	22,201.29
Purchases (Jan. 1, 1910, to Dec. 31, 1910)	27,499.26
Sales (Jan. 1, 1910, to Dec. 31, 1910)	31,400.60
Bills Receivable	297.40
Accounts Receivable (considered good)	6,710.30
Accounts Receivable (considered worthless)	320.00
Accounts Payable	8,646.45
Bills Payable	2,400.75
Real Estate (cost)	7,500.00
Fixtures (cost)	1,210.00
Traveling Expenses	662.50
Salesmen's Salaries	1,200.00
Office Salaries	720.00
Advertising	525.73
General Expense	1,524.27
Cash on hand	947.72

INVENTORIES, DEC. 31, 1910

Merchandise on hand	$26,660.27
Real Estate, valued at	7,500.00
Fixtures	1,075.00
Due J. M. Hart (bill for blank books and stationery)	67.30

Trading Account, John W. Dolson,
for period ended December 31, 1910.

Inventory, Jan. 1,				Sales			31400	60
1910,	2220.12							
Purchases	27499.26	49700	55					
Deduct								
Inventory, Dec.								
31, 1910		26660	27					
Turnover, cost of								
goods sold		23040	28					
Gross Profit on								
Trading		8360	32					
		31400	60				31400	60
Traveling Expenses	662	50		Gross Profit brought				
Advertising	525	73		down			8360	32
Salesmen Salaries	1200	—	2388	23				
Net Profit on Trading								
carried to Profit &								
Loss Account			5972	09				
			8360	32			8360	32

These are the statements called for in problem 3. Examine them very carefully. Verify all results given.

4. J. B. Altgeld and L. M. Goodman are partners, and share losses and gains in the ratio of $\frac{2}{3}$ and $\frac{1}{3}$, respectively. Prepare a trial balance, a trading account, a profit and loss account, and a balance sheet from the following data: J. B. Altgeld invested $20,079.16, and withdrew $1750; L. M. Goodman invested $10,039.58, and withdrew $1260.50. Other ledger balances were as follows: cash, $12,711.30; bills receivable, $3974.18; bills payable, $8250.50; accounts receivable, $6356.45; accounts payable, $10,976.72; mer-

Profit & Loss Account, John W. Dolson, for period ended December 31, 1910.

General Expense charged	1524.27		Net Profit brought from Trading Account		5972.09
Bill, J. M. Hart	67.30	1591.57			
Office Salaries		720.—			
Fixtures Cost	1210.—				
Inventory, Dec. 31, 1910,	1075.—	135.—			
Bad Debts	320.—	2766.57			
Balance being the Net Business Profit carried down		3205.52			
		5972.09			5972.09
John W. Dolson		3205.52	Net Business Profit		3205.52

chandise sales, $ 26,119.18; raw material, cost, $ 8416.75; plant and machinery, $ 10,500; real estate, $ 14,750; interest, credit, $ 658.61; insurance, $ 125; expense, general, $ 5505.71; advertising, $ 1375; salesmen's salaries, $ 5000; traveling expenses, $ 1908.61; office salaries, $ 1250; commissions, debit, $ 1240.25. The inventories at the close of the year were: merchandise, $ 1615.80; raw materials, $ 1250; plant and machinery, $ 9500; real estate, $ 16,000.

5. On June 30, 1911, Chester Mack had inventories as follows: Merchandise, $ 7400; real estate, $ 9500; blank books and stationery, $ 28.50. The cash balance as shown by the ledger was $ 4360. Footings of other ledger accounts were: *Debits* — Chester Mack (proprietor), $ 1960; merchandise, $ 18,200; First National Bank, $ 4608.25; real estate, $ 9000; bills receivable, $ 6420; interest and discount, $ 127.40; expense (general), $ 425.60; J. Wilson, $ 3872.75; insurance, $ 112.50; repairs, $ 87.50; office salaries, $ 2500; sales-

Balance Sheet, John W. Dolson, as at close of business December 31, 1910.

Assets			Liabilities			
Floating			**Floating**			
Cash on hand	947 72		Bills Payable			2400 75
Bills Receivable	297 40	1245 12				
Accounts Receivable			Accounts Payable			
Customers' balances		6710 30	Trade Creditors	8646 45		
			J. M. Hart Special	67 30	8713 75	
Inventory						
Merchandise	2666 27					
Fixtures	107 5—	2773 527	Capital Account			
			Investment	3091 067		
			Net Profit	3205.52		
Fixed			Less Drawings 2040 —	1165 52		
Real Estate		7500 —	Present Worth			3207619
		43190 69				43190 69

man's salary, $1800; traveling expenses, $937 60; bonds (cost), $10,500; freight (inward), $231.50; freight (outward), $318.50; merchandise returned by debtors, $297.50. *Credits* — Chester Mack (investment), $29,767.58; merchandise, $18,700; First National Bank, $2500; real estate, $250; accounts payable, $4825; bills payable, $3265; interest and discount, $192.80; interest on bonds, $500; merchandise returned to creditors, $347.50 ; J. Wilson (to be supplied by the student). Prepare a trial balance, a trading account, a profit and loss account, and a balance sheet.

· BUILDING AND LOAN ASSOCIATIONS

979. A building and loan association is a private corporation whose object is "to encourage industry, frugality, home building, and saving among its members." To accomplish this object, each member of the association subscribes for one or more shares of stock (par value from $25 to $500, according to the laws of the various states), and agrees to pay for these in weekly or monthly installments of 25¢ (a week) or $1 (a month) per share. The dues so received are loaned to some member of the association who wishes to borrow the money for the purpose of buying or building a home. No one person is allowed to borrow more than the total par value of the shares he owns. He is charged interest on the loan at the legal rate.

980. The sources of profit in an association are interest on loans, premiums on loans, fines against members who fail to pay their dues promptly, and the withdrawal of members who find it undesirable to continue in the association. Persons withdrawing have returned to them all the dues they have paid on their shares, and a portion of the profits. The remaining portion of the profits is credited to shares still in force. The book, or actual, value of a share before maturity is, therefore, somewhat greater than its withdrawal value.

981. Building and loan associations are of three classes, viz., terminating, serial, and permanent.

Shares in the **terminating association** are all dated alike, and hence all mature at the same time. A person entering the association after its organization is required to pay back dues on his shares from their date. When the dues and profits together amount to the par value of the stock (generally $100 or $200), all shares are canceled, each member, if he has not borrowed from the association, receives in cash the value of his shares, and the association ends.

The **serial plan of association** is, in effect, a number of terminating associations joined in one. That is to say, as many shares as possible dated Jan. 1, for instance, are sold. After a few meetings of the association there will be others who wish to join, but who do not wish to pay dues back to Jan. 1. Hence, a new series of shares is issued, in all respects like the first series, except that it is dated Apr. 1 or June 1, as the case may be, and

matures three or six months later than the first series. Additional
series may be issued as occasion requires, thus making the association practically permanent.

The **permanent plan of association** differs from the other plans as
follows:

1. New members may join at any time without paying back dues.
Each member's share or shares may be considered a series by itself.

2. Paid-up stock is issued.   Shares of stock are not necessarily
canceled when they mature.   If the association is able to loan the
funds, instead of canceling the stock, a certificate showing that the
shares are fully paid is issued, and the holder thereof receives his
semiannual dividends in cash.   No more dues are paid on such shares,
and they may be withdrawn at any time.

3. **Earnings** are ascertained and divided semiannually, and when
credited are subject to withdrawal the same as money payments.

982. The serial plan is the most popular.   The following problems
and discussion will, therefore, be confined to that plan of association.

983. To find the withdrawal value of a given number of shares.

X owns 20 shares in a building and loan association and has
paid dues at $1 per month per share for 6 yr. when he wishes to
withdraw.   If he is allowed profits at 4 % per annum, to what
amount is he entitled?

$72 \times \$20 = \$1440$, total amount paid in dues.

$\$1440 \times \dfrac{36\frac{1}{2}}{12} \times .04 = \$175.20$, profits.

$\$1440 + \$175.20 = \$1615.20$, withdrawal value.

On 20 shares at $1 per month per share, the total amount of dues paid in are
$1440.   The first payment of dues ($20) has been earning profits for 72 mo.,
the second payment 71 mo., the third 70 mo., and so on to the 72d payment, which has been earning profits only 1 mo.   The investment is, therefore, a decreasing arithmetical series having 72 as the first term and 1 as
the last term, and the number of terms, 72.   The sum of the series is equal to
the product of the sum of the first and last terms multiplied by half the
number of terms and equals, in this case, $[(1 + 72) \times \frac{72}{2}]$, or 2628.   The
usual method of computing the interest is to reckon it on the total dues paid
in for the average, or equated, time.   The average time is found by dividing
the total number of months, 2628, by 72, the number of months dues have
been paid, which gives 36½.   The interest on $1440 for 36½ mo. at 4% is
$175.20.   The dues paid in, $1440, plus the profits, $175.20, gives $1615.20,
the amount X can withdraw.

PROBLEMS

984. 1. A has paid dues of $1 a month per share on 18 shares in a series for 4 yr. He wishes to withdraw and is entitled to profits at 5 % per annum. How much will he receive?

2. B wishes to withdraw from his association after having paid his dues of $1 a month per share on 25 shares regularly for 10 yr. If he is entitled to 6 % profits, what sum will he receive?

3. C, after 5 yr., is unable to continue his payments of $1 a month per share on 30 shares in a building and loan association, and withdraws therefrom. There are unpaid fines* against him amounting to $3.60. If his profits are estimated at 3½ % per annum, what is the withdrawal value of his shares?

985. **To make distribution of profits.**

An association issued four series of shares as follows: First series, 400 shares, dated Jan. 1, 1908; second series, 350 shares, dated July 1, 1908; third series, 500 shares, dated Jan. 1, 1909; fourth series, 300 shares, dated July 1, 1909. The dues in each series were $1 per share per month. July 1, 1910, the entire profits were $2945.93. What was the value of a share of each series at that date?

$30 × 400 × 15½ = $186,000, first series' investment for 1 month.

$24 × 350 × 12½ = $105,000, second series' investment for 1 month.

$18 × 500 × 9½ = $85,500, third series' investment for 1 month.

$12 × 300 × 6½ = $23,400, fourth series' investment for 1 month.

$399,900, total investment for 1 month.

Share of first series is $\frac{186000}{399900}$, or $\frac{620}{1333}$ of $2945.93 = $1370.20.

Share of second series is $\frac{105000}{399900}$, or $\frac{350}{1333}$ of $2945.93 = $773.50.

Share of third series is $\frac{85500}{399900}$, or $\frac{285}{1333}$ of $2945.93 = $629.85.

Share of fourth series is $\frac{23400}{399900}$, or $\frac{7}{1333}$ of $2945.93 = $172.38.

$1370.20 ÷ 400 = $3.43, profit on 1 share of first series.

$773.50 ÷ 350 = $2.21, profit on 1 share of second series.

$629.85 ÷ 500 = $1.26, profit on 1 share of third series.

$172.38 ÷ 300 = $0.57, profit on 1 share of fourth series.

$30 + $3.43 = $33.43, value of 1 share first series.

* All unpaid fines are deducted from the amount due on shares.

$24 + $2.21 = $26.21, value of 1 share second series.

$18 + $1.26 = $19.26, value of 1 share third series.

$12 + $0.57 = $12.57, value of 1 share fourth series.

The principle that applies here is the same that applies to partnership settlement by average investment. Hence, first find the investment of each series for 1 mo. Dues of $1 a month on each of 400 shares have been paid for 30 mo. in the first series (the average time is $15\frac{1}{2}$ mo. See explanation, page 404), which makes an average investment of $30 × 400 × $15\frac{1}{2}$ = $186,000 for 1 mo. The average investments in the other series are found in the same way, making a total investment for 1 mo. of $399,900. Obviously, the share of the profits belonging to each series is in the ratio of their respective average investments, as shown in the solution. The profit per share in each series is found by dividing the entire profit belonging to each series by the number of shares in that series. The value of a share in each series is the sum of all dues paid on the share and the profit per share.

PROBLEMS

986. 1. A building and loan association has issued a new series of shares at the beginning of each year. In the first series are 500 shares; in the second series, 600 shares; in the third series, 400 shares; and in the fourth series, 500 shares. Dues in all series are $1 per month. At the end of the fourth year, the profits are $5325. Find the value of a share in each series at that time.

2. July 1, 1905, a building and loan association began business, and issued a first series of shares numbering 600. A second series of 500 shares was dated July 1, 1906. A third series of 700 shares was dated Jan. 1, 1908; and a fourth series of 800 shares was dated Jan. 1, 1909. Dues in all series were $1 per month. Find the value of a share of each series, July 1, 1910, if the profits were $9575.

987. In the association from which the following figures are taken a new series is open for subscriptions every 3 mo. Dues are 25¢ a week per share. In this statement series No. 26 has been open 704 wk., there being 43 shares in the series, Dec. 31, 1910. The total subscription paid in on series 26 is equal to 25¢ a week for 704 wk., or $176 per share, and on 43 shares, the total is $7568. During 1910, the subscriptions equal $559 (52 wk. at 25¢ on each of 43 shares). Deducting $559 from $7568 gives $7009, the total subscriptions paid in Dec. 31, 1909, which amount is earning profits during all of the year 1910. Profits for 1910 are paid on one half of the subscriptions paid in in 1910, or on $279.50, which added to $7009 gives $7288.50, the total amount on which profits are allowed for 1910 in series No. 26. The per cent of profit is found by dividing the total profit on all the series by the total subscriptions in all the series sharing in profits, as shown by the footing of the column headed "Total Profit-sharing Subscriptions, Dec. 31, 1910." The same rate is paid on all series.

988. STATEMENT SHOWING HOW PROFITS ARE DISTRIBUTED

SERIES NO.	WEEKS	SHARES	SUBSCRIPTIONS IN FULL TO DEC. 31, 1910	SUBSCRIPTIONS FOR 1910	SUBSCRIPTIONS TO DEC. 31, 1909	INCREASE IN PROFIT SHARING SUBSCRIPTION, 1910	TOTAL PROFIT SHARING SUBSCRIPTION, DEC. 31, 1910	RATE PER CENT OF PROFITS	PROFIT PER SERIES
26	704	43	7,568	559	7,009	279.50	7,288.50	$6\frac{1}{2}$	473.75
27	691	9	1,554.75	117	1,437.75	58.50	1,496.25	$6\frac{1}{2}$	97.25
28	678	68	11,526	884	10,642	442	11,084	$6\frac{1}{2}$	720.46
29	665	27	4,488.75	351	4,137.75	175.50	4,313.25	$6\frac{1}{2}$	280.36

PROBLEMS

989. 1-4. Prepare a statement like the above for series 30, 31, 32, 33, subscriptions at 25¢ a week per share having been paid for 652, 639, 626, 613, wk., respectively, the number of shares being 36, 14, 45, 76, and the rate of profit being 7 %.

5. A coöperative store, having a capital stock of 400 shares of $100 each, makes a profit on sales in one year of $12,500. The operating expenses are $4800. Find the book value of 17 shares. If there is reserved for depreciation, reserve fund, etc., $2500, find the cash dividend that can be paid on 15 shares.

EXAMINATIONS

SPEED TEST

990. Minimum time twenty minutes; maximum, one hour. Deduct one credit for each minute required beyond the minimum time.

1. Find the market value in each of the following cases:

PAR VALUE	MARKET PRICE	MARKET VALUE
$7,500	110	——
7,900	90	——
67,200	$62\frac{1}{2}$	——
475,200	$131\frac{1}{4}$	——

2. Apportion the gains and losses according to investments in the following:

A	B	C	
$5,000	$7,000	$————	$1,800 gain.
3,000	5,000	7,000	7,500 gain.
2,500	5,000	10,000	7,000 loss.
15,000	14,000	13,000	840 loss.

3. Equate the following account:

June 1, Mdse , 60 da., $6000.   July 1, Cash, $3000.
                                                  16, Cash, 1500.

4. Find the total duty on the following imports:

ARTICLE	QUANTITY	VALUE	AD. VAL. DUTY	SPECIFIC	TOTAL DUTY
Drugs	2500 lb.	£1000	10%	$1\frac{1}{4}$ ¢ a pound	————
Perfumery	250 lb.	F. 3000	50%	60 ¢ a pound	————
Hosiery	100 doz. pair	M. 500	15%	70 ¢ a dozen pair	————
Rugs	3000 sq. yd.	£500	40%	10 ¢ a square foot	————

5. Find cost of each, and the total cost of the following drafts:

FACE	RATE	COST
$1,500	50 ¢ premium	————
1,875	$1.00 discount	————
3,500	$1.50 premium	————
5,000	10 ¢ discount	————
	Total	

6. Find the total cost of the following exchange:

£ 500 @ 4.8520	————
F 10,362.5 @ 5.18$\frac{1}{8}$	————
M. 4000 @ 95$\frac{1}{4}$	————
£ 1250 @ 4.8665	————
Total	

7. Find the par value of bonds necessary to produce income as follows:

BONDS	INCOME	PAR VALUE
New York City 4$\frac{1}{2}$'s	$1,800	————
United States 3's	2,400	————
C. B. & Q. 4's	2,000	————
Sea. Air Line 5's	3,000	————

8. Add vertically, horizontally, and prove by finding the sum of the totals:

$16,157	$11,273	$10,559	——
48,287	2,380	16,272	——
473,892	8,721	145,468	——

9. Find gain or loss and per cent of gain or loss:

Cost	Selling Price
$2.50	$3.00
3.60	4.50
5.00	6.50
120.00	110.40

10. Find total interest on the following:

$1200 for 50 days at $4\frac{1}{2}\%$.
1275 for 90 days at 6%.
2500 for 72 days at 5%.
3600 for 13 days at 7%.

WRITTEN TEST

991. 1. A sight draft on Chicago for $1771 is purchased at $\frac{5}{8}\%$ premium. Find the cost.

2. Divide $5000 into four parts proportional to 8.407, 5.149, 4.132, and 2.312.

3. A lot of woolen cloth weighing 1200 lb. is invoiced in Liverpool as 1575 yd. at 9s. 6d. per yard. Find the amount of duty paid in New York at 44¢ a pound and 55% ad valorem.

4. At 4.8725, what is the cost of a bill of exchange to settle for the cloth in problem 3? At what price per yard must it be sold to gain 25%?

5. My broker buys for me 200 shares C. M. & St. P. at $121\frac{3}{4}$, brokerage $\frac{1}{8}\%$. Thirty days later, I instruct him to sell at 122. Allowing $\frac{1}{8}\%$ for selling, find my gain or loss.

6. If the cost of erecting a $28,000 building is assessed on $450,000 capital stock, how much must a stockholder pay who owns 100 shares at $100 each?

7. A and B are in partnership. Gains and losses are shared in the ratio of $\frac{3}{4}$ to A and $\frac{1}{4}$ to B. June 30, 1910, their stock is worth $6000. Each is entitled to 5% interest on his investment from Jan. 1, 1910. Find from the following trial balance each partner's present worth, June 30, 1910.

TRIAL BALANCE, JUNE 30, 1910

A, Capital		$ 10,900
A, Drawings, including interest.	$ 900	
B, Capital		5,500
B, Drawings, including interest.	255	
Creditors		7,419
Sales		53,820
Stock, Jan. 1, 1910	6,014	
Debtors (all good)	16,997	
Cash, in hand	3,489	
Bad debts	195	
Cartage	1,019	
Discount	2,509	
Fire insurance	67	
Wages	4,553	
Rent, taxes	3,244	
Repairs	202	
Salaries	1,624	
Office Expenses	271	
Purchases	36,300	

8. On the Panama Canal Bonds, interest is payable quarterly — on the 1st of February, May, August, and November, the annual rate being 2 %. Find the cost, including interest, of $ 50,000 of these bonds on the 10th of June at 99¾. (Use accurate interest.)

9. Find the equated date:

1910			1910		
Mar. 1,	Mdse., 2 mo.	$ 1500	Apr. 1,	Cash,	$ 500
Apr. 10,	" 1 mo.	750	30,	"	1000
May 1,	" Net,	1000	May 1,	Note, 2 mo.	1500

10. Find the balance due Jan. 1, 1911. Interest 6 %.

1910			1910		
Oct. 1,	Balance,	$ 1546.72	Oct. 10,	Cash,	$ 1200
30,	Mdse., 2 mo.	1268.25	Nov. 15,	Note, 1 mo.	1000
Dec. 15,	" net,	1143.88	Dec. 10,	Note, 2 mo. with int.	1200

WRITTEN TEST

NEW YORK STATE REGENTS' EXAMINATION QUESTIONS

992. 1. Find the cost of 3 tons 430 lb. of coal at $ 5.75; and 15 pieces of hemlock 6″ × 8″ × 20′ at $ 29 per M.

2. How many yards of carpet 27 in. wide will be required to carpet the floor of a room 24 ft. long and 18 ft. wide, if the strips run the long way of the room, and 6 in. are allowed on each strip for loss in matching?

3. What must be the list price of goods that cost $410 in order to make a profit of 20 %, if they are to be sold at a discount of 20 % and 10 % from the list price?

4. On May 2, 1910, J. Williams bought the following goods less a discount of 5 %, terms 2/10, n/60; 5 pieces velveteen, 42, 44, 43, 42, 45 yd. at 25 ¢ a yard. Williams sold the goods May 3 at 35 ¢ a yard, terms 1/10, n/30. He received cash for his sale on May 9, and paid for his purchase on May 10. Find his profit.

5. On an asking price of $400, which is the more favorable for the buyer, and how much: a discount of 30 %, 20 %, and 12½ %, or a discount of 40 %, 12½ %, and 10 %?

6. On July 1, 1910, a trader's stock of goods was destroyed by fire, but he saved his books of record; his goods were fully insured, and he proved his loss from the following facts: value of goods on hand Jan. 1, 1910, $21,500; purchases from Jan. 1, 1910 to July 1, $54,300; sales from Jan. 1 to July 1, $63,750. His records showed that all sales were made at an average advance of 25 % above cost. Find the value of the stock destroyed by fire.

7. A grain merchant received $22.26 for selling a quantity of grain; if he charged 2 % commission and sold the grain at $1.06 per bushel, how many bushels did he sell?

8. A 4 months' note for $875, without interest, dated May 17, 1910, was discounted at a bank July 11, 1910, at 6 %. Find the discount and the proceeds.

9. On a note for $1200 dated May 17, 1910, payable on demand with interest at 6 %, a payment of $500 was made Dec. 8, 1910. How much was due Feb. 16, 1911?

10. A city made an appropriation of $53,040 for its public schools. The assessed value of real estate was $6,710,000 and of personal property $2,130,000. What was the tax rate for school purposes? How much was Brown's school tax if his property was assessed at $12,500?

WRITTEN TEST

QUESTIONS FROM BUSINESS EDUCATORS' ASSOCIATION, CANADA

993. **1.** Find the cost of a carpet $\frac{3}{4}$ yard wide, at $1.25 per yard, for a room 22 ft. × 18 ft. if the strips run lengthwise and there is a waste of $\frac{1}{8}$ of a yard on each strip for matching the pattern.

2. An article marked $62\frac{1}{2}\%$ above cost was sold less 25 % and 20 %. If a collector was afterwards paid 20 % for collecting the account, what was the gain or loss per cent?

3. Having bought a house of A at $12\frac{1}{2}\%$ less than it cost him I spent $430 for repairs and sold it for $7293, thereby gaining 10 % on my investment. How much did the house cost A?

4. Equate the following account:

1909			1909			
Aug. 10,	To Mdse., 30 da.	$375	Sept. 3,	By Draft, 30 da.	$250
Aug. 25,	To Cash,	450	Sept. 27,	By Cash,	425
Sept. 10,	To Mdse., 2 mo.	175	Oct. 1,	By Cash,	25

5. I sold 2978 bu. wheat at $1.05 a bushel, and invested the proceeds in sugar, reserving my commission of 5 % for selling and $1\frac{1}{2}\%$ for buying, and the expenses of shipping, $53.57. How much did I invest in sugar?

6. A man invests $12,000 in 3 % stock at 75. He sells out at 90, and invests $\frac{1}{3}$ of the proceeds in $3\frac{1}{2}\%$ stock at 76 and the remainder in 5 % stock at 95. Find the change in his income.

7. I bought two houses for $11,700, paying 25 % more for one than for the other. I sold the cheaper house at 20 % profit and the higher priced one at $16\frac{2}{3}\%$ profit. What was my total gain?

8. A stock of goods was marked $22\frac{1}{2}\%$ advance on cost, but becoming damaged was sold at 20 % discount on the marked price, which caused a loss of $1186.40. Find the cost of the goods.

9. A note of $1500 dated June 20, 1908, bearing interest at 5 %, had payments indorsed as follows: Dec. 5, 1908, $300; April 2, 1909, $10; July 20, 1909, $500; Dec. 31, 1909, $400. Find the amount due June 22, 1910.

10. An invoice of woolen cloth, imported from England, was valued at £ 956 6s. If its weight was 684 lb., how much was the duty at 50 cents per pound specific, and 35 % ad valorem?

INDEX

ANSWERS TO VAN TUYL'S COMPLETE
BUSINESS ARITHMETIC

Page 12. — **1.** $57.   **2.** $160.   **3.** $289.

Page 13. — **§ 23.** **4.** $82.75.   **5.** $10,070.25.   **6.** $527.25.
§ 25. — **1.** $59.   **2.** $126.   **3.** $140.50.

Page 14. — **1.** $96.   **2.** $271.   **3.** $81.64.

Page 15. — **1.** $69.   **2.** $50.50.   **3.** $129.67.   **·** **4.** $43.75.   **5.** $224.

Page 16. — **1.** $116.   **2.** $200.50.   **3.** $170.42.   **4.** $116.   **5.** $493.17.

Page 17. — **§ 36.** **1.** $41.88.   **2.** $137.37.   **3.** $37.97.   **4.** $100.67.
5. $173.57.   **6.** $513.47.
§ 37. — **1.** $155.50.   **2.** $921.17.   **3.** $363.29.

Page 18. — **4.** $360.40.   **5.** $240.92.   **6.** $623.02.   **7.** $958.12.
8. $143.72.   **9.** $386.85.

Page 19. — **§ 39.** **1.** $21.   **2.** $88.   **3.** $44.65.
§ 41. — **1.** $140.   **2.** $177.50.   **3.** $23.   **4.** $115.08.   **5.** $118.50.
6. $198.67.   **7.** $64.29.   **8.** $630.   **9.** $1573.42.

Page 20. — **1.** $12,180.   **2.** $7825.

Page 21. — **1.** 45 yd.   **2.** 48 yd.   **3.** 120 yd.   **4.** 80 yd.   **5.** 91 yd.
6. 192 yd.   **7.** 144 yd.   **8.** 116 yd.   **9.** 112 yd.   **10.** 123 yd.   **11.** 112 yd.
12. 114 yd.   **13.** 112 yd.   **14.** 114 yd.   **15.** 144 yd.   **16.** 156 yd.   **17.** 160 yd.
18. 165 yd.   **19.** 176 yd.   **20.** 172 yd.   **21.** 160 yd.   **22.** 162 yd.
23. 68 yd.   **24.** 72 yd.   **25.** 80 yd.   **26.** 36 yd.   **27.** 62 yd.   **28.** 48 yd.
29. 42 yd.   **30.** 132 yd.

Page 22. — **1.** $35.   **2.** $130.   **3.** $62.50.   **4.** $128.77.   **5.** $340.20.
6. $542.50.   **7.** $234.19.   **8.** $310.81.   **9.** $459.50.

Page 24. — **1.** $12.14.   **2.** $75.03.   **3.** $70.44.   **4.** $199.20.   **5.** 3933.
6. 34,125.

Page 25. — **1.** $1.98 ; $2.42.   **2.** $4.38 ; $4.03.   **3.** $21.09 ; $22.78.
4. $7 ; $7.70.   **5.** $59.38 ; $64.13.   **6.** $79.75 ; $84.70.   **7.** $100 ; $90.
8. $38.75 ; $52.31.   **9.** $576.60 ; $624.65.   **10.** $1357.40 ; $1419.10.
11. $30 ; $33.75.

Page 26. — **12.** $36.25 ; $31.90.   **13.** $204.40 ; $231.26.   **14.** $81.56 ;
$76.13.   **15.** $11.42 ; $10.81.   **16.** $879.94 ; $791.94.   **17.** $538.43 ; $598.25.
18. $170.16 ; $185.63.   **19.** $52.99 ; $58.88.   **20.** $14.36 ; $17.55.   **21.** $23.94 ;
$29.93.   **22.** $45.16 ; $41.69.   **23.** $22.45 ; $21.29.   **24.** $50.02 ; $45.48.
25. $20.66 ; $23.44.   **26.** $55.74 ; $57.89.   **27.** $43.41 ; $38.20.   **28.** $161.92 ;
$158.32.   **29.** $4.25 ; $4.16.   **30.** $1.79 ; $1.75.

Page 28. — **1.** $1.30.   **2.** $2.70.   **3.** $4.   **4.** $3.36.   **5.** $3.64.   **6.** $8.
7. $6.40.   **8.** $6.60.   **9.** $14.70.   **10.** $24.   **11.** $24.   **12.** $25.   **13.** $25.20.
14. $35.20.   **15.** $45.50.   **16.** $8.   **17.** $21.   **18.** $26.40.   **19.** $38.25.
20. $60.   **21.** $75.   **22.** $144.   **23.** $12.50.   **24.** $26.80.   **25.** $40.   **26.** $63.
27. $90.   **28.** $160.   **29.** $240.   **30.** $128.   **31.** $125.   **32.** $168.   **33.** $1.75.
34. $138.   **35.** $125.   **36.** $138.60.

Page 29.—§ 68. 1. $.40. **2.** $.80. **3.** $1.50. **4.** $1.60. **5.** $4.20
6. $1.75. **7.** $7.20. **8.** $9.36. **9.** $8.96. **10.** $13. **11.** $10. **12.** $4.50.
13. $4.56. **14.** $4.48. **15.** $5.92. **16.** $7.20. **17.** $2.45. **18.** $1.20.
19. $1.20. **20.** $.70. **21.** $6.
§ 69. 1. $7.48. **2.** $28.23. **3.** $116.83.

Page 30.—4. $172.31. **5.** $22.28. **6.** $8.96. **7.** $36.13. **8.** $83.14.
9. $9.18. **10.** $13.89. **11.** $6.67. **12.** $11.48. **13.** $13.92. **14.** $17.78.
15. $47.48.

Page 36.—§ 83. 1. 43,967. **2.** 49,651. **3.** 61,105. **4.** 55,832. **5.** 55,751.
6. 44,085.
§ 84. 1. 67,892. **2.** 75,285. **3.** 58,515. **4.** 69,088. **5.** 58,126. **6.** 51,346.

Page 37.—7. 231,721,752. **8.** 252,895,949. **9.** 276,555,244. **10.** 279,012,246.
11. 343,512,628. **12.** 317,462,998. **13.** 343,052,905. **14.** 369,428,559.
15. 329,982,953. **16.** 437,567,826.
§ 85. 1. 1,214,856. **2.** 1,105,460. **3.** 1,262,010. **4.** 1,182,371.
5. 1,173,738.

Page 39.— (1905) 134,806. (1906) 136,481. (1907) 142,694. (1908) 147,846.
(1909) 149,491.

Page 40.—8. Horizontal totals: 204, 148, 215, 235. Vertical totals: 267,
197, 338. Grand total: 802. **9.** Horizontal totals: 120, 101, 162, 278. Vertical
totals: 212, 230, 219. Grand total: 661. **10.** Horizontal totals: 156, 192, 224,
158. Vertical totals: 298, 204, 228. Grand total: 730. **11.** Horizontal totals:
1822, 2553, 1927, 1901. Vertical totals: 2456, 2927, 2820. Grand total: 8203.
12. Horizontal totals: 1979, 2460, 2173, 2156. Vertical totals: 2435, 3539,
2794. Grand total: 8768. **13.** Horizontal totals: 1836, 2541, 1538, 1486.
Vertical totals: 2524, 3002, 1875. Grand total: 7401. **14.** Jan. $45,393.28,
Feb. $40,756.65, Mar. $43,984.21, Apr. $40,538.15, May $36,373.70, June
$36,495.30, July $35,184.12, Aug. $33,958.49, Sept. $38,005.62, Oct. $45,906.16,
Nov. $53,723.72, Dec. $65,264.51. Clothing, $212,127.45; Dress Goods,
$141,922.48; Gloves, $31,112.05; Hats, $53,399.38; Shoes, $77,022.55; Grand
total, $515,583.91.

Page 41.—15. (1) General tax, $2,371,259; School tax, $1,976,049; Military tax, $158,084; Public Highway tax, $790,420. (2) County: **1.** $117,644.
2. $19,058. **3.** $66,657. **4.** $186,418. **5.** $71,945. **6.** $42,465.
7. $74,330. **8.** $47,576. **9.** $63,531. **10.** $48,609. **11.** $15,630.
12. $46,188. **13.** $29,226. **14.** $74,197. **15.** $11,254. **16.** $31,584.
17. $1,424,829 **18.** $26,051. **19.** $82,962. **20.** $56,529. **21.** $123,266.
22. $157,100. **23.** $30,626. **24.** $24,179. **25.** $76,112. **26.** $548,078.
27. $10,827. **28.** $126,990. **29.** $26,824. **30.** $221,776. **31.** $547,035.
32. $65,768. **33.** $70,638. **34.** $12,753. **35.** $172,193. **36.** $162,666.
37. $217,680. **38.** $164,618. (3) Total tax, $5,295,812.

Page 45.—30. $664. **31.** $265. **32.** $1008.90. **33.** $1116.38.
34. $2543.05. **35.** $4540.50. **36.** $635.50, Gain.

Page 46.—37. $3185.75. **38.** $270.

Page 47.—1. 53. **2.** 120. **3.** 244 **4.** 1643. **5.** 1565. **6.** 1188.
7. 1122. **8.** 429. **9.** 629. **10.** 1336. **11.** 1323. **12.** 7220. **13.** Increases
in earnings: Coal traffic, $2,415,932.02; Merchandise, $583,270.99; Passengers, $131,326.82; Express, $12,124.43. Decreases in earnings: Mails,
$486.21; Miscellaneous, $16,403.58; Total earnings, 1909, $20,175,793.55,
1908, $17,050,029.08; Total increase, $3,125,764.47. Increases in expenses:
Roads, $447,986.78; Equipment, $319,887.45; Transportation, $722,656.47. De-
creases in expenses: General, $32,078.79; Total expenses, 1909, $12,129,080.33;
1908, $10,670,628.42; Total increase, $1,458,451.91; Net earnings, 1909,
$8,046,713.22; 1908, $6,379,400.66; Net increase, $1,667,312.56.

Page 51. — 1. 1419. 2. 1056. 3. 3224. 4. 1809. 5. 4399. 6. 2001. 7. 6216. 8. 4816. 9. 1184. 10. 2160. 11. 3132. 12. 7656. 13. 9114. 14. 4347. 15. 4623. 16. 5928. 17. 1911. 18. 3306. 19. 4221. 20. 6624. 21. 3956. 22. 3135. 23. 1482. 24. 1656.

Page 54. — 1. 1247. 2. 2664. 3. 1296. 4. 3108. 5. 2408. 6. 2548. 7. 4512. 8. 2576. 9. 2822. 10. 661,500. 11. 809,352. 12. 2,010,933. 13. 1,899,107. 14. 2,139,234. 15. 1,716,552. 16. 3,555,324. 17. 5,459,376. 18. 4,020,792. 19. 2304. 20. 6231. 21. 6486. 22. 5994. 23. 2378. 24. 2890. 25. 2205. 26. 4161. 27. 5293.

Page 58. — § **132.** 1. 135, rem. 12. 2. 174, rem. 41. 3. 106, rem. 51. 4. 109, rem. 31. 5. 112, rem. 70. 6. 101, rem. 68. 7. 116, rem. 8. 8. 229, rem. 31. 9. 851, rem. 18. 10. 903, rem. 25. 11. 13, rem. 3400. 12. 40, rem. 3260. 13. 59, rem. 950. 14. 68, rem. 3364. 15. 50, rem. 2689. 16. 57, rem. 5598. 17. 9, rem. 12,749. 18. 14, rem. 15,200. 19. 19, rem. 36,720. 20. 28. 21. 27.

§ **135.** 1. 11, rem. 18,139. 2. 9, rem. 9316. 3. 2, rem. 19,500. 4. 4, rem. 25,820. 5. 6, rem. 21,396. 6. 11, rem. 18,699. 7. 3, rem. 93,440. 8. 6, rem. 71,450. 9. 4, rem. 3905. 10. 10. 11. 7, rem. 27,441. 12. 10, rem. 53.

Page 61. — 1. 3, 2, 2, 2, 2. 2. 2, 2, 2, 2, 2, 2. 3. 5, 2, 2, 2, 2. 4. 5, 3, 7. 5. 2, 2, 2, 2, 3, 3. 6. 2, 2, 2, 2, 2, 5. 7. 3, 3, 3, 3, 3, 3. 8. 7, 7, 7. 9. 5, 5, 5, 3, 3. 10. 5, 5, 5, 3, 3, 3. 11. 3, 3, 2, 7, 29. 12. 2, 2, 2, 2, 7, 29. 13. 2, 2, 2, 3, 3, 37. 14. 2, 2, 2, 2, 2, 127. 15. 2, 2, 2, 2, 2, 3, 47. 16. 2, 2, 2, 2, 239. 17. 2, 2, 2, 2, 2, 151. 18. 5, 5, 5, 2, 3, 13. 19. 2, 2, 2, 3, 7, 97. 20. 2, 2, 2, 2, 2, 2, 2, 3, 67. 21. 5, 5, 5, 5, 11. 22. 5, 5, 5, 43. 23. 2, 2, 7, 11, 13. 24. 2, 2, 2, 13, 31. 25. 2, 2, 2, 2, 367. 26. 2, 2, 2, 2, 2, 2, 149. 27. 2, 2, 2, 2, 241. 28. 2, 2, 2, 2, 2, 229. 29. 3, 3, 3, 3, 3, 3, 3, 7. 30. 3, 3, 3, 3, 7, 7, 7.

Page 62. — § **148.** 1. 19. 2. 29. 3. 37. 4. 97. 5. 1. 6. 197. 7. 163. 8. 361. 9. 31. 10. 41.

Page 63. — 1. 60. 2. 90. 3. 120 4. 840. 5. 900. 6. 975. 7. 5040. 8. 3600. 9. 5760. 10. 12,600.

Page 64. — 1. 8. 2. $20\frac{1}{4}$. 3. 4 4. $1\frac{17}{17}$. 5. $\frac{27}{50}$. 7. 63 lb. 8. 72 bu. 9. 15 boys. 10. $134.40.

Page 66. — 1. $1.15. 2. 46¢. 3. $1.36. 4. $1.47. 5. 4¢. 6. 6¢. 7. $2.75 ; $75.25 8. 36¢. 9. $22.12. 10. (a) 53¢ ; (b) $90.30 ; (c) $90.

Page 71. — 1. $\frac{9}{10}, \frac{7}{10}, \frac{24}{25}$. 2. $\frac{5}{6}, \frac{1}{3}, \frac{1}{3}$. 3. $\frac{37}{39}, \frac{52}{29}, \frac{3}{4}$. 4. $\frac{3}{8}, \frac{1}{4}, \frac{1}{8}$. 5. $\frac{155}{205}$, $\frac{4}{5}, \frac{2}{3}$. 6. $\frac{2}{3}, \frac{329}{371}, \frac{34}{101}$. 7. $\frac{23}{29}, \frac{7}{6}, \frac{1}{15}$. 8. $\frac{6}{7}, \frac{4}{7}, \frac{3}{8}$. 9. $\frac{5}{8}, \frac{3}{8}, \frac{3}{5}$. 10. $\frac{9}{11}, \frac{7}{11}$, $\frac{19}{21}$. 11. $\frac{11}{13}, \frac{15}{16}, \frac{3}{4}$. 12. $\frac{2}{3}, \frac{2}{3}, \frac{2}{3}$. 13. $\frac{18}{24}$. 14. $\frac{10}{48}$. 15. $\frac{8}{16}$. 16. .050. 17. .04000. 18. $\frac{30}{36}$. 19. $\frac{40}{45}$. 20. $\frac{6}{12}, \frac{8}{12}, \frac{9}{12}, \frac{10}{12}$. 21. $\frac{24}{40}, \frac{25}{40}, \frac{30}{40}, \frac{36}{40}$. 22. $\frac{75}{100}, \frac{65}{100}, .50, .80$. 23. $\frac{15}{24}, \frac{20}{24}, \frac{18}{24}, \frac{3}{24}$. 24. $\frac{9}{48}, \frac{32}{48}, \frac{16}{48}, \frac{8}{48}$.

Page 72. — § **178.** 1. $7\frac{1}{4}$. 2. $7\frac{1}{5}$. 3. $7\frac{3}{4}$. 4. 6. 5. $7\frac{4}{5}$. 6. $11\frac{3}{8}$. 7. $9\frac{7}{8}$. 8. $17\frac{7}{8}$. 9. $17\frac{7}{8}$. 10. $27\frac{1}{4}$. 11. 22. 12. $24\frac{2}{5}$. 13. 118. 14. $16\frac{7}{8}$. 15. $1\frac{3}{4}$. 16. $16\frac{3}{10}$. 17. $2\frac{12}{25}$. 18. $3\frac{87}{100}$. 19. $48\frac{97}{100}$. 20. $2\frac{431}{1000}$. 21. $69\frac{87}{100}$. 22. $34\frac{1}{4}$. 23. $5\frac{4}{5}$. 24. $4\frac{3}{5}$. 25. $3\frac{3}{5}$. 26. $3\frac{3}{10}$. 27. $195\frac{22}{25}$. 28. 20. 29. 189. 30. 25.

§ **180.** 1. $\frac{7}{2}$. 2. $\frac{22}{8}$. 3. $\frac{34}{4}$. 4. $\frac{73}{8}$. 5. $\frac{45}{7}$. 6. $\frac{89}{9}$. 7. $\frac{43}{4}$. 8. $\frac{59}{10}$. 9. $\frac{55}{8}$. 10. $\frac{53}{7}$. 11. $\frac{71}{4}$. 12. $\frac{155}{8}$. 13. $\frac{231}{4}$. 14. $\frac{862}{9}$. 15. $\frac{231}{4}$.

Page 73. — 16. $\frac{600}{7}$. 17. $\frac{865}{8}$. 18. $\frac{218}{11}$. 19. $\frac{425}{11}$. 20. $\frac{175}{2}$. 21. $\frac{515}{4}$.
22. $\frac{1655}{7}$. 23. $\frac{1973}{7}$. 24. $\frac{4343}{9}$. 25. $\frac{2188}{5}$. 26. $\frac{7419}{8}$. 27. $\frac{5088}{7}$. 28. $\frac{2581}{7}$.
29. $\frac{15555}{16}$. 30. $\frac{12325}{6}$.

§ 182. 1. .15. 2. .625. 3. .28125. 4. .28. 5. $.26\frac{2}{3}$. 6. $.27\frac{7}{9}$. 7. $.33\frac{1}{3}$.
8. $.44\frac{4}{9}$. 9. $.36\frac{4}{11}$. 10. $.21\frac{3}{7}$. 11. .0125. 12. .075. 13. .005. 14. $.92\frac{6}{7}$.
15. $.34\frac{1}{11}$. 16. .032. 17. .72. 18. .042. 19. .00125. 20. .52.

Page 74. — 1. $\frac{4}{5}$. 2. $\frac{6}{25}$. 3. $\frac{11}{25}$. 4. $\frac{5}{8}$. 5. $\frac{5}{7}$. 6. $\frac{111}{125}$. 7. $\frac{4}{125}$.
8. $\frac{1}{800}$. 9. $\frac{1}{2000}$. 10. $\frac{1}{80}$. 11. $\frac{1}{160}$. 12. $\frac{1}{400}$. 13. $\frac{1}{800}$. 14. $\frac{3}{16}$.
15. $\frac{3}{32}$. 16. $\frac{1}{11}$. 17. $\frac{1}{14}$. 18. $\frac{1}{2000}$. 19. $\frac{7}{125}$. 20. $\frac{13}{125}$. 21. $\frac{1}{8000}$.
22. $\frac{1}{3000}$. 23. $\frac{1}{70}$. 24. $\frac{1}{150}$.

Page 75. — 1. $\frac{6}{12}, \frac{8}{12}, \frac{9}{12}$. 2. $\frac{8}{10}, \frac{9}{10}, \frac{5}{10}$. 3. $\frac{6}{8}, \frac{1}{8}, \frac{2}{8}$. 4. $\frac{25}{30}, \frac{20}{30}, \frac{3}{30}$.
5. $\frac{12}{16}, \frac{1}{16}, \frac{7}{16}$. 6. $\frac{35}{40}, \frac{36}{40}, \frac{15}{40}$. 7. $\frac{21}{24}, \frac{22}{24}, \frac{2}{24}$. 8. $\frac{10}{12}, \frac{9}{12}, \frac{8}{12}$. 9. $\frac{15}{48}$,
$\frac{14}{48}, \frac{8}{48}$. 10. $\frac{16}{18}, \frac{15}{18}, \frac{5}{18}$. 11. $\frac{20}{30}, \frac{27}{30}, \frac{14}{30}, \frac{24}{30}, \frac{2}{30}$. 12. $\frac{5}{16}, \frac{2}{16}, \frac{12}{16}, \frac{1}{16}$.
13. $\frac{30}{40}, \frac{35}{40}, \frac{20}{40}, \frac{20}{40}, \frac{10}{40}$. 14. $\frac{17}{24}, \frac{20}{24}, \frac{8}{24}, \frac{4}{24}$. 15. $\frac{15}{36}, \frac{16}{36}, \frac{3}{36}, \frac{12}{36}$. 16. $\frac{42}{48}, \frac{47}{48}, \frac{36}{48}$,
$\frac{8}{48}, \frac{20}{48}$. 17. $\frac{4}{16}, \frac{10}{16}, \frac{1}{16}, \frac{8}{16}$. 18. $\frac{42}{84}, \frac{40}{84}, \frac{27}{84}, \frac{63}{84}, \frac{24}{84}$. 19. $\frac{18}{48}, \frac{15}{48}, \frac{42}{48}, \frac{7}{48}$.
20. $\frac{9}{99}, \frac{55}{99}, \frac{21}{99}, \frac{18}{99}, \frac{66}{99}$.

Page 77. — 1. $2\frac{13}{14}$. 2. $2\frac{5}{12}$. 3. $2\frac{2}{21}$. 4. $1\frac{1}{3}$. 5. $52\frac{13}{24}$. 6. $103\frac{17}{24}$.
7. $59\frac{7}{8}$. 8. 493.308$\frac{1}{3}$. 9. 16.449. 10. $2\frac{1}{2}$. 11. $1\frac{27}{40}$. 12. $1\frac{43}{70}$.
13. $94\frac{1}{4}$. 14. $369\frac{1}{4}$. 15. 220.593$\frac{2}{3}$. 16. 161.136$\frac{2}{3}$. 17. 176.76.
18. $382\frac{2}{3}$. 19. $196\frac{7}{8}$. 20. 194.980$\frac{1}{3}$.

Page 79. — § 200. 1. $3\frac{1}{3}$. 2. $7\frac{1}{3}$. 3. $1\frac{13}{16}$. 4. $115\frac{13}{16}$. 5. $341\frac{19}{20}$.
6. 288.8545. 7. $222\frac{9}{16}$. 8. $269\frac{1}{4}$. 9. $135\frac{5}{16}$. 10. $193\frac{5}{16}$. 11. $1\frac{11}{16}$.
12. $2\frac{1}{4}$. 13. $1\frac{1}{4}$. 14. $1\frac{3}{8}$. 15. $\frac{15}{16}$. 16. $2\frac{3}{20}$. 17. $1\frac{3}{8}$. 18. $2\frac{1}{4}$.
19. $1\frac{11}{16}$. 20. $1\frac{5}{8}$. 21. $2\frac{1}{4}$. 22. $1\frac{3}{8}$. 23. $2\frac{7}{8}$. 24. $2\frac{1}{24}$. 25. $1\frac{5}{18}$.
26. $2\frac{37}{48}$. 27. $1\frac{7}{16}$. 28. $2\frac{5}{21}$. 29. $1\frac{5}{16}$. 30. $1\frac{8}{9}$.

Page 80. — 1. $\frac{1}{8}$. 2. $\frac{1}{15}$. 3. $\frac{1}{12}$. 4. $\frac{1}{16}$. 5. $\frac{3}{4}$. 6. $1\frac{1}{8}$.
7. $2\frac{3}{4}$. 8. $6\frac{1}{2}$. 9. $6\frac{1}{2}$. 10. $51\frac{5}{16}$. 11. $7\frac{3}{8}$. 12. $16\frac{2}{3}$. 13. $26\frac{11}{16}$.
14. 9. 15. 77.031$\frac{2}{3}$. 16. 30.790$\frac{3}{4}$. 17. $45\frac{25}{48}$. 18. $30\frac{89}{50}$. 19. $48\frac{11}{16}$.
20. 12.12. 21. 5.793$\frac{2}{3}$. 22. 21.705$\frac{2}{3}$. 23. 21.333$\frac{1}{3}$. 24. $60\frac{2}{3}$.
25. $42\frac{37}{40}$. 26. $39\frac{13}{20}$. 27. $15\frac{7}{17}$. 28. 27.065. 29. 24.325$\frac{1}{3}$. 30. $25\frac{2}{3}$.

Page 82. — 1. $4\frac{3}{8}$. 2. 6. 3. 4. $3\frac{3}{5}$. 5. $2\frac{5}{6}$. 6. $6\frac{6}{11}$. 7. $\frac{6}{7}$. 8. $2\frac{7}{8}$.
9. $3\frac{9}{11}$. 10. $4\frac{1}{3}$. 11. 5. 12. $4\frac{1}{3}$. 13. 10. 14. $10\frac{1}{2}$. 15. 21. 16. 45.
17. 48. 18. 40. 19. $10\frac{1}{11}$. 20. $8\frac{1}{3}$. 21. $21\frac{1}{3}$. 22. 27. 23. 28. 24. $21\frac{1}{3}$.
25. $82\frac{2}{3}$. 26. $92\frac{1}{4}$. 27. 64. 28. $11\frac{7}{8}$. 29. $60\frac{1}{2}$. 30. $73\frac{1}{4}$. 31. $4\frac{1}{16}$.
32. $47\frac{1}{2}$. 33. 63. 34. 36. 35. $31\frac{1}{4}$. 36. 45.

Page 83. — 1. 8. 2. 11.439. 3. 55.328. 4. 162. 5. 535.762. 6. 382.128.
7. $364\frac{1}{2}$. 8. 775. 9. 238.832. 10. 60.6315. 11. 609. 12. $24\frac{5}{8}$. 13. 41.168.
14. 4.895. 15. .845. 16. 8.6759. 17. 5.959. 18. $1\frac{1}{16}$. 19. 12. 20. 7.92.
21. .0138. 22. 45.85. 23. 22.035. 24. 1. 25. .8432. 26. 967.8.
27. 8.439. 28. .09637. 29. 450. 30. .64873. 31. .069478. 32. 839.02.
33. 84.8. 34. 124.89. 35. 3.4876. 36. 487.49. 37. 87.634. 38. 8563.4.
39. .7964.

Page 84. — 1. 85. 2. 56. 3. 140. 4. 129. 5. $390\frac{1}{2}$. 6. $3921\frac{2}{3}$.
7. 156. 8. $1235\frac{7}{8}$. 9. $4330\frac{2}{3}$. 10. $2628\frac{1}{4}$. 11. $4932\frac{2}{3}$. 12. $673\frac{1}{2}$.

Page 85. — 13. $464\frac{1}{2}$. 14. $2156\frac{1}{4}$. 15. $1340\frac{2}{3}$. 16. $1569\frac{2}{3}$. 17. $12105\frac{1}{8}$.
18. $14357\frac{1}{8}$. 19. $5419\frac{7}{11}$. 20. $2577\frac{1}{2}$. 21. $10,113\frac{1}{3}$. 22. $3513\frac{1}{4}$. 23. $4108\frac{2}{3}$.
24. 5880. 25. $276.96\frac{1}{4}$. 26. 27.173. 27. $148.86\frac{1}{4}$. 28. $149.27\frac{1}{4}$. 29. $7317.33\frac{1}{4}$.
30. $1716.04\frac{2}{7}$. 31. 3723.9. 32. 268.578. 33. $1143.57\frac{1}{4}$. 34. 497. 35. $282\frac{2}{3}$.
36. $1142.35\frac{1}{16}$.

Page 86. — § 217. 1. $\frac{3}{8}$. 2. $\frac{1}{2}$. 3. $\frac{1}{8}$. 4. $\frac{11}{18}$. 5. $\frac{21}{32}$. 6. $\frac{11}{18}$. 7. $\frac{21}{25}$.
8. $\frac{22}{27}$. 9. $\frac{3}{8}$. 10. $\frac{7}{8}$. 11. $\frac{7}{9}$. 12. $\frac{1}{7}$. 13. $\frac{15}{42}$. 14. $\frac{21}{28}$. 15. $\frac{22}{105}$. 16. $\frac{45}{256}$.
17. $\frac{7}{10}$. 18. $\frac{3}{8}$. 19. .3356122. 20. .0000963. 21. .00824. 22. .4241237.
23. 9.64494. 24. 855.2478. 25. 74.62784. 26. 473.473. 27. 342.17152.

§ 220. 1. $11\frac{1}{3}$. 2. $19\frac{1}{14}$. 3. $13\frac{4}{9}$. 4. $35\frac{17}{45}$. 5. $57\frac{1}{4}$. 6. $68\frac{5}{8}$. 7. $54\frac{9}{20}$.
8. $74\frac{73}{90}$. 9. $127\frac{17}{24}$. 10. $98\frac{7}{10}$. 11. $43\frac{79}{80}$. 12. $51\frac{7}{12}$.

Page 87. —13. $118\frac{1}{4}$. 14. $205\frac{33}{48}$. 15. $253\frac{7}{20}$. 16. $332\frac{2}{3}$. 17. $351\frac{45}{64}$.
18. $386\frac{33}{55}$. 19. $473\frac{11}{45}$. 20. $422\frac{3}{4}$. 21. $78\frac{9}{125}$. 22. $212\frac{33}{48}$. 23. $279\frac{5}{13}$.
24. $782\frac{7}{24}$. 25. $310\frac{31}{32}$. 26. $211\frac{11}{14}$. 27. $123\frac{3}{25}$. 28. $451\frac{7}{10}$. 29. 214. 30. 410.

§ 222. 1. $249\frac{27}{45}$. 2. $496\frac{7}{8}$. 3. $963\frac{1}{30}$. 4. $1106\frac{49}{50}$. 5. $1327\frac{3}{16}$.
6. $550\frac{1}{4}$. 7. $42\frac{3}{16}$. 8. $959\frac{3}{8}$. 9. $1571\frac{1}{2}$. 10. $2597\frac{2}{3}$. 11. $2601\frac{1}{2}$. 12. 3782.
13. $1851\frac{1}{19}$. 14. 8125. 15. $1435\frac{1}{2}$. 16. $1238\frac{23}{32}$. 17. $1037\frac{17}{25}$. 18. $2969\frac{2}{3}$.
19. $4188\frac{43}{63}$. 20. $4929\frac{49}{72}$. 21. $3331\frac{3}{13}$.

Page 88. —22. $4939\frac{1}{2}$. 23. $.2629\frac{51}{56}$. 24. $1.5501\frac{1}{4}$. 25. $1.0340\frac{37}{77}$.
26. $.23843\frac{1}{2}$. 27. $1.8438\frac{7}{10}$. 28. $1.0544\frac{4}{5}$. 29. $2.7213\frac{9}{25}$. 30. $7.3496\frac{9}{10}$.
31. $11.7363\frac{25}{64}$. 32. $13.9476\frac{4}{21}$. 33. $20\frac{1}{4}$. 34. $\$19.82$. 35. $\$21$. 36. $\$22.36$.
37. $\$30.52$. 38. $\$19.44$. 39. $\$20.97$. 40. $\$29.12$. 41. $\$45.66$.
42. $\$21.74$.

Page 90. — **§ 227.** 1. $\$1071.46$. 2. $\$12.56$. 3. $\$6905$. 4. $\$3.70$.
5. $\$84.04$. 6. $\$104$.

Page 92. —1. (1) Addison, James $\$31.92$; (2) Buell, Henry $\$26.19$;
(3) Crane, John $\$26.81$; (4) Crawford, William $\$26.47$; (5) Jones, Albert
$\$24.07$; (6) Landis, Frank $\$19.88$; (7) Money, Theodore $\$20.81$; (8) New-
ton, James $\$20.34$; (9) Orton, George $\$18.65$; (10) Rooney, Thomas $\$18.13$.
Total, $\$233.27$.

Page 93. —Total time, 65 hr. 20 min. Total wages, $\$17.97$.

Page 94. —1. $\frac{7}{8}$. 2. $\frac{2}{3}$. 3. $\frac{5}{24}$. 4. $\frac{3}{20}$. 5. $.0025$. 6. $.058$. 7. $\frac{7}{24}$.
8. $\frac{3}{16}$. 9. $\frac{1}{4}$. 10. $.242$. 11. $.0529$. 12. $.0007$. 13. $\frac{1}{100}$. 14. $\frac{5}{68}$.
15. $\frac{11}{162}$. 16. $\frac{4}{31}$.

Page 95. —17. $.0125$. 18. $.082$. 19. $.00001$. 20. $.00008$. 21. $\frac{19}{84}$.
22. $\frac{3}{25}$. 23. $\frac{11}{20}$. 24. $\frac{5}{9}$. 25. $\frac{19}{24}$. 26. $\frac{31}{192}$. 27. $\frac{4}{15}$. 28. $.0025$.
29. $.000025$. 30. $.1056$. 31. $\frac{53}{160}$. 32. $\frac{3}{80}$. 33. $\frac{7}{1024}$. 34. $\frac{9}{10}$. 35. $\frac{11}{12}$.
36. $.00002$. 37. $.009$. 38. $.0611$. 39. $.00276$. 40. $.0279$.

Page 96. —1. $\frac{3}{4}$. 2. $\frac{2}{3}$. 3. $1\frac{1}{4}$. 4. $\frac{11}{4}$. 5. $\frac{7}{10}$. 6. $2\frac{11}{32}$. 7. $6\frac{23}{32}$. 8. $23\frac{7}{8}$.
9. $7\frac{188}{189}$. 10. $13\frac{66}{301}$. 11. $12\frac{1}{253}$. 12. $42\frac{239}{240}$. 13. $59\frac{9}{32}$. 14. $79\frac{5}{96}$. 15. $86\frac{2}{15}$.
16. $327\frac{5}{96}$. 17. $451\frac{45}{44}$. 18. $71\frac{7}{216}$. 19. $41\frac{275}{276}$. 20. $97\frac{544}{545}$. 21. $132\frac{753}{784}$.
22. $917\frac{17}{72}$. 23. $52\frac{135}{152}$. 24. $88\frac{9}{16}$. 25. $59\frac{64}{65}$. 26. $138\frac{44}{45}$. 27. $112\frac{3}{160}$.
28. $482\frac{77}{85}$. 29. $64\frac{67}{72}$. 30. $193\frac{11}{24}$. 31. 29.896. 32. 49.266. 33. 2.0156.
34. 201.56. 35. 2.279. 36. 4.091. 37. 8.14. 38. 1.6218. 39. 1.2094.
40. $.287$. 41. $.77\frac{7}{8}$. 42. $1.28\frac{11}{15}$. 43. $.707\frac{1}{2}$. 44. $.88\frac{41}{99}$. 45. $7.12\frac{16}{27}$.

Page 98. —1. $\frac{8}{9}$. 2. $\frac{8}{9}$. 3. $\frac{49}{77}$. 4. $1\frac{19}{35}$. 5. $2\frac{1}{4}$. 6. $2\frac{13}{16}$. 7. $10\frac{2}{3}$. 8. $20\frac{1}{3}$.
9. $107\frac{1}{2}$.

Page 99. —10. 30. 11. $\frac{224}{225}$. 12. $\frac{5}{72}$. 13. $1\frac{1}{4}$. 14. 18. 15. $57\frac{3}{4}$. 16. 31.8.
17. 1930. 18. 309.5. 19. 20. 20. 100. 21. 8000. 22. 80. 23. $.2$. 24. 78.
25. 27.2. 26. $20,000,000$. 27. 2020.2. 28. 60.53. 29. 59.712. 30. $31,616$.
31. $25\frac{5}{8}$. 32. 49. 33. 51. 34. $133\frac{1}{2}$. 35. $65\frac{1}{4}$. 36. 40. 37. 77. 38. 34.
39. $53\frac{1}{4}$.

Page 100. —1. $19\frac{49}{296}$. 2. $57\frac{15}{32}$. 3. $50\frac{29}{156}$. 4. $11.284\frac{2}{7}$. 5. 4. 6. $21\frac{167}{187}$.
7. $393\frac{19}{29}$. 8. $106\frac{43}{69}$. 9. $481\frac{108}{125}$. 10. $21\frac{727}{271}$. 11. $13\frac{221}{488}$. 12. $12\frac{292}{297}$.
13. $28\frac{372}{375}$. 14. $24\frac{143}{148}$. 15. $22\frac{71}{174}$. 16. $11\frac{231}{254}$. 17. 75. 18. 150. 19. 112.5.
20. $59\frac{47}{130}$. 21. $112\frac{51}{51}$. 22. $103\frac{27}{47}$. 23. $16.37\frac{19}{42}$. 24. 20. 25. 30. 26. $22\frac{87}{175}$.
27. $7\frac{95}{249}$. 28. $59\frac{3}{144}$. 29. $408\frac{30}{113}$. 30. $41\frac{41}{47}$. 31. $111\frac{17}{72}$. 32. $234\frac{1}{3}$.
33. $37\frac{13}{158}$. 34. $40\frac{87}{172}$. 35. 2. 36. $39\frac{7}{132}$. 37. $62\frac{20}{45}$. 38. $59\frac{13}{102}$. 39. $5.57\frac{9}{61}$.
40. 5. 41. 241. 42. $119,370$. 43. 3000. 44. $15,720$. 45. $27,450$.

Page 108. —1. 20 ft. 2. $\frac{3}{4}$. 3. $\$440$; $\$220$. 4. $\$50$. 5. $\frac{6}{29}$.
6. $\$600$; $\$240$. 7. $\$60$. 8. 132 sheep.

Page 109. — 9. 4000 A. **10.** \$3500; \$2800. **11.** \$3000. **12.** A, 99 A; B, $74\frac{1}{4}$ A. **13.** \$10,575; \$4230. **14.** \$10,000; \$4000. **15.** \$50,000. 16. \$270; \$210. **17.** \$12,500; \$15,000. **18.** 372. **19.** \$75. **20.** $\frac{1}{9}$. **21.** \$4.50. **22.** 98. **23.** \$42. **24.** \$6; $\frac{1}{8}$.

Page 110. — **25.** \$122.31. **26.** $\frac{9}{20}$. **27.** \$3.39. **28.** \$$5\frac{1}{3}$. **29.** \$.36. **30.** $46\frac{1}{6}$. **31.** $134\frac{17}{90}$. **32.** 39. **33.** 76 wk., 2 hr. **34.** \$10. 35. \$2.01; $\frac{56}{145}$.

Page 111. — 36. \$152. **37.** \$9.89. **38.** \$11.81. **39.** \$10.73. **40.** \$9.93. **41.** \$10.34.

Page 115. — **1.** 1326 bu. **2.** $283\frac{1}{2}$ bu. **3.** \$3.63. **4.** \$.60. **5.** 81 ft. **6.** 1 yr. 4 mo. 7. 80 da. **8.** 6 da. **9.** 15 men.

Page 116. — **1.** 307 lb. 13 oz. **2.** 28 lb. 8 oz.

Page 117. — **1.** A, \$6; B, \$8. **2.** A, \$20; B, \$30. **3.** 1st, \$72; 2d, \$120; 3d, \$96. **4.** 1st, \$120; 2d, \$180; 3d, \$240. **5.** 1st, \$2160; 2d, \$2520; 3d, \$2880; 4th, \$3240. **6.** 1st, 15; 2d, 10. **7.** 1st, 20; 2d, 15. **8.** 96 ft., 120 ft., 144 ft. **9.** Each man, \$55; each boy, \$27.50. **10.** One lot, \$6000; other lot, \$8000. **11.** B, \$12; C, \$24. **12.** 1st, 96; 2d, 48; 3d, 24; 4th 12. **13.** 1st, \$1200; 2d, \$2250; 3d, \$3600. **14.** 1st son, \$16,000; 2d son, \$13,600; 3d son, \$11,200; 4th son, \$9200. **15.** A, \$1600; B, \$1800. **16.** C, \$3840; D, \$4000; E, \$1920.

Page 118. — **1.** 6,609,574,534; \$13,188.294. **2.** 169,527; 33,838; 6947.45; 688.0124. **3.** \$80.71. **4.** \$929.04. **5.** \$864.83.

Page 119. — **6.** Final sum, 10,031. **7.** 11, rem. 866. **8.** (*a*) $\frac{12}{24}$; $\frac{16}{24}$; $\frac{10}{24}$; $\frac{21}{24}$. (*b*) $\frac{27}{36}$; $\frac{30}{36}$; $\frac{28}{36}$; $\frac{10}{36}$. (*c*) $\frac{54}{96}$; $\frac{60}{96}$; $\frac{20}{96}$; $\frac{9}{96}$. (*d*) $\frac{24}{30}$; $\frac{20}{30}$; $\frac{8}{30}$; $\frac{9}{30}$. **9.** $\frac{13}{200}$; $\frac{1}{16}$; $\frac{9}{40}$; $\frac{1}{800}$. **10.** $270\frac{2}{5}$; $19\frac{73}{86}$; .0609; 54,732.5.

1. Horizontal totals, \$58,864; \$88,221; \$98,237; \$154,113. Vertical totals, \$119,044; \$188,105; \$92,286. Grand total, \$399,435. **2.** Gain on first, \$2375.63; 2d, \$3618.59; 3d, \$1,502.89; 4th, \$8716.04; 5th, \$20,806.75. Total, \$37,019.90. **3.** 510; 702; $\frac{2}{9}$; $4\frac{15}{16}$; $\frac{11}{18}$. **4.** 36; 120; 32; 60; 216.

Page 120. — **5.** $20\frac{5}{6}$; $\frac{1}{8}$. **6.** (*a*) $\frac{27}{72}$; $\frac{52}{72}$; $\frac{42}{72}$; $\frac{22}{72}$. (*b*) $\frac{42}{70}$; $\frac{40}{70}$; $\frac{55}{70}$; $\frac{63}{70}$; (*c*) $\frac{252}{945}$; $\frac{225}{945}$; $\frac{140}{945}$; $\frac{30}{945}$. (*d*) $\frac{35}{84}$; $\frac{32}{84}$; $\frac{33}{84}$; $\frac{26}{84}$. **7.** 25,222,216; 5,761,098. **8.** $16\frac{1}{2}$. **9.** \$5214.50. **10.** \$34.49.

1. \$939.30. **2.** \$90.95. **3.** \$3648. **4.** \$206.64.

Page 121. — **5.** \$1794.25. **6.** In safe, \$950.57; in bank, \$1238.79. **7.** \$247.73. **8.** 6 yr. 1 mo. 22 da. **9.** \$68.61. **10.** Total cost, \$25,626.15; Average per capita, \$24.45.

Page 136. — **1.** 301 pt. **2.** 1760 in. **3.** 33,425 lb. **4.** 570,458 sec. 5. 737,404 cu. in. **6.** 4875 cm. **7.** 5749 g. **8.** 753 l. **9.** 622,850 sq. in. **10.** \$4.25. **11.** \$13.44. **12.** \$3.50. **13.** \$75.33.

Page 137. — **§ 335.** **15.** 2 pk., 4 qt. **16.** 5 oz., 2 pwt., 20 gr. **17.** 106 sq. rd., 20 sq. yd., 1 sq. ft., 72 sq. in. **18.** 5 lots. **19.** 660 boards.

§ 337. **1.** 77 gal., 3 qt., 1 gi. **2.** 2 lb., 8 oz., 15 pwt., 9 gr. **3.** 2 da., 7 hr., 12 min., 6 sec. **4.** 1 sq. rd., 7 sq. yd., 3 sq. ft., 21 sq. in. **5.** 2 mi., 63 rd., 2 yd., 1 ft., 9 in. **6.** 3 da., 52 min. **7.** 10° 41′ 16″. **8.** 4 Hg., 8 Dg., 7 g., 9 dg., 2 cg. **9.** 5 Kl., 3 Hl., 9 Dl., 8 l., 6 dl. **10.** 5 Dm., 8 m., 3 dm., 7 cm., 1 mm.

Page 138. — **12.** $\frac{8}{9}$ yd. **13.** $.452\frac{1}{12}$ lb. **14.** $24\frac{1}{6}$ da. **15.** $\frac{9}{40}$ T. **16.** .3045 A. **17.** $\frac{9}{16}$ da. **18.** $\frac{3}{4}$ da. **19.** .4001 A.

Page 139. — **1.** 19 lb., 5 oz., 6 pwt., 16 gr. **2.** 20 lb., $9\frac{1}{7}$ oz. **3.** 26 lb., 8 oz., 6 dr., 2 sc. **4.** \$8.07. **5.** \$5.58. **6.** $13\frac{3}{4}$ pwt. **8.** 44 lb., $1\frac{1}{2}$ oz. **9.** 100 lb., 5 oz. **10.** 12.2471 Kg. **11.** 22 Kg. **12.** \$1232.65. **13.** 643 rings. **14.** 149.3 g. **15.** \$47.28.

Page 141.—1. $717.18. **2.** $1586.13. **3.** $2232.69. **4.** £140, 15s., 2d.
5. £163, 11s., 4d. **6.** £246, 11s., 8d. **7.** £377, 3s., 5d. **8.** $8022.04.
9. $8620.40. **10.** £717, 5s., 1d. **11.** $115.42. **12.** $248.01. **13.** $672.42.
14. F. 8,253.73. **15.** F. 17,283.50. **16.** F. 30,519.56. **17.** $1445.78.
18. $1249.20. **19.** F. 20,577.54. **20.** F. 24,868.22. **21.** $326.54.
22. $592.32. **23.** $927.25. **24.** M. 18,366.49. **25.** M. 15,468.49.
26. M. 7,490.55. **27.** $556.34. **28.** $791.60.

Page 142.—29. M. 21,028.72. **30.** M. 25,193.26. **31.** £349, 4s., 6d. ;
F. 8,805.53 ; M. 7140.76. **32.** $4700.75 ; F. 24,355.76 ; M. 19,751.05.
33. $2866.49 ; £589, 6d. ; M. 12,044.08. **34.** $2350.25 ; £482, 18s., 11d ;
F. 12,177.23. **35.** $1730.08 ; £355, 10s., 2d. ; M. 7269.24. **36.** $3088.76 ;
£634, 14s. ; F. 16,003.64. **37.** £2851, 17s., 11d. ; F. 71,909.27 ; M. 58,314.08.
38. $6152.49 ; F. 31,877.59 ; M. 25,850.80. **39.** $2389.19 ; £490, 18s., 11d. ;
M. 10,038.61. **40.** $7715.59 ; F. 39,976.40 ; M. 32,418.45.
1. 208 lb., 14 oz. **2.** 20 lb., 14 oz. **3.** 4½ gal. **4.** 27 gal., 3 qt., 1⅔ pt.
5. 11 qt., 1 pt., 1 gi.

Page 143.—7. 52.493 yd. **8.** 131.233 yd. **9.** 27 mi., 307 rd., 4 yd.
10. 19.31 Km. **11.** 2.776 mi. U. S. train faster. **12.** $51.09. **13.** 13 yd., 2 ft.,
10. in. **14.** 538.2 sq. yd. **16.** 11.9239 Dl. **17.** $9.12 gain. **18.** 88.905 Kg.
19. 11 ¢ a Kg. is 58 ¢ better.

Page 144.—20. 168.962¼ ares. **21.** $821.95. **22.** 45.936 bu.

Page 145.—1. 37 rd., 1 yd., 1 ft. **2.** 25 lb., 4 oz., 6 pwt., 6 gr. **3.** 45 A.,
132 sq. rd., 20 sq. yd., 7 sq. ft., 44 sq. in. **4.** £1441, 12s., 10d. **5.** 452 rd.,
4 yd. **6.** $148.05. **7.** $3.35. **8.** 16 Km., 7 Hm., 5 Dm., 7 m., 5 dm.
9. 24 Kg., 6 Hg., 9 Dg., 2 g., 1 dg. **10.** 3 Hl., 5 Dl., 5 l. **11.** 16 m., 2 dm.,
6 cm. **12.** 86 Ha., 4 ares, 63 ca.

Page 146.—14. 1 bu., 1 pk., 6 qt., 1 pt. **15.** 5 yd., 4½ in. **16.** 8 oz.
5 pwt., 19¼ gr. **17.** 8 oz., 2 dr., 1 sc., 5 gr. **18.** 8 dm., 3.6 mm. **19.** 1 Hl.,
3 l., 1 cl., 5 ml.
§ 351. **1.** 1 da., 21 hr., 38 min., 16 sec. **2.** 7 bu., 3 pk., 5 qt., 1 pt. **3.** £13,
16s., 9d., 3 far. **4.** 13 cd., 104 cu. ft. **5.** 1 Km., 6 Hm., 7 Dm., 6 m. **6.** 430
cu. m., 668 cu. dm., 484 cu. cm. **7.** 28 gal., 3 qt., 1 pt.

Page 147.—8. 22 A., 151 sq. rd., 28¼ sq. yd. **9.** 81 Hl., 8 Dl., 8 l. **11.** 4 qt.
12. 78 sq. rd., 25 sq. yd., 1 sq. ft., 126 sq. in. **13.** 3 yd., 1 ft., 2.4 in. **14.** 4 Hg.,
6 Dg. **15.** 30 sq. dm., 50 sq. cm. **16.** 759 cu. dm., 652 cu. cm., 500 cu. mm.
17. 2 ft., 5½ in. **18.** 122 sq. rd., 25 sq. yd., 1 sq. ft., 126 sq. in. **19.** 1958.9
Kg. ; $215.48.
§ 353. **1.** 41 gal., 2 qt., 2 gi. **2.** 55 da., 23 hr., 8 min., 33 sec. **3.** 3 A.,
35 sq. rd., 19 sq. yd., 7 sq. ft., 36 sq. in.

Page 148.—4. 65 mi., 270 rd., 3 yd., 2 ft., 3 in. **5.** 44 A., 22 sq. rd., 3 sq.
yd., 3 sq. ft. **6.** 22 lb., 4 oz., 5 dr., 12 gr. **7.** 58 lb., 10 oz. **8.** 86 bu., 2 qt.
9. $29.20. **10.** $157.95. **11.** 33 Kg., 4 Hg., 9 Dg., 5 g. **12.** 69 Hl., 5 Dl.,
5 l., 2 dl. **13.** 365 sq. m., 47 sq. dm., 5 sq. cm. **14.** 3846 cu. m., 211 cu. dm.,
740 cu. cm. **15.** $42.75. **16.** $1,041,360. **17.** 68,161.5 Kg. **18.** $3281.85.
19. ⅔⅔ mi. per min., 47₁₁³ mi. per hr.

Page 149.—1. 16 hr., 54 min., 33 sec. **2.** 3 lb., 11 oz., 1 dr., 1 sc., 5 gr.
3. 11 gal., 2 qt., 2 gi. **4.** 1 cu. yd., 1 cu. ft., 1267 cu. in. **5.** 33 sq. rd.,
1 sq. yd., 1 sq. ft., 13½ sq. in. **6.** 2 pk., 2 qt., 1 pt. **7.** 29 cu. m., 989 cu. dm.,
215 cu. cm., 500 cu. mm. **8.** 1 Kl., 7 Hl., 5 Dl., 1 l., 5 dl., 2 cl., 9 ml.
9. 2 Hl., 4 l. **10.** 28 sq. rd., 3 sq. yd., 6 sq. ft., 21⅔ sq. in.

Page 150.—1. 1. **2.** 4 ; 8 ; 16. **3.** 9 ; 27 ; 81. **4.** 16 ; 64 ; 256.
5. 25 ; 125 ; 625. **6.** 36 ; 216 ; 1296. **7.** 49 ; 343 ; 2401. **8.** 64 ; 512 ; 4096.
9. 81 ; 729 ; 6561. **10.** 100 ; 1000 ; 10,000. **11.** 121 ; 1331 ; 14,641.
12. 144 ; 1728 ; 20,736. **13.** 169 ; 2197 ; 28,561. **14.** 196 ; 2744 ; 38,416.

15. 225 ; 3375 ; 50,625. **16.** 256 ; 4096 ; 65,536. **17.** 289 ; 4913 ; 83,521.
18. 324 ; 5832 ; 104,976. **19.** 361 ; 6859 ; 130,321. **20.** 400 ; 8000 ; 160,000.
21. .0001 ; .000001 ; .00000001. **22.** .0004 ; .000008 ; .00000016. **23.** .0009 ;
.000027 ; .00000081. **24.** .0016 ; .000064 ; 00000256. **25.** .0025 ; .000125 ;
.00000625. **26.** .0036 ; .000216 ; .00001296. **27.** .0049 ; .000343 ; .00002401.
28. .0064 ; .000512 ; .00004096. **29.** .0081 ; .000729 ; .00006561. **30.** .01 ;
.001 ; .0001. **31.** .0121 ; .001331 ; .00014641. **32.** .0144 ; .001728 ; .00020736.
33. .0169 ; .002197 ; .00028561. **34.** .0196 ; .002744 ; .00038416. **35.** .0225 ;
.003375 ; .00050625. **36.** .0256 ; .004096 ; .00065536. **37.** .0289 ; .004913 ;
.00083521. **38.** .0324 ; .005832 ; .00104976. **39.** .0361 ; .006859 ; .00130321.
40. .04 ; .008 ; .0016. **41.** $\frac{1}{4}$; $\frac{1}{8}$; $\frac{1}{16}$. **42.** $\frac{1}{9}$; $\frac{1}{27}$; $\frac{1}{81}$. **43.** $\frac{4}{9}$; $\frac{8}{27}$; $\frac{16}{81}$.
44. $\frac{1}{16}$; $\frac{1}{64}$; $\frac{1}{256}$. **45.** $\frac{9}{16}$; $\frac{27}{64}$; $\frac{81}{256}$. **46.** $\frac{1}{25}$; $\frac{1}{125}$; $\frac{1}{625}$. **47.** $\frac{4}{25}$; $\frac{8}{125}$; $\frac{16}{625}$.
48. $\frac{9}{25}$; $\frac{27}{125}$; $\frac{81}{625}$. **49.** $\frac{16}{25}$; $\frac{64}{125}$; $\frac{256}{625}$. **50.** $\frac{2}{?}$; $\frac{125}{216}$; $\frac{125}{1296}$. **51.** $\frac{42}{64}$; $\frac{343}{512}$;
$\frac{2401}{4096}$. **52.** $\frac{25}{81}$; $\frac{125}{729}$; $\frac{625}{6561}$. **53.** $\frac{81}{256}$; $\frac{729}{4096}$; $\frac{6561}{65536}$. **54.** $\frac{25}{121}$; $\frac{125}{1331}$; $\frac{625}{14641}$.
55. $\frac{16}{49}$; $\frac{64}{343}$; $\frac{256}{2401}$. **56.** $\frac{9}{64}$; $\frac{27}{512}$; $\frac{81}{4096}$. **57.** $\frac{64}{81}$; $\frac{512}{729}$; $\frac{4096}{6561}$. **58.** $\frac{49}{144}$; $\frac{343}{1728}$;
$\frac{2401}{20736}$. **59.** $\frac{81}{100}$; $\frac{729}{1000}$; $\frac{6561}{10000}$. **60.** $\frac{100}{289}$; $\frac{1000}{4913}$; $\frac{10000}{83521}$.

Page 154.—**1.** 17. **2.** 31. **3.** 42. **4.** 53. **5.** 67. **6.** 107. **7.** 3.931.
8. 4.268. **9.** 4.25. **10.** 7.4. **11.** 12.5. **12.** 11.792. **13.** .379. **14.** .759.
15. .945. **16.** .884. **17.** .0037. **18.** 1.414. **19.** 2.828. **20.** $\frac{24}{31}$. **21.** $\frac{23}{25}$.
22. .968. **23.** .866. **24.** $\frac{1}{4}$.

Page 158.—**1.** 152 ft.; 1440 sq. ft. **2.** 58 yd.; 210 sq. yd. **3.** 72 yd.;
324 sq. yd. **4.** 76 rd.; $2\frac{1}{4}$ A. **5.** 208 rd.; $16\frac{4}{5}$ A. **6.** 105 yd.; $17\frac{9}{11}$ sq. rd.
7. $23\frac{3}{4}$ yd.; $25\frac{1}{2}$ sq. yd. **8.** $424\frac{1}{2}$ rd.; 67 A.; $41\frac{7}{8}$ sq. rd. **9.** 472 rd.; 76 A.
10. $27\frac{2}{10}$ Hm.; $44\frac{8}{10}$ sq. Hm. **11.** 66 A. **12.** 52 rd. **13.** 160 rd. **14.** $234.

Page 159.—**15.** $50. **16.** 506,880 blocks. **17.** $5.73. **18.** 294.03 sq. ft.
19. (a) $102\frac{2}{3}$ sq. yd.; (b) $110\frac{2}{3}$ sq. yd.

Page 160.—**1.** 896 sq. ft. **2.** 696 sq. yd. **3.** $7\frac{3}{5}$ A. **4.** $20\frac{5}{8}$ A. **5.** $123\frac{1}{2}$
sq. in. **6.** 185.73 sq. ft. **7.** 464.14 sq. ft. **8.** 18+ A. **9.** $14\frac{1}{3}$ A. **10.** 14.696
sq. in. **11.** $1104. **12.** 640 sq. ft. **13** 117.14 sq. yd.

Page 161.—**14.** $27\frac{1}{16}$ A. **15.** 1324 sq. ft. **16.** 18 ft. **17.** 12 yd.
18. 32 rd. **19.** 48 rd. **20.** 16 in. **21.** 54 ft. **22.** 220 yd. **23.** 80 rd.
24. 50 ft. ; 40 ft.; 30 ft.

Page 162.—**1.** 30 ft.; 216 sq. ft. **2.** $37\frac{1}{2}$ ft.; $337\frac{1}{2}$ sq. ft. **3.** $22\frac{1}{2}$ ft.;
$121\frac{1}{2}$ sq. ft. **4.** 36 yd.; 486 sq. yd. **5.** 5 ft. ; 30 sq. ft. **6.** 24 rd.; 84 sq. rd.
7. 41 ft.

Page 163.—**8.** 36 ft. **9.** 700 sq. ft. **10.** 38.4+ ft. **11.** 19.5+ ft. **12.** 1 ft.
13. 5844.8 sq. ft.

Page 164.—**1.** 44 ft.; 43.9824 ft. **2.** 88 yd. ; 87.9648 yd. **3.** 110 in. ;
109.956 in. **4.** 132 ft. ; 131.9472 ft. **5.** 143 ft. ; 142.9428 ft. **6.** 176 yd. ;
175.9296 yd. **7.** $14\frac{2}{3}$ ft. ; 14.6608 ft. **8.** $16\frac{1}{2}$ ft. ; 16.4934 ft. **9.** $20\frac{1}{8}$ ft. ;
20.1586 ft. **10.** $78\frac{1}{2}$ ft. ; 78.8018 ft. **11.** 352 rd. ; 351.8592 rd. **12.** $314\frac{3}{4}$ rd. ;
314.16 rd. **13.** $3\frac{1}{2}$ ft. ; 3.501 ft. **14.** 21 ft. ; 21.008 ft. **15.** 21 in. ; 21.008 in.
16. $10\frac{1}{2}$ rd. ; 10.504 rd. **17.** 35 yd. ; 35.014 yd. **18.** $44\frac{6}{11}$ yd. ; 44.563 yd.
19. $17\frac{1}{2}$ ft. ; 17.507 ft. **20.** 28 in. ; 28.011 in. **21.** 49 ft. ; 49.02 ft. **22.** 203
in. ; 203.081 in. **23.** 1680 ft.; 1680.672 ft. **24.** 63 in. ; 63.025 in. **25.** 6 mi. ;
$301\frac{1}{3}$ rd. **26.** 188.496 ft.

Page 166.—§ **407.** **1.** 153.9384 sq. ft. **2.** 615.7536 sq. yd. **3.** 962.115 sq. in.
4. 1385.4456 sq. ft. **5.** 1625.97435 sq. ft. **6.** 2463.0144 sq. yd. **7.** $17.1042\frac{2}{3}$ sq. ft.
8. $21.6475\frac{7}{8}$ sq. ft. **9.** $32\frac{1}{4}$+ sq. ft. **10.** $494.1529\frac{11}{14}$ sq. ft. **11.** 61 A. 92+ sq. rd.
12. 49 A. 14 sq. rd. **13.** 9.628 sq. ft. **14.** 346.632 sq. ft. **15.** 2.407 sq. ft.
16. 86.658 sq. rd. **17.** 962.885 sq. rd. **18.** 1559.705 sq. rd. **19.** 240.721 sq. ft.
20. 4.279 sq. ft. **21.** 1887.27 sq. ft. **22.** 224.94 sq. ft. **23.** 50 A. 148.72 sq. rd.
24. 21.665 sq. ft. **25.** 4.276 sq. ft. **26.** 2827.44 sq. ft.

§ **409.** 1. 4 ft.; 12.5664 ft. 2. 6 rd.; 18.8496 rd. 3. 8 ft.; 25.1328 ft.
4. 10 yd.; 31.416 yd. 5. 12 yd.; 37.6992 yd. 6. 14 rd.; 43.9824 rd.

Page 167. — § **409.** 7. 24 rd.; 75.3984 rd. 8. 11.284 rd.; 35.45 rd.
9. 22.568 rd.; 70.9 rd. 10. 14.273 rd.; 44.84 rd. 11. 55.279 rd.;
173.665 rd. 12. 320 rd. (1 mi.); 3.1416 mi. 13. 160 rd.; 1.5708 mi.
14. 451.35 rd.; 4.431 mi.

§ **410.** 1. $41.87 less. 2. 22,698.06 sq. ft. 3. 64 rd. by 80 rd.
4. 13 ft. 5. 10.39 ft. 6. 130½ sq. ft. 7. 93.528 sq. ft. 8. 13.3975 ft.
9. $233.13. 10. 1250 sq. rd.

Page 169. — 1. 252 cu. ft. 2. 72 cu. ft. 3. 55.424 cu. ft. 4. 176.715
cu. ft. 5. 237.5835 cu. ft. 6. 35.343 cu. in. 7. 280.584 cu. ft.
8. 54 cu. ft. 9. 96 cu. ft. 10. 2412.7488 cu. ft. 11. 460 T., 390⅝ lb.
12. 4 T., 859⅜ lb.

Page 170. — 13. 15 hr., 59.37 min. 15. 18 ft. 16. 10 ft. 17. 5 in.

Page 171. — 1. 168 sq. ft.; 240 sq. ft. 2. 84 sq. ft.; 108 sq. ft. 3. 96
sq. ft.; 109.856 sq. ft. 4. 141.372 sq. ft.; 180.642 sq. ft. 5. 172.788 sq. ft.;
220.305 sq. ft. 6. 47.124 sq. in.; 61.2612 sq. in. 7. 216 sq. ft.; 262.764
sq. ft. 8. 108 sq. ft.; 120 sq. ft. 9. 128 sq. ft.; 152 sq. ft.
10. 603.1872 sq. ft.; 1005.3120 sq. ft. 11. 698.13⅓ sq. yd. 12. $3.84.

Page 172. — 13. 23.08+ sq. yd. 14. 130¼ sq. ft. 16. 15 ft. 17. 5½ ft.
18. 7½ ft.

Page 173. — 1. 108 cu. ft. 2. 36.083 cu. ft. 3. 256 cu. ft. 4. 58.905
cu. ft. 5. 1884.96 cu. dm. 6. 1527.88 cu. ft.

Page 174. — 7. ⅛. 9. 4 ft. by 4 ft. 10. 12 in.

Page 175. — 1. 100 sq. ft. 2. 96 sq. ft. 3. 648 sq. ft. 4. 60 sq. m.
5. 94.248 sq. ft. 6. 219.92 sq. m. 7. 157.08 sq. ft. 8. 1507.968 sq. dm.
9. $160. 10. 20 A., 131 sq. rd., 11 sq. yd., 6½ sq. ft., area of sides; 12 A.,
124 sq. rd., 4 sq. yd., 1 sq. ft., area of base; 3,297,873 cu. yd., volume.

Page 177. — 1. 61¼ rd. 2. 67½ A. 3. 320 rd.· 4. 360 sq. ft.
5. 80 ft. 6. 80 rd. 7. $223.50.

Page 178. — 8. 1 to 4. 9. 3456 sq. ft.

Page 179. — 1. 1280 bu. 2. 113.0976 cu. in. 3. 80 ft. diam.; 80 ft. deep.
4. 180 lb.

Page 180. — 1. 2040 lb. 2. 431¼ lb. 3. 1181¼ lb.

Page 181. — 4. 54 lb., 2 oz., 2 pwt., 2.88 gr. 5. 4025 lb. 6. 4630 lb.
7. 57½ lb. 8. 730½ lb. 9. 112¼ lb. 10. 2500 T. 11. .750. 12. 160 lb.
13. 86 lb., 3½+ oz· 14. 8.

Page 183. — 1. 3 yr., 2 mo., 27 da. 2. 5 yr., 8 mo., 29 da. 3. 3 yr.,
10 mo., 17 da. 4. 2 yr., 7 mo., 18 da. 5. 4 yr., 10 mo., 29 da. 6. 2 yr.,
4 mo., 16 da. 7. 6 yr., 1 mo., 11 da. 8. 1 yr., 4 mo. 9. 2 yr., 10 mo.,
15 da. 10. 4 yr., 5 mo., 21 da. 11. 158 da.; 5 mo., 5 da. 12. 221 da.;
7 mo., 7 da. 13. 268 da.; 8 mo., 26 da. 14. 143 da.; 4 mo., 21 da.
15. 167 da.; 5 mo., 16 da. 16. 114 da.; 3 mo., 25 da. 17. 211 da.; 6 mo.,
27 da. 18. 129 da.; 4 mo., 7 da. 19. 142 da.; 4 mo., 20 da. 20. 162 da.;
5 mo., 9 da. 21. 1 yr., 9 mo., 29 da.; 1 yr., 304 da. 22. 10 mo., 28 da.;
335 da. 23. 4 yr., 5 da.; 4 yr., 6 da. 24. 1 yr., 8 mo., 1 da.; 1 yr., 243 da.
25. 2 yr., 8 mo., 20 da.; 2 yr., 263 da. 26. 10 mo., 25 da.; 329 da.

Page 184. — 27. 2 yr., 10 mo., 15 da.; 2 yr., 320 da. 28. 1 yr., 10 mo.,
21 da.; 1 yr., 325 da.

Page 186. — 1. Oct. 13; Apr. 6. 2. Oct. 18; Mar. 12. 3. June 11;
Feb. 17. 4. Oct. 11; Jan. 20. 5. Dec. 7; Jan. 21. 6. Apr. 6, 1910;
Apr. 25, 1909. 7. Mar. 28, 1910; Feb. 5, 1909. 8. May 21, 1909; Aug. 10,
1908· 9. May 1, 1909; Sept. 21, 1908. 10. June 12, 1909; Sept. 23, 1908.

11. Sept. 15, 1909 ; Oct. 14, 1908.   **12.** Aug. 20, 1908 ; Sept. 7, 1907.
13. Aug. 30, 1909 ; May 2, 1909.   **14.** July 30, 1909 ; Jan. 31, 1909.
15. Aug. 28, 1909 ; Mar. 3, 1909.   **16.** Feb. 10, 1910 ; July 23, 1909.
17. Feb. 17, 1910 ; Jan. 11, 1909.   **18.** Feb. 14, 1910 ; Feb. 21, 1909.
19. Aug. 28, 1909 ; Dec. 31, 1908.   **20.** July 18, 1910 ; Apr. 16, 1909.

Page 189. — **1.** $133.87.   **2.** $73.36.   **3.** $217.98.   **4.** $219.08.
5. $94.56.   **6.** $75\frac{7}{34}$ sq. yd.

Page 190. — **7.** 130 − sq. yd.   **8.** 34 sq. yd.   **9.** $22.88.   **10.** $1.80.
11. $3.65.   **12.** $10.   **13.** 15 rolls.

Page 191. — **14.** $34\frac{1}{3}$ yd.   **15.** $40.95.
§ **471.** **1.** 319 sq. ft.   **2.** 1591 sq. ft. ; $87.51.   **3.** $2553.60.

Page 192. — **2.** 22 ft., $9\frac{9}{16}$ in.; 21 ft., $3\frac{1}{2}$ in.; 24 ft., $7\frac{1}{6}$ in.; 29 ft., 11.8 in.; 37 ft., 2 in.

Page 193. — § **478.** **4.** 24,250 shingles.   **5.** 7068 slates.   **6.** 470 sheets.
7. 3072 slates.   **8.** 22,250 shingles.
§ **480.** **1.** $6930.   **2.** 213,840 blocks ; $73,920.

Page 194. — **3.** 675,840 stones.   **4.** $422.50.

Page 196. — **1.** $35'' \times 46''$.   **2.** 8000 volumes.   **3.** $177.05.

Page 197. — **4.** $240.   **5.** $5.50.   **6.** 16 reams.   **7.** $220.   **8.** 400 pads.
9. 40 reams.   **10.** $707.55.

Page 198. — **1.** 560.   **2.** 1056.   **3.** 960.   **4.** 960.   **5.** 7680.   **6.** 2880.

Page 199. — **7.** 15,840.   **8.** 2100.   **9.** 5750.   **10.** 5334.   **11.** 245.   **12.** 864.
13. 3024.   **14.** 7296.   **15.** 7467.   **16.** 8960.   **17.** 6840.   **18.** 6667.

Page 200. — § **494.** **1.** $49.41.   **2.** 1230 ft.
§ **496.** **1.** 99 cd.; 396 cd.   **2.** 18 cd.   **3.** $246.09.   **4.** 4 ft., $4\frac{1}{2}$ in.   **5.** 293.578 s.

Page 202. — **1.** 26,880 bu.   **2.** (a) 153.6 bu. (b) 120.96 bu.   **3.** (a) 576 bu.
(b) 453.6 bu.   **4.** (a) 320 bu. (b) 252 bu.   **5.** (a) 2764.8 bu. (b) 2177.28 bu.
6. (a) 4320 bu. (b) 3402 bu.   **7.** (a) 160 bu. (b) 126 bu.   **8.** (a) 864 bu.
(b) 680.4 bu.   **9.** (a) 144 bu. (b) 113.4 bu.   **10.** (a) 268.8 bu. (b) 211.68 bu.
11. (a) 235.2 bu. (b) 185.22 bu.   **12.** (a) 343.2 bu. (b) 270.27 bu.
13. (a) 612 bu. (b) 481.95 bu.   **14.** (a) 648 bu. (b) 510.3 bu.   **15.** (a) 26 bu.
(b) 20.475 bu.   **16.** 25 ft.   **17.** 6 ft., 8 in.   **18.** $787\frac{1}{2}$ bu.   **19.** 11.13 T.
20. (a) 578.56 bu. (b) 452.8 bu.   **21.** (a) 321.43 bu. (b) 251.55 bu.
22. (a) 2777.12 bu. (b) 2173.43 bu.   **23.** (a) 4339.24 bu. (b) 3395.99 bu.
24. (a) 160.71 bu. (b) 125.78 bu.   **25.** $277.40.   **26.** 10.8 bu. more by weight.
27. 202.9824 Hl.   **28.** 112.768 Hl.   **29.** 974.3155 Hl.   **30.** 1522.368 Hl.
31. 56.384 Hl.   **32.** 304.4736 Hl.   **33.** 50.7456 Hl.

Page 203. — **1.** (a) $3830\frac{7}{77}$ gal.   (b) 3829.76 gal.   **2.** (a) $2827\frac{7}{11}$ gal.
(b) 2827.44 gal.   **3.** (a) $86,025\frac{7}{99}$ gal.   (b) 86,020 gal.   **4.** (a) 362,880 gal.
(b) 362,854.8 gal.   **5.** (a) $1885\frac{1}{11}$ gal. (b) 1884.96 gal.   **6.** (a) $1870\frac{10}{77}$ gal.
(b) 1870 gal.   **7.** (a) $8752\frac{1}{2}\frac{6}{55}$ gal. (b) 8751.6 gal.   **8.** (a) $7714\frac{2}{7}$ gal.
(b) 7713.75 gal.   **9.** (a) $12,791\frac{7}{55}$ gal. (b) 12,790.8 gal.   **10.** (a) $12,926\frac{7}{9}\frac{8}{9}$ gal.
(b) 12,925.44 gal.   **11.** (a) 14,496.4076 l. (b) 14,496.4076 Kg.   **12.** (a) 10,702.4259 l.
(b) 10,702.4259 Kg.   **13.** (a) 325,602.904 l.   (b) 325,602.904 Kg.
14. (a) 1,373,477.989 l. (b) 1,373,477.989 Kg.   **15.** (a) 7134.9506 l.
(b) 7134.9506 Kg.   **16.** (a) 7078.324 l. (b) 7078.324 Kg.

Page 204. — **18.** $\frac{7}{10}$ gal.   **19.** 752 gal.   **20.** $1468\frac{4}{5}$ gal.   **21.** 995,225 gal.
22. 38,070 gal.   **23.** 477,708 gal.   **24.** 15,040 gal.   **25.** 4 ft.   **26.** 48 ft.

Page 205. — **2.** 480 A.   **3.** $329.85.   **4.** $832 ; $416.   **5.** 357.77 rd.
6. 640 rd. ; 480 rd. ; 320 rd.   **7.** 1600 rd.

Page 209. — **1.** $12.25 ; 36 bd. ft. ; $27.50 ; $10.80.   **2.** $415.70.   **3.** $37\frac{1}{2}$ yd.,

27 yd. **4.** 6 rolls; 7 rolls. **5.** 950,400 blocks; 620,928 blocks. **6.** 2550 lb.; 44.5787 lb. Av. **7.** 1152 bu.; 75.6 bu.

Page 210. — § **513.** **8.** 3170.1 gal. **9.** 88.184 bu.; 110.23 bu. **10.** 360 Hl. § **514.** **1.** $558.95. **2.** $52\frac{1}{2}$ mi. **3.** $78,214.45. **4.** $24,562\frac{1}{2}$ cu. ft. **5.** 14,000 cu. ft. **6.** $44.90. **7.** $128. **8.** $43.27. **9.** $108.42. **10.** $1220.40.

Page 216. — § **535.** **1.** 510. **2.** $40. **3.** $58,033.57. **4.** $12,000. **5.** $966.31. **6.** Silica, 2200 lb.; Iron and Alumina, 1098 lb.; Lime, 6150 lb.; Magnesia, 247 lb.; Sulphur anhydride, 170 lb.; Combustible materials, 85 lb.

Page 217. — **7.** $.47$\frac{3}{16}$. **8.** 29,393 A. **9.** 5 yr. **10.** $.35. **11.** $15. **12.** $6.08.

Page 219. — **1.** Boys, $45\frac{5}{11}$ %; Girls, $54\frac{6}{11}$ %. **2.** $11\frac{1}{9}$ %. **3.** 60 %. **4.** 95 %. **5.** Akron, 61.6 %; Bridgeport, 43.7 %; Camden, 24.5 %; D.C., 18.8 %; Grand Rapids, 28.6 %; Houston, 76.6 %; Reading, 21.7 %; St. Paul, 31.7 %. **6.** $66\frac{2}{3}$ %; $1231. **7.** 9 %. **8.** 1–29 da., 4.7 %; 30–59 da., 19.4 %; 60–79 da., 57.7 %; 80 da. or more, 18.2 %.

Page 220. — **9.** New York, 56.4 %; Antwerp, 81.4 %; London, 22.5 %; Hamburg, 78.8 %; Rotterdam, 86.8 %; Hongkong, 30.7 %; Liverpool, 39.7 %; Montevideo, 287.9 %; Marseilles, 67.1 %; Singapore, 67.1 %; Cardiff, 27.8 %; Kobe, 631.3 %; Genoa, 55.1 %; Buenos Ayres, 78.6 %. **10.** France, (1906) 10.2 % exc.; (1907) 12.3 % exc.; (1908) 12.1 % def.; (1909) .9 % def.; (1910) 11.9 % exc. Germany, (1906) 42.6 % def.; (1907) 37 % def.; (1908) 48.4 % def.; (1909) 38.7 % def.; (1910) 32.4 % def. Italy, (1906) 14.6 % def.; (1907) 19.4 % def.; (1908) 16.7 % def.; (1909) 16.9 % def.; (1910) 5.7 % def. Russia, (1906) 12.5 % def.; (1907) 10.5 % def.; (1908) 31.3 % def.; (1909) 31.3 % def.; (1910) 5.9 % def. United Kingdom, (1906) 64 % def.; (1907) 59.5 % def.; (1908) 67.3 % def.; (1909) 59.4 % def.; (1910) 46.3 % def.

Page 221. — **11.** (1903) 32.8 % inc.; (1904) 9.8 % inc.; (1905) 8 % dec.; (1906) 2.7 % inc.; (1907) 28.7 % inc. **12.** U.S., 83 %; Can., 12 %; Eng. 5 %. **13.** (1909) 912 T.; (1910) 1119 T.; 22.7 % inc.

Page 222. — **14.** 12,371,060 bu.; 6.6 % less than May. **15.** U.S., 63.9 %; United Kingdom, 22.1 %; Germany, 2.8 %; France, 2.8 %; all others, 8.4 %. **16.** U.S., 73 %; United Kingdom, 3 %; Switzerland, 24 %. **17.** 2.1 %.

Page 223. — **1.** $1000. **2.** $120.

Page 224. — **3.** $1500. **4.** $10,000. **5.** $7500. **6.** 150 doz. **7.** $22,947.89. **8.** $18,715.67 lia.; $1789.20 Cr. rec. **9.** 200. **10.** Lot, $2400; House, $4000. **11.** $16,825. **2.** $29,145. **13.** $3890. **14.** 4797.

Page 226. — § **543.** **1.** (1) 20 %. (2) $37\frac{1}{2}$ %. (3) $14\frac{2}{7}$ %. (4) 25 %. (5) 25 %. (6) $12\frac{1}{2}$ %. (7) $16\frac{2}{3}$ %. (8) 20 %. (9) 25 %. (10) $12\frac{1}{2}$ %. (11) 20 %. (12) 25 %. (13) $11\frac{1}{9}$ %. (14) 20 %. (15) $16\frac{2}{3}$ %. (16) $16\frac{2}{3}$ %. (17) $12\frac{1}{2}$ %. (18) 20 %. (19) $16\frac{2}{3}$ %. (20) 20 %. **2.** (1) $33\frac{1}{3}$ %. (2) $12\frac{1}{2}$ %. (3) $28\frac{4}{7}$ %. (4) $16\frac{2}{3}$ %. (5) $12\frac{1}{2}$ %. (6) 20 %. (7) 25 %. (8) 10 %. (9) 10 %. (10) $16\frac{2}{3}$ %. (11) $12\frac{1}{2}$ %. (12) $12\frac{1}{2}$ %. (13) $11\frac{1}{9}$ %. (14) $16\frac{2}{3}$ %. (15) 20 %. (16) $11\frac{1}{9}$ %. (17) $12\frac{1}{2}$ %. (18) $16\frac{2}{3}$ %. (19) $11\frac{1}{9}$ %. (20) 20 %. **3.** (1) $18.75. (2) $19.20. (3) $29.40. (4) $28.80. (5) $37.33. (6) $45. (7) $64. (8) $62.50. (9) $70. (10) $96. (11) $96. (12) $108. (13) $126. (14) $165. (15) $187.50. (16) $300. (17) $270. (18) $360. (19) $480. (20) $540. **4.** (1) $13.50. (2) $14. (3) $14.70. (4) $16. (5) $24. (6) $35.56. (7) $36. (8) $35. (9) $45. (10) $60. (11) $70. (12) $88. (13) $102.60. (14) $60. (15) $128.57. (16) $157.50. (17) $180. (18) $225. (19) $300. (20) $375. **5.** (1) 40 %. (2) $18\frac{3}{4}$ %. (3) 20 %. (4) $8\frac{1}{3}$ %. (5) $18\frac{3}{4}$ %. (6) 25 %. (7) 25 %. (8) 30 %. (9) $33\frac{1}{3}$ %. (10) $33\frac{1}{3}$ %. (11) $37\frac{1}{2}$ %. (12) 25 %. (13) $8\frac{1}{3}$ %. (14) $16\frac{2}{3}$ %. (15) $33\frac{1}{3}$ %. (16) $22\frac{2}{9}$ %. (17) 25 %. (18) $8\frac{1}{3}$ %. (19) 25 %. (20) $16\frac{2}{3}$ %. **6.** (1) 20 %. (2) 25 %. (3) 30 %. (4) $33\frac{1}{3}$ %. (5) $37\frac{1}{2}$ %. (6) 20 %. (7) $41\frac{2}{3}$ %. (8) 36 %.

(9) 30%. (10) $12\frac{1}{2}$%. (11) 10%. (12) $16\frac{2}{3}$%. (13) $33\frac{1}{3}$%. (14) 10%. (15) $26\frac{2}{3}$%. (16) $33\frac{1}{3}$%. (17) $12\frac{1}{2}$%. (18) $33\frac{1}{3}$%. (19) $12\frac{1}{2}$%. (20) 10%. **7.** (1) $44\frac{4}{9}$%. (2) $36\frac{4}{11}$%. (3) $37\frac{1}{2}$%. (4) $33\frac{1}{3}$%. (5) 30%. (6) $28\frac{8}{9}$%. (7) $35\frac{5}{7}$%. (8) 25%. (9) 28%. (10) $25\frac{25}{27}$%. (11) $27\frac{1}{12}$%. (12) 30%. (13) 20%. (14) $30\frac{5}{9}$%. (15) $31\frac{3}{7}$%. (16) $23\frac{17}{21}$%. (17) $22\frac{2}{9}$%. (18) $30\frac{5}{9}$%. (19) $23\frac{17}{21}$%. (20) $33\frac{1}{3}$%.

Page 227.—8. (1) 80%. (2) $57\frac{1}{7}$%. (3) 60%. (4) 50%. (5) $42\frac{6}{7}$%. (6) $40\frac{5}{8}$%. (7) $55\frac{5}{9}$%. (8) $33\frac{1}{3}$%. (9) $38\frac{8}{9}$%. (10) 35%. (11) $37\frac{1}{7}$%. (12) $42\frac{6}{7}$%. (13) 25%. (14) 44%. (15) $45\frac{5}{6}$%. (16) $31\frac{1}{4}$%. (17) $28\frac{4}{7}$%. (18) 44%. (19) $31\frac{1}{4}$%. 20. 50%. **9.** (1) $14.40. (2) $18.33. (3) $17.14. (4) $25. (5) $34.29. (6) $40. (7) $42. (8) $48. (9) $64.29. (10) $60.75. (11) $80. (12) $106.67. (13) $102.86. (14) $104.73. (15) $140. (16) $126. (17) $240. (18) $300. (19) $315. (20) $450. **10.** (1) $20. (2) $25.14. (3) $34.29. (4) $45. (5) $53.33. (6) $50.63. (7) $74.67. (8) $85.71. (9) $100. (10) $97.20. (11) $109.71. (12) $130.91. (13) $126.32. (14) $288. (15) $204.17. (16) $240. (17) $360. (18) $480. (19) $504. (20) $648.

Page 229.—1. 3600 lb. **2.** 72%; 8%. **3.** $1875.68. **4.** 24%. **5.** New Haven, 79.92% L.; 20.08% W. Syracuse, 97.18% L.; 2.82% W. Scranton, 98.82% L.; 1.18% W. Fall River, 85.22% L.; 14.78% W. Albany, 96.07% L.; 3.93% W. **6.** 750. **7.** 640. **8.** $1237\frac{1}{2}$ lb. **9.** 25%; $33\frac{1}{3}$%. **10.** $3000, 1st; $4500, 2d; $5250, 3d.

Page 230.—11. $59\frac{13}{37}$%. **12.** 16.39% inc. 1900–05; 19.16% inc. 1905–10; 38.68% inc. 1900–10. **13.** $3\frac{1}{3}$%. **14.** $11.64. **15.** $7,044,192,674; Manhattan, 67.35%; Bronx, 7.01%; Brooklyn, 19.93%; Queens, 4.75%; Richmond, .96%. **16.** $704,419,267.40. **17.** 5.29%. **18.** $22,250,514.90. **19.** $150,000.

Page 231.—20. $45. **21.** $8\frac{3}{4}$%. **22.** 5%. **23.** $6400. **24.** 714 bu. **25.** 20%. **26.** 750 A. **27.** $7000. **28.** $2000. **29.** $2926. **30.** 83.87.

Page 235.—1. $391.20.

Page 236.—2. $603.29. **3.** $2316.52.

Page 237.—4. $22,107.10. **5.** $870.04. **6.** $92.18. **7.** $222.25. **8.** $3493.70. **9.** $2140.28. **10.** 2d, $14.17 better.

Page 238.—11. 1st, $29.60. **12.** $872.92. **13.** $179.39.

Page 239.—§ 563. **1.** 50%. **2.** 50%. **3.** 50%. **4.** 40%. **5.** 40%. **6.** 40%. **7.** 44%. **8.** 46%. **9.** 36%. **10.** 28%. **11.** 30%. **12.** 25%. **13.** 70%. **14.** 65%. **15.** 60%. **16.** 55%. **17.** 64%. **18.** 58%. **19.** 52%. **20.** 43%. **21.** $43\frac{3}{4}$%. **22.** $40\frac{5}{8}$%. **23.** 48%. **24.** $41\frac{1}{2}$%. **25.** 51%. **26.** 37%. **27.** $43\frac{3}{4}$%. **28.** $32\frac{1}{2}$%. **29.** 25%. **30.** $23\frac{1}{2}$%. **31.** 19%. **32.** $14\frac{1}{2}$%. **33.** 24%. **34.** 65%. **35.** 70%. **36.** 64%. **37.** 72%. **38.** 76%. **39.** 80%. **40.** 70%. **41.** $22\frac{2}{9}$%. **42.** 80%. **43.** 50%. **44.** 50%. **45.** 40%. **46.** 44%. **47.** 46%. **48.** $31\frac{1}{5}$%. **49.** 55%. **50.** 60%.

§ 564. **2.** Single dis. .4%; $1.30. **3.** $288.

Page 240.—4. $1795.50. **5.** $26.60. **6.** A's $283.50. **7.** $92.16. **8.** $78.40. **9.** $270. **10.** $23\frac{7}{11}$%. **11.** $14\frac{2}{7}$%. **12.** 12.82%.

Page 242.—1. $4050 loss; $9450 S. P. **2.** $1000 gain; $3400 S. P. **3.** $.40 gain; $.25 S. P. **4.** $397.50. **5.** $.18. **6.** $.30.

Page 243.—§ 576. **7.** $8.72. **8.** $19.50 gain; $71.50 S. P. **9.** $3.50. **§ 578.** **1.** 23%. **2.** 30.9%. **3.** 20%. **4.** 17.7%. **5.** 32.7%. **6.** $33\frac{1}{3}$%. **7.** 31.1%. **8.** 20%. **9.** 15.6%. **10.** 17.5%. **11.** 25%. **12.** 21.3%. **13.** 28%. **14.** 13.4%. **15.** 20.9%. **16.** $23\frac{1}{3}$%.

Page 244.—2. $33\frac{1}{3}$%. **3.** $28\frac{4}{7}$%. **4.** 50%. **5.** $66\frac{2}{3}$%.

Page 245. — **7.** \$9000 gain; 25% gain. **8.** \$5162.50 loss; 8.6% loss. **9.** 78.6% gain. **10.** \$.36.

Page 246. — **1.** \$6000. **2.** \$72 Net Cost; \$100 List Price. **3.** \$80,000, Cost; \$68,000 S. P. **4.** \$3500; \$4200. **5.** Pineapples, \$25; oranges, \$60. **6.** \$3600; $22\frac{2}{9}$%. **7.** 17.7%.

Page 247. — § **585.** **2.** \$.30. **3.** \$54. **4.** \$4. **5.** \$4.05.

Page 248. — **6.** 20%. **7.** \$20. **8.** \$400 loss; 4% loss. **9.** \$3.43. **10.** \$48.

Page 249. — **1.** (1) \$18. (2) \$21. (3) \$30. (4) \$35. (5) \$40. (6) \$52. (7) \$75. (8) \$108. (9) \$165. (10) \$200. (11) \$276. (12) \$378. (13) \$620. (14) \$885. (15) \$1170. (16) \$1300. (17) \$1600. (18) \$2200. **2.** (1) \$14.40. (2) \$16. (3) \$18.75. (4) \$24.50. (5) \$27. (6) \$40. (7) \$56.25. (8) \$60. (9) \$112. (10) \$144. (11) \$225. (12) \$348. (13) \$485. (14) \$660. (15) \$840. (16) \$1170. (17) \$1410. (18) \$1764. **3.** (1) 25%. (2) $33\frac{1}{3}$%. (3) $16\frac{2}{3}$%. (4) $14\frac{2}{7}$%. (5) $33\frac{1}{3}$%. (6) $8\frac{1}{3}$%. (7) 40%. (8) $6\frac{1}{4}$%. (9) $12\frac{1}{2}$%. (10) 10%. (11) $6\frac{1}{4}$%. (12) 15%. (13) 5%. (14) 1%. (15) 15%. (16) 15%. (17) 20%. (18) 25%. **4.** (1) $6\frac{1}{4}$%. (2) $11\frac{1}{9}$%. (3) 50%. (4) 25%. (5) $8\frac{1}{3}$%. (6) $4\frac{1}{6}$%. (7) 20%. (8) 25%. (9) 10%. (10) 5%. (11) 10%. (12) 5%. (13) 2%. (14) 6%. (15) 8%. (16) 7%. (17) 3%. (18) 5%. **5.** (1) $12\frac{1}{2}$%. (2) $33\frac{1}{3}$%. (3) $33\frac{1}{3}$%. (4) $7\frac{1}{7}$%. (5) $16\frac{2}{3}$%. (6) $33\frac{1}{3}$%. (7) 20%. (8) $33\frac{1}{3}$%. (9) $33\frac{1}{3}$%. (10) 40%. (11) $16\frac{2}{3}$%. (12) 25%. (13) 15%. (14) 20%. (15) 20%. (16) $16\frac{2}{3}$%. (17) $3\frac{1}{3}$%. (18) $16\frac{2}{3}$%. **6.** (1) 25%. (2) $16\frac{2}{3}$%. (3) $33\frac{1}{3}$. (4) $28\frac{4}{7}$%. (5) $16\frac{2}{3}$%. (6) 25%. (7) 20%. (8) $16\frac{2}{3}$%. (9) 10%. (10) 6%. (11) 5%. (12) 10%. (13) 2%. (14) 4%. (15) 7%. (16) 12%. (17) 10%. (18) $16\frac{2}{3}$%. **7.** (1) \$32; \$36. (2) \$36; \$42. (3) \$11.25; \$15. (4) \$16; \$20. (5) \$108; \$120. (6) \$48; \$52. (7) \$125; \$150. (8) \$48; \$54. (9) \$40; \$55. (10) \$45; \$60. (11) \$100; \$115. (12) \$1080; \$1134. (13) \$104.17; \$129.17. (14) \$41.67; \$49.17. (15) \$450; \$585. (16) \$2160; \$2340. (17) \$4500; \$4800. (18) \$2025; \$2475. **8.** (1) \$10; \$9. (2) \$18; \$16. (3) \$67.50; \$56.25. (4) \$56; \$49. (5) \$12; \$9. (6) \$12; \$10. (7) \$125; \$112.50. (8) \$64; \$40. (9) \$180; \$168. (10) \$187.50; \$180. (11) \$384; \$360. (12) \$540; \$522. (13) \$333.33; \$323.33. (14) \$375; \$330. (15) \$1080; \$1008. (16) \$3360; \$3276. (17) \$750; \$705. (18) \$4500; \$4050. **9.** (1) $28\frac{4}{7}$%. (2) $33\frac{1}{3}$%. (3) $14\frac{2}{7}$%. (4) $15\frac{5}{13}$%. (5) 40%. (6) $6\frac{2}{3}$%. (7) 50%. (8) 4.9%. (9) $10\frac{10}{29}$%. (10) $7\frac{9}{13}$%. (11) $5\frac{35}{53}$%. (12) $13\frac{7}{11}$%. (13) $4\frac{6}{11}$%. (14) .84%. (15) $14\frac{2}{7}$%. (16) 14.7%. (17) 24%. (18) $27\frac{3}{11}$%. **10.** (1) $7\frac{9}{13}$%. (2) $11\frac{13}{17}$%. (3) $42\frac{6}{7}$%. (4) $25\frac{25}{27}$%. (5) $9\frac{1}{11}$%. (6) $5\frac{5}{19}$%. (7) 20%. (8) $23\frac{1}{13}$%. (9) 10%. (10) 5.05%. (11) $9\frac{11}{21}$%. (12) $5\frac{5}{19}$%. (13) 2%. (14) $5\frac{15}{17}$%. (15) 7.9%. (16) $7\frac{7}{19}$%. (17) $3\frac{7}{31}$%. (18) 5.6%.

Page 251. — **1.** \$2.40. **2.** \$3.50. **3.** $38\frac{8}{9}$%. **4.** 27.7%. **5.** 20. **6.** \$.36. **7.** \$225. **8.** $33\frac{1}{3}$%; 25%. **9.** 80%. **10.** \$21.

Page 252. — **11.** Roll-top desks, \$33.75; flat-top desks, \$18.75; china closets, \$28.13; sideboards, \$46.88; dining tables, \$18.75; dining chairs, \$15 a set. **12.** \$66.97; smoothing planes, \$1.00; screw drivers, \$.45; carriage bolts, $\frac{3}{8}''\times 2'$, \$3.73 per C.; $\frac{3}{8}''\times 2\frac{1}{2}''$, \$3.87 per C.; $\frac{1}{2}'\times 4\frac{1}{2}''$, \$4.17 per C.; machine bolts, $\frac{3}{4}''\times 6''$, \$5.04 per C. **13.** 19.7% gain. **14.** 20% inc.

Page 253. — **15.** $2\frac{7}{9}$% dec. **16.** 50% inc. **17.** \$500. **18.** 34.4%. **19.** 35.5%. **20.** 18.5% loss. **21.** 35.7% gain.

Page 254. — **22.** 51.8% gain. **23.** 16.3% gain. **24.** 26.1% gain.

Page 255. — **1.** ES.BW/NE.CY. **2.** K.OB/R.LC. **3.** O.HB/E.SC. **4.** R.WB/S.YC. **5.** S.HB/T.CY. **6.** E.KB/O.SC. **7.** SW.HB/TC.YZ. **8.** EH.BW/NA.CY. **9.** O.WB/E.YC. **10.** K.WB/R.YC. **11.** A.WB/H.YC.

12. E.WB/N.ON.   **13.** E.SB/N.SC.   **14.** E.HB/N.AC.   **15.** S.WB/O.RE.
16. EC.BW/ON.OE.   **17.** EO.BW/NL.AC.   **18.** EL.BW/ON.TA.
1. DO.KP/W.HP.   **2.** C.BK/W.SI.   **3.** A.EK/.CM.   **4.** DE.KP/W.TS.
5. DN.KP/W.IP.   **6.** DB.KP/W.RI.   **7.** DA.KP/H.SM.   **8.** OK.PQ/H.SC.
9. ON.KP/H.EH.   **10.** OL.KP/A.SM.   **11.** ZO.KP/A.SM.   **12.** ZP.BK/A.IA.
13. ZE.KP/A.PT.   **14.** EN.KP/T.PS.   **15.** NB.KP/P.AM.

Page 256. — § **595.** **16.** $12/$1.25.   **17.** $4.36/$.49.   **18.** $22.50/$2.50.
19. $37/$3.50.   **20.** $3.60/$.38.   **21.** $23.20/$2.34.   **22.** $24.40/$2.50.
23. $64/$5.   **24.** $20/$2.   **25.** $19.20/$2.17.
§ **597.** **1.** $50.   **2.** $42.   **3.** $9.   **4.** $.33 ; $1.11 ; $.78 ; $1.33 ; $1.17.

Page 257. — **5.** $23.30.   **6.** $.98 ; $1.17 ; $.23 per ft.   **7.** $4.80.   **8.** $.72.
9. $64.91.

Page 259. — **21.** $332.50 ; $16,202.50.

Page 260. — **22.** $1267.50.   **23.** $1594.50.   **24.** $2776.74.   **25.** 23.2%
gain.

Page 261. — **26.** $3808.85.   **27.** $195.05.

Page 263. — **1.** $260.40 ; $13.02.   **2.** 1150 bu.   **3.** $.88.   **4.** $.96.
5. $11,625.   **6.** $2722.50 ; 9\frac{1}{11}$%.   **7.** $891.25.

Page 264. — **8.** $40.   **9.** $1600.   **10.** 4905 bbl. ; $4.50.   **11.** $200.74 ;
2788 bbl. ; $.28.   **12.** $7369.   **13.** Net proceeds, $5187.12.   **14.** Gross
cost, $2071.01.

Page 265. — **1.** (a) $.22 ; (b) $4.80 ; (c) $70 ; (d) $3.75 ; (e) $830.07.
2. (a) 6 hr., 23 min. ; (b) 6 hr., 39 min. ; (c) 2 hr., 31 min. ; (d) 2 hr., 45 min. ;
(e) 12 hr., 48 min.   **3.** (a) 25% ; (b) 17% ; (c) 21% ; (d) 54% ; (e) 33%.
4. 80.16.

Page 266. — § **612.** **5.** 859.61 ; $4769.28 ; 76,557.12 ; $33.67.   **6.** (a) 10% ;
(b) 16$\frac{2}{3}$% ; (c) 33$\frac{1}{3}$% ; (d) 6$\frac{1}{4}$%.   **7.** (a) $3230 ; (b) $88,450 ; (c) $79,560 ;
(d) $30,399.33.   **8.** (a) $202.50 ; (b) $288 ; (c) $5.04 ; (d) $480 ; Total,
$975.54.   **9.** (1908) $247,630,809 ; (1909) $269,137,210.   **10.** Total excess,
$21,506,401.
§ **613.** **1.** $152,083.33.   **2.** 100%.   **3.** $240.   **4.** $981.25.

Page 267. — § **613.** **5.** 2550 bu.   **6.** 11$\frac{1}{8}$%.   **7.** $381.69.   **8.** 10.35% gain.
9. Net cost, $1694.79 ; $2.27 per pr. ; $2 79 per doz. ; $17.76 per doz.   **10.** 65.1%.
§ **614.** **1.** 50.9% ; 81.3%.   **2.** 24.7% decr. ; 12.5% incr.   **3.** 6.4%.

Page 268. — **5.** (1840) 2715 mi. ; 6787$\frac{1}{2}$%.   (1850) 5816 mi. ; 211.1%.   (1860)
20,349 mi. ; 237.4%.   (1870) 20,248 mi. ; 70%.   (1880) 38,556 mi. ; 78.4%.
(1890) 75,838 mi. ; 86.5%.   (1900) 29,782 mi. ; 18.2%.   (1909) 43,524 mi. ;
22.5%.   **6.** Boston, 15.6% ; New York, 39.7% ; Philadelphia, 16.9% ; Baltimore, 13.6% ; New Orleans, 7.4% ; San Francisco, 6.8%.   **7.** $723,744.67.

Page 269. — **8.** (Loans, real estate) $1,057,307.92, inc. 10.7% ; (other loans,
etc.) $12,815,363.43, inc. 32.2% ; (stocks, etc.) $3,142,308.77, inc. 43.9% ;
(due from banks) $2,494,708.68, inc. 14.4% ; (banking house, etc.) $70,082.24,
inc. 1.4% ; (other real estate) $193,595.02, inc. 15.9% ; (expenses) $370,333.73,
dec. 37.5% ; (overdrafts) $104,063.19, dec. 13% ; (profit and loss) $513.99,
inc. 101% ; (other resources) $2,260,362.40, inc. 1175.5% ; (cash on hand)
$1,107,853.18, inc. 11.1% ; (trust investments) $791,765.99, dec. 100%. Totals,
$21,875,932.72 ; 23.5% inc.

9–10.

Industry	No. Reported	No. Employed	Aggreg. Days Employed	Aggregate Wages	Per Cent Emp.	Av. Days Emp.	Av. Daily Wage	Av. Earning per Man
Building	116,755	100,304	5,376,294	$20,161,104	85.9	53.6	$3.75	$201.00
Transpor'n	60,367	53,549	4,396,373	12,155,623	88.7	82.1	2.76	226.60
Clothing and textiles	46,481	45,353	2,820,957	7,392,539	97.6	62.2	2.62	162.96
Metals and machinery	30,428	29,850	2,092,485	6,537,150	98.1	70.1	3.12	218.71
Printing and binding	23,995	23,216	1,694,768	5,943,946	96.8	73.	3.51	256.23

Page 272. — 1. $179.91.   2. $262.10.   3. $77.15.   4. $98.69.
5. $314.30.   6. $10.04.

Page 273. — 1. $.48.   2. $2.96.   3. $7.95.   4. $15.30.   5. $18.15.
6. $50.46.   7. $35.44.   8. $59.42.   9. $12.60.   10. $11.
11. $18.60.   12. $22.20.   13. $2.13.   14. $4.20.   15. $3.09.
16. $3.33.   17. $2.93.   18. $2.41.   19. $3.84.   20. $13.93.
21. $2.76.   22. $8.96.   23. $10.50.   24. $58.32.   25. $36.   26. $25.

Page 274. — 27. $3.44.   28. $5.46.   29. $30.82.   30. $23.73.
31. $39.98.   32. $.14.13.   33. $16.80.   34. $54.17.   35. $17.04.
36. $12.50.   37. $35.63.   38. $16.21.   39. $101.89.   40. $44.14.
41. $320.25.   42. $219.77.   43. $61.39.   44. $58.41.   45. $268.69.
46. $91.92.   47. $120.   48. $123.20.   49. $105.30.   50. $138.60.
51. $107.50.   52. $90.75.

Page 275. — 1. $.52 ; $.56.   2. $3.20 ; $3.45.   3. $8.61 ; $9.28.
4. $16.58 ; $17.85.   5. $19.67 ; $21.18.   6. $54.67 ; $58.87.   7. $38.40 ;
$41.35.   8. $64.37 ; $69.32.   9. $13.65 ; $14.70.   10. $11.92 ; $12.83.
11. $20.15 ; $21.70.   12. $24.05 ; $25.90.   13. $2.66 ; $1.59.   14. $5.25 ;
$3.15.   15. $3.87 ; $2.32.   16. $4.16 ; $2.50.   17. $3.66 ; $2.19.
18. $3.01 ; $1.81.   19. $4.80 ; $2.88.   20. $17.42 ; $10.45.   21. $3.45 ;
$2.07.   22. $11.20 ; $6.72.   23. $13.13 ; $7.88.   24. $72.90 ; $43.74.
25. $33 ; $30.   26. $22.92 ; $20.83.   27. $3.15 ; $2.86.   28. $5 ;
$4.55.   29. $28.25 ; $25.68.   30. $21.75 ; $19.78.   31. $36.65 ; $33.31.
32. $12.96 ; $11.78.   33. $15.40 ; $14.   34. $49.65 ; $45.14.   35. $15.62 ;
$14.20.   36. $11.46 ; $10.42.   37. $32.66 ; $29.69.   38. $14.86 ; $13.51.
39. $169.81 ; $67.93.   40. $73.56 ; $29.42.   41. $533.75 ; $213.50.
42. $366.29 ; $146.52.   43. $102.32 ; $40.93.   44. $97.35 ; $38.94.
45. $447.81 ; $179.12.   46. $153.19 ; $61.28.   47. $200 ; $80.   48. $205.33 ;
$82.13.   49. $175.50 ; $70.20.   50. $231 ; $92.40.   51. $179.17 ; $71.67.
52. $151.25 ; $60.50.

Page 276. — 1. $8.75.   2. $9.80.   3. $12.60.   4. $14.85.   5. $34.75.
6. $137.80.   7. $118.80.   8. $109.50.   9. $128.   10. $56.   11. $117.
12. $148.   13. $276.   14. $157.40.   15. $250.20.   16. $296.10.
17. $83.65.   18. $143.64.   19. $62.75.   20. $37.85.

Page 277. — 1. $49.20.   2. $297.20.   3. $477.   4. $1840.   5. $1230.
6. $307.50.   7. $4710.   8. $3228.   9. $1251.90.   10. $315.70.   11. $1056.
12. $1288.   13. $1545.60.   14. $750.75.   15. $137.51.   16. $1626.50.
17. $532.37.   18. $123.80.   19. $216.40.   20. $1847.

Page 278. — **1.** $29. **2.** $23.50. **3.** $8.30. **4.** $3.80. **5.** $9.25.
6. $27. **7.** $86. **8.** $79.50. **9.** $105.33. **10.** $5.50. **11.** $1.65.
12. $1.48. **13.** $3.55. **14.** $7.36. **15.** $6.18. **16.** $3.67. **17.** $16.65.
18. $13.20. **19.** $1.42. **20.** $1.52. **21.** $3.50. **22.** $16. **23.** $15.40.
24. $6.30. **25.** $34.10. **26.** $30. **27.** $22.10. **28.** $18.20. **29.** $3.42.
30. $4.31. **31.** $.42. **32.** $.40. **33.** $2.80. **34.** $.25. **35.** $400.
36. $275.10.

Page 279.

	Compound Time	Exact Time	Bankers' Time		Compound Time	Exact Time	Bankers Time
1.				**15.**			
Int.	$19.32	$19.67	$19.32	Int.	$172.25	$175.94	$173.48
Amt.	$439.32	$439.67	$439.32	Amt.	$7554.25	$7557.94	$7555.48
2.				**16.**			
Int.	$18.88	$19.31	$18.99	Int.	$284.90	$291.50	$286.00
Amt.	$658.88	$659.31	$658.99	Amt.	$6884.90	$6891.50	$6886.00
3.				**17.**			
Int.	$18.95	$19.32	$18.95	Int.	$140.00	$142.50	$140.00
Amt.	$578.95	$579.32	$578.95	Amt.	$7640.00	$7642.50	$7640.00
Page 280.				**18.**			
4.				Int.	$372.00	$378.67	$372.00
Int.	$29.22	$29.83	$29.22	Amt.	$8372.00	$8378.67	$8372.00
Amt.	$756 72	$757.33	$756.72	**19.**			
5.				Int.	$1174.00	$1178.00	$1174.00
Int.	$28.65	$28.97	$28.65	Amt.	$13,174.00	$13,178.00	$13,174.00
Amt.	$672.40	$672.72	$672.40	**20.**			
6.				Int.	$51.17	$51.83	$51.17
Int.	$25.95	$26.21	$26.08	Amt.	$1051.17	$1051.83	$1051.17
Amt.	$824.35	$824.61	$824.48	**21.**			
7.				Int.	$33.30	$33.60	$33.30
Int.	$53.13	$53.76	$53.13	Amt.	$933.30	$933.60	$933.30
Amt.	$1313.13	$1313.76	$1313.13	**22.**			
8.				Int.	$56.08	$56.34	$56.21
Int.	$61.88	$63.25	$61.88	Amt.	$826.08	$826.34	$826.21
Amt.	$1436.88	$1438.25	$1436.88	**23.**			
9.				Int.	$90.42	$91.08	$90.42
Int.	$61.31	$62.19	$61.31	Amt.	$1080.42	$1081.08	$1080.42
Amt.	$1821.31	$1822.19	$1821.31	**24.**			
10.				Int.	$34.78	$35.21	$34.78
Int.	$80.65	$81.57	$80.65	Amt.	$684.78	$685.21	$684.78
Amt.	$1920.65	$1921.57	$1920.65	**25.**			
11.				Int.	$104.47	$106.02	$104.47
Int.	$20.14	$20.38	$20.22	Amt.	$1964.47	$1966.02	$1964.47
Amt.	$495.94	$496.18	$496.02	**26.**			
12.				Int.	$349.80	$352.20	$349.80
Int.	$22.54	$22.80	$22.67	Amt.	$3949.80	$3952.20	$3949.80
Amt.	$791.04	$791.30	$791.17	**27.**			
13.				Int.	$126.70	$129.50	$127.40
Int.	$20.71	$20.71	$20.71	Amt.	$4326.70	$4329.50	$4327.40
Amt.	$999.01	$999.01	$999.01	**28.**			
14.				Int.	$14.65	$14.94	$14.65
Int.	$371.76	$376.75	$373.00	Amt.	$441.39	$441.68	$441.89
Amt.	$7856.76	$7861.75	$7858.00				

	Compound Time	Exact Time	Bankers' Time		Compound Time	Exact Time	Bankers' Time
29.				**30.**			
Int.	$28.49	$28.98	$28.49	Int.	$38.54	$39.03	$38.54
Amt.	$615.99	$616.48	$615.99	Amt.	$774.94	$775.43	$774.94

	Compound Time		Exact Time		Bankers' Time	
	5%	7%	5%	7%	5%	7%
31.						
Int.	$16.10	$22.54	$16.39	$22.95	$16.10	$22.54
Amt.	$436.10	$442.54	$436.39	$442.95	$436.10	$442.54
32.						
Int.	$15.73	$22.03	$16.09	$22.52	$15.82	$22.15
Amt.	$655.73	$662.03	$656.09	$662.52	$655.82	$662.15
33.						
Int.	$15.79	$22.10	$16.10	$22.54	$15.79	$22.10
Amt.	$575.79	$582.10	$576.10	$582.54	$575.79	$582.10
34.						
Int.	$24.35	$34.09	$24.86	$34.80	$24.35	$34.09
Amt.	$751.85	$761.59	$752.36	$762.30	$751.85	$761.59
35.						
Int.	$23.87	$33.42	$24.14	$33.80	$23.87	$33.42
Amt.	$667.62	$677.17	$667.89	$677.55	$667.62	$677.17
36.						
Int.	$21.62	$30.27	$21.85	$30.58	$21.73	$30.43
Amt.	$820.02	$828.67	$820.25	$828.98	$820.13	$828.83
37.						
Int.	$44.28	$61.99	$44.80	$62.72	$44.28	$61.99
Amt.	$1304.28	$1321.99	$1304.80	$1322.72	$1304.28	$1321.99
38.						
Int.	$51.56	$72.19	$52.71	$73.79	$51.56	$72.19
Amt.	$1426.56	$1447.19	$1427.71	$1448.79	$1426.56	$1447.19
39.						
Int.	$51.09	$71.52	$51.82	$72.55	$51.09	$71.52
Amt.	$1811.09	$1831.52	$1811.82	$1832.55	$1811.09	$1831.52
40.						
Int.	$67.21	$94.10	$67.98	$95.17	$67.21	$94.10
Amt.	$1907.21	$1934.10	$1907.98	$1935.17	$1907.21	$1934.10
	5½%	6½%	5½%	6½%	5½%	6½%
41.						
Int.	$18.46	$21.82	$18.68	$22.08	$18.54	$21.91
Amt.	$494.26	$497.62	$494.48	$497.88	$494.34	$497.71
42.						
Int.	$20.66	$24.42	$20.90	$24.70	$20.78	$24.56
Amt.	$789.16	$792.92	$789.40	$793.20	$789.28	$793.03
43.						
Int.	$18.98	$22.43	$18.98	$22.43	$18.98	$22.43
Amt.	$997.28	$1000.73	$997.28	$1000.73	$997.28	$1000.73

	COMPOUND TIME		EXACT TIME		BANKERS' TIME	
	5½%	6½%	5½%	6½%	5½%	6½%
44.						
Int.	$340.78	$402.73	$345.35	$408.14	$341.92	$404.09
Amt.	$7825.78	$7887.73	$7830.35	$7893.14	$7826.92	$7889.09
45.						
Int.	$157.89	$186.60	$161.28	$190.60	$159.02	$187.93
Amt.	$7539.89	$7568.60	$7543.28	$7572.60	$7541.02	$7569.93
46.						
Int.	$261.16	$308.64	$267.21	$315.79	$262.17	$309.83
Amt.	$6861.16	$6908.64	$6867.21	$6915.79	$6862.17	$6909.83
47.						
Int.	$128.33	$151.67	$130.63	$154.38	$128.33	$151.67
Amt.	$7628.33	$7651.67	$7630.63	$7654.38	$7628.33	$7651.67
48.						
Int.	$341.00	$403.00	$347.11	$410.22	$341.00	$403.00
Amt.	$8341.00	$8403.00	$8347.11	$8410.22	$8341.00	$8403.00
49.						
Int.	$1076.17	$1271.83	$1079.83	$1276.17	$1076.17	$1271.83
Amt.	$13,076.17	$13,271.83	$13,079.83	$13,276.17	$13,076.17	$13,271.83
50.						
Int.	$46.90	$55.43	$47.51	$56.15	$46.90	$55.43
Amt.	$1046.90	$1055.43	$1047.51	$1056.15	$1046.90	$1055.43

	COMPOUND TIME		EXACT TIME		BANKERS' TIME	
51.	7½%	4%	7½%	4%	7½%	4%
Int.	$41.63	$22.20	$42.00	$22.40	$41.63	$22.20
Amt.	$941.63	$922.20	$942.00	$922.40	$941.63	$922.20
52.						
Int.	$70.10	$37.39	$70.42	$37.56	$70.26	$37.47
Amt.	$840.10	$807.39	$840.42	$807.56	$840.26	$807.47
53.						
Int.	$113.03	$60.28	$113.85	$60.72	$113.03	$60.28
Amt.	$1103.03	$1050.28	$1103.85	$1050.72	$1103.03	$1050.28
54.						
Int.	$43.47	$23.18	$44.01	$23.47	$43.47	$23.18
Amt.	$693.47	$673.18	$694.01	$673.47	$693.47	$673.18
55.						
Int.	$130.59	$69.65	$132.53	$70.68	$130.59	$69.65
Amt.	$1990.59	$1929.65	$1992.53	$1930.68	$1990.59	$1929.65
56.						
Int.	$437.25	$233.20	$440.25	$234.80	$437.25	$233.20
Amt.	$4037.25	$3833.20	$4040.25	$3834.80	$4037.25	$3833.20
57.						
Int.	$158.38	$84.47	$161.88	$86.33	$159.25	$84.93
Amt.	$4358.38	$4284.47	$4361.88	$4286.33	$4359.25	$4284.93
58.						
Int.	$18.31	$9.77	$18.67	$9.96	$18.31	$9.77
Amt.	$445.05	$436.51	$445.41	$436.70	$445.05	$436.51
59.						
Int.	$35.62	$19.00	$36.23	$19.32	$35.62	$19.00
Amt.	$623.12	$606.50	$623.73	$606.82	$623.12	$606.50
60.						
Int.	$48.17	$25.69	$48.79	$26.02	$48.17	$25.69
Amt.	$784.57	$762.09	$785.19	$762.42	$784.57	$762.09

Page 281.—1. $18.25. **2.** $15.75. **3.** $26.28. **4.** $14.62. **5.** $20.80.
6. $2.56. **7.** $2.22. **8.** $8.85. **9.** $36.72. **10.** $24.63. **11.** $173.18.
12. $134.87.

Page 282.—13. $303.22. **14.** $300.55. **15.** $70.50. **16.** $238.60.
17. $13.39. **18.** $12.25. **19.** $5.26. **20.** $17.42. **21.** $32.43. **22.** $33.89.
23. $110.83. **24.** $12.05. **25.** $238.36. **26.** $12,633.56. **27.** $58,289.32;
$1,806,968.92. **28.** 2.329%.

Page 283.—1. $30.30, stock better. **2.** $151,160. **3.** $570 gain.
4. $496.03. **5.** 11.7%. **6.** $1915.91. **7.** $7000. **8.** $368.40. **9.** $9152.

Page 284.—10. $3.29. **11.** $1.80. **12.** $75. **13.** 15% gross; $15\frac{1}{3}$% net.
14. 20 yr.; 16 yr., 8 mo.; 12 yr., 6 mo.; 11 yr., 1 mo., 10 da. **15.** 4%; 7%;
10%; 12%. **16.** Cash terms; 4%-30; 5%-30; 30 da. **17.** $15\frac{5}{19}$%, $135.74.
18. $88.22.

Page 287.—1. $221.03. **2.** $376.67. **3.** $432.59. **4.** $388.10.
5. $686.64.

Page 288.—6. $1828.49. **7.** $982.33. **8.** $797.56 loss. **9.** $546.01.
10. $1875.58.

Page 290.—1. $3745.58.

Page 291.—2. $318,761.44.

3.

PERIODS	ANNUAL PAYMENT	INTEREST ON BALANCE	AMORTIZATION	PRINCIPAL UNPAID $100.00
First	$7.36	$4.00	$3.36	$96.64
Second	7.36	3.87	3.49	93.15
Third	7.36	3.72	3.64	89.51
Fourth	7.36	3.58	3.78	85.73
Fifth	7.36	3.43	3.93	81.80
Sixth	7.36	3.28	4.08	77.72
Seventh	7.36	3.11	4.25	73.47
Eighth	7.36	2.94	4.42	69.05
Ninth	7.36	2.76	4.60	64.45
Tenth	7.36	2.58	4.78	59.67
Eleventh	7.36	2.39	4.97	54.70
Twelfth	7.36	2.19	5.17	49.53
Thirteenth	7.36	1.98	5.38	44.15
Fourteenth	7.36	1.77	5.59	38.56
Fifteenth	7.36	1.55	5.81	32.75
Sixteenth	7.36	1.31	6.05	26.70
Seventeenth	7.36	1.07	6.29	20.41
Eighteenth	7.36	.82	6.54	13.87
Nineteenth	7.36	.56	6.80	7.07
Twentieth	7.36	.29	7.07	0.00

4. $95,628.43.

Page 298.

Due Date	Term of Discount	Due Date	Term of Discount
1. May 15	66 da.	14. April 7	29 da.
2. May 24	84 da.	15. June 12	39 da.
3. Sept. 10	54 da.	16. May 1	54 da.
4. Aug. 12	78 da.	17. Nov. 15	46 da.
5. Dec. 7	47 da.	18. Nov. 9	40 da.
6. Oct. 1	21 da.	19. July 11	31 da.
7. Sept. 6	23 da.	20. Aug. 16	46 da.
8. June 15	32 da.	21. March 13	34 da.
9. June 3	32 da.	22. Feb. 28	57 da.
10. May 19	49 da.	23. Feb. 28	89 da.
11. Sept. 17	85 da.	24. Jan. 7	66 da.
12. Oct. 3	57 da.	25. April 19	26 da.
13. Jan. 19	70 da.		

Page 299.

Date of Maturity	Term of Discount	Bank Discount	Proceeds
1. Dec. 1	112 da.	$33.60	$1766.40
2. Oct. 11	71 da.	23.67	1976.33
3. Aug. 11	68 da.	18.59	1621.41
4. May 23	46 da.	7.35	1142.65
5. June 2	27 da.	7.46	1982.54
6. Nov. 19	110 da.	52.62	2817.38
7. Sept. 17	47 da.	33.59	3641.41
8. Dec. 17	46 da.	72.07	9327.93
9. July 5	24 da.	25.43	8452.57
10. Nov. 13	86 da.	68.80	7131.20
11. Feb. 28	24 da.	21.63	5878.37
12. Sept. 1	18 da.	24.64	8187.36
13. Feb. 6	92 da.	91.04	4658.96
14. Dec. 30	73 da.	62.35	5062.65
15. Dec. 4	54 da.	54.00	5946.00
16. Oct. 10	48 da.	57.60	5342.40
17. Aug. 10	30 da.	28.13	4471.87
18. June 12	91 da.	47.02	3052.98
19. July 12	72 da.	7.43	405.07
20. Aug. 8	58 da.	5.94	362.81

Page 300. — 1. 212 da.   2. 334 da.   3. 122 da.   4. 306 da.
5. 184 da.   6. 184 da.

Page 301. — 7. 181 da.   8. 217 da.   9. 134 da.   10. 141 da.
11. 271 da.   12. 261 da.   13. 307 da.   14. 340 da.   15. 125 da.
16. 221 da.   17. 144 da.   18. 298 da.   19. 48 da.   20. 264 da.

Pages 302-303.

Date of Maturity	Term of Discount	Bank Discount	Collection	Proceeds
1. May 10	63 da.	$52.50	$2.50	$4945.00
2. June 15	63 da.	47.25	4.50	4448.25
3. Sept. 6	67 da.	27.92	3.13	2468.95
4. Dec. 16	36 da.	9.60	4.00	1586.40
5. Oct. 11	40 da.	11.65		1736.35
6. April 11	158 da.	30.11	3.43	1338.46
7. Feb. 15	78 da.	56.98	4.38	4321.39
8. Feb. 8	57 da.	4.17	1.32	521.31
9. June 8	87 da.	37.70		2562.30
10. Feb. 4	53 da.	4.11	1.39	552.18
11. Feb. 1	122 da.	19.19		924.56
12. March 2	26 da.	1.74	.32	319.54
13. May 30	47 da.	2.45	1.04	413.01
14. Feb. 28	56 da.	4.13	1.33	526.14
15. March 11	57 da.	28.98	3.81	3017.81

Page 304. — 1. $1399.98. 2. $1553.03. 3. $2174.75. 4. $557.91.
5. $950.12. 6. $1123.30. 7. $2890.89. 8. $3145.66. 9. $4713.33.
10. $5099.88. 11. $6394.22. 12. $4732.57.

Page 305. — 1. $500 ; $20. 2. $600 ; $18. 3. $800 ; $24.
4. $1600 ; $56.

Page 306. — 5. $1200 ; $12. 6. $1000 ; $20. 7. $1200 ; $60.
8. $4000 ; $100. 9. $900 ; $72. 10. $1000 ; $100. 11. $800.
12. $100. 13. $50. 14. $.87. 15. $3.82. 16. 15%. 17. $1470.59.
18. Alike. 19. $788.20. 20. $2542.05. 21. $7692.31.

Page 308. — 1. $991.69. 2. $2008.33. 3. $3058.21.

Page 310. — Compound Time : 1. $235.48. 2. $439.61. 3. $578.40.
4. $304.57. Exact Time : 1. $236.06. 2. $440.13. 3. $579.43.
4. $305 12.

Page 312. — 1. $3123 ; 7 yr. ; $6123.43. 2. $10,056 ; 7 yr. ; $19,721.34.
3. $2754.50 ; 6 yr. ; $5302.11. 4. $12,492 ; 6 yr. ; $24,029.09.
5. $2017.50 ; 8 yr. ; $4092.68. 6. $3248 ; 8 yr. ; $6466.47.

Page 316. — 1 5¼ mills ; $62.81. 2. $4.2919 per $1000. 3. $1,449.529.20 ;
22.123%. 4. $55.70. 5. $121.61. 6. $291. 7. $900.35. 8. $784.57.
9. Manhattan, .017579 ; Bronx, .017579 ; Brooklyn, .0181499 ; Queens, .0181079 ;
Richmond, .0187501.

Page 317. — 10. Manhattan, 68% ; Bronx, 6.8% ; Brooklyn, 19.7% ; Queens,
1.6% ; Richmond, .9%. 11. Manhattan, 67.4% ; Bronx, 6.7% ; Brooklyn,
20.2% ; Queens, 4.7% ; Richmond, 1%. 12. $221,748.17. 13. $315,251.09.
14. $301,617. 15. $1413. 16. $239.86.

Page 318. — 17. State Tax $42 50 ; School Tax $150 ; City Tax $354.69.
18. State Tax $9.54 ; School Tax $33.66 ; City Tax $81.35. 19. State Tax
$2040 : School Tax $7200 ; City Tax $17,040.40. 20. State Tax $8.50 ; School
Tax $30 ; City Tax $72.50. 21. $732.20. 22. $203.77. 23. $8251.44.
24. $39.21. 25. $483.85. 26. $306.56. 27. $124.27. 28. $68.81.

Page 320. — 29. State $450; County $795; City $2115; School $2325; Sanitary $510; Park $810; Town $195; Lake Shore $120; Total $7320. **30.** State $270; County $477; City $1269; School $1395; Sanitary $306; Park $459; Total $4176 + $41.76 (penalty) + $.19 = $4217.95. **31.** State $45; County $79.50; City $211.50; School $232.50; Sanitary $51; Total $619.50 + $18.59 + $.19 = $638.28. **32.** State $51; County $90.10; City $239.70; School $263.50; Sanitary $57.80; Park $86.70; Total $788.80 + $15.78 + $.29 = $804.87.

Page 323. — 1. $29.04. **2.** $36. **3.** $73.80. **4.** $178.75. **5.** $21.25. **6.** $64.69. **7.** $113.48. **8.** $249.08. **9.** $411.47. **10.** $237.60. **11.** $75. **12.** $67.38. **13.** $243.20. **14.** $297.

Page 324. — 1. $7800. **2.** $7307. **3.** Co. X $8595.73; Co. Y $11,397.97; Co. Z $15,113.30.

Page 325. — 4. $1600. **5.** $4583.33.

Page 326. — 1. $30. **2.** $1.44.

Page 327. — § 766. 3. $91.15. **4.** $53.63. **5.** $40; $66.67. **6.** (a) $112.50; (b) $67.50. **7.** $63.
§ 767. 1. $60. **2.** $7.80. **3.** (a) Hartford Ins. Co. $11,750; Albany Ins. Co. $17,662.50; (b) Hartford Ins. Co. $7750; Albany Ins. Co. $11,662.50. **4.** $6000. **5.** $10,044.64. **6.** $5000. **7.** $2272.73. **8.** $13,888.89.

Page 328. — 9. $358.40; $89.78.

Page 331. — 1. $149.60. **2.** $3740; $1479.44. **3.** $934.45; $3250; 26 yr., 66 da. **4.** $508; $26,191.29.

Page 334. — 1. $617.56. **2.** $730.11. **3.** $1532.77. **4.** $618.25. **5.** $730.11 **6.** $1534.44.

Page 335. — § 795. 1. $10591\frac{2}{3}\%$; 4340.4%; 7829.2%. **2.** 20.6%. **3.** $35.25. **4.** $625.
§ 796. 1. $65.50; $99.90; $310; $118.13. **2.** $344.75; $740.62; $955.20; $148.50.

Page 336. — § 796. 3. $620.67. **4.** $62.50; $139.20; $2.04; $7.70. **5.** $500; $2000; $850; $1525. **6.** $200; $500; $90; $40; Total $830. **7.** $42, farm; $337.50, house and lot; $88, city lot; $517.38, store; Total $984.88. **8.** $1.80; $2.80; $1.67; $1.50. **9–10.** Grand total 335,508.
§ 797. 1. $\frac{1}{4}\%$; $23.98.

Page 337. — 2. $48.43. **3.** $53.82. **4.** $277.50. **5.** $3764.76. **6.** $50,372.62. **7.** $165.54. **8.** $197.03. **9.** $461.59. **10.** On bldg. $7333.33; on machinery $5357.14; on stock $11,250.

Page 342. — 1. $13,868.75. **2.** $17,687.50. **3.** $7131.25. **4.** $24,478.13. **5.** $37,378.13. **6.** $47,165.63. **7.** $8062.50. **8.** $83,718.75. **9.** $70,725. **10.** $59,200 **11.** $26,550. **12.** $13,165.62. **13.** $20,650. **14.** $13,837.50. **15.** $43,687.50. **16.** $23,156.25.

Page 343. — § 816. 17. $51,312.50. **18.** $18,000. **19.** $217,500. **20.** $12,050. **21.** $56.25 gain. **22.** $150 gain. **23.** $112.50 gain. **24.** $200 gain. **25.** 0. **26.** $87.50 gain. **27.** $125 loss. **28.** $168.75 gain. **29.** $250 gain. **30.** $250 gain.
§ 818. — 1. 280. **2.** 160. **3.** 120. **4.** 400. **5.** 800.

Page 344. — 1. $2160. **2.** $840.75 increase. **3.** $1.36 gain.

Page 345. — § 820. 4. 15%; reserve fund, $106,173.41; $22,500.
§ 822. — 1. $31,290. **2.** $39,065.63. **3.** $55,171.88. **4.** $24,000.

Page 346. — § **824.** 1. 5%. 2. $5\frac{5}{7}$%. 3. Latter, $\frac{5}{11}$%. 4. Latter, $\frac{2}{45}$%.
5. Increased $14\frac{2}{7}$%.
§ **826.** 1. $120. 2. $\$174\frac{7}{8}$. 3. $\$89\frac{7}{8}$. 4. $43,750. 5. $37.50 more.

Page 351. — 1. $5446.33. 2. $6473.56. 3. $3837. 4. $13,266.
5. $19,630.56. 6. $15,716.25. 7. $20,253.33. 8. $30,559.58. 9. $43,013.33.
10. $33,318.06.

Page 353. — 1. $113.68. 2. $93.72. 3. $118.18. 4. $86.90.
5. $93.98. 6. $119.77. 7. $4\frac{3}{4}$%. 8. $4\frac{7}{8}$%. 9. $4\frac{1}{8}$%. 10. 4.9%.
11. 6%. 12. $4\frac{7}{8}$%.

Page 357. — 1. $1500.15. 2. $3876.21. 3. $4583.66. 4. $547.88.
5. $12,879.02. 6. $1350.20. 7. $15,730.27. 8. $135.37. 9. $79.

Page 358. — 10. $151.45. 11. P. O. Money Order cheapest by 11 ¢.
12. M. O. $.50 cheaper. 13. $2336.06. 14. Draft, $2.83 better.

Page 359. — 1. $780.08. 2. $1256.85. 3. $2496.87. 4. $3671.32.
5. $3855.62. 6. $4261.82. 7. $1759.20. 8. $2358.91. 9. $4150.79.
10. $7378.72. 11. $3971.48. 12. $4284.28. 13. $1558.59. 14. $5277.97.
15. $6871.46. 16. $9738.98. 17. $8490.25. 18. $4958.59. 19. $3819.77.
20. $4360.95. 21. $951.45. 22. $275.39.

Page 366. — 1. $3641.25. 2. $7310.97. 3. $4752.57. 4. $6615.21.
5. $2401.71. 6. $4725.60. 7. $3573.84. 8. $2809.86. 9. $3811.56.
10. $1957.41. 11. $5460.12. 12. $4668.68. 13. $6995.74.
14. $12,006.16. 15. $348.25. 16. $824.49. 17. $7883.37.

Page 368. — 1. $526.23. 2. $1002.70. 3. $1946.73. 4. $2457.88.
5. $1503.57. 6. $1980.83. 7. £105, 1s., 6d. 8. £304, 3s., 3d. 9. £401, 2s.
10. £509, 5s., 1d. 11. £204, 3s., 1d. 12. £209, 5s.

Page 369. — 1. 4.8510. 2. 4.8410.

Page 370. — § **895.** 3. 4.8285. 4. 4.8430. 5. $5.22\frac{1}{2} - \frac{1}{32}$. 6. $5.20 + \frac{3}{64}$.
7. $5.19\frac{3}{8} - \frac{1}{64}$. 8. $5.22\frac{1}{2} + \frac{3}{64}$.
§ **897.** 1. £722, 7s., 11d. 2. £923, 1s., 6d. 3. £1106, 17s., 11d.
4. £1539, 5s. 5. F. 12,146.56. 6. F. 18,091.72. 7. F. 33,240.
8. F. 51,625. 9. M. 12,732.10. 10. M. 16,831.03. 11. M. 5235.60.
12. M. 5879.27.

Page 371. — 14. F. 9308.11. 15. F. 11,884.41. 16. F. 11,133.20.
17. M. 15,740.16. 18. M. 18,330.88. 19. M. 31,491.14. 20. M. 5321.09.
21. F. 7500.11.

Page 372. — 1. $8464.89. 2. $2102.65. 3. $2136.79. 4. $4547.45.
5. $638.31. 6. $1120.97.

Page 373. — 1. $930. 2. $610,962.50. 3. Gold bars, $620 better.

Page 374. — § **904.** 1. $293.99 gain. 2. $41.54 gain. 3. $1096.14 gain.
§ **905.** 1. $22,442.14. 2. $4182.36. 3. F. 142,726.5. 4. $2832.46.

Page 375. — 5. $3572.45. 6. $1005. 7. £16, 5s.; M. 333.20. 8. $2948.19.
9. F. 1881. 10. M. 3056.25. 11. $2472.51. 12. £500, 5s., 2d.
13. F. .0047 gain via London. 14. $2316.03. 15. $4.8845. 16. $976,480.

Page 380. — § **934.** 1. $25.36. 2. $61.90. 3. $1100. 4. $812.50.
5. $64.80. 6. $118.80. 7. $172.80. 8. $2.24. 9. $125. 10. $1600.
§ **936.** 1. $13.20. 2. $76.05. 3. $320. 4. $948.

Page 381. — § **936.** 5. $251.90. 6. $29.60. 7. $297.50. 8. $260.50.
9. $27.80. 10. $428.45. 11. $175. 12. $178.50. 13. $181.
14. $660.10. 15. $3.60.
§ **937.** 1. $3083.44. 2. $227.75; $1.09 a ream; $.67 a box. 3. $.58.
4. $1.24. 5. $4781.43. 6. $.11\frac{1}{2}$.

Page 382. — **7.** $.95 first lot ; $1.17 second lot ; $1.33 third lot.   **8.** $1.54.
9. $.06¼.   **10.** Gain, 36.86 % on hdkfs. ; 37.77 % on ribbon.   **11.** $547.63.
12. $4860.40.

Page 383. — **1.** $24.30.   **2.** $111.   **3.** $157.50.

Page 386. — **1.** March 13.   **2.** Sept. 26.   **3.** Dec. 15.   **4.** May 22.
5. July 10.   **6.** Aug. 30.

Page 387. — **8.** May 13.   **9.** Aug. 27.   **10.** Nov. 1.

Page 390. — **1.** Oct. 1.   **2.** Feb. 4.   **3.** April 21.   **4.** June 29.

Page 392. — **1.** $694.69.   **2.** $1295.74.   **3.** $2639.51.

Page 394. — **1.** 1st, $6750 ; 2d, $9000 ; 3d, $11,250.   **2.** $8000 each.
3. $5400.   **4.** $5000.   **5.** 1st, $1200 ; 2d, $1500 ; 3d, $1800 ; Insolvent.
6. Capital, 1st, $2000 ; 2d, $3000.

Page 395. — **7.** E, $9000 ; F, $10,000.   **8.** $2500.   **9.** Capital, 1st yr.,
$11,000 ; 2d yr., $10,000 ; $5000.   **10.** A, $11,200 ; B, $12,600 ;
C, $14,000.   **11.** X, $3000 ; Y, $4800 ; Z, $4200.   **12.** D, $2000 ; E, $3000 ;
F, $4000.   **13.** Parker, $9994.95 ; Rogers, $11,422.80 ; Robinson, $12,850.65.

Page 397. — **1.** A, $8535.13 ; B, $12,008.11 ; C, 14,806.76.   **2.** C, $5200 ;
D, $3600.   **3.** E, $25,000 ; F, $29,200 ; G, $16,800.   **4.** A, $280.40 ;
B, $299.60.

Page 398. — **§ 977.**   **5.** X, $14,973.54 ; Y, $8608.46.   **6.** A $8044.74 ;
B, $13,630.26.   **7.** X, $23,802.20 ; Y, $24,939.56 ; Z, $26,258.24.
§ 978. — **1.** F, $56,900 ; G, $63,000.   **2.** D, $5641.10 ; E, $9002.18.

Page 400. — **4.** T. B., footing, $76,123.75 ; trading acct., $11,044.37 profit ;
net profit, $5072.27 ; present worth, J. B. Altgeld, $21,710.67 ; L. M. Goodman,
$10,469.84.

Page 401. — **5.** T. B. footing, $65,759.10 ; trading acct., $4662.40 profit ;
net bus. profit, $2880.70 ; P. W., $30,688.28.

Page 405. — **1.** $952.20.   **2.** $3907.50.   **3.** $1956.53.

Page 406. — **1.** 1st, $53.46 ; 2d, $39.09 ; 3d, $25.39 ; 4th, $12.36.   **2.** 1st,
$68.16 ; 2d, $53.24 ; 3d, $32.07 ; 4th, $18.76.

Page 407. — **§ 989.**   **1–4.** Profit per series, #30, $394.38 ; #31, $150.19 ;
#32, $472.50 ; #33, $780.71.   **5.** $2027.25, book value ; $195, div.
§ 990.   **1.** $8250 ; $7110 ; $42,000 ; $623,700.

Page 408. — **2.** A, $750 ; B, $1050.   A, $1500 ; B. $2500 ; C, $3500.
A, $1000 ; B, $2000 ; C, $4000.   A, $300 ; B, $280 ; C, $260.   **3.** Oct. 14.
4. $4718.50.   **5.** $1500.75 ; $1873.13 ; $3505.25 ; $4999.50 ; total, $11,878.63.
6. $11,461.63.   **7.** $40,000 ; $80,000 ; $50,000 ; $60,000.

Page 409. — **§ 990.**   **8.** Total, $733,009.   **9.** 20 % gain ; 25 % gain; 30 %
gain ; 8 % loss.   **10.** $60.73.
§ 991.   **1.** $1782.07.   **2.** $2101.75 ; $1287.25 ; $1033 ; $578.   **3.** $2530.55.
4. $3645.24 ; $4.90.   **5.** 0.   **6.** $622.22.   **7.** A, $12,831.50; B, $6235.50.

Page 410. — **8.** $49,984.59.   **9.** July 30, 1909.   **10.** $562.40.   **1.** $53.29.

Page 411. — **2.** 65½ yd.   **3.** $683.33.   **4.** $24.57.   **5.** 2d, $7 better.
6. $24,800.   **7.** 1050 bu.   **8.** $9.92 ; $865.08.   **9.** $748.59.   **10.** 60 ¢
per $100 ; $75.

Page 412. — **1.** $75.   **2.** 22 % loss.   **3.** $7085.71.   **4.** Aug. 6.
5. $2873.87.   **6.** $246.31, inc.   **7.** $2123.33.   **8.** $59,320.
9. $389.04.   **10.** $1970.90.

ADVERTISEMENTS

LESSONS IN PHYSICAL GEOGRAPHY

By CHARLES R. DRYER, M. A., F. G. S. A., Professor of Geography, Indiana State Normal School.

SIMPLICITY and accuracy constitute two of the chief merits of this text-book. The physical features of the earth are grouped according to their causal relations and the functions which they perform in the world economy. The characteristics of each group are presented by means of a typical example described in unusual detail. Many realistic exercises, including both laboratory and field work, are introduced whenever practicable in order to direct the student how to study the thing itself.

COMMERCIAL GEOGRAPHY

By HENRY GANNETT, Geographer of the United States Geological Survey and the Twelfth Census; CARL L. GARRISON, Principal of the Morgan School, Washington, D. C.; and EDWIN J. HOUSTON, Emeritus Professor of Physical Geography and Physics, Central High School, Philadelphia.

THE study of commercial geography is here made not only informative, but truly educative and worth while. While the conditions that influence commerce in every region, the cultivation of the soil, and the great commercial staples are taken up, the largest part is devoted to each country with special reference to its industries and commerce. The book abounds with diagrammatic maps, graphic diagrams, and tables of every kind, showing locations and portraying the present condition of the world's commerce.

AMERICAN BOOK COMPANY

DRYER'S
HIGH SCHOOL GEOGRAPHY
PHYSICAL, ECONOMIC, AND REGIONAL
By CHARLES REDWAY DRYER, F.G.S.A., F.R.G.S.,
Professor of Geography and Geology, State Normal
School, Terre Haute, Ind.

Parts I and II. Physical and Economic.

THIS textbook repre.ents a new departure in geography for secondary schools — the correlation of physical and commercial geography. It is an effort to afford a clear idea of the relation between the earth and man, showing both the dependence of human life upon natural conditions, and the influence of those conditions in turn upon human life.

¶ Part I is devoted to a brief account of physical geography, which forms the necessary basis, of study, only those features and processes being emphasized which have directly affected man in his progress. Each topic is treated as to its economic relations, showing how the form and present physical condition of the earth affect commerce and civilization.

¶ Part II, economic geography, reverses the point of view of the first part. Here the outlines of household management practised by the great human family are presented against the background of the natural earth already shown. By this method of treatment both physical and economic geography are made to have a double interest and value.

¶ Part III furnishes a more detailed, intimate, and graphic study of economic geography, arranged according to the different types of environment, with reference to the economic adaptations of human life. The treatment is by natural rather than by political divisions.

¶ The book contains an unusually large number and variety of maps and illustrations, which are given in close connection with the text.

AMERICAN BOOK COMPANY

OVERTON'S HYGIENE SERIES

By FRANK OVERTON, A. M., M. D., Sanitary Supervisor, Nassau and Suffolk Counties, New York State Department of Health, Author of "Applied Physiology."

Personal Hygiene (For the lower grades)   . .
General Hygiene (For the upper grammar grades)

THESE books aim to raise the individual's standard of health, and to improve the general conditions of living.
They teach the principles of modern hygiene and sanitation in such a clear, simple and friendly way as to win the confidence and coöperation of young people.
These books are devoted to spreading the gospel of cleanliness and fresh air. They show the child how to avoid not only the little annoying illnesses, but many serious diseases. They guard him against the dangers arising from impure milk and contaminated water, from flies and other disease-carrying insects. They tell what foods to eat, and explain why these are the most nourishing, from the modern standpoint of the development of heat and force in terms of calories. The importance of exercise and of plenty of sleep is made clear. Enough anatomy and physiology is taught to make plain the reasons for every rule of hygiene and sanitation.
Throughout, the books emphasize the important truth that good health is essential to the happiness and success of every man, woman and child. They show that while it is short-sighted and foolish to neglect one's own health, it is criminal to endanger the health of others; and that the health of the community is dependent upon the health of the individual. The training in right living given by these books is directly in line with many public movements, and with the steps urged by the best authorities.

AMERICAN BOOK COMPANY

(154)

FOR PENMANSHIP CLASSES

HEALEY'S FREE ARM MOVEMENT WRITING

Complete,   In three Parts, each, per doz.

The muscular movement system as applied in Healey develops ease, rapidity and legibility. It makes the pupil's writing automatic and removes all temptation to scribble or write with his fingers. . This manual teaches the standard letter forms, each preceded by thorough movement drills, and followed by word and sentence drills. The movement drills and letter forms are presented in their natural order. The course covers six school years, commencing with the third, and with its logically organized pedagogical system and due regard to motivation, puts writing on the high plane of other studies. Only blank paper is required for the work.

BARNETT'S BUSINESS PENMANSHIP

1st and 2nd Grades,   ' 5th and 6th Grades,
3rd and 4th Grades,   Advanced Grades .

A complete course for elementary schools, each of the four parts being intended for two grades. Each part consists of loose sheets of ruled paper and copy, with an instruction card, inclosed in an envelope. This system teaches form and movement together, and produces the good, plain, rapid handwriting demanded by modern life. The work is well-balanced and carefully graded, the movement exercises being related directly to the letter forms and the drill being supplied where needed.

STEADMANS' GRADED LESSONS IN WRITING

Eight Pads, each,   Teacher's Manual,
     Chart . . . .

An easy, graceful system of free-hand writing, with full play for the writer's individuality, is taught by this muscular movement course. Each pad covers a year's work. The exercises are arranged by carefully planned gradations from the simpler forms and letters to the more complex.

AMERICAN BOOK COMPANY

BRIGHAM and McFARLANE'S
ESSENTIALS OF GEOGRAPHY

By ALBERT PERRY BRIGHAM, A. M., Professor of
Geology, Colgate University, Hamilton, N. Y., and
CHARLES T. McFARLANE, Pd. D., Professor of
Geography, Teachers College, Columbia University.

FIRST BOOK .   . SECOND BOOK

Each book also published in two parts.

IN THIS series every lesson makes a vivid impression
because it shows the pupil that geography has a real and
definite connection with the everyday world he knows.
The human side of the study is continually emphasized, yet
there is no trace of fictionized geography, no irrelevant story-
matter. Up-to-date, accurate and pertinent facts give the pupil
inspiring ideas of the unity of the world and its peoples.

¶ These are the first school geographies to employ the natural
groups adopted by the United States Census Bureau,—the
groups alluded to in all government publications, magazine
articles and newspapers.

¶ For the first time in any school geography the inter-relation
of the world's life and work with physical geography is ade-
quately shown. Each main industry is described in connection with the place where it is chiefly located or developed.

¶ Over and above all other topics treated in these books is
the dual subject of industry and commerce. The industrial
side of life in every country of the world is continually kept in
the foreground. In connection with commerce, the transportation routes of the various countries of the world are fully
described and illustrated.

¶ One of the most striking features of this series is the perfection
of the maps, which were made especially for these geographies,
regardless of expense. They represent the highest art of map-
making known today. Equaling the maps in unusual merit and
charm are the illustrations, which have been reproduced from
photographs from all over the world.

AMERICAN BOOK COMPANY

Lightning Source UK Ltd.
Milton Keynes UK
UKHW021620261118
332986UK00012B/926/P

9 780243 171392